Electrical and Mechanical Services in High Rise Buildings

Design and Estimation Manual

Second Edition

INCLUDING GREEN BUILDINGS

W0230504

Electrical and Mechanical Services in High Rise Buildings

Design and Estimation Manual

Second Edition

INCLUDING GREEN BUILDINGS

AK Mittal M Tech

Practicing Consultant
New Delhi

Ex-Superintending Engineer
Central Public Works Department
(CPWD)

CBS Publishers & Distributors Pvt Ltd

New Delhi • Bengaluru • Chennai • Kochi • Mumbai • Pune
Hyderabad • Kolkata • Nagpur • Patna • Vijayawada

**Electrical and
Mechanical Services in
High Rise Buildings**

Design and Estimation Manual
Including Green Buildings

ISBN: 978-81-239-2435-9

Copyright © Author and Publishers

Second Edition: 2014

Reprint: 2015

First Edition: 2007

Published by Satish Kumar Jain and produced by Varun Jain for

CBS Publishers & Distributors Pvt Ltd

4819/XI Prahlad Street, 24 Ansari Road, Daryaganj, New Delhi 110 002, India.
Ph: 23289259, 23266861, 23266867   Website: www.cbspd.com
Fax: 011-23243014   e-mail: delhi@cbspd.com; cbspubs@airtelmail.in.
Corporate Office: 204 FIE, Industrial Area, Patparganj, Delhi 110 092
Ph: 4934 4934   Fax: 4934 4935   e-mail: publishing@cbspd.com; publicity@cbspd.com

Branches

- **Bengaluru:** Seema House 2975, 17th Cross, K.R. Road,
  Banasankari 2nd Stage, Bengaluru 560 070, Karnataka
  Ph: +91-80-26771678/79   Fax: +91-80-26771680   e-mail: bangalore@cbspd.com
- **Chennai:** No. 7, Subbaraya Street, Shenoy Nagar, Chennai 600 030, Tamil Nadu
  Ph: +91-44-26260666, 26208620   Fax: +91-44-42032115   e-mail: chennai@cbspd.com
- **Kochi:** 36/14 Kalluvilakam, Lissie Hospital Road, Kochi 682 018, Kerala
  Ph: +91-484-4059061-65   Fax: +91-484-4059065   e-mail: kochi@cbspd.com
- **Mumbai:** 83-C, Dr E Moses Road, Worli, Mumbai-400018, Maharashtra
  Ph: +91-22-24902340/41   Fax: +91-22-24902342   e-mail: mumbai@cbspd.com
- **Pune:** Bhuruk Prestige, Sr. No. 52/12/2+1+3/2 Narhe, Haveli
  (Near Katraj-Dehu Road Bypass), Pune 411 041, Maharashtra
  Ph: +91-20-64704058, 64704059, 32392277 Fax: +91-20-24300160   e-mail: pune@cbspd.com

Representatives

- **Hyderabad** 0-9885175004   · **Kolkata** 0-9831437309, 0-9051152362
- **Nagpur** 0-9021734563   · **Patna** 0-9334159340
- **Vijayawada** 0-9000660880

Printed at India Binding House, Nodia, UP

to

---

the divine mother
late Mrs Kunti Devi

and

my father
Mr UC Mittal

Foreword to the Second Edition

Today India is on the rapid path of urbanisation and industrialisation, resulting in significant growth and spread of infrastructural facilities involving the building sector. Over the years, the demand for energy, water and materials has grown enormously and need has arisen to address the minimisation of natural resources used in the construction of buildings and their associated impact on environment. With shrinkages in total available spaces, it has become necessary to devise ways and means of optimising the building construction in a more sustainable manner. Moreover, with rising awareness in stakeholders for green buildings, India is witnessing a spur in the attempts to achieve sustainable high rise buildings. In view of these concerns and constraints, it becomes imperative for every stakeholder of the construction sector to judiciously expend resources and make the best use of the latest and emerging technologies, which not only addresses national priorities but also make good business sense.

Today green buildings in India are a perfect blend of ancient architectural heritage and modern technological innovations. One has to invariably visit a green building to learn how both ancient knowledge and modern technological innovations can dovetail and create architectural splendour and majesty. With the right orientation, better energy efficiency is achieved right from the design stage. Incorporating green concepts and techniques in building sector can help address national issues like energy efficiency, conservation of natural resources, handling of consumer waste, water efficiency and reduction in fossil fuel used in commuting. Most importantly, these concepts can enhance occupants health, happiness and well-being. The building envelope should be examined closely with the help of energy simulation and daylight software. This is possible only through the interdisciplinary exchange of information and coordination. The MEP engineer should be able to collaborate with the architect by leveraging the information on solar azimuth, to arrive at an appropriate orientation of the building. This helps in the design of vertical and horizontal shades without compromising on the natural daylight penetration inside the building.

In the light of this, electrical and mechanical services have a pivotal role in ensuring that the green building project meets the desired expectations. The rich and varied technical services offered by the electrical and mechanical services can be integrated with the green design philosophy of the project with great acumen and precision. It, therefore, becomes critical to equip the technical team with contemporary tools and technologies to enable them to take informed decisions in the design and estimation of the project objectively and scientifically.

In this regard, this manual aims at bridging the gap for architects and MEP designers, between theoretical knowledge and practical methodologies. It provides an overreaching view of all MEP services with reference to national and international building codes, standards and certifications. This would be an excellent ready-reckoner for all the architects and engineers to plug in the gaps frequently encountered while designing and executing the project. MEP designers will be able to contextualise the site-related execution problems in the framework of design philosophies and techniques. Ready-to-use data tables, empirical relations and thumb rule data, step-by-step approach and examples make this book a complete compendium of information accessed and used by personnel in the construction industry.

The new chapter Green Buildings in this second edition is a very valuable addition to the original wide spectrum of information. The author has simplified and decoded the concept of the green building for comprehension and utilization by architects, engineers, builders, proprietors and end users. The alternative design strategies demonstrated by the author will enable the designers to consider and weigh different options to cut down on cost and energy requirements. They will be able to select machines at maximum energy efficiency, provide automatic controls for energy conservation and approach designs from environment-friendly perspectives. The book serves the purpose of raising consciousness towards our environmental responsibility, and facilitates the application of the concept in dynamic, creative and productive ways.

I am sure , this manual will add a great value to the green building movement and will go a long way in facilitating India emerge as one of the global leaders in green buildings by 2025.

Dr Prem C Jain

Chairman
Indian Green Building Council (IGBC)

Foreword to the First Edition

Shelter is one of the basic necessities of a human being. Buildings as such form a very common item with which everyone is intimately connected, whether it is residential or nonresidential. Buildings in the past few years have become more complex and are no more mere four walls and a roof to protect us from the adverse environment. Our dependency on services in buildings has also grown. The National Building Code, which is the leading guide document for proper design, had remained almost static from 1983 (but for a few amendments) and had not kept pace with the developments. That was primarily the provocation for this book. The National Building Code has now been revised and issued on September 16, 2005. The National Building Code 2005 identifies and lays down the need for a coordinated, integrated approach and team work in developing the building or building complex, right from the initial stage of concept design to the final completion, commissioning and later in the operation and maintenance of the building.

Building and building complexes have grown in size and complexity over the years. Among the non residential buildings, the special features in different type of buildings have also undergone a significant change. Buildings have become highly specialized based on its use. Not only there has been a quantum jump in the number of electrical and electronic gadgets that we use in our buildings, but also in the number of services, that have moved from the status of optional to essential. Today use of voice and data communication, automation, remote control, and security has become very much essential and common place. The changed or the enhanced set of components in a building, their dependency on energy, influence on environment, demand for higher levels of reliability, speedy construction, minimum demand on maintenance effort, life cycle cost concept are factors which have changed the very process of design and construction of buildings. This aspect has been identified in the NBC 2005 and Part 0 of the NBC brings out the basic need of the integrated approach at all stages, such as the design, approvals, procurement, construction, commissioning, operation and maintenance of a building. Today one is more concerned with the cost of ownership and in particular the operation cost of a building, than the cost of construction, which brings the focus on the services as the services are the active elements and have the constant influence on the day to day cost, by way of upkeep, energy, etc.

The pressure on land has also been forcing us to go for large multistoried buildings. With the high rise buildings, the requirements placed on the services has also been on the increase, both in terms of facilities and in terms of reliability. With the high rises, issues like fire protection and emergency management, which were simple needs have become complex and highly professional. Transportation within the building complex, issues related to energy and environment today control our decisions to a very great extent.

Increased use of electrical gadgets in tall buildings brings with it complexities, which did not exist in single and two storied buildings. In particular, provisions for emergencies and fire safety requirements bring in a full set of facilities to be built in an integrated form.

Developments such as IT and IT based services have had a dual effect on the construction industry and the buildings in particular. The sudden growth of IT/ITES sector in the 1995 2000 period saw a mushrooming of tall buildings which were constructed at a very fast pace. These buildings had requirement of an extremely high degree of reliability and redundancy in the services, which hitherto existed only for a few sections of buildings such as air traffic control, operating theatres, ICUs and on-line computer centres. Such requirements have influenced the design principles and the manner of integration of the architectural design with the services design. This has necessitated the interdisciplinary exchange of information essential at all stages.

A decade ago one would never bother about the operating cost of a building. What one bothered about was the initial cost. Today one may be able to afford to build a complex building but may find a problem in dealing with its operating costs and the strategy to cover the electrical and fire hazards associated with high-rise construction.

Today we are growing more and more conscious of the environment around us. Energy and environment are today guiding most of our decisions. Our building systems have to take care of the environment. Quite often the requirement for one parameter conflicts with the other. Design practices are also constantly evolving. New products, new equipment and new techniques of construction are all influencing design.

This book aims at bringing out the design needs, codal requirements, standard good practice, emerging trends in integrated approach, influence of ever rising cost of energy and the awareness regarding preserving the environment. All these aspects are looked at and the book forms a convenient source of ready reference to get concise information on all these aspects.

Apart from being a source of consolidated information from various standards, codes and good practice, the author has also put in various empirical relations and thumb rules, which are very useful in the preliminary stages of the design. While these thumb rules have been tried and tested over time, they should be used with appropriate caution and should not avoid or skip detailed design.

Technology has been changing fast and new products are coming up every day. An attempt has been made to provide the latest information available with regard to new products, which have satisfactorily been in use in recent times.

Energy is progressively becoming expensive and energy conservation is gaining importance due to the effects of energy production on environment. India has taken a big step in its construction field by constructing quite a few Green Buildings. The book covers briefly the requirements for green buildings, the concept and the key parameters on the basis of which the evaluation is made.

The book should not only be used by the designers of the services for buildings, but also by architects, whose appreciation of the space needs, maintenance requirements, safety aspects and energy considerations will make the buildings more energy efficient, user friendly, environment friendly. All building professionals, architects, structural engineers, and other design engineers also require a source of reference regarding the services to participate in the team effort for the integrated approach to produce excellence in the ultimate product.

JN Bhavani Prasad
Former Director General (Works)
Central Public Works Department
Senior Advisor, Construction
Industry Development Council

The first edition of this manual was published seven years ago. The overwhelming response and lots of technical inputs sent by the readers have prompted me to come out with this edition. This manual is basically an effort to compile as much information as is possible, related to design methodologies and ready to use reference tools about all building services, in one place so that an engineer or a designer can understand in simpler terms what otherwise would be a gigantic task to browse through volumes of codes, bye laws and other literature on specific subjects.

During these years, construction sector has seen large-scale developments and changes in construction methodology. Stand-alone buildings have been replaced with large townships whereas low rise buildings have given way to skyscrapers. The construction is taking place with computational speed. Drawings which used to be two-dimensional are now prepared with computerized three-dimensional models incorporating the coordinated schemes of all services. Concept of green buildings had just taken off in India at that time while today the total footprint of green buildings here is to the tune of three billion square feet and is expanding at a rate of 15% annually.

The first edition bridged the gap between theoretical knowledge given to undergraduate engineering students in the form of derivation of formulas, equations and small examples on one hand and the practical methods used in real-world designing using pre-tabulated data, graphs and computerized tools on the other. The compulsion to write this book arose when at the beginning of my career, I felt the need to understand these methodologies and there was none to teach them. Over a period of time, these methods were imbibed by delving deeper and deeper into all aspects of designing, making visits to manufacturers facilities and understanding the conflicts that arise between the designed schemes *versus* site realities and ease of postconstruction maintenance.

It was also felt that there are books in the market which focus only on one of all the building services in detail, e.g. there are innumerous number of books on the subject of air-conditioning and almost all of them give theoretical knowledge related to the subject. There was hardly any book which could detail out the thumb rules on estimation, space requirements for various services, heat load calculation or duct and pipe design. The design of fire alarm, lifts and sprinkler systems were considered to be scary subjects. No book would detail out the method to calculate the electrical load for a residential apartment or a commercial building and the diversities to be adopted since all the occupants may not be using all the appliances at a time and number of subjects like that. This manual was designed with the intent of imparting practical knowledge to the reader who may be a consultant in one field of services design but wants to expand in the sphere of other branches or he/she may be a fresh graduate from engineering college and wants to start his/her career in design.

In the current edition, readers will make themselves conversant with the concept of green buildings which is the need of the day to conserve environment. So far green buildings were considered to be a complex mechanism for which some special knowledge was required. But in this edition these myths have been broken and the reader will find that he/she already knew everything about green buildings but only that he/she was unable to connect with it thinking of it as a specialized subject. This volume attempts to revive interest of a common man in nature and those of designers in developing more efficient and power saving designs through various environment-friendly techniques. They will find that small steps like preserving trees or top soil from the undisturbed site before construction and reusing the same after construction, using renewable sources of power, providing bins for collecting and segregating debris and recycling waste water and sewerage can fetch additional points for their projects and make them eligible for green rating.

So far emphasis had been on designing the electrical distribution for a stand-alone building or at best a small complex, but with the development of large townships, the designers are facing the challenge to think on a large scale and understand the philosophy of developing an integrated distribution system. Similarly with the development of computational techniques, the illumination design, selection of luminaires are being done with the help of patented software developed by the manufacturers, but the basic theory of optics was lost somewhere in the light of modern tools except that taught in the engineering colleges. In this manual, efforts have been made to re-invent the basic theory so that design elements are known to the designer before using computational techniques. It is not enough to install luminaires, say of 150/250 watts sodium vapour lamps for road illumination but calculation of horizontal and vertical illuminance and their ratio which increases night visibility and reduces road accidents. This will produce more efficient and innovative designs, enable selection of light fixtures to produce appropriate lux levels even if calculation software is used to avoid complex calculations.

The scope of air-conditioning design has been widened in this edition by incorporating the basic psychrometry as many readers suggested that a link with theoretical base was necessary to comprehend the practical designing methods. Similarly, heat load calculations have been further simplified with the help of examples of real-life projects elaborating on the method to calculate heat transfer coefficients of different construction materials, selection of parameters for summer and winter load calculations and much more on the request of readers.

Pumps are an important part of any building for efficient water supply system and also to ensure sufficient supply of water for fire-fighting. An efficient mechanical ventilation system is necessary for exhausting the smoke during fire and carbon dioxide fumes from the cars parked in the basements or to induct fresh air in the living floors by the use of intake suction fans. This volume explains, with the help of practical examples, the techniques to design the fan CFM capacity and the motor ratings. Since pumps and fans are one of the most power-consuming installations in any building after air-conditioning, their selection to operate at the best efficiency, by examining various operating parameters and using them in various combinations, will help to save power. The detailed methods to avoid overdesigning have been explained in this edition.

I will appreciate and welcome suggestions and invite any information the readers would like to share with me and fraternity of engineers at my email address: mittal@cmncc.com

A K Mittal

When the idea of constructing a building is conceived, the first and the foremost requirement is to work out an estimate. A consultant or a consulting firm is usually hired to compute the preliminary estimate of the cost of the entire project. Such an estimate includes the cost of construction, air-conditioning system, diesel generating set, internal electrical installation, fire alarm and fire fighting system, substation, etc. Since the civil works are not a part of the scope of this book, only electrical and mechanical (E&M) services estimation, design and execution have been covered here.

Who benefits: It is imperative for an E&M consultant to be well versed with various parameters based on which he/she can promptly submit an estimate to his/her client. The book is also of special utility for fresh graduates of electrical and mechanical engineering. It has been written with the objective of helping fresh graduates to integrate their theoretical knowledge with the demands of practical application. The students who wish to start their career as E&M consultants will benefit tremendously from this book.

E&M consultants are required to estimate and design all the services in an upcoming project and are supposed to possess specialization in all the areas. The book covers all the services required in a building along with the latest techniques available in the market. In this way, it is a reference manual and an E&M consultant s best guide.

Brief description of the contents of this book

Section A: **Preliminary Estimation** comprises the following two chapters:

1. **Preliminary Cost Estimation: Thumb Rules**
2. **Space Requirement for Various Electrical and Mechanical Services**

Chapter 1: **Preliminary Cost Estimation: Thumb Rules** illustrates the most practical method of framing and submitting preliminary estimates. By following this approach, you beat the competition by submitting estimates in time. Also, the estimates so framed are fairly reasonable in accuracy

Chapter 2: **Space Requirement for Various Electrical and Mechanical Services** guides you to assess the space requirement for DG set room, substation, shafts for electrical rising mains, fire hydrants, fire alarm panels, wiring and intercom services, space for air-conditioning plant rooms, space above false ceiling for supply and return air ducts, etc. The calculation of space requirements is the backbone of an efficient design of E&M services and this chapter guides you to plan the space accurately as per the various Indian Standards and National Building Code.

Section B: **Air-Conditioning** is further divided into following chapters:

3. **Heat Load Calculation**
4. **Calculation of Duct Sizes**
5. **Design of Chilled Water System**
6. **Selection of Chiller Machine**

Air-conditioning system involves the maximum finances for the client and efforts of the design engineer. These chapters give step-by-step method to calculate various parameters, sizes and capacities of the equipment involved. The electrical energy is becoming costlier day-by-day. This section describes in detail the various methods and devices for saving energy.

Section C: **Other Services** comprises the following chapters:

7. **Fire Alarm System**
8. **Automatic Sprinkler System**
9. **Lifts Design: Important Factors and Traffic Constraints**
10. **Substation**
11. **Internal Wiring and Electrical Distribution System**
12. **Electric Generator**

Chapter 7: **Fire Alarm System** is about detailed method of designing fire protection system, viz. type of fire detectors for different application, design of panels.

Chapter 8: **Automatic Sprinkler System** describes the various complexities involved in the design of sprinkler system, various types of sprinklers available and their usage.

Chapter 9: **Lifts Design: Important Factors and Traffic Constraints** illustrates the complicated calculations involved in traffic analysis for designing the speed of an elevator, size and the number of lifts required. This chapter provides a systematic approach to design an efficient elevator system in a building.

Chapter 10: **Substation** pertains to the method of calculating electrical load, diversity factor, design of transformer, HT/LT panels, design of various safeties, rating of switchgears and capacitors. The calculation of capacitance required to increases the power factor of electrical supply is the most critical item in economizing the consumption of electrical energy as many electric supply companies impose heavy penalties for a poor power factor.

Chapter 11: **Internal Wiring and Electrical Distribution System** is followed by Chapter 12: **Electric Generator** which describes in detail the various types of alternators, excitation systems and controls available in the market. It also describes the method to calculate and balance the resistive and inductive load likely to be fed through DG set.

A K Mittal

Acknowledgements

This edition has been a hard work of two years involving interaction with many field engineers, consultants and manufacturers in understanding their expectation out of this book by incorporating practical aspects of design wherever the traditional methods are in conflict with the practical difficulties faced at site. The decoding of green buildings from its complexities was made possible with the help of Mr M Anand of CII and Mr Apoorv Vij of GRIHA whom I wish to thank sincerely. I wish to thank Dr PC Jain, a name synonymous with the green building concept in India, who has very kindly written the Foreword to this edition of the book.

I thank my *Karmbhoomi* CPWD for giving me all the opportunities to experiment with the design philosophies and M/s Jaiprakash Associates Ltd where I widened my horizons on large scale township planning. Special thanks to Mr Mukesh Gupta, my colleague and Senior General Manager, M/s Jaiprakash Associates Ltd (formerly Chief Engineer, Housing Board, Haryana), for extending his heartfelt cooperation during the period of our togetherness.

I wish to thank my wife Mrs Sangeeta, my daughter Prapti, and son Armaan, who have been a source of great inspiration and courage whenever I take up such assignments.

Mr YN Arjuna and Mr PS Ghuman of CBS Publishers & Distributors need a special mention here whom I thank from the bottom of my heart as they did their best to bring out this volume in the present form once again and appreciated my work. I also thank all the readers who have responded with their valuable feedback about the past edition which has made this volume even more user-friendly.

A K Mittal

Contents

Foreword to the Second Edition   vii

Foreword to the First Edition   ix

Preface to the Second Edition   xi

Preface to the First Edition   xiii

SECTION A: PRELIMINARY ESTIMATION

1. Preliminary Cost Estimation: Thumb Rules   **3–7**

1.1 Internal Electrical Installation (IEI) 3

1.2 Substation and DG Set 3

1.3 Air-Conditioning 5

1.4 Ventilation 5

1.5 Lift and Escalators 6

1.6 Fire Alarm System 6

1.7 Fire Fighting 6

1.8 Building Management System 7

1.9 Public Address, Cable TV, Security System 7

2. Space Requirement for Various Electrical and Mechanical Services   **8–16**

2.1 Shaft for Electrical Distribution 8

2.2 Air-conditioning 9

2.3 Substation and DG Sets 10

2.4 Pump Room 10

2.5 Space for LV Services 13

2.6 Size of Lift Well 13

SECTION B: AIR CONDITIONING AND MECHANICAL VENTILATION

3. Air-Conditioning Principles   **19–23**

3.1 Comfort Standards 19

3.2 Designing the HVAC System 19

3.3 Properties of Air 20

3.4 Wet Bulb Temperature and its Importance 20

3.5 Adiabatic Saturation Temperature 20

3.6 Dew Point Temperature 21

3.7 Specific Volume of Humid Air 21

3.8 Measurable Air Properties 21

3.9 Mathematical Equations 22

4. Psychrometry   **24–30**

4.1 Psychrometric Chart 24

4.2 Psychrometric Processes 25

    4.2.1 Sensible Heating/Cooling 25

    4.2.2 Cooling and Dehumidification 26

    4.2.3 Cooling with Humidification of Air 27

    4.2.4 Adiabatic Chemical Dehumidification 27

    4.2.5 Humidification by Steam Injection 27

    4.2.6 Mixing of Air Streams 27

4.3 Combination Process on Psychrometric Chart 28

    4.3.1 Heating and Humidification 28

    4.3.2 Humidification and Heating 28

    4.3.3 Dehumidification and Cooling 28

    4.3.4 Humidification and Cooling 29

    4.3.5 Dehumidification and Heating 29

    4.3.6 Double Dehumidification and Cooling 29

    4.3.7 Adiabatic Humidification and Double Heating 29

    4.3.8 Mixing, Dehumidification and Heating 29

    4.3.9 Regenerative Heat Exchange from Air to Air 30

5. Heat Load Calculation   **31–70**

5.1 Theory in Brief *31*

    5.1.1 Asymmetrical Pattern of Heat Gain *31*

    5.1.2 Components of Heat Transfer *31*

    5.1.3 Mean Temperature Difference *32*

    5.1.4 How Heat Enters a Building *33*

5.2 Step by Step Approach: Heat Load Calculation *33*

    5.2.1 Summary Table *33*

    5.2.2 Proforma *33*

6. Duct Design and Sizing   **71–81**

6.1 Economical Factors in Duct Design *72*

6.2 Duct Design *73*

    6.2.1 Friction Loss or Pressure Loss Method *75*

    6.2.2 Equal Friction Method *75*

    6.2.3 Static Regain Method *77*

    6.2.4 Concept of Equivalent Length *78*

    6.2.5 Effect of Heat Gain/Loss on Duct Design *79*

6.3 Important Considerations in Duct Design *79*

6.4 Insulating Materials *80*

6.5 Economical Considerations *80*

6.6 Modern Tools for the Calculation of Duct Size *81*

7. Chilled Water System Design   **82–100**

7.1 Regulating Devices *82*

    7.1.1 Variable Speed (Frequency) Drives *82*

    7.1.2 Throttling or Two-way Control Valve *85*

    7.1.3 Bypass or Three-way Valve *87*

7.2 Various arrangements of Chilled Water Connections *87*

    7.2.1 Chiller Piping (Evaporator Side) *87*

    7.2.2 Chiller Piping (Condenser Side) *89*

    7.2.3 Water Pipe Connections (Load Side) *91*

    7.2.4 System Water Temperature *94*

7.3 Sizing Water Pipes *95*

    7.3.1 Pressure Loss in Pipe Fittings *95*

    7.3.2 Piping System Pressure Drop *95*

    7.3.3 Calculation of Water Quantity *96*

    7.3.4 System Pipe Sizing *96*

8. Chiller Machines and Selection Criteria   **101–116**

8.1 Refrigeration Cycle *101*

8.2 Unit of Refrigeration and Coefficient of Performance *101*

8.3 Theoretical Single Stage Vapour Compression Cycle *102*

    8.3.1 Effect of Variation of Suction (Evaporation) and Discharge (Condensation) Pressures on COP *103*

    8.3.2 Effect of Subcooling Liquid Refrigerant on COP and Refrigerating Capacity *103*

    8.3.3 Effect of Suction Superheat *103*

    8.3.4 Suction-Liquid Heat Exchanger *104*

    8.3.5 Actual Vapour-Compression Cycle *104*

    8.3.6 Multistage Vapour Compression Cycle *104*

8.4 Refrigerants and their Effect on Environment *106*

8.5 Compressor *106*

    8.5.1 Reciprocating Compressors *106*

    8.5.2 Rotary Compressors *107*

    8.5.3 Centrifugal Compressors *111*

    8.5.4 Limitation of Various Types of Compressors *112*

8.6 Vapour Absorption Refrigeration *113*

8.7 Comparison of Chiller Machines *116*

9. Mechanical Ventilation Fan Design   **117–136**

9.1 Mandatory requirement of NBC *117*

9.2 Fan Engineering Basics *117*

    9.2.1 Static Pressure *119*

    9.2.2 Velocity Pressure *120*

    9.2.3 Total Pressure *121*

    9.2.4 Analysis of Pressure Changes *121*

9.3 Selection of Fans *123*

    9.3.1 Centrifugal Fans *123*

    9.3.2 Axial Flow Fans *124*

    9.3.3 Special Application Fans *125*

9.4 Fan Operating Point *126*

9.5 Fan Performance Parameters *126*

9.6 Fan Laws and their Importance *127*

9.7 Calculations of Capacity of Fans *128*

9.8 Mechanical Ventilation and Smoke Control *131*

    9.8.1 Duct Material and Fire Barrier *132*

    9.8.2 Prevent Spread of Fire Between Zones *132*

    9.8.3 Location of Fire Damper *132*

    9.8.4 False Ceiling Material *132*

    9.8.5 Exhaust Fan in AHU *133*

    9.8.6 Lift/Staircase/Emergency Lobby Ventilation *133*

    9.8.7 Basement Ventilation *134*

9.9 Temperature Rating of Fan, Motors *134*

9.10 Basic Fan Selection *135*

SECTION C: ELECTRICAL SERVICES

10. Internal Wiring and Electrical Distribution System 139–149

10.1 Point Wiring 139
    10.1.1 Definition 139
    10.1.2 Scope 139
    10.1.3 Measurement of Point Wiring 139
    10.1.4 Point Wiring for Socket Outlet Points 140
    10.1.5 Group Control Point Wiring 140
    10.1.6 Twin Control Light Point 140
    10.1.7 Multiple Controlled Call Bell Points Wiring 140
10.2 Circuit and Submain Wiring 140
10.3 System of Distribution 141
    10.3.1 Control at the Point of Entry of Supply 141
    10.3.2 Distribution 141
    10.3.3 Wiring System 142
    10.3.4 Joints in Wiring 142
10.4 Ratings of Outlets (to be adopted for Design) 142
10.5 Capacity of Circuits 142
10.6 Conformity to IE Act, IE Rules, and Standards 142
10.7 Rating of Components 143
10.8 Wiring/Cables 143
10.9 Accessories 143
10.10 Switch Box Covers 143
10.11 Fittings 143
10.12 Switchgear and Controlgear General Aspects 143
10.13 Completion Plan and Completion Certificate 144
10.14 Earthing 144
    10.14.1 Statutory Requirement 144
    10.14.2 Supply System Requirement 144
    10.14.3 Special Earthing Requirements 144
    10.14.4 Material of Earth Equipments 144
    10.14.5 Type of Earthing Systems 144
    10.14.6 Earth Bus 145
    10.14.7 Location of Earth Electrode 145
    10.14.8 Earth Resistance 146
    10.14.9 Number of Earth Electrodes 146
10.15 Protection of Building against Lightning 146
    10.15.1 Principle and Zone of Protection 146
    10.15.2 Material of Protecting Components 147
    10.15.3 Earth Termination Network 148
10.16 Illuminating the Building 148
    10.16.1 Selection of Number of Fittings 148
    10.16.2 Selection of Type of Fitting 148

11. Substation 150–181

11.1 Electrical Load estimation 150
11.2 Components of a Substation 155
    11.2.1 Capacity and Number of Transformers 155
    11.2.2 Limitation on Transformer Size due to LT Switchgear 157
    11.2.3 Incoming Supply and Fault Level 157
11.3 Developing Single Line Diagram 157
11.4 Fault Level Calculation 157
11.5 Fault withstand Capacity 158
11.6 Protection System and Fault Discrimination 159
11.7 Capacitors Rating and Selection 159
    11.7.1 Reactive Power 159
    11.7.2 Methods of Power Factor Correction 164
    11.7.3 Classification of Capacitors 164
    11.7.4 Harmonics and Transients 165
    11.7.5 Capacitor Inductor Combination 167
    11.7.6 Harmonic Filter Circuit 167
    11.7.7 Components of a Capacitor Installation 169
11.8 Bus Bar, Rising Main and Bus Trunking 172
    11.8.1 Latest Trends: Air Insulated Bus Trunking 173
    11.8.2 Sandwich Insulated Bus Trunking 174
11.9 Instrument Transformers 174
    11.9.1 Current Transformers 174
    11.9.2 Potential Transformers 176
11.10 Cables 177
11.11 Battery Bank for Trip Supply 177
11.12 Ingress Protection (IP classification) 177
11.13 Panels 177
    11.13.1 Classification of Panel 177
    11.13.2 Forms of Separation 178
    11.13.3 Type Test Criteria 178
    11.13.4 Protection Requirements 178

12. Electric Generator 182–200

12.1 Basic Design Theory 182
    12.1.1 General 182
    12.1.2 Theory 182
12.2 Machine Types 184
    12.2.1 Rotating Field Type 184
    12.2.2 Rotating Armature Type 184
12.3 Operating Principles 184
    12.3.1 Introduction 184
    12.3.2 Self Excited, Rotating Field, Brushless, AC Generator with Electronic Voltage Control 184

12.3.3 Separately Excited, Rotating Field, Brushless AC Generator with Electronic Voltage Control *185*

12.3.4 Self Excited, Rotating Field, Brushless, AC Generator with Transformer Control *186*

12.4 Power Rating *186*

    12.4.1 Efficiency and Drive Power *187*

    12.4.2 Transient *187*

    12.4.3 Temperature, Altitude, Humidity *187*

12.5 General Comments on Variety of Loads *188*

    12.5.1 Constant or Steady State Loads *189*

    12.5.2 Transient or Motor Starting Loads *189*

12.6 Motor Starting methods *190*

    12.6.1 Star-Delta Starter *190*

    12.6.2 Direct on Line Starting *191*

12.7 Load Calculation *191*

    12.7.1 Load Details *191*

    12.7.2 Load Analysis *192*

    12.7.3 Load Summation *193*

    12.7.4 Derate Factors *194*

    12.7.5 Standard Sizing *194*

    12.7.6 Power Factor Correction *194*

12.8 Parallel Operation of AC Generators *195*

    12.8.1 Introduction and Theory *195*

    12.8.2 Load Sharing *196*

    12.8.3 Setting up Procedure *197*

    12.8.4 Step by Step Setting Procedure *198*

    12.8.5 Working Procedure *199*

    12.8.6 Difficulties *200*

    12.8.7 Neutral Interconnections *200*

13. Pumping System   **201–231**

13.1 Introduction *201*

13.2 Fluid Properties *201*

13.3 Types of Flow through Pipes *201*

13.4 Hydraulic Terms and Equations *202*

    13.4.1 Hydraulic Mean Depth (*m*) *202*

    13.4.2 Bernoulli s Theorem *202*

    13.4.3 Continuity Equation *202*

    13.4.4 Loss of Head in Pipe Flow *202*

    13.4.5 Loss of Head in Valves and Fittings *204*

    13.5.6 Limitation and Comparison of Various Formulas *204*

13.5 Types of pumps *205*

    13.5.1 Centrifugal Pump *206*

    13.5.2 Positive Displacement Pumps *208*

13.6 Specific Speed, Pump Type, Pumping Power *208*

13.7 General Characteristic Curves *210*

13.8 Constant Efficiency Curve or ISO Efficiency Curve (Muschel curves) *211*

13.9 Pump Performance at Various Heads *212*

13.10 Pumps Operating in Series *212*

13.11 Pumps Operating in Parallel *213*

13.12 Throttling of Pump Discharge *214*

13.13 Net Positive Suction Head (NPSH) *215*

13.14 Water hammer *215*

13.15 Sources of Water and Methods of Pumping *216*

    13.15.1 Categories of Surface Water *216*

    13.15.2 Categories of Ground Water *216*

    13.15.3 Developing a Well *217*

    13.15.4 Tube Well Pumping Station *218*

13.16 Water Demand Calculation *219*

13.17 Types of Valves *221*

    13.17.1 Line Valves *221*

    13.17.2 Pressure Relief Valves *221*

    13.17.3 Check Valves *221*

    13.17.4 Ball Valve or Ball Float Valve *222*

13.18 Selection of a Pump *222*

13.19 Methods of Starting a Pump V_s Operating Point *226*

13.20 Achieving Efficiency and Economy in Pump Selection *227*

    13.20.1 Effect of Speed on Pump Selection *227*

    13.20.2 Operating Pumps in Parallel *229*

    13.20.3 Stop/Start Control *229*

    13.20.4 Eliminating Discharge Valve (Throttling) *230*

    13.20.5 Eliminating By-pass Control *230*

    13.20.6 Impeller Trimming *230*

    13.20.7 Smaller Steps in Energy Conservation *230*

    13.20.8 Comparison of Different Energy Conservation Techniques *231*

14. Illumination Engineering   **232–253**

14.1 Terminology *232*

14.2 Types of Lamps *235*

14.3 Understanding a Polar Curve *237*

14.4 Interpretation of Data given by Manufacturer s of Luminaires *239*

14.5 Calculation of Illumination at a Point *242*

14.6 Indoor Lighting Design *243*

14.7 Recommended Illuminance Ratio and Surface Reflections *246*

14.8 Polar Curve and Lumen Calculation *246*

14.9 Outdoor Lighting Design *248*

    14.9.1 Lumen Calculation Method *248*

    14.9.2 Level of Illumination *253*

14.10 Mounting Height and Spacing *253*

14.11 Computer Aided Lighting Designs *253*

SECTION D: OTHER SERVICES

15. Green Buildings in Simple Steps 257–265

15.1 What is a Green Building 257
15.2 Benefits of Green Buildings 258
15.3 Evaluation of a Green Building 259
    15.3.1 Procedure of Evaluation 259
15.4 Simple Steps Towards Green Building 260
    15.4.1 Site Selection and Development 260
    15.4.2 Water Conservation 262
    15.4.3 Energy Consumption, Generation and Atmosphere 262
    15.4.4 Regional, Renewable and Recycled Material 264
    15.4.5 Indoor Environment 264
    15.4.6 Innovative Designs 265

16. Fire Alarm System 266–280

16.1 Fire Alarm System Basics 266
    16.1.1 Components of a Fire Alarm System 266
    16.1.2 Type of Circuits 267
    16.1.3 Functions to be Performed by FAS 269
16.2 Stages of Fire and Detection 269
    16.2.1 Detectors for Incipient Stage 270
    16.2.2 Detectors for Smouldering Stage 271
    16.2.3 Air Sampling Smoke Detectors 271
    16.2.4 Detectors for Flame Stage 272
    16.2.5 Detectors for Heat Stage 273
16.3 Areas to be Monitored by Heat Detectors 275
16.4 Calculation of Number of Detectors 275
    16.4.1 Effect of Ceiling Height 275
    16.4.2 Supply Air Duct 275
    16.4.3 Return Air Duct 275
    16.4.4 Computer Installations 275
    16.4.5 Calculation of Number of Heat Detectors and their Spacing 275
    16.4.6 Basic Concepts 276
    16.4.7 Heat Detectors 276
    16.4.8 Minimum Number of Heat Detectors 276
    16.4.9 Smoke Detectors 276
    16.4.10 Effect of Air Dilution 276
16.5 Effect of Ceiling Construction on Heat and Smoke Detector Location 276
16.6 Designing of Control and Indicating Panels 277
16.7 Sounders 278
    16.7.1 Number of Fire Alarm Sounders 279

16.8 Power Supply Equipments 279
    16.8.1 Requirements of Power Supply to FAS 279
    16.8.2 Standby Battery Supply 279
16.9 Wiring 279
    16.9.1 Detector Circuits 279

17. Automatic Sprinkler System 281–299

17.1 Definitions 282
17.2 Type of Systems 282
17.3 Classification of Occupancies 284
    17.3.1 Light Hazard Occupancies 284
    17.3.2 Ordinary Hazard Occupancies 284
    17.3.3 Extra Hazard Occupancies 285
17.4 Sprinkler Types 285
    17.4.1 Selection on Type of Use 285
    17.4.2 Selection on Temperature Rating 287
17.5 Spacing and Area of Coverage of Sprinklers 288
    17.5.1 Upright Sprinklers 288
    17.5.2 Sidewall Sprinklers 290
17.6 Water Requirement 292
    17.6.1 Pipe Schedule Method 292
    17.6.2 Hydraulic Calculation Method 292
17.7 Hydraulic Calculation 292
17.8 Hydraulically most Demanding Area 296
17.9 Pump Capacity and Water Storage 297
17.10 Piping 298

18. Lifts Design: Important Factors and Traffic Constraints 300–315

18.1 Types of Lifts 300
18.2 Classification of Lifts 300
18.3 Essential Components of Lifts/Elevators 301
    18.3.1 Machines 301
    18.3.2 Brakes 301
    18.3.3 Ropes 301
    18.3.4 Sheave 301
    18.3.5 Divertor Pulley 301
    18.3.6 Counterweights 302
    18.3.7 Governor 302
    18.3.8 Guide 302
    18.3.9 Buffers 302
    18.3.10 Door and Door Operators 302
    18.3.11 Selector 302

18.3.12 Travelling Cables *302*

18.3.13 Hoisting Motor *302*

18.3.14 Controller *303*

18.3.15 Car *304*

18.4 Lift Design: Important Factors and Traffic Constraints *304*

    18.4.1 Human Constraints *304*

    18.4.2 Traffic Constraints *305*

18.5 Traffic Analysis *307*

    18.5.1 Positioning of Lift *307*

    18.5.2 Population in a Building *307*

    18.5.3 Calculation of Time Factors *307*

    18.5.4 Handling Capacity and Number of Lifts *308*

18.6 Power Control *313*

    18.6.1 AC Resistance Control *313*

    18.6.2 Variable Voltage Variable Frequency Drive *313*

18.7 Operating System *313*

    18.7.1 Single Automatic Operation *313*

18.7.2 Full Selective Collective *313*

18.7.3 Duplex Collective *314*

18.7.4 Group Automatic Operation *314*

18.8 Hydraulic Elevators *314*

18.9 Dimensional and Structural Requirements *314*

    18.9.1 Dimensional Requirement *314*

    18.9.2 Structural Requirement *314*

18.10 Classification According to Interiors *315*

18.11 List of IS Codes *315*

Annexures **317–358**

A Air-conditioning and Motor Starter Tables *317 341*

B Conversion Factors *342 343*

C Electrical Tables *344 348*

D List of IS Codes *349 356*

E List of International Codes *357 358*

Bibliography *359 360*

Index *361 365*

Section A

Preliminary Estimation

This section comprises of two chapters that form the framework of all the preliminary designs.

In the very beginning, a designer must know the fundamental costing of all the services being designed by him. It will be a futile exercise to go into the detailed costing calculations if at a later stage the client shelves the project.

Chapter 1

This chapter lists some basic thumb rules based on the prevailing market rates that can be applied very easily and a preliminary estimate will be ready within no time. Once the project is sanctioned and it is desired to submit the detailed costing, then one can start framing the specifications, brands to be used and prepare the actual costing of the project. The designer shall update these figures with the prevailing cost index and any price fluctuation in respect of steel or other components.

Chapter 2

Once the project has been sanctioned and the architect has prepared the preliminary plans, the space required for housing all the electrical and mechanical services is marked on the plans. This has to be done keeping in mind the facilitation of various movement patterns, population patterns at different locations, proximity with the area being served in synchronization with the various norms and bye-laws issued by local authorities from time to time.

This chapter illustrates in the form of figures, the sizes of shafts, substation and air-conditioning plant rooms, pump house and space requirements for various other electrical and mechanical services to be provided in the building. The user is, however, advised to keep in touch with the local bye-laws, National Building Code and any changes notified from time to time before attempting to issue the requirements to the architect.

Preliminary Cost Estimation: Thumb Rules

As soon as the plan to construct a building is conceived, the investor/owner/builder wants to know the total financial involvement, i.e. the cost of plot, building, electrical and mechanical services.

The architect has to be engaged who prepares preliminary sketches/models of the proposed complex. Based on these sketches, the civil/electrical/mechanical engineers/consultants prepare the cost of constructing the building and electrical/mechanical services respectively.

At this stage, the only requirement is to apprise the owner of the building about the preliminary cost so as to enable him to decide the source of financing the project and thus finally make up his mind whether to go ahead with the project or not.

Thus, it is very essential that the designers are well aware of the costing of various services prevailing in the market.

The designer at the preliminary stage has no other data except the floor area and the purpose for which the building is likely to be used. Thus, he must be aware of certain thumb rules, based on which he can submit the preliminary costing to his client.

This chapter describes these thumb rules in the most simplistic manner which will be of great help to designers.

The scope of this chapter is limited only to the costing method of electrical and mechanical services.

1.1 INTERNAL ELECTRICAL INSTALLATION (IEI)

IEI comprises of various sub components like main distribution boards (MDB), sub distribution boards (SDB), feeder pillars (FP), LT cables, rising mains, internal wiring, earthing, lightning protection and light/fan fixtures.

The MDB, SDB, FP, LT cables and rising mains are the major components of this subhead for costing

purpose and the other components can be clubbed to determine lump sum costing as below:

1.	MDB, SDB, FP	Rs. 550 per sq m
2.	LT aluminium cables, copper rising main	Rs. 550 per sq m
3.	Internal wiring with copper cables	Rs. 950 per sq m
4.	Earthing, lightning, protection, light and fans, external light fixtures and cabling	Rs. 400 per sq m
	Total IEI costing	Rs. 2450 per sq m

1.2 SUBSTATION AND DG SET

One of the major cost intensive component of a building is the substation and is called the heart of the building. The substation comprises of high tension (HT) panels, HT cabling between HT panel and transformers, low tension (LT) duct or cabling from transformers to LT panels.

In order to estimate the cost likely to be incurred on the installation of a substation, the fore most requirement is the estimation of a realistic load in kW and future projection, for say next 10–15 years.

This calculation requires the experience in assessing the usage pattern leading to calculation of diversity in demand expected at substation.

The electrical load expected in a building can be calculated on the basis of various thumb rules and practices followed as shown in Table 1.1. Since all the electrical points provided in a residence or office/commercial building do not operate at the same time, there is a usage factor that shall be incorporated while assessing the total load in kW of unit.

Similarly, all the units (residential or commercial) in a tower will have a different usage pattern, i.e. all of them

Table 1.1a Assessment of electrical load of a building

	Facility type	Individual demand load basis W/m²	Diversity at substation level	Diversity at grid level	Grid demand as percent of individual demand	Diversity for Generator load	Generator demand as percent of individual demand	Remarks
	(a)	*(b)*	*(c)*	*(d)*	*(e) = c × d*	*(f)*	*g = c × f*	
A.	University							
	1. Academic and Administration	59	60%	50%	30%	40%	24%	
	2. Residential and hostel	59	60%	50%	30%	40%	24%	
	3. Recreational	86	60%	50%	30%	40%	24%	Includes AC load
B.	Commercial	86	75%	50%	37.5%	40%	30%	Includes AC load
C.	Residential Flats							
	1. 1 BHK	5 kW	75%	50%	37.5%	40%	30%	
	2. 2 BHK	6 kW	75%	50%	37.5%	40%	30%	
	3. 3 BHK	7 kW	75%	50%	37.5%	40%	30%	
	4. 4 BHK	8 kW	75%	50%	37.5%	40%	30%	
	5. 5 BHK	9 kW	75%	50%	37.5%	40%	30%	
	Residential Plots							
	1. 150 sq. mt	15 kW	75%	50%	37.5%	40%	30%	
	2. 200 sq. mt	18 kW	75%	50%	37.5%	40%	30%	
	3. 250 sq. mt	20 kW	75%	50%	37.5%	40%	30%	
	4. 300 sq. mt	22 kW	75%	50%	37.5%	40%	30%	
	5. 350 sq. mt	24 kW	75%	50%	37.5%	40%	30%	
	6. 400 sq. mt	27 kW	75%	50%	37.5%	40%	30%	
	7. 450 sq. mt	30 kW	75%	50%	37.5%	40%	30%	

Note: Usage factor of 0.75 for non-residential and 0.5 for residential units has been considered for arriving at values in column (b).

Table 1.1b Equipment load basis for estimation

Service type	Load basis
1. Air conditioning	
– High side (AC plant equipments)	1.0 kW per ton
– Low side (AHU equipments)	0.25 kW per ton
– Mechanical ventilation	250 sq ft/kW
2. Light and power	20 W/sq mt
3. Window/split air conditioning (1 Ton per 10 square meter)	1.2 kW per ton
4. Drinking water pump (18 storey building)	15–20 kW
5. Fire fighting main and sprinkler Pump (18 storey building)	70–85 kW
6. Jockey pump	10 kW

will not be using the maximum load derived after considering the usage factor as described above. Thus there will be diversity factor to arrive at the net load expected from one or multiple number of towers on the substation. If there are a number of substations, say, in a large township, then the net demand load on grid sub-station should be derived by further multiplying the total derived load of all substations with a grid diversity factor.

After the electrical demand load on a substation has been calculated, the number of transformers shall be worked out. The usual practice is to divide the load in 2 to 5 transformers in such a way that all the transformers are loaded not more than 55–60% of their rated capacity. This is due to the reason that the best efficiency of transformers is within this loading range. This also takes care of future increase in load for the next 15–20 years. However, keeping such a high spare capacity increases the first time capital cost. Thus owners/developers usually tend to adopt 75–80% loading and reduce the capital cost.

The high rise building have a mandatory requirement of an alternate source of power, i.e. generating sets to power lifts, drinking and fire lighting pumps and other essential services. Depending on the nature of building and client requirement, owner may plan to provide generator back up only for essential services or part/full load of the building. Thus knowledge of power consumed by all electrical appliances is necessary to calculate the capacity and costing of generating sets.

The capacity of generating set can thus be worked out. The generator manufacturers recommend not to load the equipment beyond 80%. Similarly 20% reserve may be kept for future increase in load.

In case of high loads delivered by generator of capacity, say 1500 KVA, the nature of use of building may be kept in mind before deciding the capacity and number of generating sets.

Table 1.2 Emergency load on generators

S. no.	Equipment	Capacity	Power consumption (kW)
1.	Lifts (Elevator) 18 floors (all elevators to be on generator back up)	13 Passengers 16 Passengers 20 Passengers	10 14 16
2.	Fire pumps (main and sprinkler)	18 storey	70–85 kW
3.	Water supply pumps	18 storey	15–20 kW
4.	Light and fans	As required by client usually 50%	
5.	External security lights	As required by client usually 100%	
6.	Blowers of air handling units (AHU)	100%	0.25 kW/ton
7.	Ventilation fans of basement	100%	250 sq ft/kW
8.	Any other special load	As required by client	

As an example, a court building or an office building will function during day time, requiring a power back-up of 1500 KVA. But in the evening time or say at night, the power back up may be required only for few lifts, security lights and compound lights. This load may not be more than 250 KVA. In this case, there is no point in running a generator of 1500 KVA during night time. Thus, two generators of 1250 KVA and 250 KVA can be installed for selective operation. This will result in huge saving in operation cost.

The costing of substation equipments can now be projected in the preliminary sheet as below:

1. HT panels with VCB : Rs. 6,00,000 per panel
2. Transformer (oil type) : Rs. 1000 per KVA
3. Transformer (dry type) : Rs. 2700 per KVA
4. LT panel with capacitor bank (based on installed transformer capacity) : Rs. 2500 per KVA
5. Diesel generating set : Rs. 15000 per KVA
6. Gas generating set : Rs. 28,000 per kW$_e$

1.3 AIR-CONDITIONING

The estimation and calculation of air conditioning cost and load in tonnes at preliminary estimation stage is very challenging as major part of *E & M* costing comes from HVAC. The subject of HVAC comprises of number of equipments ranging from chiller machines to air handling units to ducts and grills. These equipments are basically classified into two categories:

1. **High side equipments:** These are the equipments which are installed in AC plant room. Centrifugal/reciprocating/vapour absorption chillers, condenser/chilled water pumps, pipes, electrical panels and cabling form part of high side equipments. Cooling towers though not installed in AC plant room is also considered part of high side equipments.
2. **Low side equipments:** Air handling units, control panels, control cables, chilled water plumbing from AC plant room to AHU room, 2 loop/3 loop plumbing, supply as well as return ducting, grills and diffusers, all form part of low side equipments.

When a new project is to be started, both high side and low side equipments cost is required to be calculated. But when an existing system is to be upgraded or equipments are to be replaced, then cost of individual equipments are to be calculated. The air conditioning load is to be calculated on per square feet area basis. For preliminary cost estimations, 200 square feet of air conditioned area can be covered with 1 Tonne of equipment (central AC) with a normal room height of 10 feet and top floor roof treatment so as to have maximum allowed heat conductivity of 0.12 BTU/hr ft^2 °F.

However, for window/split air conditioning units 100 square feet per tonne is the norm.

The estimated cost of air conditioning equipments can be worked out as below:

1. High side equipments : Rs. 75,000/- per tonne
2. Low side equipments : Rs. 25,000/- per tonne

1.4 VENTILATION

The buildings with basement, particularly used for car parking, substation, DG sets and air-conditioning plant room are required to be ventilated. As per the latest norms provided in NBC-2005, the minimum air changes for normal running of the building shall be twelve per hour, while during the fire time, the minimum air changes shall be thirty per hour.

The various components of a ventilation system are exhaust air blowers (to be designed for fire conditions), air washers for plant room/DG room/kitchen, ducting for basement ventilation, stair/lift lobby/lift well pressurization, etc. The costing based on per sq m area of basement is given below:

1. Exhaust air blowers : Rs. 350 per sq m
2. Air washers for plant room/DG room/kitchen : Rs. 60 per sq m
3. Ducting for basement ventilation : Rs. 160 per sq m
4. Ducting for stair/lift lobby/ lift well pressurisation : Rs. 60 per sq m
5. Grilles/diffusers/toilet exhausts : Rs. 100 per sq m

1.5 LIFT AND ESCALATORS

The lifts and escalators are used to handle the total occupancy including visiting population of a building. The number of lifts and escalators shall be determined very carefully by the designer because the efficiency of any building is dependent on the fast movement of people to their destination with least waiting time. This calculation is described in chapter on "Lift Design: Important factors and Traffic Constraints". The cost of lift and escalators shall generally be based on the configuration of vertical and horizontal movement expected out of final design, but for broader estimation purpose, the costing provision as follow can be made in preliminary estimate figures:

1. Lifts cost in lakhs of rupees for 10 storey building
   i. for 1.0 MPS : 1.25 × passenger capacity × no. of lifts
   ii. for 1.5 MPS : 1.5 × passenger capacity × no. of lifts
2. Escalator: Rs. 50,00,000 (Lump sum)

Note: Add Rs. 50,000/- for each floor increase.

Detailed calculations of number and bank of lifts can be done at a later stage when detailed estimates are required to be prepared.

1.6 FIRE ALARM SYSTEM

The fire alarm system is of three types.
1. Conventional
2. Intelligent
3. Hybrid

The use of each type varies with the use of the building, sophistication and control required to be achieved, with the final objective in mind that fire must be detected and controlled under any circumstances.

The costing of various classes of fire detection system are as under:

1. Conventional fire alarm system : Rs. 425 per sq m
2. Intelligent (addressable) fire alarm system : Rs. 800 per sq m
3. Hybrid Fire alarm system : Rs. 500 per sq m

The difference in three categories of system is explained below.

a. **Conventional fire detection and alarm system:** Smoke detectors are generally provided in all air-conditioned areas and heat detectors are provided in all non air-conditioned areas. Response indicators are provided outside the shop/room/hall. Bell-push and hooter are provided at lift lobby and staircase. A central control panel is provided in the fire control room located at ground floor which is connected to detectors, bell-pushes and hooters.

b. **Intelligent fire detection and alarm system:** Intelligent detection and alarm system is provided by using addressable detectors. In this case, each detector is monitored separately, i.e. status of fire detectors, whether it is functioning or not and to activate the alarm on detecting smoke in that area. Response indicators are not required outside the rooms as all the detectors are addressable. In case of fire, the detectors sense the fire and give an indication in the panel where the fire has taken place and at the same time an alarm is given in the entire building, or zone wise. In most of the modern building, intelligent detection and alarm system is provided, irrespective of cost as it is a latest system available and it works effectively.

In addition to this, a new detection system is available which is *very early smoke detection* analysis, which continuously analyzes the air, and smoke indicator will send signal to fire alarm panel accordingly. The speakers provided for fire alarm system can be used for public address for emergency announcement.

c. **Combination of intelligent and conventional fire detection and alarm system, i.e. hybrid system:** This system consists of microprocessor based (addressable) fire alarm control panel, communication cables, conventional photo electric smoke detector and intelligent photo electric smoke detector, conventional heat detector, response indicator, conventional manual call points, hooters, isolator module and control cabling. Some of the critical areas of the complex are provided with the addressable detectors and rest of the areas and especially basement are provided with conventional detector/devices. The monitoring of zone/intelligent devices is done at the main control panel. Detector are also provided above false ceiling as per code. System is provided with backup power from 24 volt DC storage batteries. This will facilitate to reduce the cost and achieve the required protection.

1.7 FIRE FIGHTING

The fire fighting system comprises of pumps, fire hydrants, wet riser, sprinkler system, fire extinguishers and exit signages. The costing of various items are as given below.

1. Pumps : Rs. 85 per sq m
2. Fire hydrant system : Rs. 500 per sq m
3. Sprinkler system : Rs. 700 per sq m
4. Fire extinguisher and signages exit : Rs. 35 per sq m

The water requirement for various categories of building has been codified in National Building Code (NBC–2005) which can be referred to by the designer.

1.8 BUILDING MANAGEMENT SYSTEM

The PLC based building management system has become an essential part of the public building these days. The control of functioning of air-conditioning plant, pumps, AHU's variable speed drives, substation, breakers, load management are all done by the building management system.

The expenditure on this account can be safely estimated at the rate of Rs. 5,500 per sq m.

1.9 PUBLIC ADDRESS, CABLE TV, SECURITY SYSTEM

These equipments are required to be provided in all parts of the building including basement. The cost of providing public address, cable TV, security system and music system can be taken as Rs. 200 per sq m of gross area.

The above thumb rules have been given as a matter of convenience only for designers. These rules shall be applied with caution as costing parameters will vary with the usage of building, occupancy, dollar to rupee conversion rate and of course the cost index issued by Government of India from time to time.

The designer or user is advised to exercise their wisdom as all applications are different in nature and vary with the type and usage of the building. They may have to add or modify certain parameters based on the type of building, facilities to be provided like percentage of power back up as per client requirement.

Space Requirement for Various Electrical and Mechanical Services

Once the preliminary cost estimation is over, and the client/owner has given a final nod to go ahead with the project, the designs and plans of the buildings are prepared. The architect is required to be informed about the requirement of space for various electrical and mechanical services to be provided in the buildings.

Unless these spaces are decided carefully, the efficient working of equipments may not be feasible. Simultaneously, the space demanded should not be very large as every sq m of land costs and is major cost of project.

This chapter explains in detail the space requirement based on experience and as defined in various codes and bye-laws for this purpose.

2.1 SHAFT FOR ELECTRICAL DISTRIBUTION

Nowadays, multistoried buildings are fed with electricity through rising mains, which is basically an enclosed chamber housing bus-bars for phases and neutral. Nowadays with the advent of technology, the compact rising main which is very small in size compared to the conventional ones, are provided. These rising mains about 150 mm wide at the ground floor are connected to a suitable capacity switching unit about 450 mm wide. Thus rising main should not be run adjoining side wall of the shaft.

As seen in the Fig. 2.1, the rising main has to pass through a shaft all throughout the height of the building. The shaft shall be wide enough to house the rising mains, electrical distribution board and sub-distribution board as per the scheme. The ground floor shall have enough space to house the switching unit usually called end feed unit, which in itself occupies a very large space. This is because the switching unit itself is very large and large size cable termination, bending of cable needs sufficient space.

Generally, there are two rising mains, i.e. one for light and fans and another for power and air-conditioning load (say AHU or high side equipments).

Hence, the following sizes should be recommended for electrical rising mains.

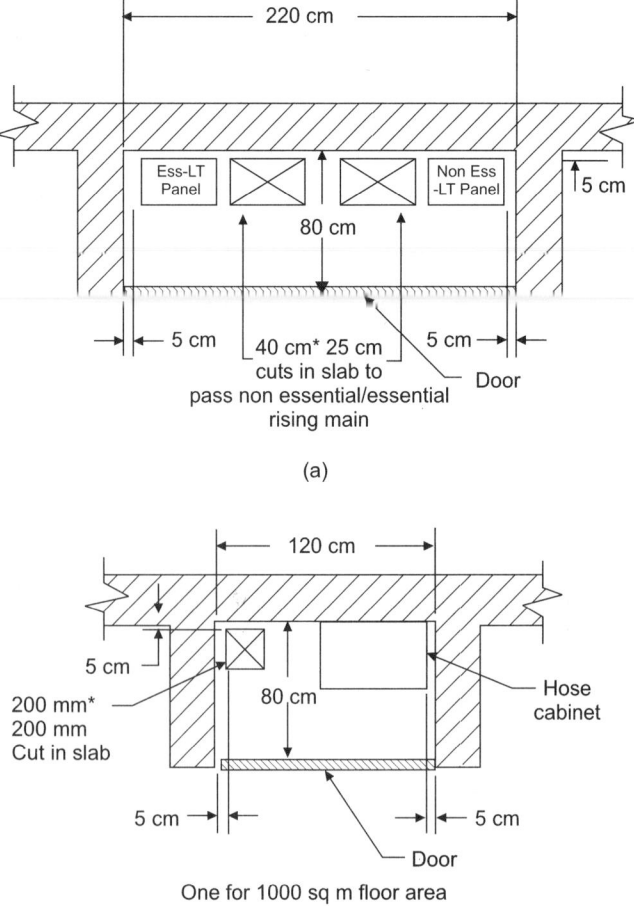

(a)

One for 1000 sq m floor area

(b)

Fig. 2.1 (a) Electrical rising main shaft, (b) Wet riser shaft

Width of One Shaft

One rising main   :   1.2 m wide   ⎫
Two rising mains   :   2.2 m wide   ⎬ both 80 cm. deep
                                                ⎭

2.2 AIR-CONDITIONING

The air-conditioning equipments comprise of high side as well as low side. The space requirement for installation of various equipments of both type of systems is discussed below.

(a) High Side

In an air-conditioning system, high side equipments comprise of chiller machine, condenser water pump (water cooled system) or blowers (air cooled system), chilled water pump, cooling tower, electrical panel.

The general arrangement and the number of equipments will decide the space requirement. A sample AC plant room is shown in Fig. 2.2 alongwith dimensions. The actual requirement may be worked out depending on the capacity of equipments and their numbers. However, as a general guideline:

Size of central plant room: 5 sq m + 0.5 sq m per ton of refrigeration.

(Clear height not less than 5.00 m below soffit of beam).

The height of the AC plant room shall not be less than 5.00 m as considerable height of the room is occupied in laying chilled water pipes and condenser water pipes as well as cable trays.

Since the cooling towers are required to be located in an open space, sufficient space for cooling tower either on roof top or on ground, where sufficient air movement is available shall be marked as shown in Fig. 2.3.

Space requirement for cooling tower
Natural draft   :   0.40 sq m per TR.
Forced draft   :   0.15 sq m per TR.

(b) Low Side

In the low side system, air handling unit, supply air duct as well as return air duct consume most of the space. The building height shall not be less than 3.6–3.9 m, so that the false ceiling height above floor level does not reduce below 2500 mm after accommodating depth of ducts.

This much floor height is very essential, if the central air-conditioning with ducted air supply is to be provided. However, if fan coil unit is to be provided, then the floor height of 3.10 m may be sufficient as duct of small size for fresh air will only be required. No return air duct or open space for return air need to be left.

The air handling unit shall be provided in an enclosed room and preferably in the center of the floor to be air-conditioned so that ducts of smaller size run on both

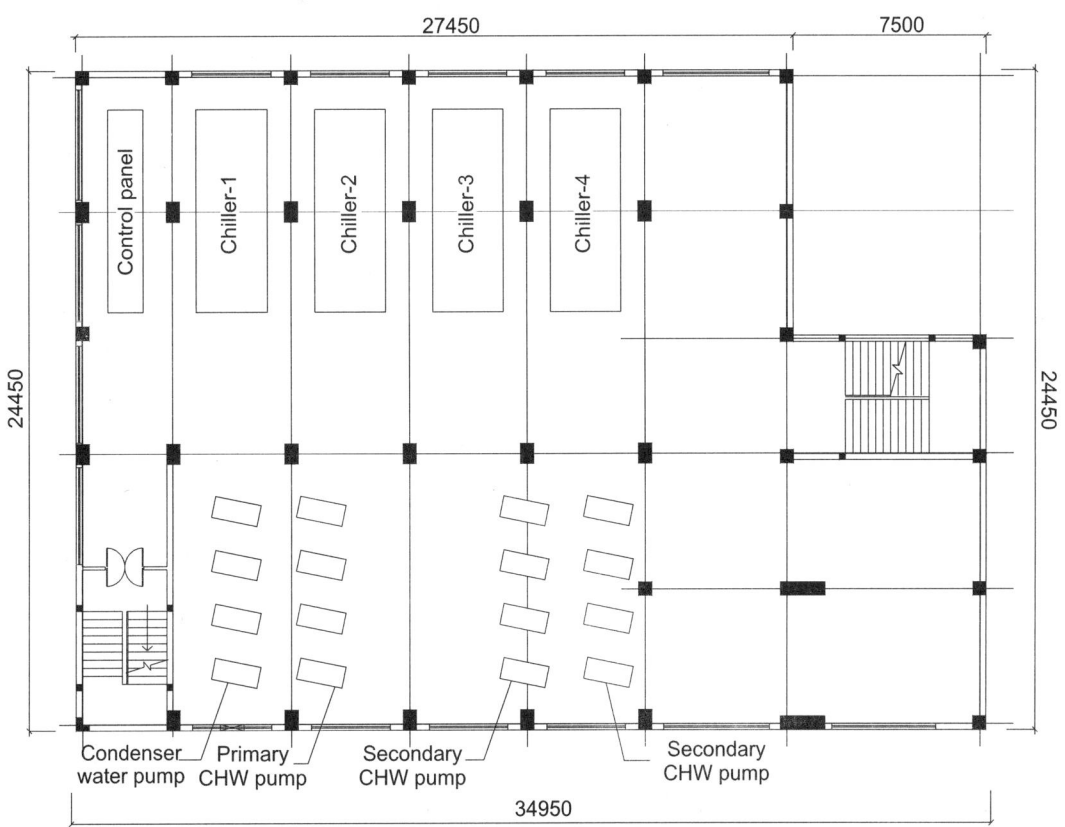

Fig. 2.2 Space required for chiller and pumps

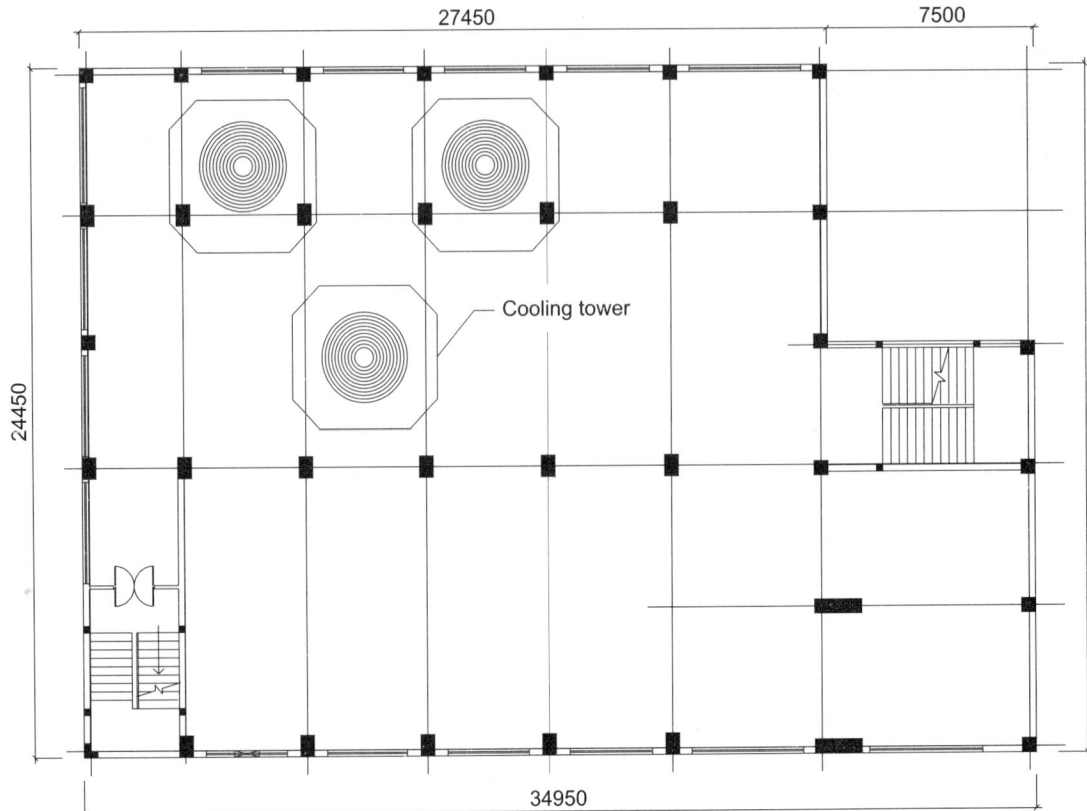

Fig. 2.3 Space required for cooling tower

the sides compared to the size of duct, when AHU room is provided on one side of the corridor.

The size of weather maker room (AHU Room): 5 sq m + 0.5 sq m per ton.

2.3 SUBSTATION AND DG SETS

The space requirement for substation and DG set shall depend on the capacity and number of equipments to be provided. A general layout diagram of a typical substation, AC plant cum DG set room is shown in Fig. 2.4.

The height of the substation/DG set room shall be 4.5 m to allow heat to be dissipated, for allowing space for lifting and lowering of equipments and for laying of overhead cable, trays, etc.

A rolling shutter of 2.5 m width and 3 m height shall be provided.

As a thumb rule, the space mentioned in Table 2.1a can be earmarked for substation and DG set.

Additional area that is required for one generator is shown in Table 2.1b.

The clear height required for the generating set room shall be a minimum of 3.6 m up to 100 kW capacity and 4.5 m for higher capacities.

2.4 PUMP ROOM

The fire fighting pump set and water supply pump set shall be provided in the same pump room, so that they can draw water from the adjoining underground tank (Fig. 2.5).

Table 2.1a   Substation area

Substation with transformer capacity of	Total transformer room area required	Total substation area required i/c HV/MV panel transformers but without generators	Suggested minimum face width
2* 500 kVa	36.00 sq m	130.00 sq m	145 m
3* 500 kVa	54.00 sq m	172.00 sq m	190 m
2* 800 kVa	39.00 sq m	135.00 sq m	145 m
3* 800 kVa	58.00 sq m	181.00 sq m	190 m
2* 1000 kVa	39.00 sqm	149.00 sq m	145 m
3* 1000 kVa	58.00 sq m	197.00 sq m	190 m

Note: The clear height required for substation equipments shall be a minimum of 3.6 m.

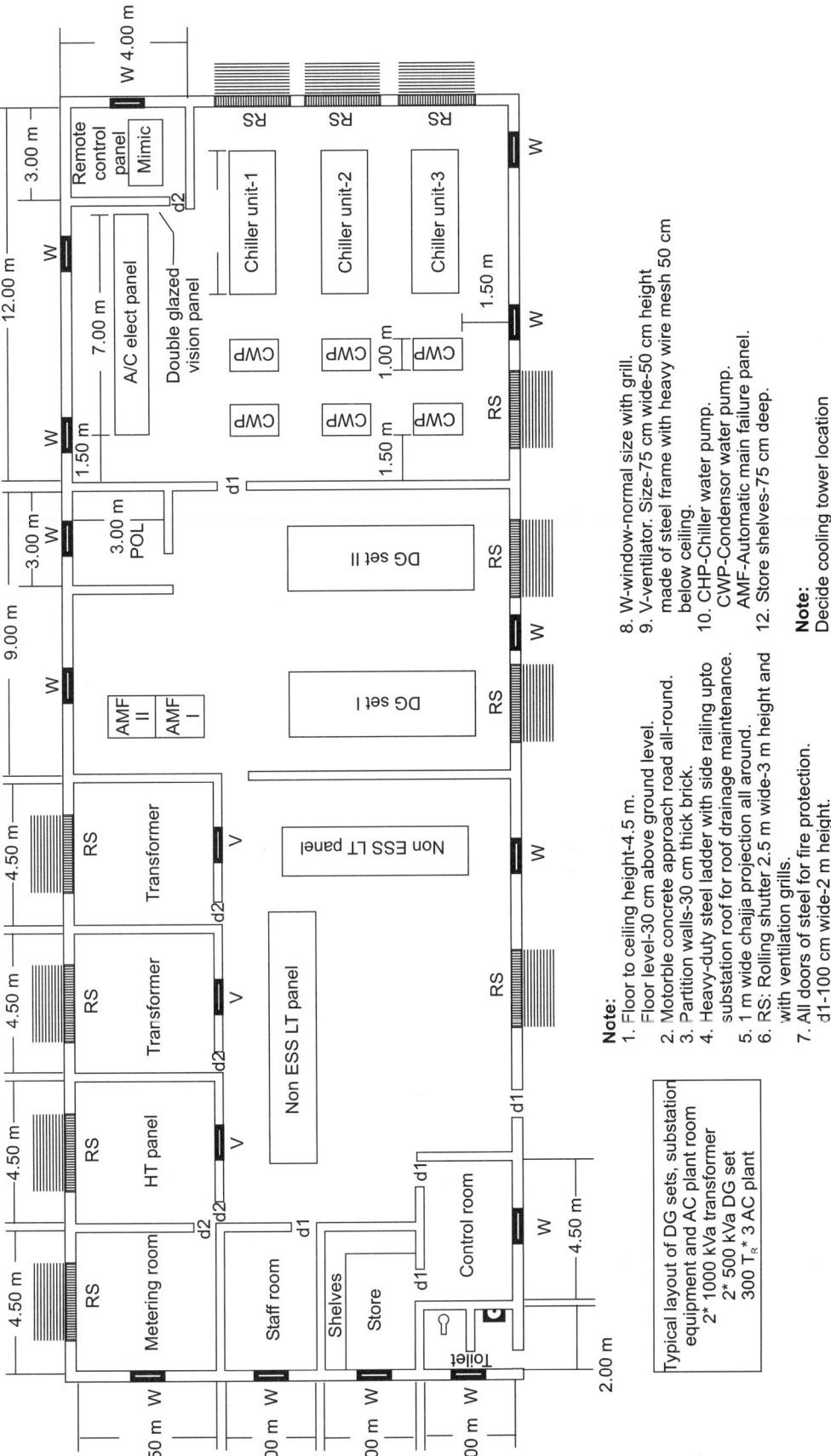

Note:
1. Floor to ceiling height-4.5 m.
   Floor level-30 cm above ground level.
2. Motorble concrete approach road all-round.
3. Partition walls-30 cm thick brick.
4. Heavy-duty steel ladder with side railing upto substation roof for roof drainage maintenance.
5. 1 m wide chajja projection all around.
6. RS: Rolling shutter 2.5 m wide-3 m height and with ventilation grills.
7. All doors of steel for fire protection.
   d1-100 cm wide-2 m height.
   d2-75 cm wide-1.8 m height.
8. W-window-normal size with grill.
9. V-ventilator. Size-75 cm wide-50 cm height made of steel frame with heavy wire mesh 50 cm below ceiling.
10. CHP-Chiller water pump.
    CWP-Condensor water pump.
    AMF-Automatic main failure panel.
12. Store shelves-75 cm deep.

Note:
Decide cooling tower location

Typical layout of DG sets, substation equipment and AC plant room
2* 1000 kVa transformer
2* 500 kVa DG set
300 T$_R$* 3 AC plant

Fig. 2.4 A layout of a typical substation AC plant cum DG set room

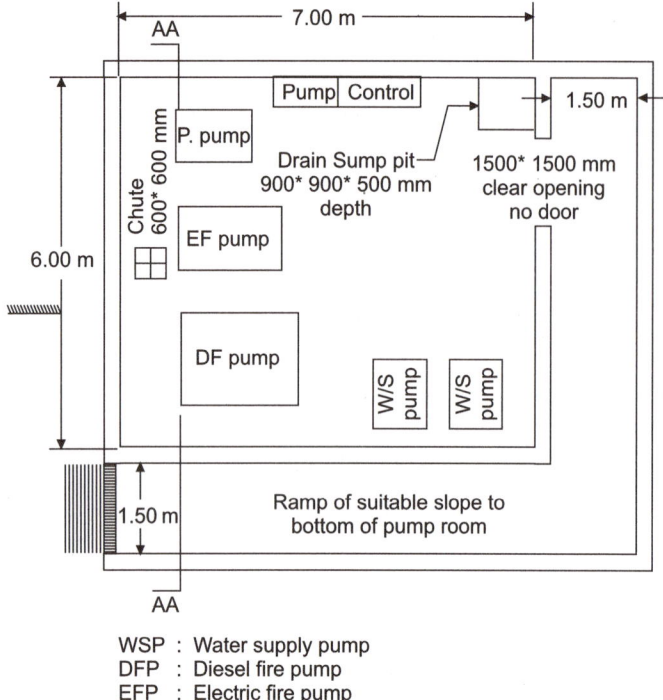

WSP : Water supply pump
DFP : Diesel fire pump
EFP : Electric fire pump
PP : Pressurization pump

Fig. 2.5 Fire fighting and drinking supply pump room

The pump shall have net positive suction head (NPSH), i.e. the suction of the pump shall always be immersed in water.

The water tank of adequate capacity for fire fighting shall be provided adjacent to pump house. The drinking water tank shall get the water from overflow of fire fighting tank so that fire fighting tank is always full (Fig. 2.6).

Table 2.1b Space for DG set

Capacity	Area
25 kW	56.00 sq m
48 kW	56.00 sq m
100 kW	65.00 sq m
150 kW	72.00 sq m
248 kW	100.00 sq m

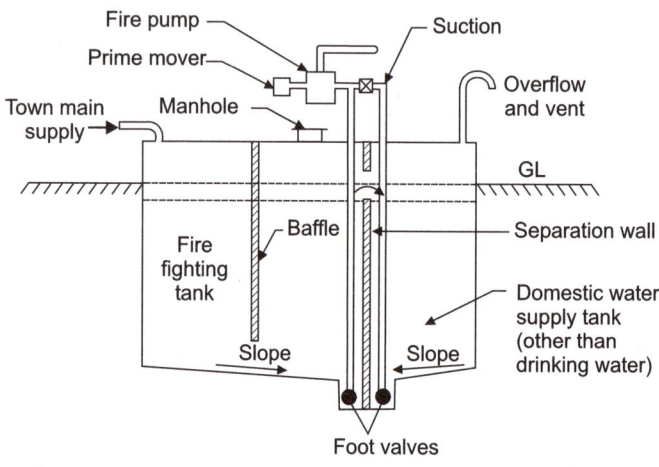

(a) With negative suction

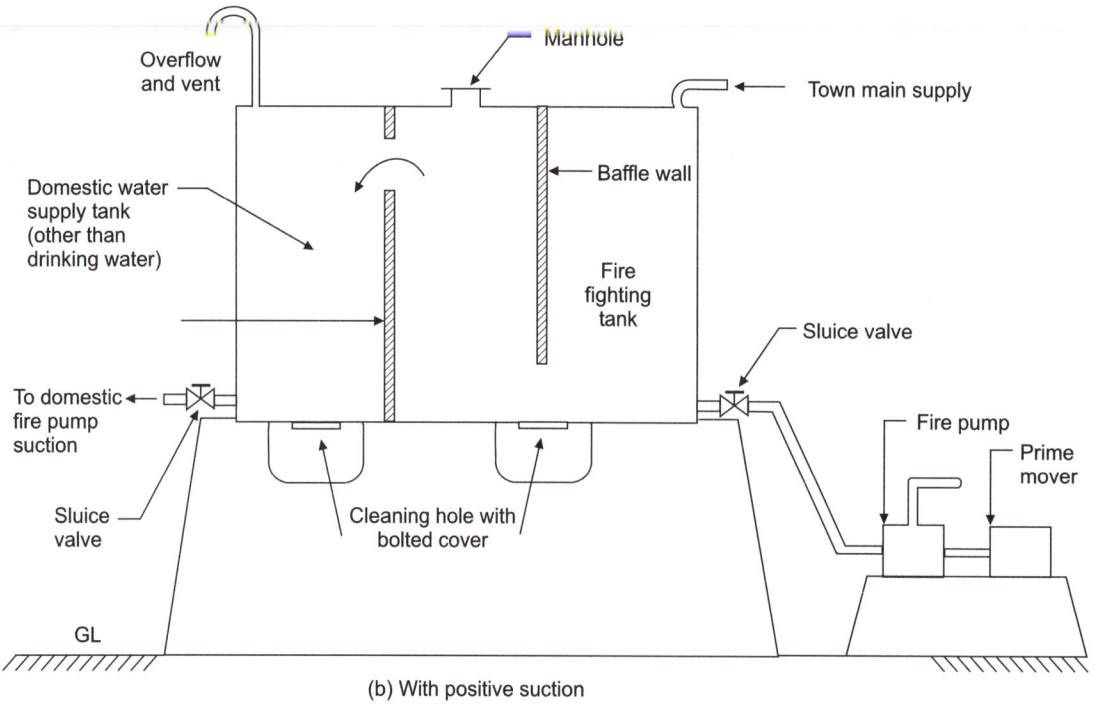

(b) With positive suction

Fig. 2.6 Typical arrangement for providing combined fire fighting and domestic water storage tank

Water requirement for fire protection with wet riser/down comer system is given in Table 2.2a, b (extracted from NBC 2005).

Table 2.2a Residential buildings

	UG water storage tank static (in litres)	Terrace tank (in litres)
Above 15 m up to 30 m	50,000	10,000
Above 30 m up to 45 m	1,00,000	20,000
Above 45 m	2,00,000	40,000

Table 2.2b Business buildings

	UG water storage tank static (in litres)	Terrace tank (in litres)
Above 15 m up to 30 m	1,00,000 50,000 lts if covered area in GF is less than 300 sq m)	20,000
Above 30 m up to 45 m	2,00,000	20,000
Above 45 m	2,50,000	50,000

The shaft of wet riser/fire hydrant shall not be less than 1.2 m wide and 0.8 m deep. The number of such shafts shall be one for every 1000 sq m of floor area. The location of shaft shall be so decided that one shaft shall serve maximum 30 m distance (Fig. 2.1).

2.5 SPACE FOR LV SERVICES

The general space requirement for telephone and BMS control room and shaft is shown in Fig. 2.7.

The size of the fire control room/telephone/BMS: 12 sq m each.

Shaft for telephone/fire alarm/computer cabling: 0.6 × 0.3 m each.

2.6 SIZE OF LIFT WELL

The knowledge of lift shaft dimensions is very important for a successful designer as mistakes in other services can be adjusted to a certain extent by minor modifications, but the lift well once made for a particular design of lift, cannot be modified without undertaking major changes in the structure.

The recommended dimensions of the lift well for various lift applications are shown in Figs. 2.8–2.11. For details, one may refer to NBC-2005 and IS 14665.

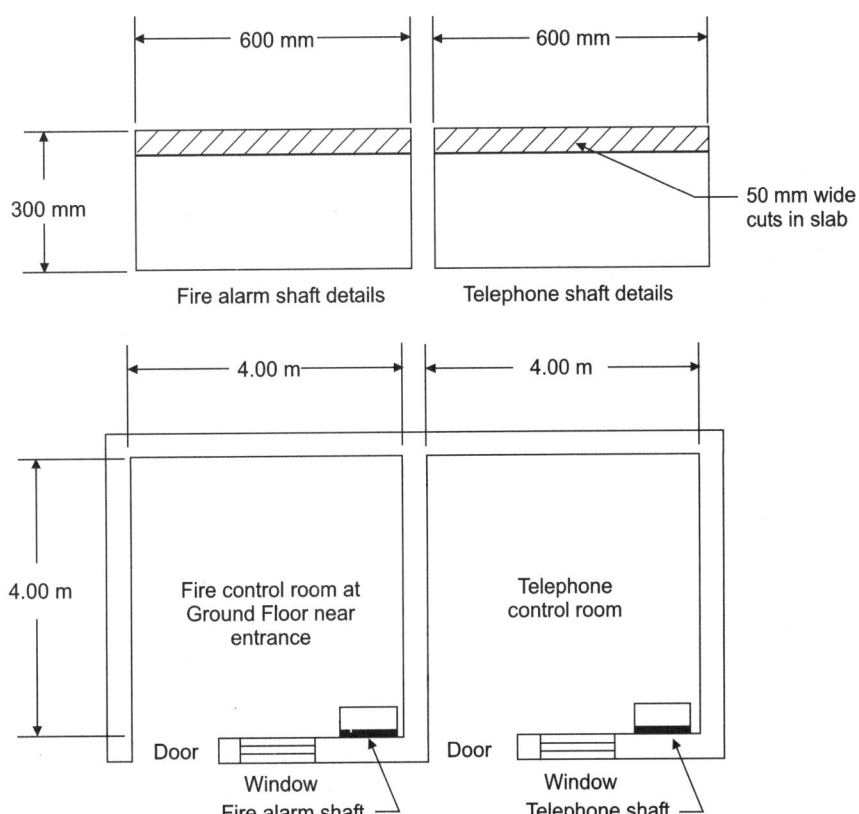

Fig. 2.7 Telephone and fire control room

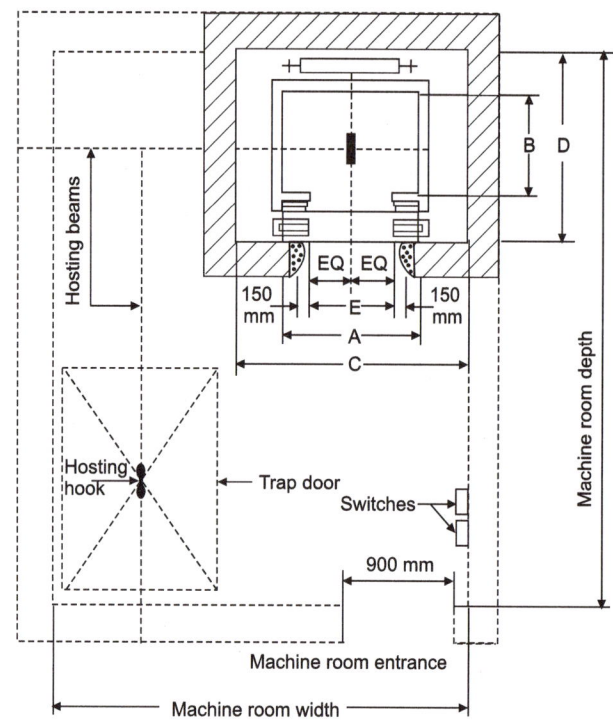

Plans						
Load		Cab inside		Lift well		Entrance
Persons	kg	A	B	C	D	E
4	272	1100	700	1900	1300	700 (min)
6	408	1100	1000	1900	1700	700 (min)
8	544	1300	1100	1900	1900	800
10	680	1300	1350	1900	2100	800
13	884	2000	1100	2500	1900	900
16	1088	2000	1300	2500	2100	1000
20	1360	2000	1500	2500	2400	1000

* All Dimensions in millimetres.

Fig. 2.8 Recommended dimensions of passenger lifts

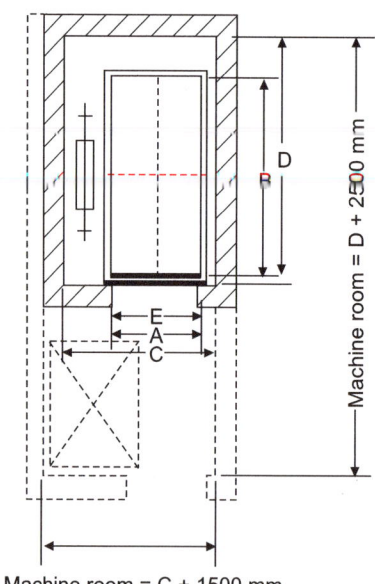

Load		Cab inside		Lift well		Entrance
Persons	kg	A	B	C	D	E
15	1020	1000	2400	1800	3000	800
20	1360	1300	2400	2200	3000	1200
26	1768	1600	2400	2400	3000	1200

* All Dimensions in millimetres.

Fig. 2.9 Recommended dimensions of goods lifts (for speeds up to 1.5 m/s)

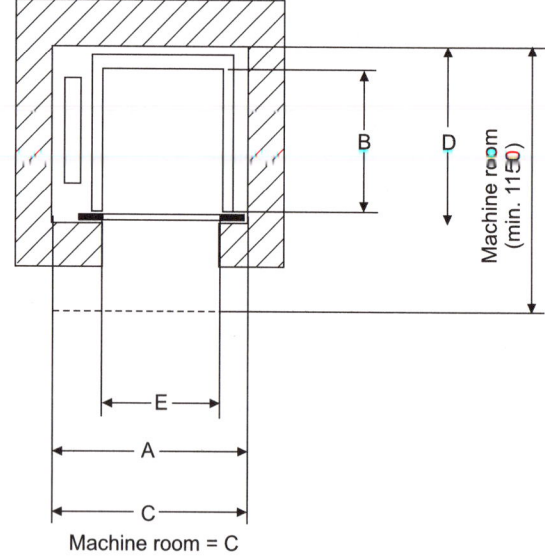

Load	Cab inside			Lift well		Entrance
kg	A	B	H	C	D	E
700	700	700	800	1200	900	700
800	800	800	900	1300	1000	800
900	900	900	1000	1400	1100	900
1000	1000	1000	1000	1500	1200	1000

* All Dimensions in millimetres.

Fig. 2.10 Recommended dimensions of hospital lifts (for speeds up to 1.5 m/s)

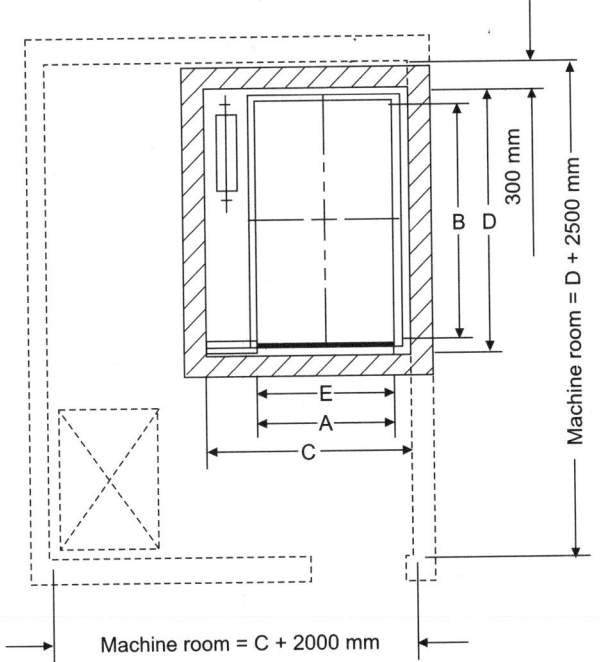

| Load | Cab inside | | Lift well | | Entrance |
kg	A	B	C	D	E
500	1100	1200	1900	1500	1100
1000	1400	1800	2300	2100	1400
1500	1700	2000	2600	2300	1700
2000	1700	2500	2600	2800	1700
2500	2000	2500	2900	2800	2000
3000	2000	3000	2900	3300	2000
4000	2000	3000	3400	3300	2500
5000	2500	3600	3400	3900	2500

* All Dimensions in millimetres.

Fig. 2.11 Recommended dimensions of service lifts (for speeds up to 0.5 m/s)

Section B

Air Conditioning and Mechanical Ventilation

This section has been composed with the intent of providing reader a first hand experience in designing each and every aspect of HVAC system. The HVAC design is a complicated affair on account of multiple parameters of air to be treated to achieve a set of comfort conditions, various patterns of geographical needs of human beings and the demand to conserve energy by a set of design elements, to be introduced in the system, like energy recovery wheels, variable frequency drives etc etc.

This chapter has been divided into seven chapters as below:

Chapter 3: Air Conditioning Principles

This chapter summarises various air conditioning principles learnt in graduation and post graduation, the output result of various derivations, description of various terms used in HVAC so as to refresh the knowledge gained in previous years and also to serve as a ready-to-use reference whenever needed.

Chapter 4: Psychrometry

Psychrometry is the branch of science which deals with determination of physical and thermodynamic properties of air or gas-vapour mixture. This branch of science is of more interest in HVAC as it deals with mixture of water vapour and air for comfort conditioning. In this chapter, the psychrometric chart, describing various properties of air vapour mixture, has been explained. The treatment of air, to achieve a set of comfort conditions, require passing the air through multiple processes, based on air conditioning principles.

This chapter describes these processes in detail and their representation on psychrometric chart.

Chapter 5: Heat Load Calculation

Having understood the principles of air conditioning and behavior of air when passed through various processes, the calculation of heat load or cooling load to achieve comfort conditions has been explained in this chapter. Ready to use tables, proformas have been given and explained with the help of real life example. Selection of parameters for various latitudes, direction, correction factors, building mass calculation etc. have been explained in great details.

Chapter 6: Duct Design and Sizing

After heat load to be removed from a room has been calculated, the next step is to calculate the quantity of cold air to be circulated through each room, calculation of duct sizes so as to prevent circulation noise of air inside duct, least friction and temperature gain from the unconditioned area through which it passes. All this has been explained with the help of examples and easy to use practical methods.

Chapter 7: Chilled Water System Design

The water temperature is lowered in chiller and this chilled water is then taken to various buildings and floor air handling units (AHUs) through pipes and taken back through pipes. In the process, it has to encounter frictional resistance, temperature gain and has to pass through heat exchanger, cooling towers etc. This chapter describes various forms of loops, intake and out-take connections from main headers and size calculation strategies.

Chapter 8: Chiller Machines and Selection Criterias

The water is cooled in various types of chiller machines based on various air conditioning principles. Vapour compression and vapour absorption techniques have been explained here. Centrifugal, screw, rotary compression machines are some forms of chiller machines based on the above principles.

Chapter 9: Mechanical Ventilation Design

Mechanical ventilation of building floors, particularly basements, has been made mandatory to prevent smoke build up during fire. During normal operation, this system works to maintain the level of oxygen, carbon monoxide and carbon dioxide. This chapter describes in detail the NBC requirements, accepted principles, the calculation procedure to derive CFM and accordingly selection of fan, sizing of inlet and exhaust duct with the help of examples.

3

Air Conditioning Principles

To an average person, air conditioning simply means the cooling of air. However, from the perspective of HVAC designer, this definition is neither useful nor accurate. The definition in the technical terms, instead is

"Process of treating air in an internal environment to establish and maintain required standards of temperature, humidity, cleanliness and motion".

Most air conditioning systems are used either for human comfort or for process control. From life experiences it is already established that air conditioning enhances our comfort and work efficiency. Air conditioning is also used to provide controlled conditions required by some processes say manufacturing of life saving drugs and other scientific processes. Textile printing, photographic processing, computer rooms are some of the applications which require a controlled temperature and humidity.

3.1 COMFORT STANDARDS

Studies of the conditions that affect human comfort have led to the development of recommended indoor air conditions for comfort, published in American Society of Heating, Refrigeration and Air Conditioning Engineers (ASHRAE) manuals. One of these results is shown in Fig. 3.1.

The shaded region in this figure is called comfort zone. This zone depicts the region of the air temperature and relative humidity (RH), where at least 80% of the occupants will find the environment comfortable.

3.2 DESIGNING THE HVAC SYSTEM

The design of HVAC system for a large building project is an extremely complex task. It calls for experience of designer and involves scores of parameters. The manual calculation takes time though these days very advanced software are also available. The design of an HVAC system involves:

i. Calculation of heating and cooling load requirement of a building in terms of TR and CFM.
ii. Calculation of duct and pipe sizes.
iii. Selection of the type and sizes of equipments (low side equipments).

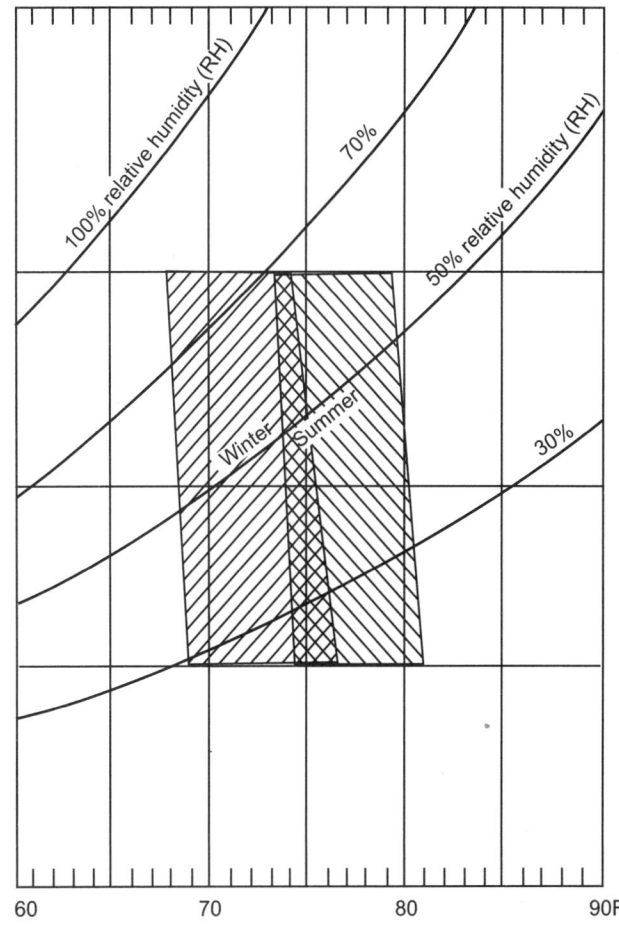

Fig. 3.1 Comfort zones of indoor air temperatures and relative humidity

iv. Selection of the type and capacities of equipment (high side equipments).

v. Selection of type of chilled/ hot water piping arrangement.

vi. Selection of control equipment to achieve the energy efficiency.

3.3 PROPERTIES OF AIR

The knowledge of fundamental laws on mixture of gases led to the understanding of properties of air and water content associated with it. It is the water content in the air, that remains in air in the form of vapour or suspended droplets, that is the cause of concern for an air conditioning engineer. Secondly, the temperature of air and the total heat content has to be managed, to provide comfortable condition, for comfort of human or process in an industry. It is thus important to study the properties of mixture of air and water vapour. Psychrometry is the subject which studies these properties of air and water vapour.

Different terms that are commonly used in psychrometry are explained below:

Dry air: The dry air is the mixture of nitrogen and oxygen in the ratio of 79% and 21% respectively (neglecting small percentages of either gases). Molecular weight of dry air is taken as 29 approximately.

Moist air: It is a mixture of dry air and water vapour. The quantity of water vapour varies from zero to maximum (at saturation) according to temperature of air.

Water vapour: The moisture or water content present in air in the form of vapour is called water vapour. The air is considered to be saturated when it holds the maximum amount of water vapour that it can hold at a given temperature.

Dry bulb temperature: The temperature of air measured by an ordinary thermometer is called dry bulb temperature.

Wet bulb temperature: The temperature measured by a thermometer when its bulb is covered with a wet cloth and is exposed to a current of moving air. The difference between dry bulb and wet bulb temperature is known as wet bulb depression which becomes zero when air is finally saturated. Wet bulb temperature is a measure of total enthalpy of air.

Sensible heat of air: The quantity of heat that can be measured by measuring the dry bulb temperature of the air.

Latent heat of air: The quantity of heat present in the air due to vapour content present in it at saturation temperature of water.

Total heat of air: The total heat of humid air is the sum of sensible heat of the dry air as well as sensible and latent heat of water vapour associated with dry air.

3.4 WET BULB TEMPERATURE AND ITS IMPORTANCE

As described earlier, when a stream of air is passed over a wet bulb thermometer, the moisture contained in the cloth evaporates. The evaporation causes lowering of temperature measured by dry bulb temperature at the rate which equals the rate at which evaporation occurs. When the temperature measured by wet bulb thermometer has stabilized, then the heat released by thermometer bulb for evaporation equals the heat given by air to thermometer by convection as shown below (Refer Fig. 3.2):

H_c is the heat given by air to thermometer by convection

H_v is the heat going back to air by evaporation.

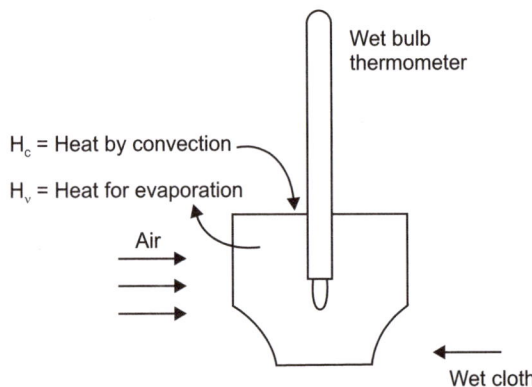

Fig. 3.2 Measurement of wet bulb temperature

The heat that causes evaporation is the sensible heat of air. During evaporation, this heat gets converted into latent heat of vaporization and comes back to air, thus maintaining total heat of air constant. Thus wet bulb temperature is a measure of total heat of air. The evaporation rate depends on the condition of air passing over it. If the air is dry, evaporation will be fast and correspondingly drop in temperature measured by wet bulb thermometer will be appreciable and vice versa. Thus wet bulb temperature is also a measure of degree of saturation of air or relative humidity. Air with 100% relative humidity will have no drop in temperature as there will be no evaporation.

3.5 ADIABATIC SATURATION TEMPERATURE

Also known as thermodynamic wet bulb temperature, it is that temperature at which air is brought to saturated condition adiabatically by evaporating the water into flowing air.

Consider the moist air flowing through a perfectly insulated chamber as shown. The water evaporates and gets carried away with flowing stream of air. The temperature of outlet air at equilibrium state is equal to water temperature t_w but lower than inlet dry bulb temperature and greater than dew point temperature of entering air. This equilibrium temperature T_{db2} (equal to t_w) is known as adiabatic saturation temperature and process is known as adiabatic saturation process (Fig. 3.3).

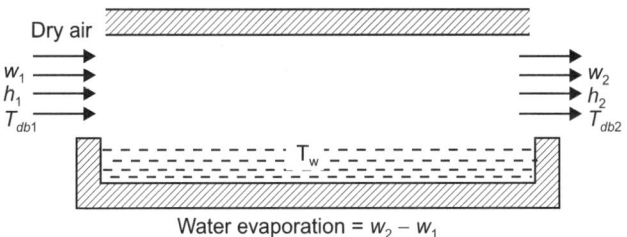

Water evaporation = $w_2 - w_1$

Fig. 3.3 Adiabatic saturation of air

Enthalpy equation becomes:
$$h_1 + (w_2 - w_1) = h_2$$

(Assumed that leaving air is fully saturated).

The wet bulb temperature is slightly different from adiabatic saturation temperature as ideal adiabatic saturation of air is not possible but for all practical purposes, both the temperatures are taken to be equal.

3.6 DEW POINT TEMPERATURE

Do you remember, when as a child, you used to draw pictures on a fogged window pane in winter or in rainy season. When the temperature of the glass becomes too low. The water vapour from moist air gets condensed over the glass. This happens when the glass temperature is lower than the saturation temperature corresponding to partial pressure of water vapour in air. This temperature is called dew point temperature.

Let the glass surface temperature is 30° F. If the condition of air is such that its dew point temperature is 32° F. Then the air will condense immediately on coming in contact with glass as the glass temperature is lower than dew point temperature. But if the dew point temperature of air is 28°F, then it will not condense on a glass at temperature of 30° F.

The dew point temperature can be located on psychrometric chart by drawing a horizontal line from the point of condition of air (Fig. 3.4).

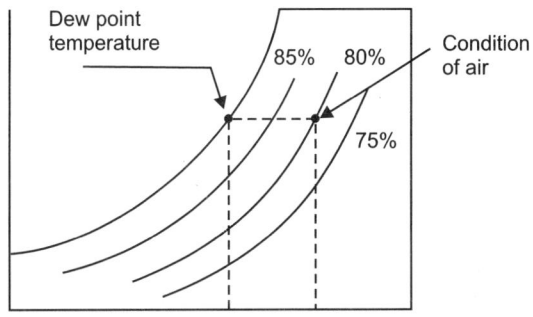

Fig. 3.4 Locating dew point temperature

3.7 SPECIFIC VOLUME OF HUMID AIR

If the partial pressure of water vapour in the air is known, then the specific volume of moist air can be calculated from partial pressure as

$$v_{sv} = \frac{R_v T}{P_v}$$

where

v_{sv} = specific volume of water vapour at partial pressure P_v

T = dry bulb temperature of air

Similarly

$$V_{sa} = \frac{R_a T}{P_a}$$

where $V_{sa} = V_{sv}$ and it is also the specific volume of humid air at pressure P_t according to Dalton's law of partial pressure

where $P_t = P_v + P_a$

3.8 MEASURABLE AIR PROPERTIES

The properties of air that can be measured directly are dry bulb temperature and wet bulb temperature. Other properties required for air conditioning calculation like enthalpy, specific volume, relative humidity and specific humidity are calculated for different dry bulb and wet bulb temperatures.

$$\text{Specific humidity of saturated air} = \frac{0.622\,P_{vs}}{P_t - P_{vs}} \quad \ldots (1)$$

P_{vs} = Steam table partial vapour pressure at a given dry bulb temperature

Total enthalpy of saturated air:
$$h = h_a + wh_v$$

where, h = total enthalpy of moist air.

h_a = enthalpy of dry air.

wh_v = enthalpy of water vapour associated with unit quantity of dry air.

w = mass of water vapour in unit mass of dry air or simply specific humidity.

$$h = c_{pa}(T_{db} - 32) + \omega\left[c_{pw}T_{dp} + \left(h_{fg}\right)_{dp} + c_{pv}\left(T_{db} - T_{dp}\right)\right]$$
$$\ldots (2)$$

where, c_{pa} = specific heat of dry air

c_{pw} = specific heat of liquid water

T_{db} and T_{dp} = Dry bulb and dew point temperature of air

$\left(h_{fg}\right)_{db}$ = latent heat of vaporization at DBT.

c_{pv} = specific heat of superheated vapour.

Under saturated conditions of air
$$T_{db} = T_{dp}$$

Thus

$$h = c_{pa}(T_{db} - 32) + w\left[c_{pw}T_{db} + \left(h_{fg}\right)_{db} + c_{pv}(T_{db} - T_{db})\right]$$

… FPS notation (Adopt zero for MKS instead of 32)

$$= T_{db}(c_{pa} + wc_{pw}) - 32c_{pa} + w\left(h_{fg}\right)_{db}$$

$$= c_{pm}T_{db} - 32c_{pa} + w\left(h_{fg}\right)_{db} \qquad \dots (3)$$

where c_{pm} = mean specific heat of humid air

3.9 MATHEMATICAL EQUATIONS

The basic design equation for HVAC calculations are given below. The derivation of these equations is beyond the scope of this book as the same can be seen in text books. The purpose of listing them here is only to refresh the knowledge gained in undergraduate courses.

The HVAC design is popularly carried out in FPS system where heat load is calculated in British Thermal Units (BTU).

1. Conversion of Farhenheit to Celsius

$$C = \frac{F - 32}{1.8}$$

where C = temperature in °C
           F = temperature in °F

2. Sensible heat

$$Q_s = m \times c \times TD$$

where Q_s = rate of sensible heat added to or removed from substance in BTU/hr
    m = mass rate flow of medium lb/hr.
    c = specific heat of medium in BTU/lb °F
      = 0.244 for air
    TD = temperature differential across which heat flows in °F.

Specific volume of moist air = 13.5 ft³/lb at 70° F db and 50% RH
Substituting values

$$Q_s = \left(\frac{1}{13.5} \times CFM\right) \times 0.244 \times TD$$

$$= 1.08\ CFM \times TD$$

3. Enthalpy equation:

$$Q = m(h_2 - h_1)$$

where Q = rate of heat added or removed from substance, BTU/hr
m = mass flow rate of medium, lb/hr
$h_2 - h_1$ = specific enthalpy change of medium, BTU/lb

4. Latent heat equation:

$$Q_L = mc(w_0 - w_i)$$

where
$w_0 - w_i$ = difference in outdoor and indoor humidity in grains of water/lb of dry air
Q_L, m, c are as defined earlier

$$m = \frac{60}{\text{specific vol.}\ \left(ft^3/lb\right)} \times \frac{q}{g_r}$$

where q = average heat required to be removed to condense one pound of water vapour from air (BTU/lb/ft³)
= 1076 BTU
g_r = grains of water per pound of air = 7000 (lb of vapour/lb of air)

$$m = \frac{60}{13.5} \times \frac{1076}{7000} = 0.68$$

$$Q_L = 0.68 \times CFM \times (w_0 - w_1)$$

5. Ideal gas law
$$Pv = MRT$$
where
    P = pressure, lb/ft² absolute
    v = volume, ft³
    M = weight of gas, lb
    R = a gas constant
    T = absolute temperature, degree F

6. Heat equation
$$E_{ch} = E_{in} - E_{out}$$
where
    E_{ch} = Change in stored energy
    E_{in} = heat supplied by heating system
    E_{out} = heat removed by the heating system

7. Heat conduction

$$Q = \frac{1}{R} \times A \times T_D$$

where
    Q = heat transfer rate, BTU/hr
    R = Thermal resistance of material, hr-ft²-°F/BTU
    A = surface area through which heat flows, ft²
    TD = Temperature difference across which heat flows, °F.

8. Heat convection

$$Q = \frac{L}{K} \times T_D$$

$$= \frac{T_D}{C}$$

where
    C = Conductance, BTU/hr-ft²-°F per inch of thickness and is inverse of resistance of material.

$$= \frac{K}{L}$$

    K = Conductivity, BTU/hr-ft²-°F, per inch of thickness
    L = Thickness of material, inch

9. Overall heat transfer coefficient
$$Q = U \times A \times TD$$

Q = heat transfer rate, BTU/hr

where U = Overall heat transfer coefficient BTU/hr-ft^2-°F

$$= \frac{1}{R_0}$$

R_0 = overall resistance to heat flow

= $R_1 + R_2 + R_3 \dots$ etc.

A = surface area through which heat flows, ft^2

TD = temperature difference, °F.

10. Humidity ratio:

This is the ratio of mass of water vapour per unit mass of dry air to the mass of water vapour in same mass of dry air when saturated at same temperature.

$$w = \frac{m_w}{m_a}$$

where

w = humidity ratio, lb of water vapour per unit lb of dry air

m_w = mass of water vapour (lb) in unit mass of dry air.

m_a = mass of water vapour (lb) in unit mass of dry air when saturated at same temperature.

11. Relative humidity:

This is the ratio of mass of water vapour per unit volume of dry air to the mass in same volume of dry air when saturated at same temperature.

$$R_H = \frac{M_w}{M_a}$$

where RH = Relative humidity, %

M_w = mass of water vapour (lb) per unit volume of dry air.

M_a = mass of water vapour (lb) per unit volume of dry air when saturated at same temperature.

12. Enthalpy:

Enthalpy of atmospheric air is the sum of individual enthalpies of the dry air and water vapour.

$$h = 0.24t + w(1061 + 0.45t)$$

where, h = enthalpy of moist air, BTU/lb of dry air

t = dry bulb temperature of air, °F.

w = humidity ratio

13. Psychrometric equations:

$$RSH = 1.08 \times CFM_{sa} \times (t_{rm} - t_{sa})$$
$$RLH = 0.68 \times CFM_{sa} \times (w_{rm} - w_{sa})$$
$$TSH = 1.08 \times CFM_{da} \times (T_{edb} - T_{ldb})$$
$$TLH = 0.68 \times CFM_{da} \times (w_{ea} - w_{la})$$
$$ERSH = 1.08 \times CFM_{da} \times (t_{rm} - t_{adp}) \times (1 - BF)$$

$$ERLH = 0.68 \times CFM_{da} \times (w_{rm} - w_{adp}) \times (1 - BF)$$
$$ERTH = 4.45 \times CFM_{da} \times (h_{rm} - h_{adp}) \times (1 - BF)$$

$$RSHF = \frac{RSH}{RSH + RLH}$$

$$= \frac{RSH}{\text{Room Total Heat}}$$

$$= \frac{RSH}{RTH}$$

$$GSHF = \frac{TSH}{TSH + TLH}$$

$$= \frac{TSH}{GTH}$$

$$\text{Bypass Factor } (BF) = \frac{t_{edb} - t_{adp}}{t_{ldb} - t_{adp}}$$

$$= \frac{w_{la} - w_{adp}}{w_{ea} - w_{adp}}$$

$$= \frac{h_{la} - h_{adp}}{h_{ea} - h_{adp}}$$

where,

CFM_{da} = cubic feet per minute of dry air

T_{edb} = dry bulb temperature of entering air, °F

T_{ldb} = dry bulb temperature of leaving air, °F

w_{ea} = entering air moisture content (grains)

w_{la} = leaving air moisture content (grains)

T_{rm} = room dry bulb temperature, °F

T_{sa} = supply air dry bulb temperature, °F

T_{adp} = apparatus dry bulb temperature, °F

w_{rm} = room air moisture content (grains)

w_{adp} = moisture content at apparatus dew point (grains)

h_{rm} = enthalpy of air at room conditions, BTU

h_{adp} = enthalpy at apparatus dew point, BTU

h_{la} = enthalpy of leaving air, BTU

h_{ea} = enthalpy of entering air, BTU

RSH = Room sensible heat, BTU

RLH = Room latent heat, BTU

TSH = Total sensible heat, BTU

TLH = Total sensible heat, BTU

$ERSH$ = Effective room sensible heat, BTU

$ELSH$ = Effective room latent heat, BTU

$ERTH$ = Effective room total heat, BTU

$RSHF$ = Room sensible heat facor

$GSHF$ = Grand sensible heat factor

4

Psychrometry

Psychrometric chart shows the inter-relation of various directly measurable properties and derived properties of air on a single X–Y chart, rendering it very easy to calculate any of the properties of air if few others are known. It is also possible to show various process lines on the chart which helps an air conditioning engineer to calculate the end property of air after it has passed through the process without going into com-plicated mathematical calculation.

4.1 PSYCHROMETRIC CHART

Psychrometric chart is prepared for a definite baro-metric pressure. The dry bulb temperature is taken along abscissa while ordinate represents the specific humidity on the right hand side of the chart.

a. **Constant relative humidity lines:** The specific humidity at various dry bulb temperature is obtained from psychrometric table when the air is fully saturated. These points are marked on the chart and joined. The curve so obtained is 100% saturation or 100% relative humidity line (Fig. 4.1). The verticals from point a and b show the

wet bulb temperature or dry bulb temperature or dew point temperature as they are same at 100% saturation of air. If the vertical from a and b and similarly other points on a saturation curve are divided into 5 equal parts and the respective parts are joined together, we get relative humidity lines of 20%, 40%, 60%, 80% saturation as shown.

b. **Constant specific volume lines:** The perfect gas equation is used to construct specific volume lines. The volume of dry air per unit mass of dry air or volume of water vapour per unit mass of dry air at their individual partial pressures gives the specific volume of air vapour mixture.

$$\text{Thus, } V_{su} = \frac{R_a T}{P_a} = \frac{R_a T}{P_t - P_v}$$

If specific volume line of 30 cubic feet/pound is to be constructed then from the above equation at various values of T, we can find corresponding values of P_v and thereafter corresponding value of w by using specific volume equation described earlier. Thus, constant specific volume line can be constructed (Fig. 4.1).

c. **Constant enthalpy line:** The constant enthalpy line is drawn with the help of equation-3 described earlier. For constructing enthalpy line of 750 BTU, substitute arbitrary valves of T_{db} in the equation-3 and obtain respective values of w. All the points are marked on the psychrometric chart. When these points are joined, the resulting line is constant enthalpy line (Fig. 4.1).

d. **Constant wet bulb temperature line:** The wet bulb temperature is a measure of enthalpy of moist air (explained earlier). Thus constant enthalpy lines also represent constant wet bulb temperature line. Mark a point on saturation curve say at 50 °F dry bulb temperature. If the

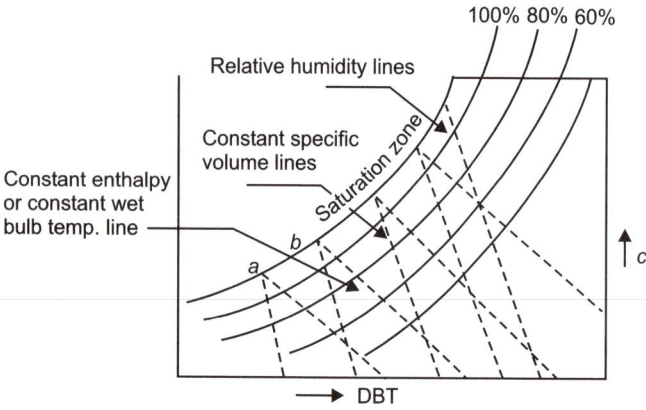

Fig. 4.1 Psychrometric chart

path is traced through this point along constant enthalpy line, this line also represents constant wet bulb temperature line and the WBT will be 50 °F all along the line whatever be the DBT (Fig. 4.1).

4.2 PSYCHROMETRIC PROCESSES

As described earlier, psychrometric chart represents various properties of air on a single chart. It is thus obvious that making any change in any of the property of air is bound to effect other properies, i.e. the condition of air undergoes a change and will be represented by another point on the psychrometric chart. This change in property of air from one point to another is effected by some kind of process say heating, cooling, humidification, etc. and can be represented by a line on the psychrometric chart.

If the initial and final state of air can be represented with the help of any two parameters for each point, then the other parameter can be found from the chart. Similarly, if initial state is known along with the process through which air is to pass then the final state can be easily located on the psychrometric chart. This is done without going into complex equation solving for their process. Some common process which are performed on air in the air conditioning applications are described here and marked on psychrometric chart.

4.2.1 Sensible heating/cooling

Heating or cooling of air without addition or removal of moisture content from air is known as sensible cooling. The heating can be achieved by passing the air through heating coils say of electric resistance or hot water or steam variety. Same way sensible cooling can be performed by passing air over brine coil, etc.

The process line is a horizontal line with arrow pointing rightwards in the case of heating, i.e. increase of DBT and vice versa is the case of cooling (Fig. 4.2a, b).

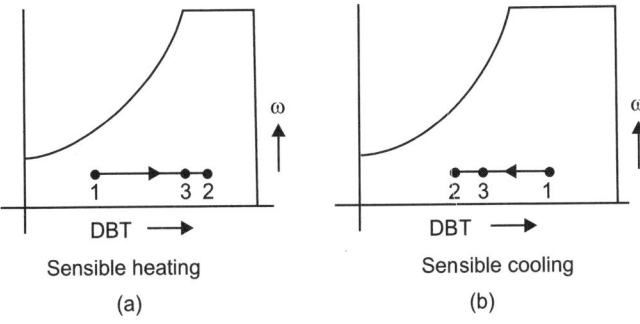

Fig. 4.2 Sensible heating and cooling process

Quantity of heat added is expressed by the equation:

$$Q_s = c_{pa}(T_2 - T_1) + \omega c_{pv}(T_2 - T_1)$$
$$= c_{pa} + \omega c_{pv}(T_2 - T_1)$$
$$= c_{pm}(T_2 - T_1)$$

Similarly quantity of heat removed is
$$Q_s = c_{pm}(T_1 - T_2)$$

As a matter of fact, it is impossible to achieve temperature T_2, as all of the air does not come into contact with the coil and some is by passed and the result is temp T_3 which is less than T_2 in heating and more than T_2 in case of cooling of air. This by passed air affects the air conditioning calculation very crucially and its estimation and calculation is very important in air conditioning design.

If out of one pounds of air passing through the coil, B pounds gets by passed, then only $(1 - B)$ pounds will reach temperature T_2 while B pounds at temperature T_1 will mix up with heated air at temperature T_2 thus creating a mixed air at temperature T_3.

Balancing the enthalpies, we get

$$B c_{pm} T_1 + (1 - B) c_{pm} T_2 = 1 \times c_{pm} T_3$$

$$B = \frac{T_3 - T_2}{T_1 - T_2} \text{ for sensible cooling}$$

or $\quad B = \dfrac{T_2 - T_3}{T_2 - T_1}$ for sensible heating

where B is known as the by pass factor of the coil.

The factor $(1 - B)$ is the quantity of air coming in direct contact with heating surface and is thus also termed as efficiency of coil.

On psychrometric chart, the condition of air entering coil is represented by point 1, ideal condition by point 2 while mixed air condition is represented by point 3. By pass factor is the ratio of difference in temperature $(T_2 - T_3)$ to $(T_2 - T_1)$ which is also the distance between points 2 – 3 and points 2 – 1.

Thus by pass factor can be represented as

$$B = \frac{\text{Distance } 2-3}{\text{Distance } 2-1}$$

The sensible heat given by coil is (Fig. 4.3)
$$Q_s = A_s \, U \, (LMTD)$$
where, U = overall heat transfer coefficient
$LMTD$ = log mean temperature difference
A_s = surface area of coil

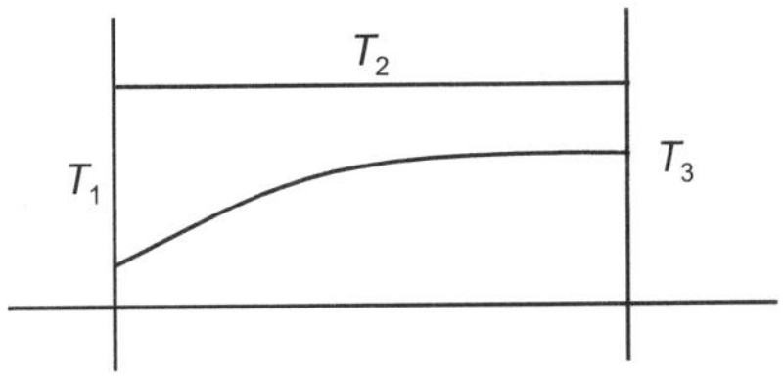

Fig. 4.3 Log mean temperature difference

$$LMTD = \frac{T_3 - T_1}{\log_e\left(\dfrac{T_2 - T_1}{T_2 - T_3}\right)}$$

$$= \frac{T_3 - T_1}{\log_e\left(\frac{1}{B}\right)}$$

Thus $Q_s = \dfrac{A_s U(T_3 - T_1)}{\log_e\left(\frac{1}{B}\right)}$

The heat is transferred to mass m of air in pound per second raising the temperature of air from T_1 to T_3.

Equating, we get $m_a c_{pm}(T_3 - T_1) = \dfrac{A_s U(T_3 - T_1)}{\log_e\left(\frac{1}{B}\right)}$

4.2.2 Cooling and dehumidification

In air conditioning, the cooling of air with high moisture content is achieved by passing the air through cooling coil having temperature below the dew point of air. This causes condensation of moisture present in the air and also reduces the temperature of the air. The removal of moisture from the air is called dehumidification. In the central air conditioning system, the AHU (Air Handling Unit) cooling coil temperature called the apparatus dew point temperature (ADP) has to be carefully selected and should be below the DPT of air. Similarly in window split or *VRV* system, the ADP of indoor unit has to be carefully selected.

As shown in the psychrometric chart Fig. 4.4 air at condition 1 passes over the cooling coil maintained at ADP represented by point 2. Under ideal conditions, the condition of air coming out of coil is same as point 2, but no cooling coil is 100% efficient, thus, the condition of air will be slightly different and towards the right side of point 2 on process line and represented by point 3 whose location will depend on the bypass factor and the design of cooling coil.

T_1 = Temperature of air entering the coil
T_2 = temperature of cooling coil
T_3 = Temperature of air coming out of coil
T_4 = Dew point temperature of air

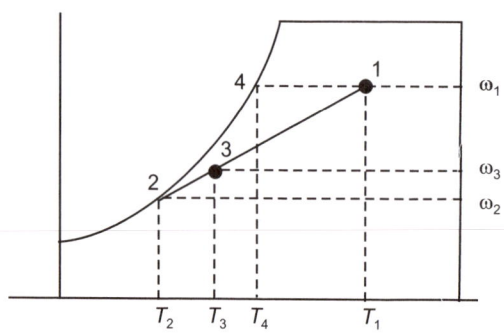

Fig. 4.4 Cooling and dehumidification

Bypass factor of cooling coil is given by:

$$B = \frac{h_3 - h_2}{h_1 - h_2} = \frac{T_3 - T_2}{T_1 - T_2} = \frac{w_3 - w_2}{w_1 - w_2}$$

Since T_2 = ADP of cooling coil

$$B = \frac{T_3 - ADP}{T_1 - ADP}$$

Since psychrometric chart represents initial and final states of air independent of the process followed, it can be assumed that air followed the path 1–5 and then 5–3 as shown (Fig. 4.5).

Thus, the process 1–5 removes latent heat from the air in the form of condensed vapours while 5–3 removes sensible heat by reducing DBT.

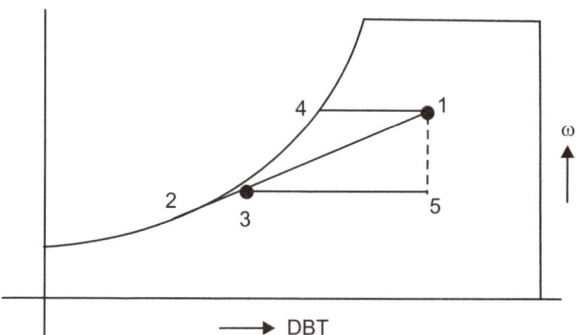

Fig. 4.5 Process as combination of sensible and latent heat

Total heat removed from the air is given by

$$Q_t = Q_L + Q_s$$
$$= (h_1 - h_5) + (h_5 - h_3)$$

The ratio $\dfrac{Q_S}{Q_t}$ is called sensible heat factor (SHF) or sensible heat ratio (SHR) and is given by:

$$SHR = \frac{Q_S}{Q_S + Q_L}$$

SHR is very important in air conditioning calculations since this ratio fixes the slope of line 1–3 on psychrometric chart. If initial condition and *SHR* are given for any process, the process line can be drawn. On the psychrometric chart, along the right hand side ordinate, a scale is given marking various slopes of *SHR* on it. This mark and initial condition can be joined and extended further to draw process line.
Capacity in Tons or Tons of refrigeration (TR):

$$TR = \frac{(h_1 - h_3) \times m_a}{12000}$$

Where h_1, h_3 are the initial and final enthalpies of air in BTU/hr/pound of air while m_a is the mass of air in pounds per hour

4.2.3 Cooling with humidification of air

Simple desert cooler used in India is an astonishing example of reducing the temperature of air by spraying water into the flowing air. As the air is sucked by the fan from grilles, the water sprayed through nozzle gets evaporated due to heat supplied by air. This causes water to convert into vapour thus increasing the moisture content of air and in turn reducing the temperature of air due to heat being absorbed by water. The total enthalpy of the air remaining constant as energy is conserved but only converted from sensible to latent (Fig. 4.6). The only extra quantity of heat carried by air is that of water droplets entrained by outgoing air which could not be converted to vapour.

$$Q = c_{pw}(\omega_2 - \omega_1)(T_3 - T_\omega)$$

where T_ω is the temperature of water sprayed.

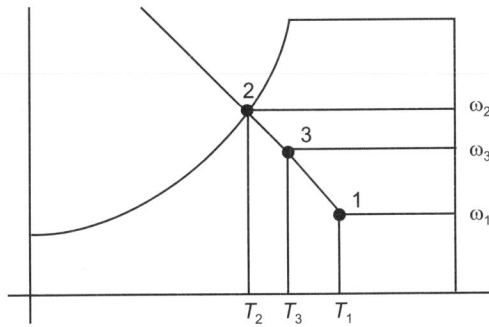

Fig. 4.6 Cooling and humidification

But this quantity is negligible and is not considered for calculation purposes. This process is also called adiabatic humidification because the total energy remains constant. This process is very suitable for achieving comfort temperature of air where the condition of air is dry with high temperature. This causes more water to evaporate thus increasing moisture content and reducing temperature of air. But there is a limit to which DBT of air can be reduced and this cannot be below the *WBT* of entering air. The effectiveness of spray chamber is defined by:

$$\varepsilon = \frac{T_1 - T_3}{T_1 - T_2}$$

4.2.4 Adiabatic chemical dehumidification

When high moisture air is passed through a chemical bed like silica gel which has the tendency to absorb moisture, part of water vapour gets absorbed. The latent heat is released and is absorbed by the flowing air, and is converted into the sensible heat and temperature of outgoing air is increased maintaining total enthalpy of air constant (Fig. 4.7).

The effectiveness of dehumidifier is

$$\varepsilon = \frac{T_3 - T_1}{T_2 - T_1} = \frac{\omega_1 - \omega_3}{\omega_1 - \omega_2}$$

where T_2 is the ideal temperature at which air can come out.

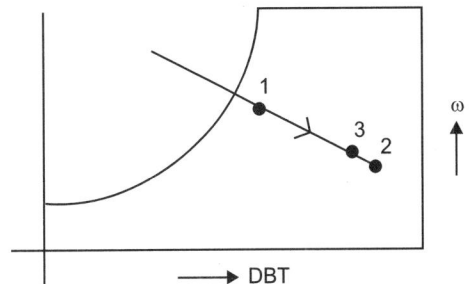

Fig. 4.7 Adiabatic chemical dehumidification

4.2.5 Humidification by steam injection

Sona bath is a perfect example of humidification by steam injection. The steam is injected into the air to increase the moisture as well as temperature of air. The hot and humid air opens the pores of the skin which help in circulation of blood and removal of dust and dirt from the pores. The psychrometric process is shown (Fig. 4.8) wherein air at low moisture content and low temperature is mixed with steam injected through nozzles. The lowest possible enthalpy of steam is at 100° C at atmospheric pressure when it is fully dry and saturated.

The amount of steam sprayed = $\omega_2 - \omega_1$

Final condition of air with enthalpy balance.

$$\omega_2 = \omega_1 + m_s$$

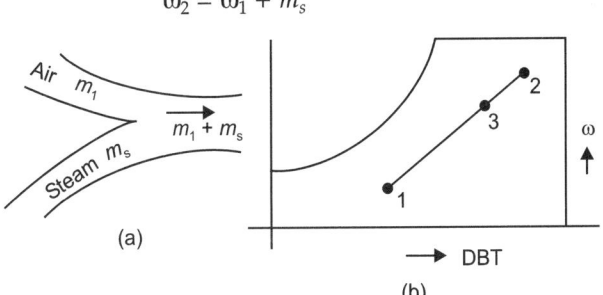

Fig. 4.8 Humidification by steam injection

where m_s is the quantity of steam added per pound of dry air

$$h_2 = h_1 + m_s h_s$$

where h_s is the enthalpy of steam per pound of steam injected.

4.2.6 Mixing of air streams

When two air streams having different relative humidity and DBT are mixed, then the condition of resultant air is as shown (Fig. 4.9). By balancing the enthalpies we get

$$m_1 h_1 + m_2 h_2 = (m_1 + m_2)h_3$$
$$m_1 \omega_1 + m_2 \omega_2 = (m_1 + m_2)\omega_3$$

On solving the above equation, we get

$$\frac{m_1}{m_2} = \frac{h_3 - h_2}{h_1 - h_3} = \frac{w_3 - w_2}{w_1 - w_3}$$

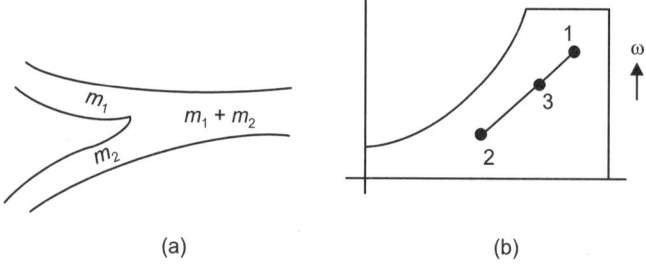

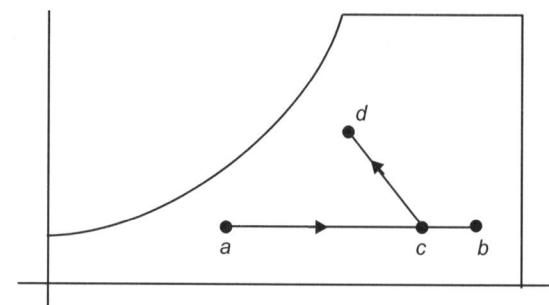

Fig. 4.9 Mixing of air streams

Since the enthalpy and humidity ratio are linear lines on psychrometric chart, the condition of resultant point 3 can be marked on the chart in the proportion of m_1 and m_2 such that

$$\frac{m_1}{m_2} = \frac{\text{distance } 2-3}{\text{distance } 1-3}$$

The point 3 shifts towards that point which has more quantity associate with it.

4.3 COMBINATION PROCESS ON PSYCHROMETRIC CHART

The air conditioning involves treatment of air through various processes before being supplied for comfort condition or industrial application. At times, when air is treated to control one parameter say RH, the process through which air is passed alters other parameters, e.g. reduction in RH through condensation will result in excessive cooling of air which cannot be supplied directly. Now this air should be treated through some other process to increase temperature to desired level.

Sometimes, air is passed through exchanger, heat recovery wheels to extract heat from the air being thrown into atmosphere.

All these processes can be represented on a psychrometric chart if some conditions and process lines are known and the final state of air can be easily worked out.

Some combination processes and their marking on psychrometric chart are shown here:

4.3.1 Heating and humidification

This combination process is required in winters when temperature and humidity both are low outside. Sensible heating increases the temperature but simultaneously reduces humidity. Adiabatic humidification process can be used to increase the relative humidity though it reduces the DBT again slightly (Fig. 4.10).

Process a–b is sensible heating

Procesd c–d is adiabatic humidification

Fig. 4.10 Heating and humidification

4.3.2 Humidification and heating

This process is used when temperature of outdoor air is reasonable but air is very dry and needs humidification. The air is passed through an air washer that increases humidity but lowers temperature also. The air shall then be heated to bring back to the same or nearby temperature (Fig. 4.11).

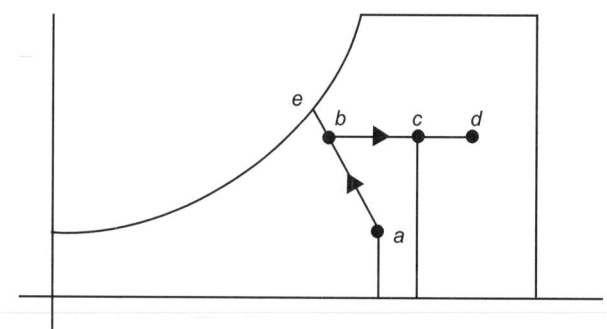

Fig. 4.11 Humidification and heating

4.3.3 Dehumidification and cooling

When the outdoor air is high in RH and high in temperature as well, then dehumidification and cooling must be performed. Air is passed through chemical dehumidifier which has the tendency to raise the temperature (as latent heat of water absorbed is released to air) and thus needs cooling by passing the air through cooling coil (Fig. 4.12).

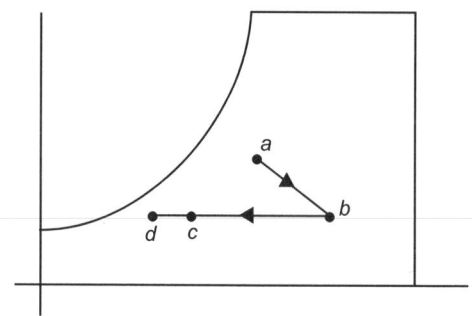

Fig. 4.12 Dehumidification and cooling

4.3.4 Humidification and cooling

This process is adopted when outdoor air is high in temperature and low in relative humidity. The air is first passed through a humidifier thus raising the RH and lowering the temperature. This process is further amplified by passing the air over a cooling coil.

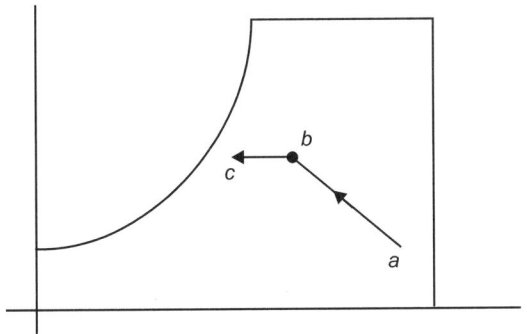

Fig. 4.13 Humidification and cooling

4.3.5 Dehumidification and heating

Another variant of process (c) described above is to first achieve dehumidification by condensation over a cooling coil and then bringing RH and DBT to desired level by sensible heating (Fig. 4.14). It may not be always possible to use chemical dehumidifier due to cost and maintenance point of view. This method is generally used in air conditioning applications. In this process, temperature of air after dehumidification is reduced as against chemical dehumidification where temperature is further increased.

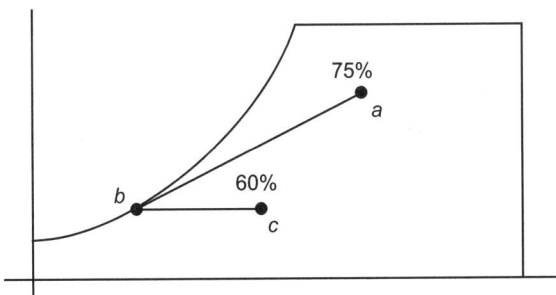

Fig. 4.14 Dehumidification and heating

4.3.6 Double dehumidification and cooling

The air is passed over a cooling coil and then through a chemical dehumidifier to reduce the water content in the air. Double process is required due to limitation imposed on a single air conditioning equipment. This process is normally used for special applications requiring low temperature and low RH. The rise in temperature due to chemical dehumidification is reduced by sensible cooling (Fig. 4.15).

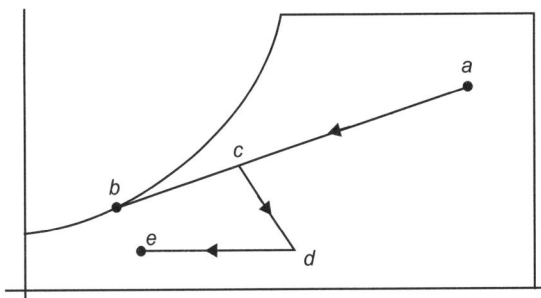

Fig. 4.15 Double dehumidification and cooling

4.3.7 Adiabatic humidification and double heating

In this process, low temperature air is brought to high temperature without much change in RH of outdoor air. The air is first heated through sensible heating process. The decrease in RH is compensated by adiabatic humidification (Fig. 4.16). This lowers the temperature which is brought to supply air temperature by second stage of sensible heating.

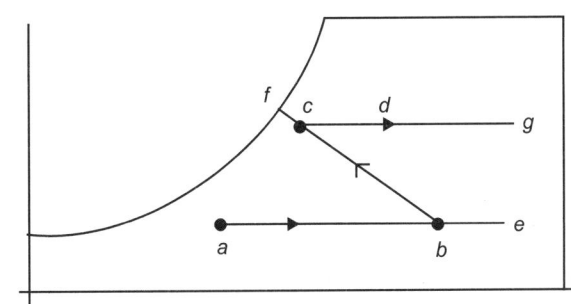

Fig. 4.16 Adiabatic humidification and double heating

4.3.8 Mixing, dehumidification and heating

In a recirculated system, the return air is not exhausted into atmosphere but mixed with small quantity of fresh air taken from atmosphere for maintaining oxygen level. Thus two streams of air at different properties are mixed and dehumidified. The lowering of temperature is offset by sensible heating (Fig. 4.17).

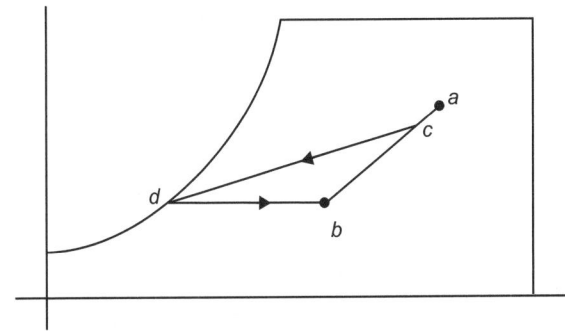

Fig. 4.17 Mixing, dehumidification and heating

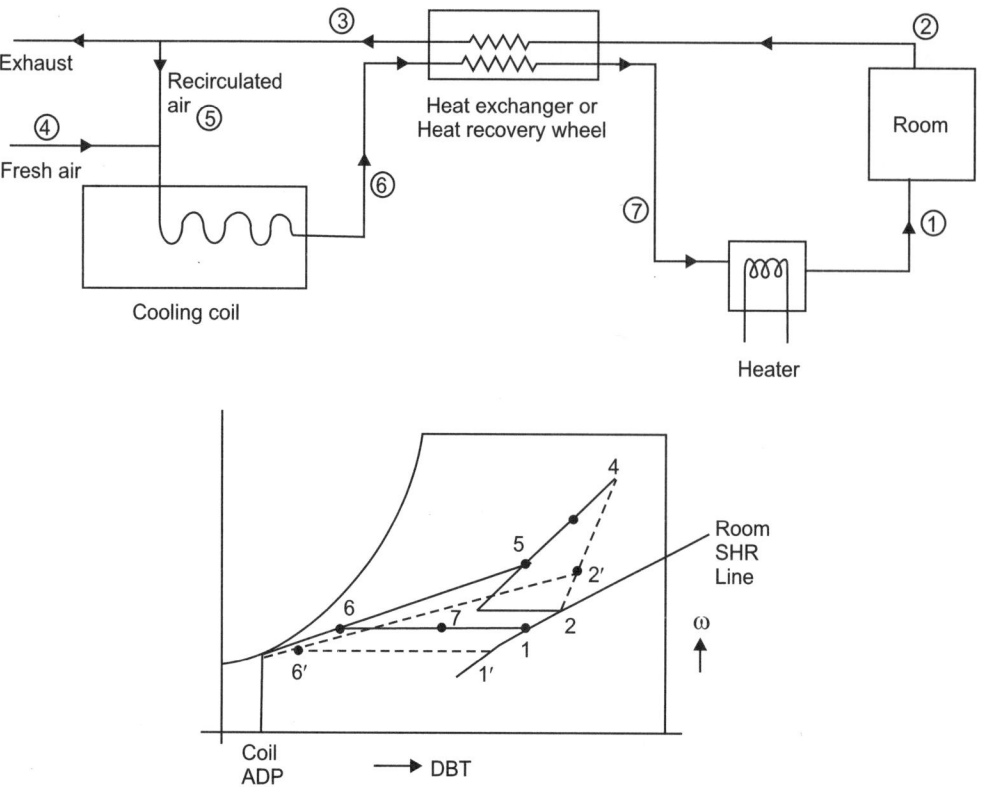

Fig. 4.18   Regenerative heat exchange process

4.3.9 Regenerative heat exchange from air to air

The cases explained above have an inherent deficiency associated with them, i.e. wastage of energy to maintain humidity. First the air has been cooled by passing it through low temperature cooling coil and then heated to bring it to the desired temperature. In both processes, the energy has been consumed.

The modern method is to heat the air in an air to air heat exchanger which eliminates the requirement of heater completely. Here the dehumidified chilled air

coming out of cooling coil extracts heat from the return air without mixing. The air gets heated sensibly and in turn cools the return air thus reducing the load on cooling coil also. Process 7–1 may be used if heat from return air is not sufficient (Fig. 4.18). The process is shown below and explained in psychrometric chart as well. The dotted chart is for process without heat exchanger. Saving in energy is = saving in cooling coil + saving in heater = $[(h_2' - h_6') - (h_5 - h_6)] + [(h_1' - h_6') - (h_1 - h_6)]$.

5

Heat Load Calculation

The objective of this chapter is to learn how to determine the amount of cooling or heating required to keep the rooms in a building comfortable in summer, monsoon and winter. As a first step in designing the air-conditioning system, the calculation process for determining cooling/heating load requirement of the building must be learnt by heart.

This chapter describes step-by-step methodology to calculate the load of the building, various factors to be kept in mind, various parameters and formulas as well as the data tables which shall be used for calculations.

After studying this chapter, you will be able to

1. Calculate the heat load in summer and monsoon.
2. Calculate the pre-heat and reheat required to achieve the desired conditions.
3. Prepare the ground work for further designing of low side of air-conditioning equipments.

5.1 THEORY IN BRIEF

To calculate the heat load in a building, it is necessary to understand the behaviour of various materials towards heat, and also the various methods of heat transfer in a building. Though you have already read these things in your college level, it is necessary to recapitulate them once again before we discuss the step-by-step approach to calculate the amount of heat load to be removed from a building to achieve the desired conditions.

5.1.1 Asymmetrical Pattern of Heat Gain

The amount of heat that must be removed is not always equal to the amount of heat received at any given point of time. This difference is due to the tendency of materials like bricks, concrete and glass to absorb heat inside them and release after some interval on the other side. Figure 5.1 shows this effect. The thermal storage and time lag causes the cooling load to be different at

different points of time and is different from instantaneous heat gain as shown in this figure. Hence, cooling load is the rate, at which heat must be removed at any point of time. This is why air-conditioning load is defined in terms of tons of refrigeration or BTU/hr.

Due to this time lag capacity of materials, the load of building may not be as high during day time as it may be during night. This is why, those installations, which are not required to be air-conditioned in night, like office buildings, are very hot in the next day morning hours.

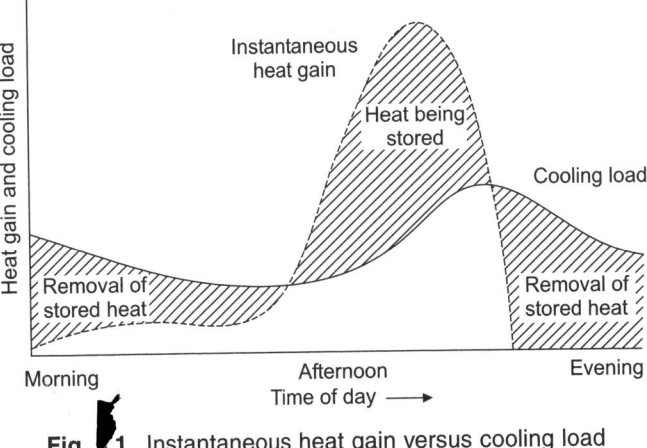

Fig. 5.1 Instantaneous heat gain versus cooling load

5.1.2 Components of Heat Transfer

We all know that solar gain is the major cause of heat build-up inside the building. This heat enters the room in the following forms.

i. Radiation

The radiation heat enters through glass irrespective of whether heat is being conducted through the glass, from the space to the outside or from the outside into the space.

The light energy has wavelength from fraction of microns to many kilometers. But, heat is generated by waves of wavelength 0.4 to 40 micron called *visible light* and infrared rays, which get absorbed by substances and generate heat in the process of absorption.

The energy emitted by a body is described by the law called *Stefan-Boltzman law*.

$$E = C \left(\frac{T}{100}\right)^4$$

Where

E = The amount of energy emitted by a body.

T = Temperature of body emitting energy.

C = A constant.

ii. Conduction

The property of a substance to absorb heat at a higher temperature and transmit it to the other side at a lower temperature is called *conduction*.

The rate of heat transfer by conduction varies with the physical property, i.e. thermal conductivity and the thickness of the substance. It is expressed as follow.

$$Q = \frac{T_1 - T_4}{\frac{X_1}{K_1 A} + \frac{X_2}{K_2 A} + \frac{X_3}{K_3 A}} \quad \text{(according to Fig. 5.2).}$$

where

Q = Rate of heat transfer.

T_1 = Temperature of heat surface (inlet face of heat).

T_4 = Temperature of cold surface (discharge face of heat).

X_1, X_2, X_3 = Thickness of different layers which are conducting.

K_1, K_2, K_3 = Thermal conductivity of different layers.

A = Area of conducting surface.

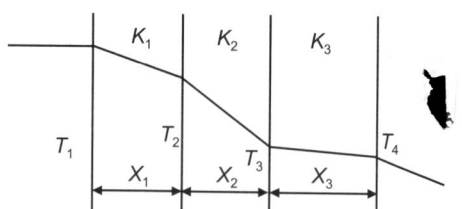

Fig. 5.2 Heat flow by conduction through composite walls

iii. Convection

Convection is the process of heat transfer from a medium of fluid. This fluid can be air, gas or liquid. Higher the turbulent flow of the fluid, higher the heat transfer rate. As shown in Fig. 5.3, the heat from hot fluid at temperature T_i is first transferred to solid surface

by convection, then the heat is transferred through the solid barrier by conduction and then the heat is given on the other side by convection through the fluid. The heat flow is defined as below.

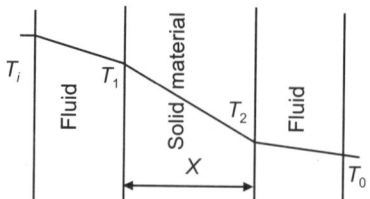

Fig. 5.3 Heat flow by convection

$$Q = H_i A (T_i - T_1) = \frac{KA (T_1 - T_2)}{X} = H_o A (T_2 - T_o)$$

where

H_i and H_o = Inside and outside heat transfer coefficients respectively.

$$Q = UA (T_i - T_o)$$

Where

U = Overall heat transfer coefficient and is a combination of conduction and convection coefficients.

5.1.3 Mean Temperature Difference

All the above discussions are based on the assumption that a constant temperature exists on two sides of surface. However, this condition does not exist always. Both hot and cold surfaces undergo a temperature change, hence a mean temperature difference must be used to compute the heat transfer. As shown in Fig. 5.4 a, b, i.e. parallel and counter flow heat exchange, the equation can be defined as

$$Q = UA (T_m)$$

T_m = Mean temperature difference

$$= \frac{Q_i - Q_o}{\log_e (Q_i / Q_o)}$$

$$\left.\begin{array}{l} Q_i = T_{hi} - T_{ci} \\ Q_o = T_{ho} - T_{co} \end{array}\right\} \text{for parallel flow}$$

$$\left.\begin{array}{l} Q_i = T_{hi} - T_{co} \\ Q_o = T_{ho} - T_{ci} \end{array}\right\} \text{for counter flow}$$

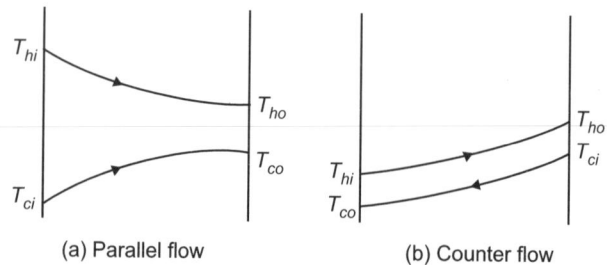

(a) Parallel flow   (b) Counter flow

Fig. 5.4 Mean temperature difference

5.1.4 How Heat Enters a Building

As shown in Fig. 5.5, the heat enters the room from different sources and from different mediums as below.

   i. *Conduction:* Through exterior walls, roof and glass, through interior partitions, ceilings, floors.

   ii. *Radiation:* Through glass.

   iii. *Convection:* Through infiltration or ventilation.

   iv. *Light electrical energy:* Through electrical lights and equipments.

   v. *Metabolic heat:* Heat from people.

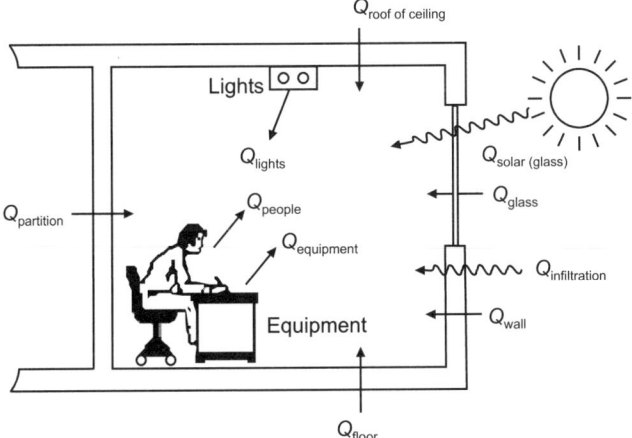

Fig. 5.5 Room heat gain components, Q

The air conditioning system has to remove two types of loads known as sensible and latent heat load. There are various sources which contribute to both kind of loads which a designer should know. In addition, it is important to understand the location of contributory heat load as the location has an important bearing in calculating its effect on room load or on air conditioning system.

1. Sensible heat load sources:

1. Heat flow from exterior walls, ceiling, floor, windows, doors, glass due to temperature difference on two sides.
2. Solar radiation load transmitted directly through glass in windows and ventilators.
3. Solar radiation load absorbed by wall and roof and later transmitted inside the room.
4. Metabolic heat generated by occupants.
5. Equipment heat
6. Outdoor air infiltration through cracks in doors, windows etc and their frequent opening.
7. Heat gain by ducts carrying supply air to room.
8. Heat transfer from unconditioned rooms to conditioned rooms through partition.

2. Latent heat load sources:

1. Metabolic Heat load from occupants.
2. Heat load from cooking food or stored material.

3. Moisture penetration through walls due to high vapour pressure.
4. Heat load due to infiltrated air.

   The above sources add heat in the room being air conditioned. However there are other heat sources which though do not add load in room but in the conditioner room i.e. AHU (air handling unit) and must be removed by air conditioning system. These are as below:

   1. Fresh air taken inside AHU.
   2. Heat given by fan before cooling coil. If some fan is installed after cooling coil, its effect will be considered on room load.
   3. Heat gain by air in the return duct.

Thus total heat load on the system is the sum of

1. Sensible heat
2. Latent heat
3. Other loads

To calculate the heat inflow, the procedure of calculation has been explained in heat load calculation sheet enclosed (courtesy: carrier corporation). Various factors affect the heat inflow in a system, e.g. weight of building walls and floors, material used and its insulation and reflective values, area of glass, internal/external venetian blinds, other shading devices.

All these factors effect the heat entering/leaving the building. These factors have been tabulated in the chapter and used with the help of an example to enable designers to effectively use them.

5.2 STEP BY STEP APPROACH: HEAT LOAD CALCULATION

5.2.1 Summary table

After the layout plan marked with dimensions, partition walls (AC or non AC) of adjoining areas, windows has been made (Fig. 5.6), make a summary table for the area of walls, windows, floor, partition, their material, thickness and density. The calculation in this table will be very useful whenever the designer has to revert back to calculations due to revision in design by architect or change in material of wall etc (Refer Table 5.1).

5.2.2 Proforma

Now that all the key parameters related to building have been tabulated, the heat load proforma (Table 5.2 and 5.3) will help to calculate the heat load, dehumidified CFM requirement, sensible Heat Ratio and other parameters required for designing air conditioning system.

   i. Fill the name of project, zone of the building, floor number and date on which calculation is being done so that subsequent revisions can be identified easily.

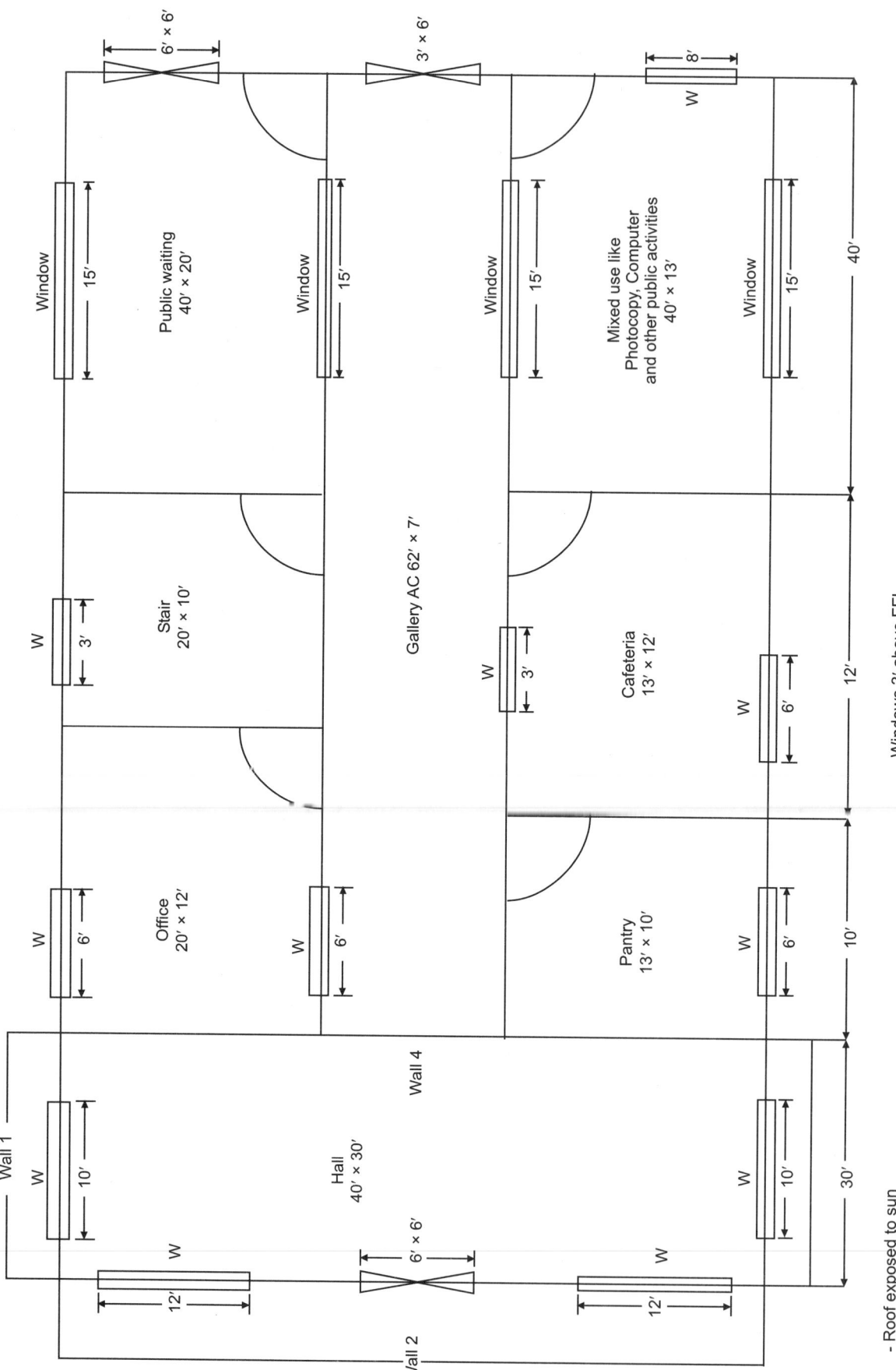

Fig. 5.6 Heat load of a typical building

Table 5.1 Summary of glass and wall areas

Name of Area	Size Sq ft	Wall-1				Wall-2				Wall-3				Wall-4				Area of partition (Non AC)	Total inside wall area
		Direction	Area Sq ft	Window area (glass) 6 mm	Net wall area	Direction	Area Sq ft	Window area (glass) 6 mm	Net wall area	Direction	Area Sq ft	Window area (glass) 6 mm	Net wall area	Direction	Area Sq ft	Window area (glass) 6 mm	Net wall area		
Hall	1200	NE	300	45	255	NW	400	144	256	SW	300	45	255	SE	400		400	130	400
Office	240	NE	120	27	93	NW	200		200	SW	120	27	93	SE	200		200	200	200
Public waiting	800	NE	400	67.5	332.5	NW	200		200	SW	400	67.5	332.5	SE	200	36	164	200	200
Cafeteria	156	NE	120	13.5	106.5	NW	130		130	SW	120	27	93	SE	130		130	130	130
Mixed use	520	NE	400	67.5	332.5	NW	130		130	SW	400	67.5	332.5	SE	130	36	94	130	130
Gallery	434	NE	620											SE	70	18	52	100	1240
Total	3350				1120				916				1106				1040	760	2300
Floor	3350																		
Ceiling	3350																		

Notes: 1. Outer walls 9" brick with 25 mm plaster on outer face and 15 mm on inner face. Internal and partition walls 4.5" with 15 mm plaster on both faces.
2. Brick work of density 118 lb/cubic ft and plaster 112 lb/cubic ft
3. Slab (RCC) thickness 6–8" with 1.5" flooring.
4. Slab density 156 lb/cubic ft and flooring density 137 lb/ft
5. Glass thickness 6 mm with density 156 lb/cubic ft and U value 1.13 BTU/hr/sq ft/F°

Table 5.2 (i) Heat load proforma (Summer)

	Job name	**Sample**				Date		Rev	0
	Space for	**Hall**				**Estimated by:**			
	Floor area	**Typical**				Estimate for: **Summer**			
	Height 10 ft								
	Size 1200 sq ft, Vol. 12000 cubic ft								

Part-I — Solar gain glass / Heat gain / Part-VII

Item	Area (Sq ft)	Sun gain (Btu/h sq ft)	Factor	Btu/hour
N - Glass	0	18	0.56	0
NE - Glass	45	12	0.56	302
E - Glass	0	12	0.56	0
SE - Glass	0	12	0.56	0
S - Glass	0	12	0.56	0
SW - Glass	45	90	0.56	2268
W - Glass	0	161	0.56	0
NW - Glass	144	130	0.56	10483
Sky light				0

Part-VII

Condition	DB(°F)	WB(°F)	%RH	DP(°F)	GR/LB
Outside	110	75	20		74
Room	75		60		68
Difference	35	XXXX	XXXX	XXXX	6

Part-VIII

Outside air (ventilation)
25 People X	5	CFM/Person	125
1200 Sq ft	0.06	CFM/Sq ft	72
		CFM ventilation	197

Eff sensible heat factor (ESHF)	= 0.93
Indicated ADP	= 59.3 °F
Selected ADP	= 55 °F
Dehum temp rise	= 17.00 °F
Dehumdified CFM	= **4296**

Part-II — Solar and trans gain walls and roof

Item	Area (Sq ft)	Eq temp diff (°F)	U (Btu/h sq ft)	Btu/hour
N-Wall	0	24.5	0.31	0
NE-Wall	255	30.5	0.31	2411
E-Wall	0	38.5	0.31	0
SE-Wall	400	38.5	0.31	4774
S-Wall	0	36.5	0.31	0
SW-Wall	255	34.5	0.31	2727
W-Wall	0	32.5	0.31	0
NW-Wall	256	26.5	0.31	2103
Roof sun	1200	52.5	0.23	14490
Roof insulated	0	0	0.12	0

Notes

Occupancy =	25 Nos.
Lighting =	1.0 W/Sq ft
Eq. Load =	0.5 W/Sq ft
Height =	10.0 FT

Part-III — Trans Gain except walls and roof

Item	Area (Sq ft)	Temp.diff. (°F)	U (Btu/h sq ft)	Btu/hour
All Glass	234	35	1.13	9255
Partition	130	20	0.35	910
Ceiling	0	20	0.4	0
Floor	**1200**	20	0.32	7680

Part-IV — Internal Heat Gain

			Btu/hour
People	25 Nos X	245	6125
Light	1200.0 W X 1.25	3.41	5115
Eq. Load	600 W X	3.41	2046

Internal Heat - People

Sensible heat	= 245
Latent heat	= 205

Room sensible Heat (RSH)	**70690**

Part-V

Supply	Heat gain from unconditioned space(%) 5.0			3534
duct	safety factor (%)		5.0	3534

Outside air

CFM	°F	BF	Factor	
197	35	0.15	0.68	1117

Effective room sensible heat (ERSH)	**78876**

Actual TR = 8.00

Latent heat

People	25	Nos X	205	
		Room latent heat (RLH)		5125
Supply duct leakage loss + safety factor %	5.0			256

Outside air

CFM	GR/LB	BF	Factor	
197	6	0.15	0.68	121

Effective room latent heat (ERLH)	**5502**
Effective room total heat (ERTH)	**84377**

Part-VI

Outside air heat (sensible)

CFM	°F	1–BF	Factor	
197	35	0.85	1.08	6330

Outside air heat (sensible)

CFM	GR/LB	1–BF	Factor	
197	6	0.85	0.68	683
			Heat sub total	91390

Return duct			
heat+ pump + Dehum & Pipe losses (%)	5.0		4570

TR	**8.00**	**Grand total heat**	**95960**

Check figures

BTU/hr/Sq ft =	80.0
CFM/Sq ft =	3.58
Sq ft/TR =	150
CFM/TR =	537
Total CFM =	4296

Table 5.2 (ii) Heat load proforma (Summer)

	Job name	**Sample**								
	Space for	**Office**				Date		Rev	0	
	Floor area	**Typical**				**Estimated by:**				
	Height 10 ft					Estimate for: **Summer**				
	Size 240 sq ft, Vol. 2400 cubic ft									

Part-I — Solar gain glass

Item	Area (Sq ft)	Sun gain (Btu/h sq ft)	Factor	Btu/hour (Heat gain)
N - Glass	0	18	0.56	0
NE - Glass	27	12	0.56	181
E - Glass	0	12	0.56	0
SE - Glass	0	12	0.56	0
S - Glass	0	12	0.56	0
SW - Glass	27	90	0.56	1361
W - Glass	0	161	0.56	0
NW - Glass	0	130	0.56	0
Sky light				0

Part-II — Solar and trans gain walls and roof

Item	Area (Sq ft)	Eq temp diff (°F)	U (Btu/h sq ft)	Btu/hour
N-Wall	0	24.5	0.31	0
NE-Wall	93	30.5	0.31	879
E-Wall	0	38.5	0.31	0
SE-Wall	200	38.5	0.31	2387
S-Wall	0	36.5	0.31	0
SW-Wall	93	34.5	0.31	995
W-Wall	0	32.5	0.31	0
NW-Wall	200	26.5	0.31	1643
Roof sun	240	52.5	0.23	2898
Roof insulated	0	0	0.12	0

Part-III — Trans Gain except walls and roof

Item	Area (Sq ft)	Temp.diff. (°F)	U (Btu/h sq ft)	Btu/hour
All Glass	54	35	1.13	2136
Partition	200	20	0.35	1400
Ceiling	0	20	0.4	0
Floor	**240**	20	0.32	1536

Part-IV — Internal Heat Gain

				Btu/hour
People	2 Nos X		245	490
Light	240.0 W X 1.25		3.41	1023
Eq. Load	120 W X		3.41	409

Room sensible Heat (RSH)				**17338**

Part-V

Supply Heat gain from unconditioned space(%)	5.0			867
duct safety factor (%)			5.0	867
Outside air				
CFM	°F	BF	Factor	
24	35	0.15	1.08	138

Effective room sensible heat (ERSH)				**19210**

Latent heat

People	2	Nos X	205	
		Room latent heat (RLH)		410
Supply duct leakage loss + safety factor %		5.0		21
Outside air				
CFM	GR/LB	BF	Factor	
24	6	0.15	0.68	15

Effective room latent heat (ERLH)				**445**
Effective room total heat (ERTH)				**19656**

Part-VI

Outside air heat (sensible)				
CFM	°F	1–BF	Factor	
24	35	0.85	1.08	784
Outside air heat (latent)				
CFM	GR/LB	1–BF	Factor	
24	6	0.85	0.68	85
		Heat sub total		20524
Return duct				
heat + pump + Dehum & Pipe losses (%)		5.0		1026

TR	**1.80**	**Grand total heat**		**21550**

Part-VII

Condition	DB(°F)	WB(°F)	%RH	DP(°F)	GR/LB
Outside	110	75	20		74
Room	75		60		68
Difference	35	XXXX	XXXX	XXXX	6

Part-VIII

Outside air (ventilation)			
2 People X	5	CFM/Person	10
240 Sq ft	0.06	CFM/Sq ft	14
		CFM ventilation	24

Eff sensible heat factor (ESHF) =	0.93
Indicated ADP =	59.9 °F
Selected ADP =	55 °F
Dehum temp rise =	17.00 °F
Dehumdified CFM =	**1046**

Notes

Occupancy =	25 Nos.
Lighting =	1.0 W/Sq ft
Eq. Load =	0.5 W/Sq ft
Height =	10.0 FT

Internal Heat - People

Sensible heat	= 245
Latent heat	= 205

Actual TR = 1.80

Check figures

BTU/hr/Sq ft =	89.8
CFM/Sq ft =	4.36
Sq ft/TR =	134
CFM/TR =	583
Total CFM =	1046

Table 5.2 (iii) Heat load proforma (Summer)

1	Job name	**Sample**		Date	Rev 0
2	Space for	**Public waiting**		**Estimated by:**	
3	Floor area	**Typical**		Estimate for: **Summer**	
4	Height 10 ft				
	Size 800 sq ft, Vol. 8000 cubic ft				

Part-I — Solar gain glass / Heat gain

	Item	Area	Sun gain	Factor	Btu/hour
6		(Sq ft)	(Btu/h sq ft)		
8	N - Glass	0	18	0.56	0
9	NE - Glass	67.5	12	0.56	454
10	E - Glass	0	12	0.56	0
11	SE - Glass	0	12	0.56	0
12	S - Glass	0	12	0.56	0
13	SW - Glass	67.5	90	0.56	3402
14	W - Glass	0	161	0.56	0
15	NW - Glass	0	130	0.56	0
16	Sky light				0

Part-II — Solar and trans gain walls and roof

	Item	Area	Eq temp diff	U	
19		(Sq ft)	(°F)	(Btu/h sq ft)	
20	N-Wall	0	24.5	0.31	0
21	NE-Wall	332.5	30.5	0.31	3144
22	E-Wall	0	38.5	0.31	0
23	SE-Wall	164	38.5	0.31	* 1957
24	S-Wall	0	36.5	0.31	0
25	SW-Wall	332.5	34.5	0.31	3556
26	W-Wall	0	32.5	0.31	0
27	NW-Wall	200	26.5	0.31	1643
28	Roof sun	800	52.5	0.23	9660
29	Roof insulated	0	0	0.12	0

Part-III — Trans Gain except walls and roof

	Item	Area	Temp. diff.	U	
32		(Sq ft)	(°F)	(Btu/h sq ft)	
33	All Glass	171	35	1.13	6763
34	Partition	200	20	0.35	1400
35	Ceiling	0	20	0.4	0
36	Floor	**800**	20	0.32	5120

Part-IV — Internal Heat Gain

38	People	10 Nos X		245	2450
39	Light	800.0 W X 1.25		3.41	3410
40	Eq. Load	400 W X		3.41	1364

41	**Room sensible Heat (RSH)**		**44565**

Part-V

42	Supply Heat gain from unconditioned space(%)	5.0			2228
43	duct	safety factor (%)	5.0		2228
44	Outside air				
45	CFM	°F	BF	Factor	
46	98	35	0.15	1.08	556
47	**Effective room sensible heat (ERSH)**				**49577**

48	Latent heat				
49	People	10	Nos X	205	
50			Room latent heat (RLH)		2050
51	Supply duct leakage loss + safety factor %	5.0			103
52	Outside air				
53	CFM	GR/LB	BF	Factor	
54	98	6	0.15	0.68	60
55	**Effective room latent heat (ERLH)**				**2212**
56	**Effective room total heat (ERTH)**				**51789**

Part-VI

57	Outside air heat (sensible)				
58	CFM	°F	1–BF	Factor	
59	98	35	0.85	1.08	3149
60	Outside air heat (latent)				
61	CFM	GR/LB	1–BF	Factor	
62	98	6	0.85	0.68	340
				Heat sub total	55278
63	Return duct				
64	heat + pump + Dehum & Pipe losses (%)	5.0			2764
65	**TR**	**4.84**	**Grand total heat**		**58042**

Part-VII

Condition	DB(°F)	WB(°F)	%RH	DP(°F)	GR/LB
Outside	110	75	20		74
Room	75		60		68
Difference	35	XXXX	XXXX	XXXX	6

Part-VIII

Outside air (ventilation)				
10 People X	5	CFM/Person	50	
800 Sq ft	0.06	CFM/Sq ft	48	
		CFM ventilation	98	

Eff sensible heat factor (ESHF)	=	0.96
Indicated ADP	=	59.6 °F
Selected ADP	=	55 °F
Dehum temp rise	=	17.00 °F
Dehumdified CFM	=	**2700**

Notes

Occupancy =	25 Nos.
Lighting =	1.0 W/Sq ft
Eq. Load =	0.5 W/Sq ft
Height =	10.0 FT

Internal Heat - People

Sensible heat	= 245
Latent heat	= 205

Actual TR = 4.84

Check figures

BTU/hr/Sq ft =	72.6
CFM/Sq ft =	3.38
Sq ft/TR =	165
CFM/TR =	558
Total CFM =	**2700**

Table 5.2 (iv) Heat load proforma (Summer)

1		Job name	**Sample**					Date			Rev	0
2		Space for	**Cafeteria**					**Estimated by:**				
3		Floor area Height 10 ft	**Typical**									
4		Size 156 sq ft, Vol. 1560 cubic ft						Estimate for: **Summer**				

Part-I — Solar gain glass / Heat gain

		Item	Area (Sq ft)	Sun gain (Btu/h sq ft)	Factor	Btu/hour
5		Solar gain glass				Heat gain
6		Item	Area	Sun gain	Factor	Btu/hour
7			(Sq ft)	(Btu/h sq ft)		
8		N - Glass	0	18	0.56	0
9		NE - Glass	13.5	12	0.56	91
10		E - Glass	0	12	0.56	0
11		SE - Glass	0	12	0.56	0
12		S - Glass	0	12	0.56	0
13		SW - Glass	27	90	0.56	1361
14		W - Glass	0	161	0.56	0
15		NW - Glass	0	130	0.56	0
16		Sky light				0

Part-II — Solar and trans gain walls and roof

		Item	Area (Sq ft)	Eq temp diff (°F)	U (Btu/h sq ft)	Btu/hour
17		Solar and trans gain walls and roof				
18		Item	Area	Eq temp diff	U	
19			(Sq ft)	(°F)	(Btu/h sq ft)	
20		N-Wall	0	24.5	0.31	0
21		NE-Wall	106.5	30.5	0.31	1007
22		E-Wall	0	38.5	0.31	0
23		SE-Wall	130	38.5	0.31	1552
24		S-Wall	0	36.5	0.31	0
25		SW-Wall	93	34.5	0.31	995
26		W-Wall	0	32.5	0.31	0
27		NW-Wall	130	26.5	0.31	1068
28		Roof sun	156	52.5	0.23	1884
29		Roof insulated	0	0	0.12	0

Part-III — Trans Gain except walls and roof

		Item	Area (Sq ft)	Temp. diff. (°F)	U (Btu/h sq ft)	Btu/hour
30		Trans Gain except walls and roof				
31		Item	Area	Temp. diff.	U	
32			(Sq ft)	(°F)	(Btu/h sq ft)	
33		All Glass	40.5	35	1.13	1602
34		Partition	130	20	0.35	910
35		Ceiling	0	20	0.4	0
36		Floor	**156**	20	0.32	998

Part-IV — Internal Heat Gain

		Item		Factor	Btu/hour
37		Internal Heat Gain			
38		People	10 Nos X	245	2450
39		Light	156.0 W X 1.25	3.41	665
40		Eq. Load	78 W X	3.41	266

Part-V

						Btu/hour
41		Room sensible Heat (RSH)				**14847**
42		Supply Heat gain from unconditioned space(%) 5.0				742
43		duct	safety factor(%)		5.0	742
44		Outside air				
45		CFM	°F	BF	Factor	
46		59	35	0.15	1.08	337
47		**Effective room sensible heat (ERSH)**				**16669**

						Btu/hour
48		Latent heat				
49		People	10 Nos X		205	
50			Room latent heat (RLH)			2050
51		Supply duct leakage loss + safety factor%		5.0		103
52		Outside air				
53		CFM	GR/LB	BF	Factor	
54		59	6	0.15	0.68	36
55		**Effective room latent heat (ERLH)**				**2189**
56		**Effective room total heat (ERTH)**				**18858**

Part-VI

						Btu/hour
57		Outside air heat (sensible)				
58		CFM	°F	1–BF	Factor	
59		59	35	0.85	1.08	1907
60		Outside air heat (latent)				
61		CFM	GR/LB	1–BF	Factor	
62		59	6	0.85	0.68	206
62				Heat sub total		20971
63		Return duct				
64		heat + pump + Dehum & Pipe losses (%) 5.0				1049
65		**TR**	**1.83**	**Grand total heat**		**22019**

Part-VII

Condition	DB(°F)	WB(°F)	%RH	DP(°F)	GR/LB
Outside	110	75	20		74
Room	75		60		68
Difference	35	XXXX	XXXX	XXXX	6

Part-VIII

Outside air (ventilation)

10 People X	5	CFM/Person	50
156 Sq ft	0.06	CFM/Sq ft	9
		CFM ventilation	59

Eff sensible heat factor (ESHF)	=	0.88
Indicated ADP	=	58.7°F
Selected ADP	=	55 °F
Dehum temp rise	=	17.00 °F
Dehumdified CFM	=	**908**

Notes

Occupancy =	25 Nos.
Lighting =	1.0 W/Sq ft
Eq. Load =	0.5 W/Sq ft
Height =	10.0 FT

Internal Heat - People

Sensible heat	= 245
Latent heat	= 205

Actual TR = 1.83

Check figures

BTU/hr/Sq ft =	141.1
CFM/Sq ft =	5.82
Sq ft/TR =	85
CFM/TR =	495
Total CFM =	**908**

Table 5.2 (v) Heat load proforma (Summer)

1 Job name **Sample**	
2 Space for **Mixed use**	Date Rev 0
3 Floor area **Typical**	**Estimated by:**
4 Height 10 ft	
Size 520 sq ft, Vol. 5200 cubic ft	Estimate for: **Summer**

Solar gain glass / Heat gain — **Part-VII**

Item	Area	Sun gain	Factor	Btu/hour		Condition	DB(°F)	WB(°F)	%RH	DP(°F)	GR/LB
	(Sq ft)	(Btu/h sq ft)				Outside	110	75	20		74
N - Glass	0	18	0.56	0		Room	75		60		68
NE - Glass	67.5	12	0.56	454		Difference	35	XXXX	XXXX	XXXX	6
E - Glass	0	12	0.56	0							
SE - Glass	36	12	0.56	242		**Part-VIII**					
S - Glass	0	12	0.56	0		Outside air (ventilation)					
SW - Glass	67.5	90	0.56	3402		8 People X	5	CFM/Person			40
W - Glass	0	161	0.56	0		520 Sq ft	0.06	CFM/Sq ft			31
NW - Glass	0	130	0.56	0				CFM ventilation			71
Sky light				0							

Part-I (rows 6–16)

Solar and trans gain walls and roof (Part-II)

Item	Area	Eq temp diff	U			
	(Sq ft)	(°F)	(Btu/h sq ft)		Eff sensible heat factor (ESHF)	= 0.96
N-Wall	0	24.5	0.31	0	Indicated ADP	= 59.7 °F
NE-Wall	332.5	30.5	0.31	3144	Selected ADP	= 55 °F
E-Wall	0	38.5	0.31	0	Dehum temp rise	= 17.00 °F
SE-Wall	94	38.5	0.31	1122	Dehumidified CFM	**= 2084**
S-Wall	0	36.5	0.31	0		
SW-Wall	332.5	34.5	0.31	3556		
W-Wall	0	32.5	0.31	0		
NW-Wall	130	26.5	0.31	1068		
Roof sun	520	52.5	0.23	6279		
Roof insulated	0	0	0.12	0		

Trans Gain except walls and roof (Part-III)

Item	Area	Temp. diff.	U		**Notes**	
	(Sq ft)	(°F)	(Btu/h sq ft)		Occupancy =	25 Nos.
All Glass	171	35	1.13	6763	Lighting =	1.0 W/Sq ft
Partition	0	20	0.35	0	Eq. Load =	0.5 W/Sq ft
Ceiling	0	20	0.4	0	Height =	10.0 FT
Floor	**520**	20	0.32	3328		

Internal Heat Gain (Part-IV)

					Internal Heat - People	
People	0 Nos X		245	1900	Sensible heat	= 245
Light	520.0 W X 1.25		3.41	2217	Latent heat	= 205
Eq. Load	260 W X		3.41	887		

Room sensible Heat (RSH)	**34420**

Part-V

Supply Heat gain from unconditioned space(%) 5.0				1721
duct safety factor (%)			5.0	1721
Outside air				
CFM	°F	BF	Factor	
71	35	0.15	1.08	404

Actual TR = 3.72

Effective room sensible heat (ERSH)	**38266**

Latent heat

People	8	Nos X	205	
		Room latent heat (RLH)		1640
Supply duct leakage loss + safety factor % 5.0				82
Outside air				
CFM	GR/LB	BF	Factor	
71	6	0.15	0.68	44

Effective room latent heat (ERLH)	**1766**
Effective room total heat (ERTH)	**40032**

Part-VI

Outside air heat (sensible)				
CFM	°F	1–BF	Factor	
71	35	0.85	1.08	2288
Outside air heat (latent)				
CFM	GR/LB	1–BF	Factor	
71	6	0.85	0.68	247
			Heat sub total	42566
Return duct				
heat + pump + Dehum & Pipe losses (%) 5.0				2128
TR	**3.72**	**Grand total heat**		**44695**

Check figures

BTU/hr/Sq ft =	86.0
CFM/Sq ft =	4.01
Sq ft/TR =	140
CFM/TR =	560
Total CFM =	2084

Table 5.2 (vi) Heat load proforma (Summer)

1	Job name	**Sample**		
2	Space for	**Gallery**	Date	Rev 0
3	Floor area	**Typical**	**Estimated by:**	
	Height 10 ft			
4	Size 434 sq ft, Vol. 4340 cubic ft		Estimate for: **Summer**	

Part-I — Solar gain glass

Item	Area (Sq ft)	Sun gain (Btu/h sq ft)	Factor	Btu/hour
N - Glass	0	18	0.56	0
NE - Glass	0	12	0.56	0
E - Glass	0	12	0.56	0
SE - Glass	18	12	0.56	121
S - Glass	0	12	0.56	0
SW - Glass	0	90	0.56	0
W - Glass	0	161	0.56	0
NW - Glass	0	130	0.56	0
Sky light				0

Part-VII

Condition	DB(°F)	WB(°F)	%RH	DP(°F)	GR/LB
Outside	110	75	20		74
Room	75	60			68
Difference	35	XXXX	XXXX	XXXX	6

Part-VIII — Outside air (ventilation)

5 People X	5	CFM/Person	25
434 Sq ft	0.06	CFM/Sq ft	26
		CFM ventilation	51

Eff sensible heat factor (ESHF) = 0.93
Indicated ADP = 59.3 °F
Selected ADP = 55 °F
Dehum temp rise = 17.00 °F
Dehumdified CFM = **854**

Part-II — Solar and trans gain walls and roof

Item	Area (Sq ft)	Eq temp diff (°F)	U (Btu/h sq ft)	Btu/hour
N-Wall	0	24.5	0.31	0
NE-Wall	0	30.5	0.31	0
E-Wall	0	38.5	0.31	0
SE-Wall	52	38.5	0.31	621
S-Wall	0	36.5	0.31	0
SW-Wall	0	34.5	0.31	0
W-Wall	0	32.5	0.31	0
NW-Wall	0	26.5	0.31	0
Roof sun	434	52.5	0.23	5241
Roof insulated	0	0	0.12	0

Part-III — Trans Gain except walls and roof

Item	Area (Sq ft)	Temp.diff. (°F)	U (Btu/h sq ft)	Btu/hour
All Glass	18	35	1.13	712
Partition	100	20	0.35	700
Ceiling	0	20	0.4	0
Floor	**434**	20	0.32	2778

Notes

Occupancy = 25 Nos.
Lighting = 1.0 W/Sq ft
Eq. Load = 0.5 W/Sq ft
Height = 10.0 FT

Internal Heat - People

Sensible heat = 245
Latent heat = 205

Part-IV — Internal Heat Gain

People	5 Nos X	245	1225
Light	434.0 W X 1.25	3.41	1850
Eq. Load	217 W X	3.41	740

Room sensible Heat (RSH)		**13987**

Part-V

Supply Heat gain from unconditioned space(%) 5.0				699
duct	safety factor(%)		5.0	699
Outside air				
CFM	°F	BF	Factor	
51	35	0.15	1.08	289

Effective room sensible heat (ERSH)		**15675**

Actual TR = 1.63

Latent heat

People	5	Nos X	205
		Room latent heat (RLH)	1025
Supply duct leakage loss + safety factor%	5.0		51
Outside air			
CFM	GR/LB	BF	Factor
51	6	0.15	0.68	31

Effective room latent heat (ERLH)		**1107**
Effective room total heat (ERTH)		**16782**

Part-VI

Outside air heat (sensible)				
CFM	°F	1–BF	Factor	
51	35	0.85	1.08	1640
Outside air heat (latent)				
CFM	GR/LB	1–BF	Factor	
51	6	0.85	0.68	177
			Heat sub total	18599
Return duct				
heat + pump + Dehum & Pipe losses (%) 5.0				930
TR	**1.63**	**Grand total heat**		**19529**

Check figures

BTU/hr/Sq ft = 45.0
CFM/Sq ft = 1.97
Sq ft/TR = 267
CFM/TR = 525
Total CFM = **854**

Table 5.3 (i) Heat load proforma (monsoon)

Job name	**Sample**
Space for	**Hall**
Floor area	**Typical**
Height 10 ft	
Size 1200 sq ft, Vol. 12000 cubic ft	

Date   Rev   0
Estimated by:
Estimate for: **Monsoon**

Part-I — Solar gain glass

Item	Area	Sun gain	Factor	Btu/hour
	(Sq ft)	(Btu/h sq ft)		
N - Glass	0	14	0.56	0
NE - Glass	45	12	0.56	302
E - Glass	0	12	0.56	0
SE - Glass	0	12	0.56	0
S - Glass	0	12	0.56	0
SW - Glass	45	100	0.56	2520
W - Glass	0	164	0.56	0
NW - Glass	144	123	0.56	9919
Sky light				0

Part-II — Solar and trans gain walls and roof

Item	Area	Eq temp diff	U	
	(Sq ft)	(°F)	(Btu/h sq ft)	
N-Wall	0	14.5	0.31	0
NE-Wall	255	20.5	0.31	1621
E-Wall	0	28.5	0.31	0
SE-Wall	400	28.5	0.31	3534
S-Wall	0	26.5	0.31	0
SW-Wall	255	24.5	0.31	1937
W-Wall	0	22.5	0.31	0
NW-Wall	256	16.5	0.31	1309
Roof sun	1200	42.5	0.23	11730
Roof insulated	0	0	0.12	0

Part-III — Trans Gain except walls and roof

Item	Area	Temp. diff.	U	
	(Sq ft)	(°F)	(Btu/h sq ft)	
All Glass	234	20	1.13	5288
Partition	130	20	0.35	910
Ceiling	0	20	0.4	0
Floor	**1200**	20	0.32	7680

Part-IV — Internal Heat Gain

People	25 Nos X		245	6125
Light	1200.0 W X 1.25		3.41	5115
Eq. Load	600 W X		3.41	2046

Room sensible Heat (RSH)	**60036**

Part-V

Supply Heat gain from unconditioned space(%) 5.0				3002
duct safety factor (%)			5.0	3002
Outside air				
CFM	°F	BF	Factor	
197	20	0.15	1.08	638
Effective room sensible heat (ERSH)				**66678**

Latent heat

People	25	Nos X	205	
		Room latent heat (RLH)		5125
Supply duct leakage loss + safety factor %		5.0		256
Outside air				
CFM	GR/LB	BF	Factor	
197	122	0.15	0.68	2451
Effective room latent heat (ERLH)				**7833**
Effective room total heat (ERTH)				**74511**

Part-VI

Outside air heat (sensible)				
CFM	°F	1–BF	Factor	
197	20	0.85	1.08	3617
Outside air heat (latent)				
CFM	GR/LB	1–BF	Factor	
197	122	0.85	0.68	13892
			Heat sub total	92019
Return duct				
heat + pump + Dehum & Pipe losses (%) 5.0				4601
TR	**8.05**	**Grand total heat**		**96620**

Part-VII

Condition	DB(°F)	WB(°F)	%RH	DP(°F)	GR/LB
Outside	95	75	75		190
Room	75		60		68
Difference	20	XXXX	XXXX	XXXX	122

Part-VIII

Outside air (ventilation)			
25 People X	5	CFM/Person	125
1200 Sq ft	0.06	CFM/Sq ft	72
		CFM ventilation	197

Eff sensible heat factor (ESHF)	=	0.89
Indicated ADP	=	58.8 °F
Selected ADP	=	55 °F
Dehum temp rise	=	17.00 °F
Dehumidified CFM	=	**3632**

Notes

Occupancy =	25 Nos.
Lighting =	1.0 W/Sq ft
Eq. Load =	0.5 W/Sq ft
Height =	10.0 FT

Internal Heat - People

Sensible heat	= 245
Latent heat	= 205

Actual TR = 8.05

Check figures

BTU/hr/Sq ft =	80.5
CFM/Sq ft =	3.03
Sq ft/TR =	149
CFM/TR =	451
Total CFM =	3632

Table 5.3 (ii) Heat load proforma (monsoon)

Job name	**Sample**				Date		Rev	0
Space for	**Cafeteria**				**Estimated by:**			
Floor area	**Typical**				Estimate for: **Monsoon**			
Height 10 ft								
Size 240 sq ft, Vol. 2400 cubic ft								

Part-I — Solar gain glass / Heat gain

Item	Area	Sun gain	Factor	Btu/hour
	(Sq ft)	(Btu/h sq ft)		
N - Glass	0	14	0.56	0
NE - Glass	27	12	0.56	161
E - Glass	0	12	0.56	0
SE - Glass	0	12	0.56	0
S - Glass	0	12	0.56	0
SW - Glass	27	100	0.56	1512
W - Glass	0	164	0.56	0
NW - Glass	0	123	0.56	0
Sky light				0

Part-II — Solar and trans gain walls and roof

Item	Area	Eq temp diff	U	
	(Sq ft)	(°F)	(Btu/h sq ft)	
N-Wall	0	14.5	0.31	0
NE-Wall	93	20.5	0.31	591
E-Wall	0	28.5	0.31	0
SE-Wall	200	28.5	0.31	1767
S-Wall	0	26.5	0.31	0
SW-Wall	93	24.5	0.31	705
W-Wall	0	22.5	0.31	0
NW-Wall	200	16.5	0.31	1023
Roof sun	240	42.5	0.23	2356
Roof insulated	0	0	0.12	0

Part-III — Trans Gain except walls and roof

Item	Area	Temp. diff.	U	
	(Sq ft)	(°F)	(Btu/h sq ft)	
All Glass	54	20	1.13	1220
Partition	200	20	0.35	1400
Ceiling	0	20	0.4	0
Floor	**240**	20	0.32	1536

Part-IV — Internal Heat Gain

People	2 Nos X		245	490
Light	240.0 W X 1.25		3.41	1023
Eq. Load	120 W X		3.41	409

Part-V

Room sensible Heat (RSH)				**14205**
Supply duct	Heat gain from unconditioned space(%) 5.0			710
	safety factor (%)		5.0	710
Outside air				
CFM	°F	BF	Factor	
24	20	0.15	1.08	79
Effective room sensible heat (ERSH)				**15705**

Latent heat				
People	2	Nos X	205	
		Room latent heat (RLH)		410
Supply duct leakage loss + safety factor %	5.0			21
Outside air				
CFM	GR/LB	BF	Factor	
24	20	0.15	1.08	304
Effective room latent heat (ERLH)				**734**
Effective room total heat (ERTH)				**16439**

Part-VI

Outside air heat (sensible)				
CFM	°F	1–BF	Factor	
24	20	0.85	1.08	448
Outside air heat (latent)				
CFM	GR/LB	1–BF	Factor	
24	122	0.85	0.68	1721
			Heat sub total	18608
Return duct				
heat + pump + Dehum & Pipe losses (%)	5.0			930
TR	**1.63**	**Grand total heat**		**19538**

Part-VII

Condition	DB(°F)	WB(°F)	%RH	DP(°F)	GR/LB
Outside	95	75	75		190
Room	75	60			68
Difference	20	XXXX	XXXX	XXXX	122

Part-VIII

Outside air (ventilation)

10 People X	5	CFM/Person	10
800 Sq ft	0.06	CFM/Sq ft	14
		CFM ventilation	24

Eff sensible heat factor (ESHF)	=	0.96
Indicated ADP	=	59.6 °F
Selected ADP	=	55 °F
Dehum temp rise	=	17.00 °F
Dehumdified CFM	=	**855**

Notes

Occupancy =	25 Nos.
Lighting =	1.0 W/Sq ft
Eq. Load =	0.5 W/Sq ft
Height =	10.0 FT

Internal Heat - People

Sensible heat	= 245
Latent heat	= 205

Actual TR = 1.63

Check figures

BTU/hr/Sq ft =	81.4
CFM/Sq ft =	3.56
Sq ft/TR =	147
CFM/TR =	525
Total CFM =	**855**

Table 5.3 (iii) Heat load proforma (monsoon)

Job name **Sample**
Space for **Public waiting**
Floor area **Typical**
Height 10 ft
Size 800 sq ft, Vol. 8000 cubic ft

Date   Rev   0
Estimated by:
Estimate for: **Monsoon**

Part-I — Solar gain glass

Item	Area (Sq ft)	Sun gain (Btu/h sq ft)	Factor	Heat gain (Btu/hour)
N - Glass	0	14	0.56	0
NE - Glass	67.5	12	0.56	454
E - Glass	0	12	0.56	0
SE - Glass	36	12	0.56	242
S - Glass	0	12	0.56	0
SW - Glass	67.5	100	0.56	3780
W - Glass	0	164	0.56	0
NW - Glass	0	123	0.56	0
Sky light				0

Part-II — Solar and trans gain walls and roof

Item	Area (Sq ft)	Eq temp diff (°F)	U (Btu/h sq ft)	Btu/hour
N-Wall	0	14.5	0.31	0
NE-Wall	332.5	20.5	0.31	2113
E-Wall	0	28.5	0.31	0
SE-Wall	164	28.5	0.31	1449
S-Wall	0	26.5	0.31	0
SW-Wall	332.5	24.5	0.31	2525
W-Wall	0	22.5	0.31	0
NW-Wall	200	16.5	0.31	1023
Roof sun	800	42.5	0.23	7820
Roof insulated	0		0.12	0

Part-III — Trans Gain except walls and roof

Item	Area (Sq ft)	Temp. diff. (°F)	U (Btu/h sq ft)	Btu/hour
All Glass	171	20	1.13	3865
Partition	200	20	0.35	1400
Ceiling	0	20	0.4	0
Floor	**800**	20	0.32	5120

Part-IV — Internal Heat Gain

				Btu/hour
People	10 Nos X		245	2450
Light	800.0 W X 1.25		3.41	3410
Eq. Load	400 W X		3.41	1364

Room sensible Heat (RSH)	**37014**

Part-V

Supply duct	Heat gain from unconditioned space (%) 5.0	1851
	safety factor (%) 5.0	1851

Outside air

CFM	°F	BF	Factor	
98	20	0.15	1.08	318

Effective room sensible heat (ERSH)	**41033**

Latent heat

People	10	Nos X	205	
		Room latent heat (RLH)		2050
Supply duct leakage loss + safety factor % 5.0				103

Outside air

CFM	GR/LB	BF	Factor	
98	20	0.15	1.08	1220

Effective room latent heat (ERLH)	**3372**
Effective room total heat (ERTH)	**44405**

Part-VI

Outside air heat (sensible)

CFM	°F	1–BF	Factor	
98	20	0.85	1.08	1799

Outside air heat (latent)

CFM	GR/LB	1–BF	Factor	
98	122	0.85	0.68	6911
		Heat sub total		53115

Return duct
heat + pump + Dehum & Pipe losses (%) 5.0   2656

TR	**4.65**	**Grand total heat**	**55771**

Part-VII

Condition	DB(°F)	WB(°F)	%RH	DP(°F)	GR/LB
Outside	95	75	75		190
Room	75		60		68
Difference	20	XXXX	XXXX	XXXX	122

Part-VIII

Outside air (ventilation)

10 People X	5	CFM/Person	50
800 Sq ft	0.06	CFM/Sq ft	48
		CFM ventilation	98

Eff sensible heat factor (ESHF)	=	0.92
Indicated ADP	=	59.2 °F
Selected ADP	=	55 °F
Dehum temp rise	=	17.00 °F
Dehumdified CFM	=	**2235**

Notes

Occupancy =	25 Nos.
Lighting =	1.0 W/Sq ft
Eq. Load =	0.5 W/Sq ft
Height =	10.0 FT

Internal Heat - People

Sensible heat	= 245
Latent heat	= 205

Actual TR = 4.65

Check figures

BTU/hr/Sq ft =	69.7
CFM/Sq ft =	2.79
Sq ft/TR =	172
CFM/TR =	481
Total CFM =	**2235**

Table 5.3 (iv) Heat load proforma (monsoon)

Job name	**Sample**			
Space for	**Cafeteria**		Date	Rev 0
Floor area	**Typical**		**Estimated by:**	
Height 10 ft			Estimate for: **Monsoon**	
Size 156 sq ft, Vol. 1560 cubic ft				

Part-I — Solar gain glass / Heat gain

	Item	Area	Sun gain	Factor	Btu/hour
		(Sq ft)	(Btu/h sq ft)		
	N - Glass	0	14	0.56	0
	NE - Glass	13.5	12	0.56	91
	E - Glass	0	12	0.56	0
	SE - Glass	0	12	0.56	0
	S - Glass	0	12	0.56	0
	SW - Glass	27	100	0.56	1512
	W - Glass	0	164	0.56	0
	NW - Glass	0	123	0.56	0
	Sky light				0

Part-VII

Condition	DB(°F)	WB(°F)	%RH	DP(°F)	GR/LB
Outside	95	75	75		190
Room	75		60		68
Difference	20	XXXX	XXXX	XXXX	122

Part-VIII

Outside air (ventilation)

10 People X	5	CFM/Person	50
156 Sq ft	0.06	CFM/Sq ft	9
		CFM ventilation	59

Eff sensible heat factor (ESHF)   =   0.83
Indicated ADP   =   57.9 °F
Selected ADP   =   55 °F
Dehum temp rise   =   17.00 °F
Dehumidified CFM   =   **761**

Part-II — Solar and trans gain walls and roof

	Item	Area	Eq temp diff	U	
		(Sq ft)	(°F)	(Btu/h sq ft)	
	N-Wall	0	14.5	0.31	0
	NE-Wall	106.5	20.5	0.31	677
	E-Wall	0	28.5	0.31	0
	SE-Wall	130	28.5	0.31	1149
	S-Wall	0	26.5	0.31	0
	SW-Wall	93	24.5	0.31	706
	W-Wall	0	22.5	0.31	0
	NW-Wall	130	16.5	0.31	665
	Roof sun	156	42.5	0.23	1525
	Roof insulated	0	0	0.12	0

Part-III — Trans Gain except walls and roof

	Item	Area	Temp. diff.	U	
		(Sq ft)	(°F)	(Btu/h sq ft)	
	All Glass	40.5	20	1.13	915
	Partition	130	20	0.35	910
	Ceiling	0	20	0.4	0
	Floor	**156**	20	0.32	998

Notes

Occupancy =	25 Nos.
Lighting =	1.0 W/Sq ft
Eq. Load =	0.5 W/Sq ft
Height =	10.0 FT

Part-IV — Internal Heat Gain

People	10 Nos X	245	2450
Light	156.0 W X 1.25	3.41	665
Eq. Load	78 W X	3.41	266

Internal Heat - People

Sensible heat	= 245
Latent heat	= 205

Room sensible Heat (RSH)	**37014**

Part-V

Supply Heat gain from unconditioned space(%) 5.0				626
duct	safety factor (%)		5.0	626

Outside air

CFM	°F	BF	Factor	
59	20	0.15	1.08	192

Actual TR = 1.94

Effective room sensible heat (ERSH)	**13974**

Latent heat

People	10 Nos X	205	
		Room latent heat (RLH)	2050
Supply duct leakage loss + safety factor % 5.0			103

Outside air

CFM	GR/LB	BF	Factor	
59	20	0.15	1.08	739

Effective room latent heat (ERLH)	**2891**
Effective room total heat (ERTH)	**16865**

Part-VI

Outside air heat (sensible)

CFM	°F	1–BF	Factor	
59	20	0.85	1.08	1090

Outside air heat (latent)

CFM	GR/LB	1–BF	Factor	
59	122	0.85	0.68	4186
			Heat sub total	22141

Return duct heat + pump + Dehum & Pipe losses (%) 5.0	1107

Check figures

BTU/hr/Sq ft =	149.0
CFM/Sq ft =	4.88
Sq ft/TR =	81
CFM/TR =	393
Total CFM =	**761**

TR	1.94	**Grand total heat**	**23248**

Table 5.3 (v) Heat load proforma (monsoon)

1	Job name	**Sample**			
2	Space for	**Mixed use**		Date	Rev 0
3	Floor area	**Typical**		**Estimated by:**	
4	Height 10 ft			Estimate for: **Monsoon**	
5	Size 520 sq ft, Vol. 5200 cubic ft				

Part-I — Solar gain glass

	Item	Area	Sun gain	Factor	Heat gain Btu/hour
6		(Sq ft)	(Btu/h sq ft)		
7					
8	N - Glass	0	14	0.56	0
9	NE - Glass	67.5	12	0.56	454
10	E - Glass	0	12	0.56	0
11	SE - Glass	36	12	0.56	242
12	S - Glass	0	12	0.56	0
13	SW - Glass	67.5	100	0.56	3780
14	W - Glass	0	164	0.56	0
15	NW - Glass	0	123	0.56	0
16	Sky light				0

Part-II — Solar and trans gain walls and roof

	Item	Area	Eq temp diff	U	
18		(Sq ft)	(°F)	(Btu/h sq ft)	
20	N-Wall	0	14.5	0.31	0
21	NE-Wall	332.5	20.5	0.31	2113
22	E-Wall	0	28.5	0.31	0
23	SE-Wall	94	28.5	0.31	830
24	S-Wall	0	26.5	0.31	0
25	SW-Wall	332.5	24.5	0.31	2525
26	W-Wall	0	22.5	0.31	0
27	NW-Wall	130	16.5	0.31	665
28	Roof sun	520	42.5	0.23	5083
29	Roof insulated	0	0	0.12	0

Part-III — Trans Gain except walls and roof

	Item	Area	Temp.diff.	U	
31		(Sq ft)	(°F)	(Btu/h sq ft)	
33	All Glass	171	20	1.13	3865
34	Partition	0	20	0.35	0
35	Ceiling	0	20	0.4	0
36	Floor	**520**	20	0.32	3328

Part-IV — Internal Heat Gain

38	People	0 Nos)(		245	1960
39	Light	520.0 W X 1.25		3.41	2217
40	Eq. Load	260 W X		3.41	887

41	**Room sensible Heat (RSH)**	**27948**

Part-V

42	Supply duct — Heat gain from unconditioned space(%) 5.0			1397
43	safety factor (%) 5.0			1397
44	Outside air			
45	CFM	°F	BF	Factor
46	71	20	0.15	1.08 — 231
47	**Effective room sensible heat (ERSH)**			**30974**

Latent heat

49	People	8 Nos X	205	
50	Room latent heat (RLH)			1640
51	Supply duct leakage loss + safety factor % 5.0			82
52	Outside air			
53	CFM	GR/LB	BF	Factor
54	71	20	0.15	1.08 — 886
55	**Effective room latent heat (ERLH)**			**2608**
56	**Effective room total heat (ERTH)**			**33582**

Part-VI

57	Outside air heat (sensible)			
58	CFM	°F	1–BF	Factor
59	71	20	0.85	1.08 — 1307
60	Outside air heat (latent)			
61	CFM	GR/LB	1–BF	Factor
62	71	122	0.85	0.68 — 5021
62	Heat sub total			39910
63	Return duct			
64	heat + pump + Dehum & Pipe losses (%) 5.0			1995
65	**TR 3.49**	**Grand total heat**		**41905**

Part-VII

Condition	DB(°F)	WB(°F)	%RH	DP(°F)	GR/LB
Outside	95	75	75		190
Room	75		60		68
Difference	20	XXXX	XXXX	XXXX	122

Part-VIII

Outside air (ventilation)

8 People X	5	CFM/Person	40	
520 Sq ft	0.06	CFM/Sq ft	31	
		CFM ventilation	31	

Eff sensible heat factor (ESHF)	=	0.92
Indicated ADP	=	59.2 °F
Selected ADP	=	55 °F
Dehum temp rise	=	17.00 °F
Dehumdified CFM	=	**1687**

Notes

Occupancy =	25 Nos.
Lighting =	1.0 W/Sq ft
Eq. Load =	0.5 W/Sq ft
Height =	10.0 FT

Internal Heat - People

Sensible heat	= 245
Latent heat	= 205

Actual TR = 3.49

Check figures

BTU/hr/Sq ft =	80.6
CFM/Sq ft =	3.24
Sq ft/TR =	149
CFM/TR =	483
Total CFM =	**1687**

Table 5.3 (vi) Heat load proforma (monsoon)

1		Job name	**Sample**								
2		Space for	**Gallery**			Date		Rev	0		
3		Floor area	**Typical**			**Estimated by:**					
		Height 10 ft									
4		Size 434 sq ft, Vol. 4340 cubic ft				Estimate for: **Monsoon**					

Part-I — Solar gain glass

Item	Area (Sq ft)	Sun gain (Btu/h sq ft)	Factor	Heat gain Btu/hour
N - Glass	0	14	0.56	0
NE - Glass	0	12	0.56	0
E - Glass	0	12	0.56	0
SE - Glass	18	12	0.56	121
S - Glass	0	12	0.56	0
SW - Glass	0	100	0.56	0
W - Glass	0	164	0.56	0
NW - Glass	0	123	0.56	0
Sky light				0

Part-II — Solar and trans gain walls and roof

Item	Area (Sq ft)	Eq temp diff (°F)	U (Btu/h sq ft)	Btu/hour
N-Wall	0	14.5	0.31	0
NE-Wall	0	20.5	0.31	0
E-Wall	0	28.5	0.31	0
SE-Wall	52	28.5	0.31	459
S-Wall	0	26.5	0.31	0
SW-Wall	0	24.5	0.31	0
W-Wall	0	22.5	0.31	0
NW-Wall	0	16.5	0.31	0
Roof sun	434	42.5	0.23	4242
Roof insulated	0	0	0.12	0

Part-III — Trans Gain except walls and roof

Item	Area (Sq ft)	Temp. diff. (°F)	U (Btu/h sq ft)	Btu/hour
All Glass	18	20	1.13	407
Partition	100	20	0.35	700
Ceiling	0	20	0.4	0
Floor	**434**	20	0.32	2778

Part-IV — Internal Heat Gain

			Btu/hour
People	5 Nos X	245	1225
Light	434.0 W X 1.25	3.41	1850
Eq. Load	217 W X	3.41	740

Room sensible Heat (RSH)	**12522**

Part-V

				Btu/hour
Supply duct	Heat gain from unconditioned space(%) 5.0			626
	safety factor (%)		5.0	626
Outside air				
CFM	°F	BF	Factor	
51	20	0.15	1.08	165

Effective room sensible heat (ERSH)	**13940**

Latent heat				Btu/hour
People	5	Nos X	205	
		Room latent heat (RLH)		1025
Supply duct leakage loss + safety factor % 5.0				51
Outside air				
CFM	GR/LB	BF	Factor	
51	122	0.15	1.08	635

Effective room latent heat (ERLH)	**1711**
Effective room total heat (ERTH)	**15651**

Part-VI

				Btu/hour
Outside air heat (sensible)				
CFM	°F	1–BF	Factor	
51	20	0.85	1.08	937
Outside air heat (latent)				
CFM	GR/LB	1–BF	Factor	
51	122	0.85	0.68	3599
			Heat sub total	20187
Return duct				
heat + pump + Dehum & Pipe losses (%) 5.0				1009
TR	**1.77**	**Grand total heat**		**21197**

Part-VII

Condition	DB(°F)	WB(°F)	%RH	DP(°F)	GR/LB
Outside	95	75	75		190
Room	75		60		68
Difference	20	XXXX	XXXX	XXXX	122

Part-VIII

Outside air (ventilation)

5 People X	5	CFM/Person	25
434 Sq ft	0.06	CFM/Sq ft	26
		CFM ventilation	51

Eff sensible heat factor (ESHF)	=	0.89
Indicated ADP	=	58.8 °F
Selected ADP	=	55 °F
Dehum temp rise	=	17.00 °F
Dehumidified CFM	=	**759**

Notes

Occupancy =	25 Nos.
Lighting =	1.0 W/Sq ft
Eq. Load =	0.5 W/Sq ft
Height =	10.0 FT

Internal Heat - People

Sensible heat	= 245
Latent heat	= 205

Actual TR = 1.77

Check figures

BTU/hr/Sq ft =	48.8
CFM/Sq ft =	1.75
Sq ft/TR =	246
CFM/TR =	430
Total CFM =	759

Table 5.4a Design conditions for comfort

	Summer (Temperature in °C)					Winter (Temperature °C)			
Sr. no.	Optimum condition		Maximum condition		Sr. no.	Optimum conditions		Maximum condition	
	Dry-bulb	Wet-bulb	Dry-bulb	Wet-bulb		Dry-bulb	Wet-bulb	Dry-bulb	Wet -bulb
1.	23.3	19.4	25.9	21.8	1.	21.4	17.8	18.3	15.0
2.	23.9	18.4	26.1	21.6	2.	21.7	17.3	18.9	13.4
3.	24.4	17.6	26.7	20.9	3.	22.2	16.4	19.4	12.0
4.	25.0	16.8	27.2	20.1	4.	22.8	15.3	19.7	12.0
5.	25.6	16.0	27.8	19.4	5.	23.3	14.4	–	–
6.	26.1	15.2	28.3	18.8	6.	23.6	13.4	–	–
7.	–	–	28.9	18.1					
8.	–	–	29.4	17.6					

Table 5.4b Recommended inside design temperature for winter

Place	°C	Place	°C	Place	°C
Art gallieries	20	Operation theatres	18–21	**Hotels**	
Assembly hall	18	Wards	18	Bedroom (standard)	22
Bar	18	Sport pavilions	21	Bedroom (luxury)	20
Canteens	20	Ware houses	16	Public rooms	21
Churches	18	Laboratories	20	**Schools and Colleges**	
Flats/Houses		Law courts	20	Classrooms	18
Living rooms	21	Libraries	20	Lecture rooms	18
Bedrooms	18	**Offices**		**Shop**	
Bathrooms	22	General	20	Small	18
Entrance hall	16	Private	20	Large	18
Hospital		Police station	18	**Swimming Baths**	
Corridors	16			Changing rooms	22
Offices	20	Restaurant	18	Bath hall	26

Table 5.4c Outdoor design data

Place	Summer				Monsoon				Winter				Lat
	DB °C	WB °C	RH %	Month	DB °C	WB °C	RH %	Month	DB °C	WB °C	RH %	Month	
Agra	42.2	23.9	35.2	May	35.6	28.3	58	July	8.9	6.1	67	Jan.	27.10
Ahmedabad	43.3	25.6	24	May	32.2	29.4	82	July	15.6	10.6	60	Dec.	23.02
Ahmed Nagar	42.2	23.9	20	May	38.3	31.1	60	June	10.0	6.1	55	Jan.	19.05
Ajmer	42.2	23.3	20	May	33.9	27.2	60	Aug.	7.2	3.9	59	Jan.	26.27
Aligarh	42.2	23.9	20	May	35.6	28.3	58	July	8.9	6.1	67	Jan.	27.53
Allahabad	43.3	24.4	22	May	35.6	28.3	58	July	8.9	7.8	87	Jan.	25.27
Ambala	43.3	23.9	20	June	35.0	26.7	52	July	7.2	5.0	70	Jan.	30.23
Assansol	42.2	25.6	26	May	32.2	30.0	85	July	11.1	6.7	50	Jan.	23.41
Aurangabad	41.7	24.4	24	May	32.2	26.7	65	July	12.8	8.9	60	Dec.	19.53
Bangalore	35.6	25.6	45	April	27.8	25.6	82	Sep.	14.4	12.2	78	Jan.	12.58
Baroda	43.3	25.6	24	May	31.1	26.1	68	July	15.6	6.1	58	Jan.	22.18
Belgaum	37.8	25.0	35	April	27.8	25.0	80	July	14.4	11.7	71	Jan.	15.51
Bellary	40.6	25.0	28	May	34.4	25.6	50	Sep.	18.3	15.0	70	Dec.	15.09

(Contd.)

Table 5.4c *(Contd.)*

Place	Summer				Monsoon				Winter				Lat
	DB °C	WB °C	RH %	Month	DB °C	WB °C	RH %	Month	DB °C	WB °C	RH %	Month	
Bhopal	41.1	22.8	20	May	33.3	28.3	69	July	7.2	3.3	50	Jan.	23.16
Bhubaneswar	37.8	27.8	46	May	32.2	30.0	85	Aug.	13.3	8.9	55	Jan.	20.15
Calicut	35.4	27.8	55	May	29.4	27.8	88	July	22.2	18.3	69	Jan.	11.15
Chanda	46.1	23.9	15	May	31.1	26.7	70	July	12.8	9.4	63	Dec.	19.58
Cocanada	39.4	27.8	41	May	28.3	26.7	88	Oct.	18.3	13.9	60	Jan.	16.57
Cochin	35.0	27.8	58	April	29.4	27.8	88	Jan.	22.2	18.3	69	Jan.	9.58
Coimbatore	36.7	24.4	37	May	27.8	23.9	72	Oct.	18.3	13.9	60	Dec.	11.00
Cuttack	40.6	27.8	37	May	32.2	30.6	85	Aug.	13.3	8.9	55	Jan.	20.00
Dehradun	40.6	23.9	25	June	32.2	26.7	65	Aug.	5.6	3.3	70	Jan.	30.19
Dibrugarh	32.2	25.6	59	June	27.8	25.6	84	July	11.1	10.6	95	Jan.	27.28
Durgapur	42.2	25.6	26	May	32.2	30.0	85	July	11.1	6.7	50	Jan.	23.14
Gauhati	32.2	25.6	59	April	31.1	27.8	78	July	11.1	8.3	69	Jan.	26.11
Gaya	43.3	23.3	18	May	32.2	28.9	78	Aug.	10.0	6.7	60	Jan.	24.40
Goa	32.2	27.8	70	April	28.9	27.2	86	Jan.	18.3	15.6	72	Jan.	15.25
Hyderabad	41.1	25.6	28	May	29.4	27.2	82	Sep.	12.8	8.9	60	Dec.	17.86
Indore	41.1	25.0	28	May	32.2	27.8	70	July	10.0	7.2	65	Jan.	22.43
Jaipur	43.3	23.9	20	May	35.0	25.6	48	Aug.	7.8	5.0	65	Jan.	26.27
Jamnagar	37.8	26.7	43	May	29.4	27.2	82	July	12.8	9.4	65	Dec.	22.29
Jamshedpur	43.3	25.6	87	May	32.2	28.9	78	Aug.	10.0	7.8	75	Dec.	22.49
Hansi	43.9	24.4	20	May	36.1	28.9	58	July	8.9	7.2	80	Jan.	25.27
Jodhpur	43.3	25.0	23	May	35.8	26.7	52	Aug.	7.8	3.9	52	Jan.	26.18
Jabalpur	42.8	23.9	22	May	33.9	28.9	70	July	7.2	5.0	75	Dec.	23.16
Kanpur	42.8	25.0	23	May	36.1	28.9	58	July	7.2	5.6	80	Jan.	26.26
Kathmandu	29.4	23.9	63	May	24.4	21.7	79	July	7.2	5.6	78	Jan.	27.42
Kolkata	37.8	28.3	49	May	32.2	30.0	85	July	13.3	8.9	55	Dec.	22.32
Kota	45.0	23.9	17	May	37.2	26.7	43	July	7.2	5.0	70	Jan.	25.11
Kurnool	42.2	26.7	30	May	33.9	27.2	60	Sep.	15.6	11.7	62	Dec.	15.50
Lucknow	42.8	26.1	26	May	34.4	28.3	64	Aug.	8.9	6.1	67	Jan.	26.52
Chennai	39.4	27.8	41	May	28.3	26.7	88	Nov.	18.3	13.9	60	Jan.	13.04
Madurai	38.3	25.6	36	May	34.4	25.6	50	Oct.	20.0	16.7	72	Jan.	9.55
Mangalore	35.6	27.8	55	May	29.4	27.8	88	July	21.1	17.8	72	Jan.	12.52
Merrut	43.3	23.9	20	May	35.0	28.3	60	Aug.	7.2	5.0	70	Jan.	28.35
Mumbai	35.0	28.3	60	April	29.4	27.8	88	July	18.3	14.4	65	Jan.	18.54
Mysore	37.8	26.6	38	April	29.4	25.0	69	Oct.	18.9	15.6	70	Dec.	12.18
Nagpur	44.4	24.4	18	May	29.4	27.2	82	July	15.6	11.1	58	Dec.	21.09
Nellore	42.2	28.3	35	May	34.4	27.8	60	Nov.	18.3	13.9	60	Jan.	14.27
New Delhi	43.3	25.6	20	May	35.0	28.3	60	Aug.	7.2	5.0	70	Jan.	28.35
Ootacamund	22.8	15.6	45	May	18.9	14.4	62	July, Oct.	3.3	0.6	60	Dec.	11.24
Patna	42.2	25.6	26	May	32.2	28.9	78	July, Oct.	10.0	6.7	60	Dec.	25.37
Poona	40.0	24.4	28	April	28.3	26.1	82	July	10.0	5.6	50	Dec.	18.32
Raichur	41.7	26.1	29	May	33.9	27.8	62	Sep.	15.6	12.2	69	Dec.	16.12
Raipur	43.3	25.0	22	May	33.3	28.3	68	July	10.0	7.2	70	Dec.	21.14
Ranchi	37.8	27.8	46	May	28.9	25.6	76	July	8.9	5.6	60	Jan.	23.23
Rourkela	43.9	25.6	87	May	30.6	27.8	80	July	12.2	6.7	41	Dec.	21.28
Salem	39.4	26.7	38	May	33.9	25.0	50	Oct.	18.3	15.0	70	Jan.	11.39
Shillong	29.4	21.7	50	April	23.3	21.1	82	June	3.3	0.0	50	Jan.	25.34
Sholapur	42.2	25.0	25	May	33.3	27.8	65	Sep.	12.8	8.3	55	Jan.	17.40
Shree Nagar	33.3	9.4	–	–	–	–	–	–	6.7	–	88	Jan.	34.00
Thana-Kalyan	37.8	27.8	47	April	30.0	28.3	86	July	18.3	14.4	65	Jan.	18.54
Trichrapoly	40.0	27.2	38	May	35.0	27.2	55	Oct.	21.1	17.8	72	Jan.	10.40
Trivendram	33.3	26.7	59	March	29.4	26.7	80	June	22.2	18.3	69	Jan.	8.29
Vijayawada	43.3	28.3	32	May	34.4	27.8	60	Oct.	12.8	9.4	65	Dec.	16.33
Visakhapattnam	33.3	27.3	64	May	30.6	27.8	80	Oct.	18.3	13.9	60	Jan.	17.42

Table 5.5 Occupant load

Sl. no. (1)	Group of occupancy (2)	Occupant load floor area in M^2/person (3)
(i)	Residential (A)	12.5
(ii)	Educational (B)	4
(iii)	Institutional (C)	15*
(iv)	Assembly (D)	
(a)	With fixed or loose seats and dance floors	0.6*
(b)	Without seating facilities including dining rooms	1.5
(v)	Mercantile (F)	
(a)	Street floor and sales basement	3
(b)	Upper sale floors	6
(vi)	Business and industrial (E and G)	10
(vii)	Storage (H)	30
(viii)	Hazardous (J)	10

* Occupant load in dormitory portions of homes for the aged, orphanages, insane asylums, etc. where sleeping accommodation is provided shall be calculated at not less than 7.5 m² gross floor area/person.

The gross floor area shall include, in addition to the main assembly room or space, any occupied connecting room or space in the same storey or in the stories above or below, where entrance is common to such rooms and spaces and they are available for use by the occupants of the assembly place. No deductions shall be made in the gross area for corridors, closets or other subdivisions; they shall include all space serving the particular assembly occupancy.

ii. Fill area and height of room to calculate the volume of premise.

iii. Fill indoor and outdoor design conditions. The indoor comfort design conditions will vary from country to country based on the lifestyle, type of activity like office work or residential environment, threatre, dance hall, gymnasium etc. It will also depend on whether air circulators will be used or not as temperature can be kept little higher if fans are used. The recommended indoor design conditions are given in Table 5.4a and b. The outdoor design conditions for a city or town of India are given in Table 5.4c.
The indoor condition adopted here is 75° F DBT with 60% RH and 69 grains/lb. The outdoor conditions for Delhi are 110°F DBT and 75°F WBT with 74 grain/lb. Irrespective of type of weather, the design conditions for indoor remains same.

iv. Ventilation air requirement is the sum of air required for the total occupancy and the air, required based on area of the room.
i.e. $V = n \times v_p + A \times v_a$
where V = ventilation air requirement in CFM
n = no of person to be calculated from Table 5.5 as per NBC
v_p = ventilation air required per person as per ASHRAE (Refer Table 5.6)
A = area of room in square feet
v_a = ventilation air requirement based an area criteria as per ASHRAE

ASHRAE standard 62-2001 has defined the air requirements v_p and v_a. Some of the values for common applications are given in Table 5.6.

Table 5.6 Minimum ventilation rates (as per ASHRAE)

Occupancy Category	People Outdoor air rate CFM/Person	Area outdoor air rate CFM/Sq ft
Lecture Hall (Fixed Seats)	7.5	0.06
Art Class room	10	0.18
Computer lab	10	0.12
Music/ Theater/ Dance	10	0.06
Multi Use Assembly	7.5	0.06
Restaurant dining room	7.5	0.18
Cafeteria/ Fast food dining	7.5	0.18
Conference/ Meeting	5	0.06
Corridors		0.06
Bedroom/ Living room	5	0.06
Office room	5	0.06
Reception area	5	0.06
Auditorium Seating	5	0.06
Courtroom	5	0.06
Libraries	5	0.12
Museums/ galleries	7.5	0.06
Mall Common Areas	7.5	0.06
Supermarket	7.5	0.06
Gym/ Stadium (play area)		0.30
Spectator area	7.5	0.06
Stages/ Studios	10	0.06

Since this is an office space, the number of occupants as per Table 5.5 will be

$$n = \frac{120}{10} = 12$$

The ventilation air requirement is thus
V = 12 × 5 + 1200 × 0.06
= 132 CFM

v. **Solar Gain from glass (Part-I):** This is the heat transmitted or radiated into the room due to solar radiations falling on the glass. The peak solar heat gain is the maximum amount of heat falling on the glass in a particular direction and at particular point of time. The peak solar heat gain of one room may not be at the same time as that of the other room. Thus total maximum solar heat gain of the building will be at a time when combined load of all the rooms is maximum.

To understand this point clearly, let's take one building located in New Delhi, i.e. 30°N latitude. At 30° N latitude, the peak solar heat gain occurs on 21st June as 161 BTU/hr/sq ft (Table 5.7).

This value is to be multiplied with various factors as below to arrive at actual heat load entering in the building;
Heat gain = Peak solar heat gain
        × Storage load factor at desired time and required number of hours of operation (Table 5.8)
        × Overall factor with/without shading device (Table 5.9a)
        × area (sq feet).

However, this method does not account for solar heat gain from other directions. Also the solar heat gain is not constant all throughout the day. A calculation with the above method gives the peak value of solar heat gain. Also the value calculated above is for one direction only. If the peak values entering from other directions are also added,

then the total cooling load requirement will be much higher than the peak value so calculated. But these peaks do not occur at the same time. Thus we must calculate the time, when the building cooling load is at its peak. This can be done as follow for a building in New Delhi.

Direction	Solar 3.00 PM	Heat Gain 4.00 PM	BTU/nr/sq ft 5.00 PM
N	14	18	29
NE	14	12	10
E	14	12	10
SE	14	12	10
S	14	12	10
SW	90	90	75
W	143	161	156
NW	97	130	139
Total	400	447	439

Thus the peak occurs at 4.00 PM. Hence all calculations will be carried out for 4.00 PM. Though this method is generally adopted in all manual calculations, but it will be more accurate if each room is examined at different hours of day. It may so happen that one room may peak at 3.00 PM while another at 4.00 PM or 5.00 PM. But manual calculations become cumbersome and hence overall heat load is calculated at 4.00 PM. But modern day softwares give accurate results and the overall heat load is based on such calculations.

From Table 5.7, the solar gain column of heat load proforma is filled up for 21st June at 4.00 PM. Similarly for monsoon proforma, similar check for determining at which maximum load occurs is carried out from Table 5.7 for 23rd July row and values are extended.

This column represents the radiation heat load transferred from glass to the conditioned room.

Table 5.7 Solar heat gain through ordinary glass—Btu/(hr) (sq ft)

30 °N latitude		AM						Sun time	PM						30 °S latitude	
Time of year	Exposure	6	7	8	9	10	11	Noon	1	2	3	4	5	6	Exposure	Time of year
	North	33	29	18	14	14	14	14	14	14	14	18	29	33	South	
	Northeast	105	139	130	97	55	19	14	14	14	14	12	10	5	Southeast	
	East	108	156	161	143	98	44	14	14	14	14	12	10	5	East	
	Southeast	42	75	90	90	73	44	17	14	14	14	12	10	5	Northeast	
June 21	South	5	10	12	14	15	19	21	19	15	14	12	10	5	North	Dec. 22
	Southwest	5	10	12	14	14	14	17	44	73	90	90	75	42	Northwest	
	West	5	10	12	14	14	14	14	44	98	143	161	156	108	West	
	Northwest	5	10	12	14	14	14	14	19	55	97	130	139	105	Southwest	
	Horizontal	19	61	131	180	217	240	250	240	217	180	131	61	19	Horizontal	

(Contd.)

Table 5.7 Solar heat gain through ordinary glass—Btu/(hr) (sq ft) (*Contd.*)

30 °N latitude		AM						Sun time			PM				30 °S latitude	
Time of year	Exposure	6	7	8	9	10	11	Noon	1	2	3	4	5	6	Exposure	Time of year
	North	22	20	14	13	14	14	14	14	14	13	14	20	22	South	
	Northeast	93	131	123	89	46	16	14	14	14	13	12	9	4	Southeast	
	East	100	155	164	145	99	44	14	14	14	13	12	9	4	East	
	Southeast	42	82	100	100	83	53	22	14	14	13	12	9	4	Northeast	Jan. 21
July 23	South	4	9	12	14	20	27	30	27	20	14	12	9	4	North	and
and	Southwest	4	9	12	13	14	14	14	53	83	100	100	82	42	Northwest	Nov. 21
May 21	West	4	9	12	13	14	14	14	44	99	145	164	155	100	West	
	Northwest	4	9	12	13	14	14	14	16	46	89	123	131	93	Southwest	
	Horizontal	15	66	123	176	214	236	246	236	214	176	123	66	15	Horizontal	
	North	6	8	11	13	13	14	14	14	13	13	11	8	6	South	
	Northeast	55	108	100	66	27	14	14	14	13	13	11	8	2	Southeast	
	East	66	147	165	148	102	46	14	14	13	13	11	8	2	East	
	Southeast	37	98	127	129	112	82	39	15	13	13	11	8	2	Northeast	
Aug. 24	South	2	8	13	27	47	58	63	58	47	27	13	8	2	North	Feb. 20
and	Southeast	2	8	11	13	13	15	39	82	112	129	127	98	37	Northwest	and
April 20	West	2	8	11	13	13	14	14	45	102	148	165	147	66	West	Oct. 23
	Northwest	2	8	11	13	13	14	14	14	27	66	106	108	55	Southwest	
	Horizontal	6	47	107	161	200	225	235	225	200	161	107	47	6	Horizontal	
	North	0	5	10	12	13	14	14	14	13	12	10	5	0	South	
	Northeast	0	74	90	40	15	14	14	14	13	12	10	5	0	Southeast	
	East	0	124	158	144	103	48	14	14	13	12	10	5	0	East	
	Southeast	0	98	131	152	141	113	67	25	13	12	10	5	0	Northeast	
	South	0	9	18	60	82	98	105	98	82	60	18	9	0	North	Mar. 22
Sept. 22	Southwest	0	5	10	12	13	25	67	113	141	152	131	98	0	Northwest	and
and	West	0	5	10	12	13	14	14	48	103	144	158	124	0	West	Sept. 22
Mar. 22	Northwest	0	5	10	12	13	14	14	14	15	40	90	74	0	Southwest	
	Horizontal	0	25	81	135	179	202	212	202	179	135	81	25	0	Horizontal	
	North	0	3	8	11	12	13	14	13	12	11	8	3	0	South	
	Northeast	0	33	39	18	12	13	14	13	12	11	8	3	0	Southeast	
	East	0	79	135	132	94	43	14	13	12	11	8	3	0	East	
	Southeast	0	73	142	163	199	136	92	47	15	11	8	3	0	Northeast	
Oct. 20	South	0	18	57	92	121	139	145	139	121	92	57	18	0	North	April 20
and	Southwest	0	3	8	11	15	47	92	136	159	163	142	73	0	Northwest	and
Feb. 20	West	0	3	8	11	12	13	14	43	94	132	135	79	0	West	Aug. 24
	Northwest	0	3	8	11	12	13	14	13	12	18	39	33	0	Southwest	
	Horizontal	0	6	49	100	143	171	179	171	143	100	49	6	0	Horizontal	
	North	0	1	6	9	11	12	12	12	11	9	6	1	0	South	
	Northeast	0	8	16	9	11	12	12	12	11	9	6	1	0	Southeast	
	East	0	27	109	116	83	35	12	12	11	9	6	1	0	East	
	Southeast	0	28	127	161	162	143	104	64	23	9	6	1	0	Northeast	
Nov. 21	South	0	10	68	109	137	154	159	154	137	109	68	10	0	North	May 21
and	Southwest	0	1	6	9	23	64	104	143	162	161	122	28	0	Northwest	and
Jan. 21	West	0	1	6	9	11	12	12	35	83	116	109	27	0	West	July 23
	Northwest	0	1	6	9	11	12	12	12	11	9	16	8	0	Southwest	
	Horizontal	0	2	27	71	109	136	145	136	109	71	27	2	0	Horizontal	
	North	0	0	4	9	11	12	12	12	11	9	4	0	0	South	
	Northeast	0	0	10	9	11	12	12	12	11	9	4	0	0	Southeast	
	East	0	0	92	105	80	32	12	12	11	9	4	0	0	East	
	Southeast	0	0	114	157	162	143	108	72	28	9	4	0	0	Northeast	
Dec. 22	South	0	0	64	113	142	159	163	159	142	113	64	0	0	North	June 21
	Southwest	0	0	4	9	28	72	108	143	162	157	114	0	0	Northwest	
	West	0	0	4	9	11	12	12	32	80	105	92	0	0	West	
	Northwest	0	0	4	9	11	12	12	12	11	9	10	0	0	Southwest	
	Horizontal	0	0	19	60	97	122	131	122	97	60	19	0	0	Horizontal	

* For 10°, 20° and 40 °N altitude, refer Table A.31, A.32, A.33 in Annexure A of the book.
Courtesy: Abstracted from *Handbook of Air-conditioning System Design* with permission from M/S Carrier Corporation.

Table 5.8 Storage load factors, solar heat gain through glass with internal shade 24 hour operation, constant space temperature

Exposure (North lat)	Weight (lb per sq of floor area)	6	7	8	9	10	11	12	1	2	3	4	5	6	7	8	9	10	11	12	1	2	3	4	5	Exposure (South lat)
				AM							PM								AM							
Northeast	150 and over	.47	.58	.54	.42	.27	.21	.20	.19	.18	.17	.16	.14	.12	.09	.08	.07	.06	.06	.05	.05	.04	.04	.04	.03	
	100	.48	.60	.57	.46	.30	.24	.20	.19	.17	.16	.15	.13	.11	.08	.07	.06	.05	.05	.04	.04	.03	.03	.02	.02	Southeast
	30	.55	.76	.73	.58	.36	.24	.19	.17	.15	.13	.12	.11	.07	.04	.02	.02	.01	.01	0	0	0	0	0	0	
East	150 and over	.39	.56	.62	.59	.49	.33	.23	.21	.20	.18	.17	.15	.12	.10	.09	.08	.08	.07	.06	05	.05	.05	.04	.04	
	100	.40	.58	.65	.63	.52	.35	.24	.22	.20	.18	.16	.14	.12	.09	.08	.07	.06	.05	.05	04	.04	.03	.03	.02	East
	30	.46	.70	.80	.79	.64	.42	.25	.19	.16	.14	.11	.09	.07	.04	.02	.02	.01	.01	0	0	0	0	0	0	
Southeast	150 and over	.04	.28	.47	.59	.64	.62	.53	.41	.27	.24	.21	.19	.16	.14	.12	.11	.10	.09	.08	.07	.06	.06	.05	.05	
	100	.03	.28	.47	.61	.67	.65	.57	.44	.29	.24	.21	.18	.15	.12	.10	.09	.08	.07	.06	.05	.05	.04	.04	.03	Northeast
	30	0	.30	.57	.75	.84	.81	.69	.50	.30	.20	.17	.13	.09	.05	.04	.03	.02	.01	0	0	0	0	0	0	
South	150 and over	.06	.06	.23	.38	.51	.60	.66	.67	.64	.59	.42	.24	.22	.19	.17	.15	.13	.12	.11	.10	.09	.08	.07	.07	
	100	.04	.04	.22	.38	.52	.63	.70	.71	.69	.59	.45	.26	.22	.18	.16	.13	.12	.10	.09	.08	.07	.06	.06	.05	North
	30	.10	.21	.43	.63	.77	.86	.88	.82	.56	.50	.24	.16	.11	.08	.05	.04	.02	.02	.01	.01	0	0	0	0	
Southwest	150 and over	.08	.08	.09	.10	.11	.24	.39	.53	.63	.66	.61	.47	.23	.19	.18	.16	.14	.13	.11	.10	.09	.08	.08	.07	
	100	.07	.08	.08	.08	.10	.24	.40	.55	.66	.70	.64	.50	.26	.20	.17	.15	.13	.11	.10	.09	.08	.07	.06	.05	Northwest
	30	.03	.04	.06	.07	.09	.23	.47	.67	.81	.86	.79	.60	.26	.17	.12	.08	.05	.04	.03	.02	.01	.01	0	0	
West	150 and over	.08	.09	.09	.10	.10	.10	.10	.18	.36	.52	.63	.65	.55	.22	.19	.17	.15	.14	.12	.11	.10	.09	.08	.07	
	100	.07	.08	.08	.09	.09	.09	.09	.18	.36	.54	.66	.68	.60	.25	.20	.17	.15	.13	.11	.10	.08	.07	.06	.05	West
	30	.03	.04	.06	.07	.08	.08	.08	.19	.42	.65	.81	.85	.74	.30	.19	.13	.09	.06	.05	.03	.02	.02	.01	0	
Northwest	150 and over	.08	.09	.10	.10	.10	.10	.10	.10	.16	.33	.49	.61	.60	.19	.17	.15	.13	.12	.10	.09	.08	.08	.07	.06	
	100	.07	.08	.09	.09	.10	.10	.10	.10	.16	.34	.52	.65	.64	.23	.18	.15	.12	.11	.09	.08	.07	.06	.06	.05	Southwest
	30	.03	.05	.07	.08	.09	.09	.10	.10	.17	.39	.63	.80	.79	.28	.18	.12	.09	.06	.04	.03	.02	.02	.01	0	
North and Shade	150 and over	.08	.37	.67	.71	.74	.76	.79	.81	.83	.84	.86	.87	.88	.29	.26	.23	.20	.19	.17	.15	.14	.12	.11	.10	South and Shade
	100	.06	.31	.61	.72	.76	.79	.81	.83	.85	.87	.88	.90	.91	.30	.26	.22	.19	.16	.15	.13	.12	.10	.09	.08	
	30	0	.25	.74	.83	.88	.91	.94	.96	.96	.98	.98	.99	.99	.26	.17	.12	.08	.05	.04	.03	.02	.01	.01	.01	

* For 16 hour and 12 hour operation, refer Table A.34, A.35 in Annexure A of the book.

* Reprinted with permission of M/S Carrier Corporation from *Handbook of Air-conditioning System Design*.

Table 5.9a Overall factors for solar heat gain through glass with and without shading devices, outdoor wind velocity – 5 mph, angle of incidence – 30°, shading devices fully covering window

Type of glass	Glass factor no shade	Inside venetian blind 45° horizontal or vertical or roller shade			Outside venetian blind 45° horizontal slats		Outside shading screen 17° horizontal slats		Outside awning venetian sides and top	
		Light colour	Medium colour	Dark colour	Light colour	Light on outside dark on inside	Medium colour	Dark colour	Light colour	Medium or dark colour
Ordinary glass	1.00	.56	.65	.75	.15	.13	.22	.15	.20	.25
Regular plate (1/4″)	.94	.56	.65	.74	.14	.12	.21	.14	.19	.24
Heat Absorbing Glass										
40 to 48 % absorbing	.80	.56	.62	.72	.12	.11	.18	.12	.16	.20
48 to 56 % absorbing	.73	.53	.59	.62	.11	.10	.16	.11	.15	.18
56 to 70 % absorbing	.62	.51	.54	.56	.10	.10	.14	.10	.12	.16
Double pane										
Ordinary glass	.90	.54	.61	.67	.14	.12	.20	.14	.18	.22
Regular plate	.80	.52	.59	.65	.12	.11	.18	.12	.16	.20

(Contd.)

Table 5.9a (*Contd.*)

Type of glass	Glass factor no shade	Inside venetian blind 45° horizontal or vertical or roller shade			Outside venetian blind 45° horizontal slats		Outside shading screen 17° horizontal slats		Outside awning venetian sides and top	
		Light colour	Medium colour	Dark colour	Light colour	Light on outside dark on inside	Medium colour	Dark colour	Light colour	Medium or dark colour
48 to 56 % absorbing outside; ordinary glass inside	.52	.36	.39	.43	.10	.10	.11	.10	.10	.13
48 to 56 % absorbing outside; regular plate inside	.50	.36	.39	.43	.10	.10	.11	.10	.10	.12
Triple pane										
Ordinary glass	.83	.48	.56	.64	.12	.11	.18	.12	.16	.20
Regular plate	.69	.47	.52	.57	.10	.10	.15	.10	.14	.17
Painted Glass										
Light colour	.28									
Medium colour	.39									
Dark colour	.50									
Stained Glass										
Amber colour	.70									
Dark red	.56									
Dark blue	.60									
Dark green	.32									
Grayed green	.46									
Light opalescent	.43									
Dark opalescent	.37									

* Reprinted with permission of M/S Carrier Corporation from *Handbook of Air-conditioning System Design*.

Table 5.9b Transmission coefficient U-windows, sky lights, doors (BTU/hr/sq ft/°F temperature difference)

Glass

		Vertical Glass						Horizontal Glass			
	Single	Double			Triple			Single		Double	
Air Spce (thickness, inches)		1/4″	1/2″	3/4–4″	1/4″	1/2″	3/4–4″	Summer	Winter	Summer	Winter
– Without storm windows	1.3	0.61	0.55	0.53	0.41	0.36	0.34	0.86	1.40	0.50	0.70
– With storm windows	0.54							0.43	0.64		

Doors

Nominal thickness of wood (inches)	U (exposed door)	U (with storm door)
1	0.69	0.35
1 1/4	0.59	0.32
1 1/2	0.52	0.30
1 3/4	0.51	0.30
2	0.46	0.28
2 1/2	0.38	0.25
3	0.33	0.23
Glass (3/4″ Hercutite)	1.05	0.43

* Reprinted with permission from Carrier Corporation from Handbook of Air-conditioning System Design.

The ordinary glass absorbs a small portion of solar heat (5-6%) and reflects or transmits the rest. The transmitted or reflected radiation varies with angle of incidence. At low angles of incidence, 86–87% is transmitted and 8–9% is reflected.

The total solar heat gain to the conditioned space is thus:

Total solar heat gain = Amount transmitted by glass due to diffused radiation + 40% of amount absorbed by glass.

The values for 30° latitudes for each month of year and for each hour of day are given in Table 5.7. The heat load proforma Part-I is filled from here, e.g. at 4.00 PM on 21st June. The above values do not include transmission of heat across the glass caused by a temperature difference between outdoor and inside air which is accounted for in Part-III of proforma.

The heat transmitted inside gets reduced by shades, venetian blinds provided inside or outside and the colour of the blinds. Thus a multiplication factor from Table 5.9a is to be applied. For a regular plate (6 mm or 1/4 inch) with light coloured inside venetian blind, a factor of 0.56 is selected from the table and filled in Heat Load Proforma.

vi. **Solar and transmission gain through wall and roof:** Part-II of heat load proforma is the amount of heat absorbed by various outside walls exposed to sun which gets transmitted inside the room. The heat transmission to the conditioned space gets delayed by as much as 8-12 hours due to heat being absorbed by walls. Secondly solar heat is variable throughout the day and will result in unsteady state heat flow. This unsteady state causes difficulty as each individual situation is difficult to examine. This issue is solved by the concept of equivalent temperature difference which takes into account effects caused by variable solar radiation, outdoor temperature, variety of constructions, time of day and location of building. The heat flow through the structure is thus calculated as

$$Q = UA\,\Delta T$$

where Q = heat flow BTU/hr
U = Transmission coefficient, BTU/hr/ sq ft/°F temperature difference
A = Area of surface, sq ft
ΔT = equivalent temperature difference, °F

Table 5.10a, b mentions the equivalent temperature difference for sunlit and shaded walls and roofs respectively for 40° North latitude, outdoor daily range of DBT 20°F, maximum outdoor temperature of 95°F DB, indoor design temperature of 80°F. However, if any of the parameters are different from those mentioned here, correction is applied with the help of Table 5.10c. The transmission coefficient for various materials and type of construction are given in Table 5.11, 5.12, 5.13a, b.

In the example problem, Part-II of heat load proforma is filled by entering the net area of outer walls (after subtracting glass area).

Equivalent temperature difference is calculated from Table 5.10a.

In order to use the tables mentioned above, calculate the weight of wall per sq ft. This is done as below:

$$\text{Weight of wall/sq ft} = \frac{A_w(t_b \times d_b + t_p \times d_p)}{A_w}$$

where A_w = net area of wall
t_b, t_p = thickness of brick and plaster respectively
d_b, d_p = density of brick and plaster respectively

Thus

$$\text{Weight/sq ft} = \frac{9'}{12} \times 118 + \frac{1.5'}{12} \times 112$$
$$= 102.5 \text{ lb/sq ft}$$

Now, select equivalent temperature difference from Table 5.10a, at 4.00 PM for weight of wall to be 100 lb/sq ft, the parameters are:

NE	10	SW	14
E	18	W	12
SE	18	NW	6
S	16	N	4

Further, the maximum outdoor temperature in Delhi in summer is 110°F and 20% RH while the minimum average temperature is 91°F. Thus daily range is 19°F. The correction is applied to equivalent temperature difference, from Table 5.10c, by interpolating for the row of (110°F - 75°F) 35°F and column of 19°F. The correction factor to be applied is + 20.5. The effective equivalent temperature.

	Summer	Monsoon
NE	10 + 20.5 = 30.5	10 + 10.5 = 20.5
E	18 + 20.5 = 38.5	18 + 10.5 = 28.5
SE	18 + 20.5 = 38.5	18 + 10.5 = 8.5
S	16 + 20.5 = 36.5	16 + 10.5 = 20.5
SW	14 + 20.5 = 34.5	14 + 10.5 = 24.5
W	12 + 20.5 = 32.5	12 + 10.5 = 22.5
NW	6 + 20.5 = 26.5	6 + 10.5 = 16.5
N	4 + 20.5 = 24.5	4 + 10.5 = 14.5

Table 5.10a Equivalent temperature difference (°F) for dark coloured, sunlit and shaded walls based on dark coloured walls; 95 °F db outdoor design temperature; constant 80 °F db room temperature; 20 °F daily range; 24-hour operation; July and 40° North Lat

Exposure	Weight of wall (lb/sq ft)	6	7	8	9	10	11	12	1	2	3	4	5	6	7	8	9	10	11	12	1	2	3	4	5
				AM								PM											AM		
Northeast	20	5	15	22	23	24	19	14	13	12	13	14	14	14	12	10	8	6	4	2	0	−2	−3	−4	−2
	60	−1	−2	−2	5	24	22	20	15	10	11	12	13	14	13	12	11	10	8	6	4	2	1	0	−1
	100	4	3	4	4	4	10	16	15	14	12	10	11	12	12	12	11	10	9	8	7	6	6	5	5
	140	5	5	6	6	6	6	6	10	14	16	14	12	10	10	10	10	10	10	10	9	9	8	7	7
East	20	1	17	30	33	36	35	32	20	12	13	14	14	14	12	10	8	6	4	2	0	−1	−2	−3	−3
	60	−1	−1	0	21	30	31	31	19	14	13	12	13	14	13	12	11	10	8	5	4	3	1	1	0
	100	5	5	6	8	14	20	24	25	24	20	18	16	14	14	14	13	12	11	10	9	8	7	7	6
	140	11	10	10	9	8	9	10	15	18	19	18	17	16	14	12	15	14	14	14	13	13	12	10	12
Southeast	20	10	6	13	19	26	27	28	26	24	19	16	15	14	12	10	8	6	4	2	0	−1	−1	−2	−2
	60	1	1	0	13	20	24	28	26	25	21	18	15	14	13	12	11	10	8	6	5	4	3	3	2
	100	7	7	6	6	6	11	16	17	18	19	18	16	14	13	12	11	10	10	10	9	9	8	8	7
	140	9	8	8	8	8	7	6	11	14	15	16	18	16	15	14	13	12	12	12	11	11	10	10	9
South	20	−1	−2	−4	1	4	14	22	27	30	28	26	20	16	12	10	7	6	3	2	1	1	0	0	−1
	60	−1	−3	−4	−3	−2	7	12	20	24	25	26	23	20	15	12	10	8	6	4	2	1	1	0	−1
	100	4	4	2	2	2	3	4	8	12	15	16	18	18	15	14	11	10	9	8	8	7	6	6	5
	140	7	6	6	5	4	4	4	4	4	7	10	13	14	15	16	16	14	12	10	10	9	9	8	7
Southwest	20	−2	−4	−4	−2	0	4	6	19	26	34	40	41	42	30	24	12	6	4	2	1	1	0	−1	−1
	60	2	1	0	0	0	1	2	8	12	21	32	36	36	35	34	20	10	7	6	5	4	4	3	3
	100	7	5	6	5	4	5	6	7	8	12	14	19	22	23	24	23	22	15	10	10	9	9	8	7
	140	8	8	8	8	8	7	6	6	6	7	8	9	10	15	18	19	20	13	8	8	8	8	8	8
West	20	−2	−3	−4	−2	0	3	6	14	20	32	40	45	48	34	22	14	8	5	2	1	0	0	−1	−1
	60	2	1	0	0	0	2	4	7	10	19	26	34	40	41	36	28	16	10	6	5	4	3	3	2
	100	7	7	6	6	6	6	6	7	8	10	12	17	20	25	28	27	26	19	14	12	11	10	9	8
	140	8	11	10	9	8	8	8	9	10	10	10	11	12	14	16	21	22	23	22	20	18	16	15	13
Northwest	20	−3	−4	−4	−2	0	3	6	10	12	19	24	33	40	37	34	18	6	4	2	0	−1	−1	−2	−2
	60	−2	−3	−4	−3	−2	0	2	6	8	10	12	21	30	31	32	21	12	8	6	4	3	1	0	−1
	100	5	4	4	4	4	4	4	4	5	6	9	12	17	20	21	22	14	8	7	7	6	6	5	
	140	8	7	6	6	6	6	6	6	6	6	6	7	8	9	10	14	18	19	20	16	13	11	10	9
North (Shade)	20	−3	−3	−4	−3	−2	1	4	8	10	12	14	13	12	10	0	6	4	2	0	0	−1	−1	−2	−2
	60	−3	−3	−4	−3	−2	−1	0	3	6	8	10	11	12	2	12	10	8	6	4	2	1	0	−1	−2
	100	1	1	0	0	0	0	0	1	2	3	4	5	5	5	8	7	6	5	4	3	3	2	2	1
	140	1	1	0	0	0	0	0	0	0	1	2	3	4	5	6	7	8	7	6	4	3	2	2	1

* Reprinted with permission of M/S Carrier Corporation from *Handbook of Air-conditioning System Design*.

Table 5.10b Equivalent temperature difference (°F) for dark coloured, sunlit and shaded roofs based on 95 °F db outdoor design temperature; constant 80 °F DB room temperature; 20 °F daily range; 24-hour operation; July and 40° North Lat

Condition	Weight of roof (lb/sq ft)	AM							PM												AM				
		6	7	8	9	10	11	12	1	2	3	4	5	6	7	8	9	10	11	12	1	2	3	4	5
Exposed to sun	10	-4	-6	-7	-5	-1	7	15	24	32	38	43	46	45	41	35	28	22	16	10	7	3	1	-1	-3
	20	0	-1	-2	-1	2	9	16	23	30	36	41	43	43	40	35	30	25	20	15	12	8	6	4	2
	40	4	3	2	3	6	10	16	23	28	33	38	40	41	39	35	32	28	24	20	17	13	11	9	6
	60	9	8	6	7	8	11	16	22	27	31	35	38	39	38	36	34	31	28	25	22	18	16	13	11
	80	13	12	11	11	12	13	16	22	26	28	32	35	37	37	35	34	34	32	30	27	23	20	18	14
Covered with water	20	-5	-2	0	2	4	10	16	19	22	30	18	16	14	12	10	6	2	1	1	-1	-2	-3	-4	-5
	40	-3	-2	-1	-1	0	5	10	13	15	15	16	15	15	14	12	10	7	5	3	1	-1	-2	-3	-3
	60	-1	-2	-2	-2	-2	2	5	7	10	12	14	15	16	15	14	12	10	8	6	4	3	2	1	0
Sprayed	20	-4	-2	0	2	4	8	12	15	18	17	16	15	14	12	10	6	2	1	0	-1	-2	-2	-3	-3
	40	-2	-2	-1	-1	0	2	5	9	13	14	14	14	14	13	12	9	7	5	3	1	0	0	-1	-1
	60	-1	-2	-2	-2	-2	0	2	5	8	10	12	13	14	13	12	11	10	8	6	4	2	1	0	-1
Shaded	20	-5	-5	-4	-2	0	2	6	9	12	13	14	13	12	10	8	5	2	1	0	-1	-3	-4	-5	-5
	40	-5	-5	-4	-3	-2	0	2	5	8	10	12	13	12	11	10	8	6	4	2	0	-1	-3	-4	-5
	60	-3	-3	-2	-2	-2	-1	0	2	4	6	8	9	10	10	10	9	8	6	4	2	1	0	-1	-2

* Reprinted with permission of M/S Carrier Corporation from *Handbook of Air-conditioning System Design.*

Table 5.10c Corrections to equivalent temperatures (°F)

Outdoor design for month at 3 pm minus room temperature (°F)	Daily range (°F)																
	8	10	12	14	16	18	20	22	24	26	28	30	32	34	36	38	40
-30	-39	-40	-41	-42	-43	-44	-45	-46	-47	-48	-49	-50	-51	-52	-53	-54	-55
-20	-29	-30	-31	-32	-33	-34	-35	-36	-37	-38	-39	-40	-41	-42	-43	-44	-45
-10	-19	-20	-21	-22	-23	-24	-25	-26	-27	-28	-29	-30	-31	-32	-33	-34	-35
0	-9	-10	-11	-12	-13	-14	-15	-16	-17	-18	-19	-20	-21	-22	-23	-24	-25
5	-4	-5	-6	-7	-8	-9	-10	-11	-12	-13	-14	-15	-16	-17	-18	-19	-20
10	1	0	-1	-2	-3	-4	-5	-6	-7	-8	-9	-10	-11	-12	-13	-14	-15
15	6	5	4	3	2	1	0	-1	-2	-3	-4	-5	-6	-7	-8	-9	-10
20	11	10	9	8	7	6	5	4	3	2	1	0	-1	-2	-3	-4	-5
25	16	15	14	13	12	11	10	9	8	7	6	5	4	3	2	1	0
30	21	20	19	18	17	16	·15	14	13	12	11	10	9	8	7	6	5
35	26	25	24	23	22	21	20	19	18	17	16	15	14	13	12	11	10
40	31	30	29	28	27	26	25	24	23	22	21	20	19	18	17	16	15

* Reprinted with permission of M/S Carrier Corporation from *Handbook of Air-conditioning System Design.*

The above matrix shows similar calculations for monsoon as well with outdoor temperature of 95°F at 75% RH and minimum average temperature of 80°F. The heat load proforma for monsoon is enclosed.

The transmission coefficient for 9" thick masonary wall with 1.5" plaster, (0.75" an each side) from Table 5.11 are 0.45 and 0.31 BTU/hr/sq ft/°F for 8" brick and 0.75" plaster respectively. Though for 9" brick and 1.5" plaster, this value will be still lower.

Fill up these values in Part-II of proforma to calculate solar and transmission gain through wall. Similarly, values for roof can be calculated as below:

$$\text{Weght of roof/sq ft} = \frac{A_r \times (t_s \times d_s + t_f \times d_f)}{A_r}$$

where A_r = Area of roof/floor

t_s, t_f = thickness of slab, flooring respectively

d_s, d_f = density of slab, flooring respectively

Table 5.11 Transmission coefficient U-Masonry walls for summer and winter Btu/(h) (sq ft) (°F Temperature difference) All numbers in parentheses indicate weight per sq ft. Total weight per sq ft is sum of wall and finishes

Exterior finish		Thickness (inches) weight (lb per sq ft)	None	3/8" Gypsum board (Plaster board) (2)	Interior finish							
					5/8" Plaster on wall		Metal lath plastered on furring		3/8" Gypsum or wood lath plastered on furring		Insulating board plain or plastered on furring	
					Sand agg (6)	Lt Wt agg (3)	3/4" Sand plaster (7)	3/4" Lt Wt plaster (3)	1/2" Sand plaster (7)	1/2" Lt Wt plaster (2)	1/2" board (2)	1" board (4)
Solid Brick	Face and common	8 (87)	.48	.41	.45	.41	.31	.28	.29	.27	.22	.16
		2 (123)	.35	.31	.33	.30	.25	.23	.23	.22	.19	.14
		16 (173)	.27	.25	.26	.25	.21	.19	.20	.19	.16	.13
	Common only	8 (80)	.41	.36	.39	.35	.28	.26	.26	.25	.21	.15
		12 (120)	.31	.28	.30	.27	.23	.22	.22	.21	.18	.14
		16 (160)	.25	.23	.24	.23	.19	.18	.18	.18	.16	.12
Stone		8 (100)	.67	.55	.63	.53	.39	.34	.35	.32	.26	.18
		12 (150)	.55	.47	.52	.46	.34	.31	.31	.29	.24	.17
		16 (200)	.47	.41	.45	.40	.31	.28	.28	.27	.22	.16
		24 (300)	.36	.32	.35	.32	.26	.24	.24	.23	.19	.15
Adobe- blocks or brick		8 (26)	.34	.30	.32	.30	.25	.23	.23	.22	.18	.12
		12 (40)	.25	.23	.24	.23	.20	.18	.18	.18	.15	.14
Poured concrete	140 lb/cu ft	6 (70)	.75	.55	.69	.58	.41	.36	.37	.34	.27	.18
		8 (93)	.67	.49	.63	.53	.39	.34	.35	.32	.26	.17
		10 (117)	.61	.44	.57	.49	.36	.32	.33	.31	.25	.17
		12 (140)	.55	.40	.52	.45	.34	.31	.31	.29	.24	.16
	80 lb/cu ft	6 (40)	.31	.28	.30	.27	.23	.21	.22	.21	.18	.14
		8 (53)	.25	.23	.24	.23	.19	.18	.18	.18	.16	.12
		10 (66)	.21	.19	.20	.19	.17	.16	.15	.14	.14	.11
		12 (80)	.18	.17	.17	.15	.15	.14	.14	.14	.12	.10
	30 lb/cu ft	6 (15)	.13	.13	.13	.13	.12	.11	.11	.11	.13	.09
		8 (20)	.10	.10	.10	.10	.09	.09	.09	.09	.10	.07
		10 (25)	.08	.08	.08	.08	.08	.07	.08	.07	.08	.06
		12 (30)	.07	.07	.07	.07	.07	.07	.06	.06	.07	.06
Hollow concrete blocks	Sand and Grevel agg	8 (43)	.52	.44	.48	.43	.33	.29	.30	.28	.23	.17
		12 (63)	.47	.41	.45	.40	.31	.28	.28	.27	.22	.16
	Cinder agg	8 (37)	.39	.35	.37	.34	.27	.25	.25	.24	.20	.15
		12 (53)	.36	.33	.35	.32	.26	.24	.23	.23	.19	.15
	Lt wt agg	8 (32)	.35	.32	.34	.31	.26	.23	.24	.22	.19	.15
		12 (43)	.32	.29	.31	.28	.24	.22	.22	.21	.18	.14
Stucco on hollow clay tile		8 (39)	.36	.32	.34	.32	.26	.24	.24	.23	.19	.15
		10 (44)	.32	.29	.31	.28	.23	.22	.22	.21	.18	.14
		12 (49)	.29	.27	.28	.26	.22	.20	.21	.20	.17	.13

* Reprinted with permission of M/S Carrier Corporation from *Handbook of Air-conditioning System Design*.

Table 5.12 Transmission coefficient U-masonry walls and partitions for summer and winter Btu/(hr) (sq ft) (°F Temperature difference). All numbers in parentheses indicate weight per sq ft. Total weight per sq ft is sum of component materials

Exterior finish	Sheathing	None	3/4″ wood panel	3/8″ gypsum board	Metal lath plastered		3/8″ gypsum or wood lath plastered or		Insulating board plain plastered	
			(2)	(plaster board) plaster (2)	3/4″ sand plaster (7)	3/4″ L1 Wt plaster (3)	1/2″ sand plaster (7)	1/2″ Lt Wt (2)	1/2″ board (2)	1″ board (4)
1″ stucco 10	None, building paper	.91	.33	.42	.45	.39	.40	.37	.29	.20
or asbestos	5/16″ plywood (1) or ½ Gyp (2)	.68	.30	.37	.40	.35	.36	.33	.26	.19
cement siding (1)	25/32″ wood and bldg. paper (2)	.48	.25	.30	.31	.28	.29	.27	.22	.17
or asphalt	1/2″ insulating board (2)	.42	.23	.27	.29	.26	.27	.25	.21	.16
roll siding (2)	25/32″ insulating board (3)	.32	.20	.23	.24	.22	.22	.21	.18	.14
4″ face brick	None, building paper	.73	.30	.37	.40	.35	.36	.33	.26	.19
veneer (43) or	5/16″ plywood (1) or 1/2″ Gyp (2)	.57	.28	.33	.36	.32	.32	.30	.24	.18
3/8″ plywood (1)	25/32″ wood and bldg. paper (2)	.42	.23	.27	.29	.26	.27	.25	.21	.16
or asphalt	1/2″ insulating board (2)	.38	.22	.25	.27	.25	.25	.24	.20	.15
siding (2)	25/32″ insulating board (3)	.30	.19	.21	.22	.21	.21	.20	.17	.14
Wood siding (3)	None, building paper	.57	.27	.33	.35	.31	.32	.30	.24	.18
or	5/16″ plywood (1) or 1/2″ Gyp (2)	.48	.25	.30	.31	.28	.29	.27	.22	.17
Wood shingles (2)	25/32″ wood and bldg. paper	.36	.22	.25	.26	.24	.24	.23	.19	.15
or 3/4″ wood	1/2″ insulating board (2)	.33	.20	.23	.24	.22	.23	.22	.18	.14
panels (3)	25/32″ insulating board (3)	.27	.18	.20	.21	.19	.19	.19	.16	.13
Wood shingles	None, building paper	.43	.24	.28	.29	.27	.27	.25	.21	.16
over 5/16″ insul	5/16″ plywood (1) or 1/2″ Gyp (2)	.38	.22	.25	.27	.24	.25	.23	.19	.15
backer board (3)	25/32″ wood and bldg. paper	.30	.19	.22	.23	.21	.21	.20	.17	.14
or asphalt	1/2″ insulating board (2)	.28	.18	.20	.21	.20	.20	.19	.16	.13
Insulated siding (4)	25/32″ insulating board (3)	.23	.16	.18	.18	.17	.18	.17	.15	.12
Single partition (Finish on one side only)			.43	.60	.67	.55	.57	.50	.36	.23

* Reprinted with permission of M/S Carrier Corporation from *Handbook of Air-conditioning System Design*.

Table 5.13a Transmission coefficient U–ceiling and floor, (heat flow up)* based on still air both sides, Btu/(hr) (sq ft) (°F Temperature difference) All numbers in parentheses indicate weight per sq ft. Total weight per sq ft is sum of ceiling and floor

			Masonry ceiling											
			Not furred		Suspended or furred									
Floor	Concrete subfloor	Thickness (inches) and weight	None or 1/2″ sand plaster	1/2″ Lt Wt plaster (3)	Acoustical title glued		Metal lath plastered		3/8″ gypsum or wood lath plastered		Insulating board plain or 1/2″ sand agg plastered		Acoustical tile on furring 3/8″ gypsum	
		(lb per sq ft)	(5)		1/2″ tile (1)	3/4″ tile (1)	3/4″ sand plaster (7)	3/4″ Lt Wt plaster (3)	1/2″ sand plaster (5)	1/2″ Lt Wt plaster (2)	1/2″ board (2)	1/2″ board (4)	1/2″ tile (1)	3/4″ tile (1)
		2 (19)	.70	.53	.38	.31	.43	.38	.44	.41	.26	.19	.28	.24
None or		4 (39)	.63	.49	.36	.30	.41	.36	.41	.38	.25	.18	.26	.23
1/8″	Sand agg	6 (59)	.57	.45	.34	.28	.38	.34	.39	.36	.24	.18	.25	.22
linoleum		8 (79)	.52	.42	.32	.27	.36	.32	.37	.34	.23	.17	.24	.21
or		10 (99)	.48	.39	.31	.26	.34	.31	.35	.32	.23	.17	.23	.21
floor tile	Lt Wt agg	2 (15)	.48	.39	.31	.26	.34	.31	.35	.32	.23	.17	.23	.21
	80 lb/ft³	4 (28)	.35	.30	.25	.22	.27	.25	.27	.26	.19	.15	.20	.18
		6 (41)	.27	.24	.21	.18	.22	.21	.22	.21	.17	.13	.17	.15

(Contd.)

Table 5.13a (*Contd.*)

Floor	Concrete subfloor	Thickness (inches) and weight (lb per sq ft)	None or 1/2" sand plaster (5)	1/2" Lt Wt plaster (3)	Acoustical title glued 1/2" tile (1)	3/4" tile (1)	Metal lath plastered 3/4" sand plaster (7)	3/4" Lt Wt plaster (3)	3/8" gypsum or wood lath plastered 1/2" sand plaster (5)	1/2" Lt Wt plaster (2)	Insulating board plain or 1/2" sand agg plastered 1/2" board (2)	1/2" board (4)	Acoustical tile on furring 3/8" gypsum 1/2" tile (1)	3/4" tile (1)
13/16" wood block on slab	Sand agg	2 (20)	.47	.39	.30	.26	.33	.30	.33	.40	.22	.17	.23	.20
		4 (40)	.44	.36	.29	.25	.31	.28	.32	.38	.22	.16	.22	.20
		6 (60)	.41	.34	.28	.24	.30	.27	.30	.36	.21	.16	.22	.19
		8 (80)	.38	.33	.26	.23	.28	.26	.29	.34	.20	.15	.21	.19
		10 (100)	.36	.31	.25	.22	.27	.25	.27	.32	.19	.15	.20	.18
	Lt Wt agg 80 lb/ft³	2 (16)	.36	.31	.25	.22	.26	.25	.27	.32	.19	.15	.20	.18
		4 (29)	.28	.25	.21	.19	.22	.21	.23	.26	.17	.13	.17	.16
		6 (42)	.23	.21	.18	.16	.19	.18	.19	.21	.15	.12	.15	.14
Floor tile 1/8" linoleum on 5/8" plywood on 2" × 2" sleepers	Sand agg	2 (22)	.32	.28	.23	.21	.31	.28	.32	.30	.18	.14	.18	.17
		4 (42)	.31	.27	.23	.20	.30	.27	.30	.28	.18	.14	.18	.17
		6 (62)	.29	.26	.22	.19	.28	.26	.29	.27	.17	.14	.18	.16
		8 (82)	.28	.25	.21	.19	.27	.25	.27	.26	.17	.13	.17	.16
		10 (102)	.27	.24	.20	.18	.26	.24	.26	.25	.16	.13	.17	.15
	Lt Wt agg 80 lb/ft³	2 (19)	.27	.24	.20	.18	.26	.24	.26	.25	.16	.13	.17	.15
		4 (31)	.22	.20	.17	.16	.22	.20	.22	.21	.14	.12	.15	.14
		6 (44)	.19	.17	.15	.14	.18	.17	.19	.18	.13	.11	.13	.12
3/4" hardwood on 25/32" subfloor on 2" × 2" sleepers	Sand agg	2 (24)	.26	.23	.20	.18	.25	.23	.25	.24	.16	.13	.16	.15
		4 (44)	.25	.22	.19	.17	.24	.22	.24	.23	.16	.13	.16	.15
		6 (64)	.24	.21	.19	.17	.23	.21	.23	.22	.15	.12	.16	.14
		8 (84)	.23	.21	.18	.16	.22	.21	.22	.21	.15	.12	.15	.14
		10 (104)	.22	.20	.17	.16	.21	.20	.22	.21	.14	.12	.15	.14
	Lt Wt agg 80 lb/ft³	2 (20)	.22	.20	.17	.16	.21	.20	.22	.21	.14	.12	.15	.14
		4 (33)	.19	.17	.15	.14	.18	.17	.18	.18	.13	.11	.13	.12
		6 (46)	.16	.15	.14	.13	.16	.15	.16	.16	.12	.099	.12	.11

* Lower flooor unconditioned, upper floor air-conditioned.

* Reprinted with permission of M/S Carrier Corporation from *Handbook of Air-conditioning System Design*.

Table 5.13b Transmission Coefficient U-Ceiling and Floor, (Heat Flow Down)* Based on Still Air Both Sides, Btu/(hr) (sq ft) (°F Temperature Difference) All Numbers in Parentheses Indicate Weight per sq ft. Total Weight per sq ft is Sum of Ceiling and Floor

Floor	Concrete subfloor	Thickness (inches) and weight (lb per sq ft)	None or 1/2" sand plaster (5)	1/2" Lt Wt plaster (3)	Acoustical title glued 1/2" tile (1)	3/4" tile (1)	Metal lath plastered 3/4" sand plaster (7)	3/4" Lt Wt plaster (3)	3/8" gypsum or wood lath plastered 1/2" sand plaster (5)	1/2" Lt Wt plaster (2)	Insulating board plain or 1/2" sand agg plastered 1/2" board (2)	1/2" board (4)	Acoustical tile on furring 3/8" gypsum 1/2" tile (1)	3/4" tile (1)
None or 1/8" linoleum or floor tile	Sand agg	2 (19)	.48	.43	.31	.26	.32	.29	.30	.28	.23	.17	.23	.20
		4 (39)	.44	.40	.30	.25	.31	.28	.28	.27	.22	.17	.22	.20
		6 (59)	.41	.37	.28	.24	.29	.27	.27	.26	.21	.16	.22	.19
		8 (79)	.39	.35	.27	.23	.28	.26	.26	.25	.21	.16	.21	.19
		10 (99)	.36	.34	.26	.22	.27	.25	.25	.24	.20	.15	.20	.18
	Lt Wt agg	2 (15)	.36	.34	.26	.22	.27	.25	.25	.24	.20	.15	.20	.18

(*Contd.*)

Table 5.13b (Contd.)

Floor	Concrete subfloor	Thickness (inches) and weight (lb per sq ft)	None or 1/2" sand plaster (5)	1/2" Lt Wt plaster (3)	Acoustical title glued 1/2" tile (1)	Acoustical title glued 3/4" tile (1)	Metal lath plastered 3/4" sand plaster (7)	Metal lath plastered 3/4" Lt Wt plaster (3)	3/8" gypsum or wood lath plastered 1/2" sand plaster (5)	3/8" gypsum or wood lath plastered 1/2" Lt Wt plaster (2)	Insulating board plain or 1/2" sand agg plastered 1/2" board (2)	Insulating board plain or 1/2" sand agg plastered 1/2" board (4)	Acoustical tile on furring 3/8" gypsum 1/2" tile (1)	Acoustical tile on furring 3/8" gypsum 3/4" tile (1)
	80 lb/ft³	4 (28)	.29	.26	.21	.19	.22	.21	.21	.20	.17	.14	.17	.16
		6 (41)	.23	.22	.18	.17	.19	.18	.18	.17	.15	.13	.15	.14
13/16" wood block on slab	Sand agg	2 (20)	.36	.33	.25	.22	.26	.24	.24	.23	.20	.15	.20	.18
		4 (40)	.33	.31	.24	.21	.25	.23	.23	.22	.19	.15	.19	.17
		6 (60)	.32	.29	.23	.21	.24	.22	.22	.21	.18	.15	.18	.17
		8 (80)	.30	.28	.23	.20	.23	.22	.22	.21	.18	.14	.18	.16
		10 (100)	.29	.27	.22	.19	.22	.21	.21	.20	.17	.14	.17	.16
	Lt Wt agg 80 lb/ft³	2 (16)	.29	.27	.22	.19	.22	.21	.21	.20	.17	.14	.17	.16
		4 (29)	.23	.22	.19	.17	.19	.18	.18	.17	.15	.13	.15	.14
		6 (42)	.20	.19	.16	.15	.16	.16	.16	.15	.14	.11	.14	.13
Floor tile 1/8" linoleum on 5/8" plywood on 2" × 2" sleepers	Sand agg	2 (22)	.33	.31	.24	.21	.25	.23	.23	.22	.19	.15	.20	.17
		4 (42)	.32	.29	.23	.21	.24	.22	.22	.21	.18	.15	.19	.17
		6 (62)	.30	.28	.23	.20	.23	.21	.22	.21	.18	.14	.18	.16
		8 (82)	.29	.27	.22	.19	.22	.21	.21	.20	.17	.14	.18	.16
		10 (102)	.28	.26	.21	.19	.21	.20	.20	.19	.17	.13	.17	.15
	Lt Wt agg 80 lb/ft³	2 (19)	.28	.26	.21	.19	.21	.20	.20	.19	.17	.13	.17	.15
		4 (31)	.22	.21	.18	.16	.18	.17	.17	.17	.15	.12	.15	.14
		6 (44)	.19	.18	.16	.14	.16	.15	.15	.15	.13	.11	.14	.13
3/4" hardwood on 25/32" subfloor on 2" × 2" Sleepers	Sand agg	2 (24)	.26	.25	.20	.18	.20	.20	.20	.19	.16	.13	.17	.15
		4 (44)	.25	.24	.20	.18	.20	.19	.19	.18	.16	.13	.16	.15
		6 (64)	.24	.23	.19	.17	.19	.18	.19	.18	.15	.13	.16	.14
		8 (84)	.23	.22	.19	.17	.19	.18	.18	.17	.15	.12	.15	.14
		10 (104)	.22	.21	.18	.16	.18	.17	.17	.17	.14	.12	.15	.14
	Lt Wt agg 80 lb/ft³	2 (20)	.22	.21	.18	.16	.18	.17	.17	.17	.14	.12	.15	.14
		4 (33)	.19	.18	.16	.14	.16	.15	.15	.15	.13	.11	.13	.12
		6 (46)	.16	.16	.14	.13	.14	.14	.14	.13	.12	.10	.12	.11

* Lower flooor air-conditioned, upper floor unconditioned.

* Reprinted with permission of M/S Carrier Corporation from *Handbook of Air-conditioning System Design*.

$$= \frac{8}{12} \times 156 + \frac{1.5}{12} \times 137$$

$$= 121.6 \text{ lb/sq ft}$$

From Table 5.10b, for 80 lb/sq ft weight of roof at 4.00 PM, the equivalent temperature difference is 32°F. The correction of 20.5 is applied to make it 32 + 20.5 = 52.5°F. Similarly the U value for ceiling with heat flow down (summer) from Table 5.13b and with 3/4" accoustical tile glued to ceiling for 8" slab thickness will be 0.23. Fill these values in roof exposed to sun column.

If insulation is applied on the sun facing side of the ceiling, then appropriate value can be filled in "Roof Insulated" column, otherwise it is to be left blank.

vii. **Conduction or transmission gain except wall and roof:** Part III of proforma accounts for heat transmitted to conditioned space due to temperature difference on two sides of the glass (i.e. difference in atmosphere temperature and temperature of conditioned room). It also accounts for heat transfer due to temperature difference existing due to adjoining non conditioned room. The temperature of non conditioned room is usually taken as 20°F higher than that of conditioned room (some designers take this value as 15°F). Similarly if there is a non conditioned floor above or below the conditioned room, the same method is applied. Thus fill the Part-III of proforma as below:

All glass area (exposed to atmosphere) of hall = 45 + 144 + 45 = 234 sq ft

Temperature difference = 110 − 75 = 35° F

U value for 6 mm thick glass (Table 5.9b) = 1.13

Partition wall adjoining hall (non ac Pantry) = 13′ × 10′ = 130 sq ft

U value for Partition wall (4.5″ thick) with 3/4″ light plaster = 0.35 BTU/hr/sq ft/°F temperature difference.

Temperature difference = 20°F

No value to be entered for ceiling (non AC) as this example is of a single storey building.

Floor: There is no floor below the room, but ground temperature is always constant and is taken as unconditioned space thus transmitting heat upwards.

Area of floor = 1200 sq ft

Temperature difference = 20°F

U value (for 8″ slab and 1.5 ″ flooring) = 0.32 (for 3/4″ light weight plaster)

Fill these values in proforma.

viii **Internal heat gain (Part-IV):** Generally occupancy in a room is known by experience or can be calculated by NBC guidelines in Table 5.5. The sensible heat is 245 for office worker at 75°F room temperature (Table 5.14).

Similarly, light load is calculated at 1 watt/sq ft of floor area and increased by 25% for ballast losses and multiplied with 3.41 to convert watts into BTU.

Equipment load is taken as 0.5 w/sq ft or as per actual.

ix **Room sensible heat (Part-V):** Row 42 of proforma require very detailed calculations, hence the sensible heat obtained is multiplied with a suitable percentage factor based on experience, say 5% for supply duct heat as it passes through unconditioned space, another 5% for duct leakage loss.

Table 5.14 Heat gain from people

Degree of activity	Typical application	Metabolic rate (Adult male) Btu/hr	Average adjusted metabolic rate Btu/hr	Room dry-bulb temperature									
				82 °F Btu/hr		80 °F Btu/hr		78 °F Btu/hr		75 °F Btu/hr		70 °F Btu/hr	
				Sensible	Latent	Sensible	Latent	Sensible	Latent	Sensible	Latent	Sensible	Latent
Seated at rest	Theater, grade school	390	350	175	175	195	155	210	140	230	120	260	90
Seated, very light work	High school	450	400	180	220	195	205	215	185	240	160	275	125
Officer worker	Offices, hotels, apts., college	475	450	180	270	200	250	215	235	245	205	285	165
Standing, walking slowly	Dept., retail, or variety store	550	450	180	270	200	250	215	235	245	205	285	165
Walking, seated	Drug store	550	500	180	320	200	300	220	280	255	245	290	210
Standing, walking slowly	Bank	550	500	180	320	200	300	220	280	255	245	290	210
Sedentary work	Restaurant	500	550	190	360	220	330	240	310	280	270	320	230
Light bench work	Factory, light work	800	750	190	560	220	530	245	505	295	455	365	385
Moderate dancing	Dance hall	900	850	220	630	245	605	275	575	325	525	400	450
Walking, 3 mph	Factory, fairly heavy work	1000	1000	270	730	300	700	330	670	380	620	460	540
Heavy	Bowling alley	1500	1450	450	1000	465	985	485	965	525	925	605	845

* Reprinted with permission of M/S Carrier Corporation from *Handbook of Air-conditioning System Design.*

Row 43 is also safety factor to account for in filteration air due to door opening, gaps in windows/doors and frames. This factor is again a matter of experience, though detailed calculations can be carried out as per ASHRAE guidelines. For this example, factor of 5% is taken.
Part of outside air which does not come in contact with cooling coil in AHU gets by passed and adds heat load in the room. A by pass factor say 0.15 multiplied with factor 0.68 and total CFM gives sensible heat load added to the room. Similarly latent heat added to room on this account is calculated in row 49–54.

x. **Outside air heat on equipments (Part-VI):** The heat load added by outside air on AHU or conditioning equipment is worked out. This is the air which actually comes in contact with cooling coil, i.e. (1-B F). This adds sensible and latent heat on the system capacity. Similarly fan motor and pump and any leakage of outdoor air from outside before cooling coil add load on system and can be taken as 5%.

xi. **Design conditions (Part-VII):** The condition of outdoor and indoor air requirement are filled here.

xii. **Ventilation CFM (Part-VIII):** The ventilation CFM as per ASHRAE standard 62-2001 (Table 5.6) are filled here.

This part also calculates Effective sensible heat factor.

$$ESHF = \frac{\text{Effective Room Sensible Heat}}{\text{Effective Room Total Heat}}$$

The indicated ADP can be determined from Psychrometric chart as the point where SHF line crosses the 100% saturation curve or it can be selected from Table 5.15. In table, select the value corresponding to room DB of 75°F at 60% RH with ESHF as calculated, say, 0.93. With interpolation we get ADP of 59.33. The equipment manufacture's are supplying equipments with an ADP of 50–55°F. Thus we have to select the ADP as per market availability.

The dehumidified CFM is the quantity of air required to satisfy the room condition by absorbing total room load, it is also same as the air dehumidified in cooling coil. (Refer Fig. 5.7) Thus for hall, the Dehumidified CFM is calculated as below:

$$\text{Dehumidified CFM} = \frac{ERSH}{1.08\left(T_{RM} - T_{adp}\right) \times (1 - BF)}$$

$$= \frac{78876}{1.08(75 - 55)(1 - 0.15)}$$

$$= 4296$$

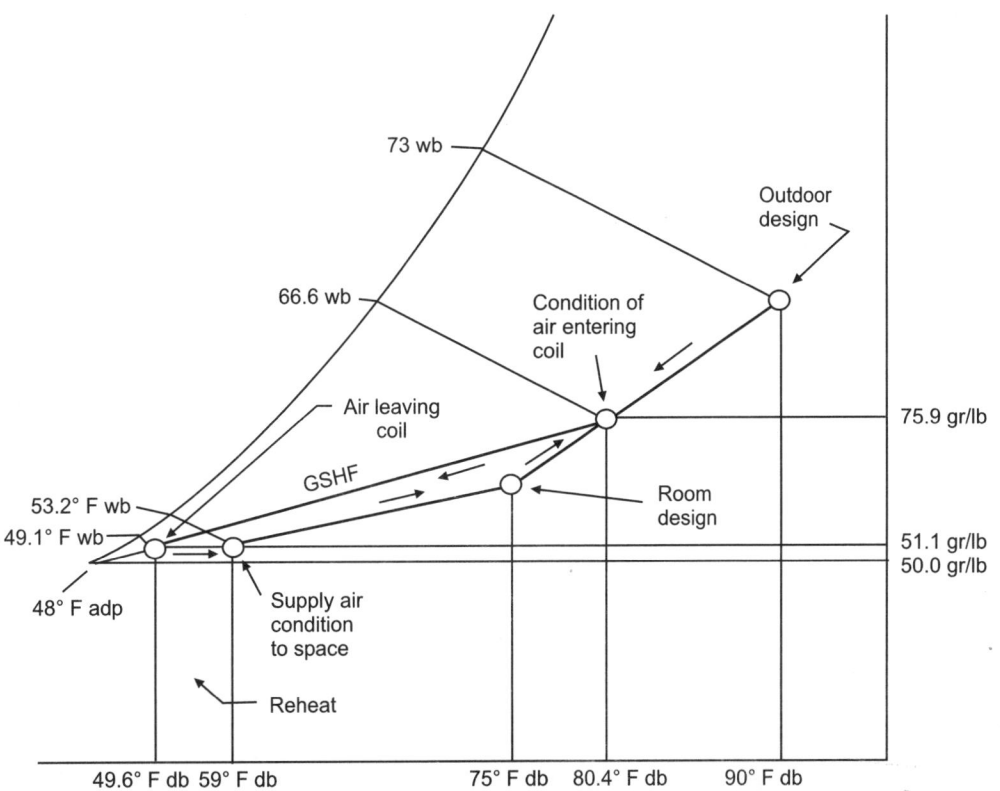

Fig. 5.7 Calculation of dehumidified air quantity

Table 5.15 Apparatus dew points

DB (F)	RH (%)	WB (F)	W (gr/lb)											
		Room conditions			*Effective sensible heat factor and aparatus dew point*									
	20	62.7	42.0	ESHF	1.00	.96	.92	.90	.88	.86	.84	.82	.81
				ADP	43.5	41	39	37	35	32	29	24	22
	25	65.1	52.7	ESHF	1.00	.96	.92	.88	.84	.82	.80	.78	.75
				ADP	49.6	48	46	44	41	39	36	32	22
	30	67.3	63.6	ESHF	1.00	.92	.87	.83	.80	.76	.74	.72	.70
				ADP	54.5	52	50	48	46	42	38	34	24
	35	69.3	74.2	ESHF	1.00	.92	.85	.81	.76	.73	.71	.69	.66
				ADP	58.8	57	55	53	50	48	45	42	33
	40	71.2	84.8	ESHF	1.00	.92	.83	.78	.74	.69	.66	.63	.62
				ADP	62.4	61	59	57	55	52	48	44	40
90	45	73.0	95.5	ESHF	1.00	.92	.82	.76	.70	.66	.62	.60	.58
				ADP	65.8	65	63	61	59	56	52	49	43
	50	74.9	106.4	ESHF	1.00	.92	.78	.68	.64	.60	.58	.56	.54
				ADP	68.9	68	66	67	61	58	56	53	47
	55	76.7	117.5	ESHF	1.00	.92	.76	.68	.64	.57	.54	.52	.50
				ADP	71.6	71	69	67	66	62	59	57	50
	60	78.4	128.4	ESHF	1.00	.86	.68	.60	.56	.52	.50	.48	.46
				ADP	74.2	73	71	69	67	64	62	59	50
	65	80.0	139.6	ESHF	1.00	.75	.68	.62	.55	.50	.47	.45	.43
				ADP	76.8	75	74	73	71	69	66	64	.59
	70	81.6	151.0	ESHF	1.00	.78	.66	.60	.52	.47	.43	.41	.39
				ADP	79.0	78	77	76	74	72	69	66	58
	20	59.6	35.8	ESHF	1.00	.98	.95	.92	.90	.88	.87	.86	.84
				ADP	39.4	38	36	34	32	30	28	26	22
	25	61.7	44.8	ESHF	1.00	.98	.93	.90	.86	.84	.82	.80	.78
				ADP	45.2	44	42	40	37	35	32	28	20
	30	63.7	54.1	ESHF	1.00	.94	.89	.85	.81	.79	.77	.75	.73
				ADP	50.2	48	46	44	40	38	35	31	22
	35	65.5	62.9	ESHF	1.00	.92	.86	.82	.78	.74	.72	.70	.69
				ADP	54.1	52	50	48	45	41	38	32	27
	40	67.4	71.7	ESHF	1.00	.92	.84	.79	.76	.73	.69	.67	.65
				ADP	57.9	56	54	52	50	48	44	40	32
85	45	69.1	81.1	ESHF	1.00	.92	.83	.77	.72	.68	.64	.62	.61
				ADP	61.2	60	58	56	54	51	46	41	.36
	50	70.8	90.1	ESHF	1.00	.92	.80	.73	.68	.64	.61	.59	.57
				ADP	64.2	63	61	59	57	54	51	48	.39
	55	72.3	99.4	ESHF	1.00	.92	.83	.73	.67	.60	.57	.56	.54
				ADP	66.9	66	65	63	61	57	54	52	.39
	60	73.9	108.8	ESHF	1.00	.92	.76	.67	.61	.56	.54	.52	.54
				ADP	69.5	69	67	65	63	60	58	55	.47
	65	75.5	118.2	ESHF	1.00	.88	.69	.61	.56	.53	.50	.48	.47
				ADP	71.9	71	69	67	65	63	61	58	54
	70	77.0	127.6	ESHF	1.00	.81	.63	.55	.51	.49	.47	.45	.43
				ADP	74.0	73	71	69	67	66	64	62	55
	35	63.3	57.0	ESHF	1.00	.92	.88	.84	.80	.76	.74	.72	.71
				ADP	51.6	49	48	46	43	39	36	31	27
	40	65.0	65.1	ESHF	1.00	.90	.87	.82	.78	.74	.74	.69	.67
				ADP	55.2	53	52	50	48	45	45	38	31

(Contd.)

Table 5.15 (*Contd.*)

DB (F)	RH (%)	WB (F)	W (gr/lb)					Effective sensible heat factor and aparatus dew point					
	45	66.7	73.5	ESHF	1.00	.91	.87	.80	.75	.72	.68	.65	.63
				ADP	58.5	57	56	54	52	50	46	41	53
	50	68.3	81.9	ESHF	1.00	.90	.80	.74	.70	.64	.62	.60	.59
				ADP	61.5	60	58	56	54	50	47	42	37
82	55	69.8	90.2	ESHF	1.00	.90	.83	.74	.68	.64	.61	.58	.56
				ADP	64.2	63	62	60	58	56	54	50	44
	60	71.3	98.5	ESHF	1.00	.92	.76	.68	.63	.59	.56	.54	.52
				ADP	66.7	66	64	62	60	58	55	52	44
	65	72.8	107.0	ESHF	1.00	.86	.71	.63	.58	.54	.52	.51	.49
				ADP	69.1	68	66	64	62	60	58	56	51
	70	74.2	115.5	ESHF	1.00	.80	.71	.65	.60	.54	.51	.48	.46
				ADP	71.2	70	69	68	67	65	63	60	56
	35	62.5	55.2	ESHF	1.00	.94	.89	.84	.81	.77	.75	.73	.71
				ADP	50.8	49	47	45	43	39	36	32	21
	40	64.2	63.2	ESHF	1.00	.94	.87	.82	.78	.75	.72	.69	.67
				ADP	54.4	53	51	49	47	45	41	36	23
	45	65.9	71.2	ESHF	1.00	.96	.91	.83	.78	.74	.70	.67	.64
				ADP	57.6	57	56	54	.52	50	47	43	36
	50	67.5	79.0	ESHF	1.00	.90	.84	.80	.74	.70	.66	.62	.60
				ADP	60.5	59	58	57	.55	53	50	45	38
	55	69.0	87.4	ESHF	1.00	.90	.77	.71	.66	.62	.60	.58	.56
				ADP	63.2	62	60	58	56	53	51	47	35
	60	70.5	95.4	ESHF	1.00	.92	.77	.68	.63	.59	.56	.54	.53
				ADP	65.8	65	63	61	59	56	53	50	46
	65	71.9	103.7	ESHF	1.00	.85	.76	.71	.66	.60	.56	.52	.50
				ADP	68.2	67	66	.65	64	62	60	56	52
	70	73.3	111.9	ESHF	1.00	.80	.71	.61	.55	.52	.48	.47	.46
				ADP	70.3	69	68	66	64	62	58	56	52
	20	56.4	30.4	ESHF	1.00	.98	.95	.93	.91	.89	.88	.87	.86
				ADP	35.4	34	32	30	28	26	24	22	20
	25	58.3	38.0	ESHF	1.00	.96	.93	.90	.88	.86	.84	.82	.81
				ADP	40.9	39	37	35	33	31	28	24	21
	30	60.0	45.6	ESHF	1.00	.96	.91	.88	.85	.83	.80	.78	.76
				ADP	45.7	44	42	40	38	36	32	28	21
	35	61.8	53.5	ESHF	1.00	.94	.88	.85	.82	.79	.77	.73	.72
				ADP	49.8	48	46	44	42	40	37	29	24
	40	63.5	61.2	ESHF	1.00	.94	.90	.84	.80	.76	.73	.70	.68
				ADP	53.5	52	51	49	47	44	41	36	28
80	45	65.1	68.9	ESHF	1.00	.96	.87	.81	.76	.73	.70	.67	.64
				ADP	56.8	56	54	52	50	48	45	41	31
	50	66.7	76.7	ESHF	1.00	.89	.80	.74	.70	.66	.64	.62	.61
				ADP	59.7	58	56	54	52	49	46	42	39
	55	68.2	84.6	ESHF	1.00	.89	.82	.74	.69	.65	.61	.59	.57
				ADP	62.3	61	60	58	56	54	50	47	40

(*Contd.*)

Table 5.15 (*Contd.*)

DB (F)	RH (%)	WB (F)	W (gr/lb)		Effective sensible heat factor and aparatus dew point								
	60	69.6	92.3	ESHF	1.00	.91	.83	.72	.66	.62	.59	.57	.54
				ADP	64.8	64	63	61	59	57	55	.53	47
81	65	71.1	100.4	ESHF	1.00	.85	.76	.71	.63	.58	.55	.52	.50
				ADP	67.2	66	65	64	62	60	58	54	47
	70	72.4	108.3	ESHF	1.00	.78	.71	.65	.61	.55	.52	.49	.47
				ADP	69.4	68	67	66	65	63	61	58	53
	35	61.0	51.5	ESHF	1.00	.96	.91	.89	.85	.82	.78	.75	.73
				ADP	48.9	48	46	45	43	41	37	32	26
	40	62.7	59.2	ESHF	1.00	.97	.90	.84	.80	.76	.74	.71	.69
				ADP	52.7	52	50	48	46	43	41	36	29
	45	64.3	66.7	ESHF	1.00	.91	.83	.78	.75	.72	.70	.67	.65
				ADP	55.9	54	52	50	48	46	44	39	32
	50	65.9	74.2	ESHF	1.00	.89	.80	.75	.71	.68	.66	.63	.61
				ADP	58.9	57	55	53	51	49	47	42	33
79	55	67.4	81.9	ESHF	1.00	.96	.82	.74	.69	.66	.63	.60	.58
				ADP	61.4	61	59	57	55	53	51	47	41
	60	68.8	89.3	ESHF	1.00	.90	.76	.69	.64	.61	.57	.55	.54
				ADP	63.9	63	61	59	57	55	51	47	41
	65	70.2	97.0	ESHF	1.00	.84	.71	.64	.59	.56	.54	.52	.51
				ADP	66.3	65	63	61	59	57	55	51	48
	70	71.6	104.8	ESHF	1.00	.81	.71	.65	.58	.54	.52	.50	.48
				ADP	68.5	67	66	65	63	61	59	57	53
	35	60.3	50.0	ESHF	1.00	.96	.91	.87	.83	.79	.77	.75	.73
				ADP	48.2	47	45	43	41	37	35	31	22
	40	61.9	57.3	ESHF	1.00	.93	.87	.82	.79	.77	.73	.71	.69
				ADP	51.7	50	48	46	44	42	38	34	25
	45	63.5	64.6	ESHF	1.00	.95	.86	.81	.76	.74	.70	.68	.66
				ADP	55.0	54	52	50	48	46	42	.39	34
78	50	65.0	71.9	ESHF	1.00	.94	.83	.76	.73	.70	.67	.64	.62
				ADP	57.9	57	55	53	51	49	47	.42	36
	55	66.6	79.2	ESHF	1.00	.96	.83	.75	.70	.65	.62	.60	.59
				ADP	60.5	60	58	56	54	51	48	44	41
	60	67.9	86.4	ESHF	1.00	.90	.82	.76	.69	.64	.60	.57	.55
				ADP	63.0	62	61	60	58	56	53	49	42
	65	69.3	93.8	ESHF	1.00	.85	.77	.71	.67	.62	.58	.54	.52
				ADP	65.2	64	63	62	61	59	57	53	48
	70	70.6	101.2	ESHF	1.00	.71	.66	.62	.59	.55	.52	.50	.48
				ADP	67.5	65	64	63	62	60	58	55	48
	35	59.6	48.3	ESHF	1.00	.96	.91	.87	.83	.79	.77	.75	.74
				ADP	47.3	46	44	42	40	36	33	28	24
	40	61.2	55.5	ESHF	1.00	.96	.89	.84	.81	.78	.76	.73	.70
				ADP	50.9	50	48	46	44	42	40	36	27
	45	62.7	61.4	ESHF	1.00	.94	.86	.81	.77	.74	.72	.69	.66
				ADP	54.1	53	51	49	47	45	43	39	29

(*Contd.*)

Table 5.15 (*Contd.*)

DB (F)	RH (%)	WB (F)	W (gr/lb)										
	Room conditions				*Effective sensible heat factor and aparatus dew point*								
77	50	64.2	69.7	ESHF	1.00	.94	.84	.77	.73	.70	.68	.65	.63
				ADP	57.0	56	54	52	50	48	46	42	37
	55	65.6	76.6	ESHF	1.00	.95	.83	.75	.70	.67	.63	.61	.59
				ADP	59.6	59	57	55	53	51	48	44	37
	60	67.1	83.6	ESHF	1.00	.89	.82	.77	.73	.67	.62	.58	.56
				ADP	62.0	61	60	59	59	56	53	48	43
	65	68.5	90.8	ESHF	1.00	.84	.72	.64	.60	.57	.55	.54	.53
				ADP	64.4	63	61	59	57	55	53	51	48
	70	69.8	97.9	ESHF	1.00	.79	.66	.60	.55	.53	.51	.50	.49
				ADP	66.5	65	63	61	59	57	55	53	49
	35	58.9	46.7	ESHF	1.00	.96	.91	.87	.84	.81	.79	.77	.74
				ADP	46.3	45	43	41	39	37	34	31	21
	40	60.4	53.7	ESHF	1.00	.96	.89	.84	.81	.78	.76	.72	.70
				ADP	49.9	49	47	45	43	41	39	32	22
	45	61.9	60.4	ESHF	1.00	.94	.86	.81	.77	.74	.71	.69	.67
				ADP	53.2	52	50	48	46	44	40	37	31
76	50	63.4	67.4	ESHF	1.00	.93	.83	.77	.73	.69	.67	.65	.63
				ADP	56.2	55	53	51	49	46	43	40	32
	55	64.9	74.0	ESHF	1.00	.94	.82	.75	.70	.67	.65	.62	.60
				ADP	58.7	58	56	54	52	50	48	44	38
	60	66.2	80.9	ESHF	1.00	.90	.77	.70	.66	.62	.60	.58	.57
				ADP	61.1	60	58	56	54	52	49	46	43
	65	67.6	87.6	ESHF	1.00	.84	.72	.65	.61	.58	.56	.54	.53
				ADP	63.4	62	60	58	56	54	52	48	43
	70	68.9	94.6	ESHF	1.00	.80	.67	.60	.56	.54	.52	.51	.50
				ADP	65.5	64	62	60	58	56	54	52	49
	20	53.2	25.7	ESHF	1.00	.98	.96	.94	.92	.90	.89		
				ADP	31.5	30	28	26	24	22	20		
	25	54.8	32.1	ESHF	1.00	.95	.92	.90	.88	.86	.84		
				ADP	36.9	34	32	30	28	25	21		
	30	56.5	38.5	ESHF	1.00	.97	.93	.90	.87	.85	.82	.80	.79
				ADP	41.4	40	38	36	34	32	28	24	20
	35	58.1	45.2	ESHF	1.00	.96	.91	.87	.84	.80	.78	.76	.75
				ADP	45.5	44	42	40	38	34	31	27	22
	40	59.6	51.8	ESHF	1.00	.96	.89	.84	.81	.79	.76	.73	71
				ADP	49.1	48	46	44	42	40	37	32	24
75	45	61.1	58.2	ESHF	1.00	.94	.87	.81	.77	.75	.72	.69	.67
				ADP	52.2	51	49	47	45	43	40	35	21
	50	62.6	65.0	ESHF	1.00	.92	.84	.78	.74	.71	.69	.66	.64
				ADP	55.2	54	52	50	48	46	44	40	34
	55	64.0	71.5	ESHF	1.00	.94	.87	.78	.73	.69	.65	.63	.61
				ADP	57.8	57	56	54	52	50	47	44	39
	60	65.3	77.9	ESHF	1.00	.90	.77	.71	.66	.63	.61	.59	.58
				ADP	60.1	59	57	55	53	51	49	46	43

(*Contd.*)

Table 5.15 (*Contd.*)

DB (F)	RH (%)	WB (F)	W (gr/lb)					Effective sensible heat factor and aparatus dew point				
	65	66.7	84.8	ESHF	1.00	.84	.72	.65	.61	.59	.57	.55	.54
				ADP	62.4	61	59	57	55	53	51	48	44
	70	68.0	91.2	ESHF	1.00	.80	.73	.68	.61	.57	.54	.52	.51
				ADP	64.5	63	62	61	59	57	55	52	49
	35	55.9	40.8	ESHF	1.00	.98	.93	.89	.86	.83	.81	.79	.77
				ADP	42.8	42	40	38	36	34	31	28	22
	40	57.3	46.7	ESHF	1.00	.95	.92	.87	.84	.81	.77	.75	.73
				ADP	46.3	45	44	42	40	38	34	30	23
	45	58.7	52.7	ESHF	1.00	.94	.87	.82	.79	.76	.74	.71	.69
				ADP	49.5	48	46	44	42	40	38	32	22
72	50	60.1	58.8	ESHF	1.00	.92	.88	.81	.77	.73	.70	.68	.66
				ADP	52.4	51	50	48	46	43	40	37	30
	55	61.4	64.4	ESHF	1.00	.93	.83	.77	.72	.68	.66	.64	.63
				ADP	54.9	54	52	50	48	45	42	39	36
	60	62.7	70.2	ESHF	1.00	.89	.79	.72	.68	.65	.63	.61	.60
				ADP	57.3	56	54	52	50	48	46	42	39
	65	64.0	76.3	ESHF	1.00	.84	.73	.67	.63	.61	.59	.58	
				ADP	59.5	58	56	54	52	50	48	47	
	70	65.2	82.3	ESHF	1.00	.80	.69	.62	.59	.56	.54	.53	
				ADP	61.6	60	58	56	54	51	48	44	
	20	49.9	21.6	ESHF	1.00	.98	96.	.94	.93				
				ADP	27.6	26	24	22	21				
	25	51.5	27.0	ESHF	1.00	.97	.94	.92	.90	.88			
				ADP	33.7	31	29	27	25	22			
	00	50.0	00.0	ECIIF	1.00	.08	.04	.01	.88	.86	.84	.82	
				ADP	37.1	36	34	32	30	27	25	20	
	35	54.4	38.0	ESHF	1.00	.97	.93	.89	.86	.84	.82	.80	.78
				ADP	41.1	40	38	36	34	32	30	27	22
	40	55.8	43.5	ESHF	1.00	.95	.90	.86	.83	.80	.78	.76	.74
				ADP	44.5	43	41	39	37	35	32	29	22
	45	57.1	49.1	ESHF	1.00	.93	.87	.82	.79	.77	.75	.73	.71
				ADP	47.7	46	44	42	40	38	36	33	27
	50	58.5	54.8	ESHF	1.00	.92	.84	.80	.76	.74	.71	.69	.67
				ADP	50.5	49	47	45	43	41	38	35	25
70	55	59.7	60.1	ESHF	1.00	.93	.83	.77	.73	.71	.68	.66	.64
				ADP	53.1	52	50	48	46	44	42	38	32
	60	60.9	65.5	ESHF	1.00	.89	.79	.73	.69	.66	.64	.62	.61
				ADP	55.4	54	52	50	48	46	43	40	36
	65	62.2	71.1	ESHF	1.00	.93	.78	.71	.66	.63	.61	.59	.58
				ADP	57.7	57	55	53	51	49	47	44	40
	70	63.4	76.9	ESHF	1.00	.90	.74	.66	.61	.59	.57	.56	.55
				ADP	59.8	59	57	55	53	51	49	47	45
	75	64.5	82.5	ESHF	1.00	.88	.70	.62	.57	.55	.53	.52	.51
				ADP	61.7	61	59	57	55	53	51	49	44
	80	65.7	88.0	ESHF	1.00	.87	.73	.65	.60	.54	.51	.49	.48

(*Contd.*)

Table 5.15 *(Contd.)*

Room conditions				Effective sensible heat factor and aparatus dew point									
DB (F)	RH (%)	WB (F)	W (gr/lb)										
				ADP	63.5	63	62	61	60	58	56	53	49
	85	66.8	93.7	ESHF	1.00	.71	.56	.52	.50	.48	.47	.46	.45
				ADP	65.3	64	62	61	60	59	58	57	54
	90	67.9	99.3	ESHF	1.00	.66	.56	.50	.47	.45	.43	.42	.41
				ADP	66.9	66	65	64	63	62	61	60	56
	95	69.0	105.0	ESHF	1.00	.60	.47	.42	.39	.38	.37		
				ADP	68.5	68	67	66	65	64	62		
	60	56.6	55.0	ESHF	1.00	.95	.84	.77	.73	.70	.68	.66	.65
				ADP	50.6	50	48	46	46	42	39	36	34
	65	57.7	59.7	ESHF	1.00	.92	.85	.80	.80	.69	.66	.64	.62
				ADP	52.9	52	51	50	50	46	44	41	37
	70	58.9	64.5	ESHF	1.00	.89	.80	.76	.76	.65	.62	.60	.58
				ADP	55.0	54	53	52	52	48	46	43	37
	75	59.9	69.2	ESHF	1.00	.88	.78	.72	.72	.61	.58	.56	.55
				ADP	56.9	56	55	54	54	50	48	45	41
65	80	51.0	73.8	ESHF	1.00	.75	.68	.63	.63	.58	.55	.53	.52
				ADP	58.7	57	56	55	55	53	51	48	46
	85	62.0	78.6	ESHF	1.00	.71	.63	.58	.58	.52	.50	.49	
				ADP	60.3	59	58	57	57	54	52	50	
	90	63.0	83.2	ESHF	1.00	.70	.58	.53	.53	.48	.46	.45	
				ADP	61.9	61	60	59	59	57	55	53	
	95	64.0	88.0	ESHF	1.00	.69	.51	.46	.46	.42	.41		
				ADP	63.5	63	62	61	61	59	58		
	60	52.3	46.2	ESHF	1.00	.94	.89	.81	.77	.74	.72	.70	.68
				ADP	46.0	45	44	42	40	38	36	34	28
	65	53.3	50.0	ESHF	1.00	.91	.86	.78	.74	.70	.69	.67	.65
				ADP	48.1	47	46	44	42	40	39	36	31
	70	54.3	53.9	ESHF	1.00	.89	.83	.74	.70	.67	.65	.63	.62
				ADP	50.1	49	48	46	44	42	40	37	34
60	75	55.3	57.8	ESHF	1.00	.79	.74	.71	.68	.64	.62	.60	.59
				ADP	52.0	50	49	48	47	45	43	40	37
	80	56.3	61.7	ESHF	1.00	.85	.76	.70	.66	.61	.59	.57	.56
				ADP	53.8	53	52	51	50	48	46	44	41
	85	57.2	65.5	ESHF	1.00	.75	.67	.63	.57	.56	.54	.53	
				ADP	55.4	54	53	52	50	49	47	45	
	90	58.2	69.4	ESHF	1.00	.72	.62	.57	.54	.52	.50	.49	
				ADP	57.0	56	55	54	53	52	50	47	
	95	59.1	73.5	ESHF	1.00	.69	.55	.49	.47	.46	.45		
				ADP	58.5	58	57	56	55	54	52		
	60	47.9	38.4	ESHF	1.00	.93	.89	.85	.80	.77	.75	.73	.71
				ADP	41.3	40	39	38	36	34	32	29	23
	65	48.8	41.4	ESHF	1.00	.91	.86	.83	.78	.74	.72	.70	.68
				ADP	43.3	42	41	40	38	36	34	31	24
	70	49.7	44.6	ESHF	1.00	.90	.84	.80	.74	.71	.69	.67	.66
				ADP	45.2	44	43	42	40	38	36	33	31

(Contd.)

Table 5.15 (*Contd.*)

Room conditions					Effective sensible heat factor and aparatus dew point								
DB (F)	RH (%)	WB (F)	W (gr/lb)										
	75	50.6	48.0	ESHF	1.00	.89	.82	.74	.69	.66	.65	.64	.63
				ADP	47.1	46	45	43	41	39	37	36	34
55	80	51.5	51.2	ESHF	1.00	.88	.79	.74	.67	.64	.62	.61	.60
				ADP	48.8	48	47	46	44	42	40	39	37
	85	52.4	54.5	ESHF	1.00	.77	.70	.66	.63	.60	.58	.57	
				ADP	50.4	49	48	47	46	44	42	40	
	90	53.2	57.7	ESHF	1.00	.76	.67	.61	.58	.55	.54	.53	
				ADP	52.0	51	50	49	48	46	44	41	
	95	54.2	61.2	ESHF	1.00	.69	.58	.54	.51	.49			
				ADP	53.6	53	52	51	50	48			

* Reprinted with permission of M/S Carrier Corporation from *Handbook of Air-conditioning System Design*.

Thus it can be seen that with decrease in the value of selected ADP, the dehumidified CFM also decreases. However, ADP cannot be decreased beyond a value as it will cause lowering the temperature of air to be supplied to room and thus requiring costly heating equipments and power consumption.

The heat load in tons of refrigeration is:

$$TR = \frac{GTH}{12000}$$

$$= \frac{95960}{12000}$$

$$= 7.99 \text{ TR}$$

Having calculated the TR of one hall, the TR requirement of other rooms should also be calculated working on the same principles, the tabulated summary of all the rooms is given in Table 5.16.

Table 5.16 Summary of TR and dehumidified CFM

	Summer		Monsoon		Adopted	
	TR	CFM	TR	CFM	TR	CFM
Hall	8.00	4296	8.05	3632	8.05	4296
Office	1.80	1046	1.63	855	1.80	1046
Public waiting	4.84	2700	4.65	2235	4.84	2700
Cafeteria	1.83	908	1.94	761	1.94	908
Mixed use	3.72	2084	3.49	1687	3.72	2084
Gallery	1.63	854	1.77	759	1.77	854
Total	21.82	11888	21.53	6929	22.12	11888

However, it is to be remembered that in monsoon or rainy season, air is highly humidified (upto 75% RH). The latent heat load is very high which adds to total heat and increase the TR calculation. In summer, it is sensible head which is high and thus more CFM is required to take this load.

The values for the example have been calculated for monsoon also and summarised in the table.

Heat load in Monsoon

During raining season, the relative humidity is very high. The high outdoor temperature, though comparatively lower than summer, combined with high RH is a cause of discomfort. Both RH and temperature of outdoor air must be lowered.

The conventional method was to pass the outdoor air first through heating coils, thus reducing RH and increasing the DB temperature. This puts additional load on the air conditioning equipment as not only the outdoor air heat is to be removed but additional energy consumption is taking place in heating coils.

With the advancement in HVAC field, the heating of air is achieved by heat recovery wheel (HRW) wherein the return air from the conditioned space is passed through HRW which in turn gets cooled and in the other compartment of heat exchanger, heats the outdoor air which has been cooled by cooling coil. The deficiency in heat is supplied by heating coil already illustrated in Fig. 4.18.

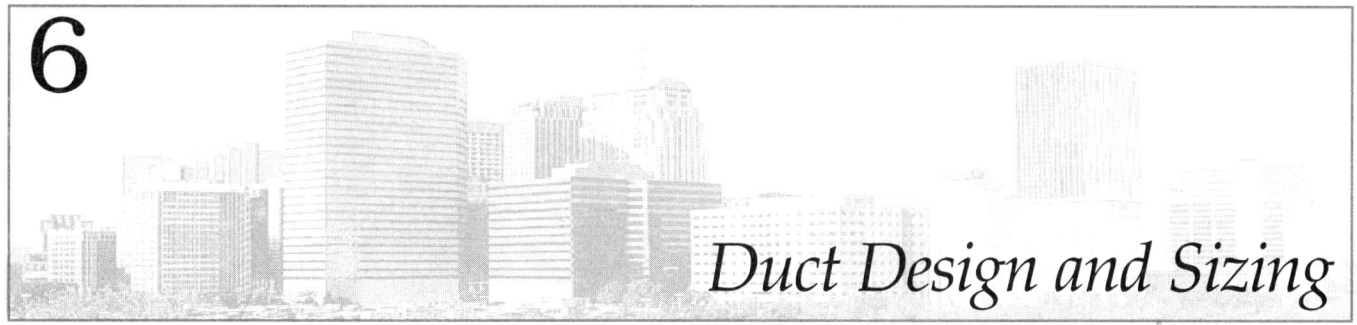

6

Duct Design and Sizing

The calculation of size of duct depends on the velocity and pressure of the air to be maintained inside the duct. The velocity and pressure conditions are decided by the comfort conditions required inside the air-conditioned area. However, by changing the various combinations of temperature, velocity, and pressure, we can obtain similar conditions. These variations are helpful in those situations, e.g. when due to space restrictions, the duct size has to be reduced. The increased velocity under such circumstances helps in maintaining the desired conditions. There may be other situations, when these parameters may have to be changed. Hence, a complete understanding of the parameters governing duct design is very important.

Velocity

Based on the requirements of air-conditioning design, there are basically two types of systems.

a. Low velocity, i.e. between 1200 to 2500 FPM.

b. High velocity, i.e. above 2500 FPM.

The return air systems are to be designed as low velocity systems in which case lower values of velocity shall be adopted.

How to select air velocity? This is a matter of experience and comfort level and type of application. The recommended supply and return velocities commonly used for various applications are listed in Table 6.1.

Table 6.1 Recommended maximum duct velocities for low velocity systems (FPM)

Application	Controlling factor-noise generation main ducts	Controlling factor-duct friction			
		Main ducts		Branch ducts	
		Supply	Return	Supply	Return
Residence	600	1000	800	600	600
Apartments Hotel bedrooms Hospital bedrooms	1000	1500	1300	1200	1000
Private offices Directors rooms Libraries	1200	2000	1500	1600	1200
Theatres Auditoriums	800	1300	1100	1000	800
General offices High class restaurants High class stores/banks	1500	2000	1500	1600	1200
Average stores /cafeterias	1800	2000	1500	1600	1200
Industrial	2500	3000	1800	2200	1500

Higher duct velocity result in lower size of ducts, but higher operating cost as motor and fan size increase and consume more electricity. Vibration and sound level may require acoustic treatment, and hence higher first initial cost. Lower duct velocity will increase the size of duct but operating cost, noise, vibration will reduce. Thus, designer has to exercise a balance between the two factors for the most efficient design. The acceptable range of velocities at various points, equipments are listed in Table 6.2 to 6.4.

Pressure

Based on the pressure to be maintained, the system can be divided into three categories.

1. Low pressure up to 3.75" Wg are used with class I fans.
2. Medium pressure from 3.75" to 6.75" Wg are used with class II fans.
3. High pressure from 6.75" to 12.75" Wg are used with class III fans.

The pressure range includes all losses through AHU, duct work and air terminal.

6.1 ECONOMICAL FACTORS IN DUCT DESIGN

It is very essential to understand the economical factors which influence the determination of duct size. The duct has to pass through corridors and above the false

Table 6.2 Occupied zone room air velocities (3 feet above floor level)

Room air velocity (FPM)	Reaction	Recommended application
0–16	Complaints about stagnant air	None
25	Ideal design–favourable	All commercial applications
25–50	Probably favourable but 50 FPM is approaching maximum tolerable velocity for seated persons	All commercial applications
65	Unfavourable–light papers are blown off a desk	
75	Upper limit for people moving about slowly–favourable	Retail and dept. store
75–300	Some factory air-conditioning installations –favourable	Factory air-conditioning higher velocities for spot cooling

Table 6.3 Typical design velocities for HVAC components

Duct element	Face velocity, FPM	Duct element	Face velocity, FPM
a. Louvers		– Pleated media (intermediate efficiency)	Up to 750
– Intake 7000 CFM and greater	400	– Hepa	250
– Exhaust 5000 CFM and greater	500	ii. Renewable media filters	
b. Filters		– Moving-curtain viscous impingement	500
i. Panel filters		– Moving-curtain dry-media	200
– Viscous impingement	400	– Electronic air cleaners Ionizing type	150 to 350
– Dry-type, extended-surface Flat (low efficiency)	Same as Duct velocity		

Table 6.4 Recommended outlet velocities

Application	Terminal velocity (FPM)	Application	Terminal velocity (FPM)
Broadcast studios	300–500	Private offices, acoustically treated	500–750
Residence	500–750	Private offices, not treated	500–800
Apartments	500–750	Motion picture theaters	1000
Churches	500–750	General offices	500–750
Hotel bedrooms	500–750	Dept. stores, upper floors	1500
Legitimate theaters	500–750	Dept. stores, main floor	2000

ceiling. An abnormal depth of duct will require more room above the false ceiling. This will eventually require higher floor height which is an economical burden on total cost.

There are various factors, which can help in reducing the first installation cost of the duct system vis-à-vis operating cost and are given below for the guidance of design engineer.

a. Higher the aspect ratio, higher the surface area resulting in higher heat loss/gain. For the same surface area, higher aspect ratio means more weight, and thus more cost of duct and insulation material. Generally aspect ratio shall not be more than 4 and aspect ratio of 1:1 is the most ideal.

$$\text{Aspect ratio} = \frac{\text{Long side}}{\text{Short side}}$$

The round duct is the best as it requires least material and gives least economical burden. However, round duct uses more space (depth below ceiling) compared to rectangular duct for the same air handling capacity.

b. While calculating the duct size, one must determine the round equivalent duct size for a given friction rate and CFM handling capacity. For a given round duct size, there could be many possibilities of rectangular ducts with various aspects ratios. The equivalent rectangular duct section should be selected from Table 6.5, so that dimensions with least friction are used, otherwise operating cost will be more and effectiveness in terms of velocity and pressure will be less.

c. Duct size reduction or expansion is needed wherever there are tap-offs for the area to be air-conditioned, or air obstruction like beam is encountered along the run of the duct. Such changes in dimensions (if cross-section area remains same) should be so gradual that a minimum slope of 1:7 is provided. However, a maximum slope of 1:4 shall not be exceeded.

d. To avoid condensation on the ducts, it is necessary that the room dew point temperature corresponding to the return air temperature shall be lower than the duct surface temperature. Table 6.8 lists the maximum difference in supply air temperature and room dew point temperature to avoid condensation on ducts.

6.2 DUCT DESIGN

Any successful duct design requires that minimum loss of heat, minimum frictional resistance to air flow, i.e. minimum pressure drop in the system should take place. The air should be supplied to the area to be air-conditioned at right temperature and pressure. Also, the duct size should be economical and occupy less space.

To design the duct system, it is important to understand the following concepts of static pressure and velocity pressure.

The total pressure of a flowing fluid is defined as.

$H_t = H_s + H_v$
H_t = Total pressure.
H_s = Static pressure.
H_v = Velocity pressure.

The static pressure is the pressure the fluid has at rest. The velocity pressure is defined by kinetic energy equation.

$$H_v = \frac{v^2}{2g}$$

v = velocity in FPM

$$= \frac{v}{60} \text{ FPS}$$

g = gravitational constant = 32.2 ft/sec^2

Also 1 inch of water = 69 ft of air

Substituting values in kinetic energy equation, we get

$$H_v = \left(\frac{v}{60}\right)^2 \times \frac{1}{2 \times 32.2 \times 69.6} \text{ inch water}$$

$$= \left(\frac{v}{4000}\right)^2 \text{ inch water}$$

The velocity pressure concept is helpful in measuring velocities and flow rates in piping and ducts. If the velocity pressure can be measured, the velocity can be found by solving the above equation and vice versa. The pressure can be measured using manometer as shown in Fig. 6.1 or by using a pitot tube.

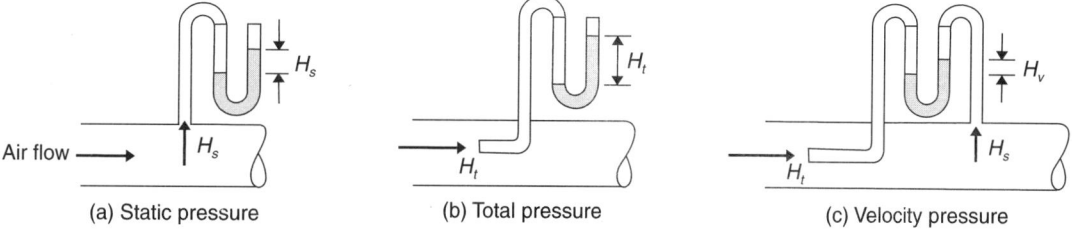

(a) Static pressure	(b) Total pressure	(c) Velocity pressure

Fig. 6.1 Measurement of pressure using manometer

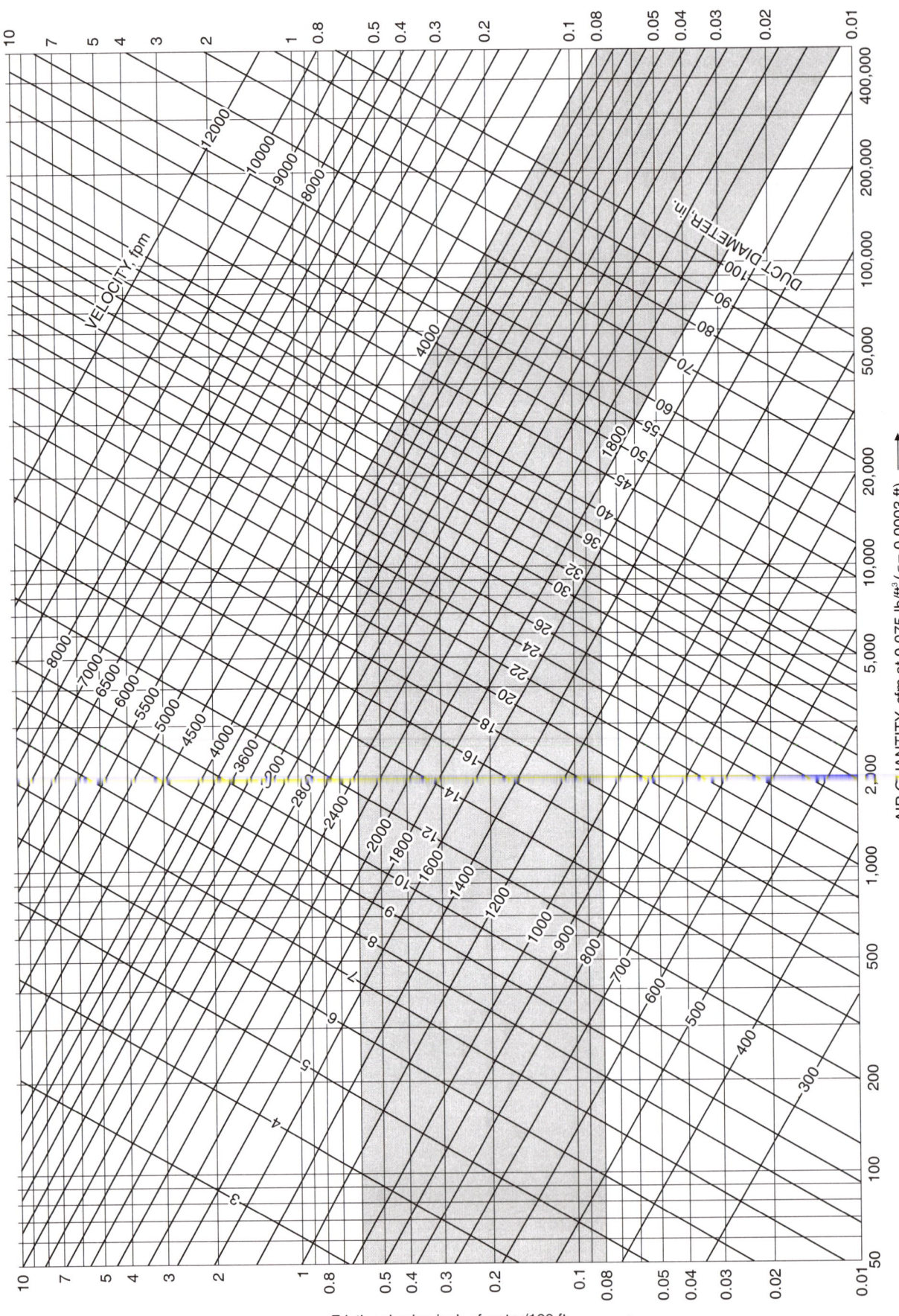

AIR QUANTITY, cfm at 0.075 lb/ft³ (ε= 0.0003 ft) ⟶

Frictional value inch of water/100 ft ⟶

Fig. 6.2 Duct design-friction chart

The static pressure can be increased in the direction of flow by reducing the velocity as the total pressure remains constant, i.e. velocity energy is converted to static and vice versa.

To design a duct system, the total pressure loss likely to be encountered in the air flow due to friction of duct material or bends, fittings and equipments shall be calculated. There are two systems of calculation of pressure loss.

6.2.1 Friction Loss or Pressure Loss Method

The loss in pressure due to friction in the duct not only affects the supply pressure but also the power of the fan drive in the AHU room. The loss due to friction depends on the following factors.

a. Air velocity.
b. Duct size and aspect ratio.
c. Roughness of surface.
d. Length of duct.

A friction chart at Fig. 6.2 shows graphically the relationship among various factors as listed above. Using this friction chart, one can find the friction loss for given parameters or if friction loss is fixed then one can find the velocity or equivalent diameter of duct. Figure 6.3 shows the graph for converting round duct to equivalent rectangular duct.

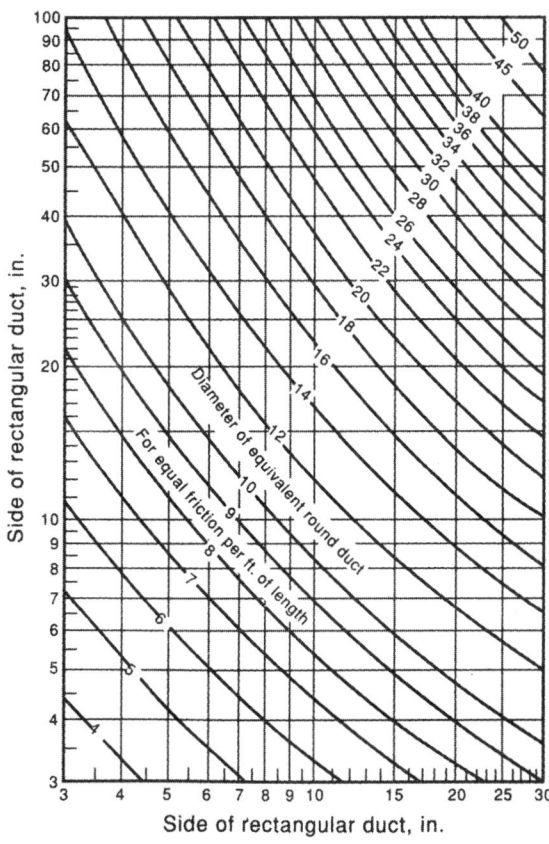

Fig. 6.3 Equivalent round duct size

Table 6.5 gives the conversion of round duct to equivalent rectangular ducts for equal friction. Thus, step-by-step procedure to design the duct system is as follow.

a. Calculate the CFM requirement of the portion for which duct design is required.

b. Calculate equivalent area of duct which is

$$= \frac{\text{Volume (CFM)}}{\text{Velocity (FPM)}}$$

c. By using friction chart, diameter of round duct is selected and then

$$\text{Area of round duct} = \frac{\pi (\text{dia.})^2}{4}$$

d. After the equivalent area is calculated, Table 6.5 can be used to calculate the dimensions of duct, so that the friction of rectangular duct is minimum and equal to, what would have been with round duct. Remember, arbitrarily choosing duct dimensions for same equivalent area may result in higher frictional loss.

e. This procedure shall be followed with all sections of the duct as the CFM goes on changing.

6.2.2 Equal Friction Method

In this method, the same value of friction loss per unit length of duct is used to size each section of duct in the system. A higher friction loss results in smaller ducts as velocity also increases but results in higher fan operating costs. Hence, a balance must be struck between the two. Return air ducts can be sized by this method in the manner similar to supply air duct. Return air ducts are usually designed as a low velocity system even if supply ducts are of high velocity type.

The equal friction method of designing ducts is quite simple and probably the most popular one. For systems that do not have long distances between the outlets at the beginning and end of the system, this method works quite satisfactorily. However, if the case is reverse then difficulties in balancing the flow rates and noise is encountered as the outlets nearer to fan are over pressurised.

Example

Find the size of each duct section for the system shown in Fig. 6.4 using the equal friction design method.

Use rectangular ducts. The system serves a public building.

Solution

1. Sum up the CFM's backward from last outlet, to find the CFM in each duct sections. The results are shown in Table 6.6.

Table 6.5 Circular equivalents of rectangular ducts for equal friction and capacity (Dimensions in inches)

Side rectangular duct	4.0	4.5	5.0	5.5	6.0	6.5	7.0	7.5	8.0	9.0	10.0	11.0	12.0	13.0	14.0	15.0	16.0
3.0	3.8	4.0	4.2	4.4	4.6	4.7	4.9	5.1	5.2	5.5	5.7	6.0	6.2	6.4	6.6	6.8	7.0
4.0	4.4	4.6	4.9	5.1	5.3	5.5	5.7	5.9	6.1	6.4	6.7	7.0	7.3	7.6	7.8	8.1	8.3
4.5	4.6	4.9	5.2	5.4	5.7	5.9	6.1	6.3	6.5	6.9	7.2	7.5	7.8	8.1	8.4	8.6	8.8
5.0	4.9	5.2	5.5	5.7	6.0	6.2	6.4	6.7	6.9	7.3	7.6	8.0	8.3	8.6	8.9	9.1	9.4
5.5	5.1	5.4	5.7	6.0	6.3	6.5	6.8	7.0	7.2	7.6	8.0	8.4	8.7	9.0	9.3	9.6	9.9

Side rectangular duct	6	7	8	9	10	11	12	13	14	15	16	17	18	19	20	22	24	26	28	30
6	6.6																			
7	7.1	7.7																		
8	7.6	8.2	8.7																	
9	8.0	8.7	9.3	9.8																
10	8.4	9.1	9.8	10.4	10.9															
11	8.8	9.5	10.2	10.9	11.5	12.0														
12	9.1	9.9	10.7	11.3	12.0	12.6	13.1													
13	9.5	10.3	11.1	11.8	12.4	13.1	13.7	14.2												
14	9.8	10.7	11.5	12.2	12.9	13.5	14.2	14.7	15.3											
15	10.1	11.0	11.8	12.6	13.3	14.0	14.6	15.3	15.8	16.4										
16	10.4	11.3	12.2	13.0	13.7	14.4	15.1	15.7	16.4	16.9	17.5									
17	10.7	11.6	12.5	13.4	14.1	14.9	15.6	16.2	16.8	17.4	18.0	18.6								
18	11.0	11.9	12.9	13.7	14.5	15.3	16.0	16.7	17.3	17.9	18.5	19.1	19.7							
19	11.2	12.2	13.2	14.1	14.9	15.7	16.4	17.1	17.8	18.4	19.0	19.6	20.2	20.8						
20	11.5	12.5	13.5	14.4	15.2	16.0	16.8	17.5	18.2	18.9	19.5	20.1	20.7	21.3	21.9					
22	12.0	13.0	14.1	15.0	15.9	16.8	17.6	18.3	19.1	19.8	20.4	21.1	21.7	22.3	22.9	24.0				
24	12.4	13.5	14.6	15.6	16.5	17.4	18.3	19.1	19.9	20.6	21.3	22.0	22.7	23.3	23.9	25.1	26.2			
26	12.8	14.0	15.1	16.2	17.1	18.1	19.0	19.8	20.6	21.4	22.1	22.9	23.5	24.2	24.9	26.1	27.3	28.4		
28	13.2	14.5	15.6	16.7	17.7	18.7	19.6	20.5	21.3	22.1	22.9	23.7	24.4	25.1	25.8	27.1	28.3	29.5	30.6	
30	13.6	14.9	16.1	17.2	18.3	19.3	20.2	21.1	22.0	22.9	23.7	24.4	25.2	25.9	26.6	28.0	29.3	30.5	31.7	32.8
32	14.0	15.3	16.5	17.7	18.8	19.8	20.8	21.8	22.7	23.5	24.4	25.2	26.0	26.7	27.5	28.9	30.2	31.5	32.7	33.9
34	14.4	15.7	17.0	18.2	19.3	20.4	21.4	22.4	23.3	24.2	25.1	25.9	26.7	27.5	28.3	29.7	31.0	32.4	33.7	34.9
36	14.7	16.1	17.4	18.6	19.8	20.9	21.9	22.9	23.9	24.8	25.7	26.6	27.4	28.2	29.0	30.5	32.0	33.3	34.6	35.9
38	15.0	16.5	17.8	19.0	20.2	21.4	22.4	23.5	24.5	25.4	26.4	27.2	28.1	28.9	29.8	31.3	32.8	34.2	35.6	36.8
40	15.3	16.8	18.2	19.5	20.7	21.8	22.9	24.0	25.0	26.0	27.0	27.9	28.8	29.6	30.5	32.1	33.6	35.1	36.4	37.8
42	15.6	17.1	18.5	19.9	21.1	22.3	23.4	24.5	25.6	26.6	27.6	28.5	29.4	30.3	31.2	32.8	34.4	35.9	37.3	38.7
44	15.9	17.5	18.9	20.3	21.5	22.7	23.9	25.0	26.1	27.1	28.1	29.1	30.0	30.9	31.8	33.5	35.1	36.7	38.1	39.2
46	16.2	17.8	19.3	20.6	21.9	23.2	24.4	25.5	26.6	27.7	28.7	29.7	30.6	31.6	32.5	34.2	35.9	37.4	38.9	40.4
48	16.5	18.1	19.6	21.0	22.3	23.6	24.8	26.0	27.1	28.2	29.2	30.2	31.2	32.2	33.1	34.9	36.6	38.2	39.7	41.2
50	16.8	18.4	19.9	21.4	22.7	24.0	25.2	26.4	27.6	28.7	29.8	30.8	31.8	32.8	33.7	35.5	37.2	38.9	40.5	42.0
52	17.1	18.7	20.2	21.7	23.1	24.4	25.7	26.9	28.0	29.2	30.3	31.3	32.3	33.3	34.3	36.2	37.9	39.6	41.2	42.8
54	17.3	19.0	20.6	22.0	23.5	24.8	26.1	27.3	28.5	29.7	30.8	31.8	32.9	33.9	34.9	36.8	38.6	40.3	41.9	43.5
56	17.6	19.3	20.9	22.4	23.8	25.2	26.5	27.7	28.9	30.1	31.2	32.3	33.4	34.4	35.4	37.4	39.2	41.0	42.7	44.3
58	17.8	19.5	21.2	22.7	24.2	25.5	26.9	28.2	29.4	30.6	31.7	32.8	33.9	35.0	36.0	38.0	39.8	41.6	43.3	45.0
60	18.1	19.8	21.5	23.0	24.5	25.9	27.3	28.6	29.8	31.0	32.2	33.3	34.4	35.5	36.5	38.5	40.4	42.3	44.0	45.7
62		20.1	21.7	23.3	24.8	26.3	27.6	28.9	30.2	31.5	32.6	33.8	34.9	36.0	37.1	39.1	41.0	42.9	44.7	46.4
64		20.3	22.0	23.6	25.1	26.6	28.0	29.3	30.6	31.9	33.1	34.3	35.4	36.5	37.6	39.6	41.6	43.5	45.3	47.1
66		20.6	22.3	23.9	25.5	26.9	28.4	29.7	31.0	32.3	33.5	34.7	35.9	37.0	38.1	40.2	42.2	44.1	46.0	47.7
68		20.8	22.6	24.2	25.8	27.3	28.7	30.1	31.4	32.7	33.9	35.2	36.3	37.5	38.6	40.7	42.8	44.7	46.6	48.4
70		21.1	22.6	24.5	26.1	27.6	29.1	30.4	31.8	33.1	34.4	35.6	36.8	37.9	39.1	41.2	43.3	45.3	47.2	49.0
72			23.1	24.8	26.4	27.9	29.4	30.8	32.2	33.5	34.8	36.0	37.2	38.4	39.5	41.7	43.8	45.8	47.8	49.6
74			23.3	25.1	26.7	28.2	29.7	31.2	32.5	33.9	35.2	36.4	37.7	38.8	40.0	42.2	44.4	46.4	48.4	50.3
76			23.6	25.3	27.0	28.5	30.0	31.5	32.9	34.3	35.6	36.8	38.1	39.3	40.5	42.7	44.9	47.0	48.9	50.9
78			23.8	25.6	27.3	28.8	30.4	31.8	33.3	34.6	36.0	37.2	38.5	39.7	40.9	43.2	45.4	47.5	49.5	51.4
80			24.1	25.8	27.5	29.1	30.7	32.2	33.6	35.0	36.3	37.6	38.9	40.2	41.4	43.7	45.9	48.0	50.1	52.0
82				26.1	27.8	29.4	31.0	32.5	34.0	35.4	36.7	38.0	39.3	40.6	41.8	44.1	46.4	48.5	50.6	52.6
84				26.4	28.1	29.7	31.3	32.8	34.3	35.7	37.1	38.4	39.7	41.0	42.2	44.6	46.9	49.0	51.1	53.2
86				26.6	28.3	30.0	31.6	33.1	34.6	36.1	37.4	38.8	40.1	41.4	42.6	45.0	47.3	49.6	51.7	53.7
88				26.9	28.6	30.3	31.9	33.4	34.9	36.4	37.8	39.2	40.5	41.8	43.1	45.5	47.8	50.0	52.2	54.3
90				27.1	28.9	30.6	32.2	33.8	35.3	36.7	38.2	39.5	40.9	42.2	43.5	45.9	48.3	50.5	52.7	54.8

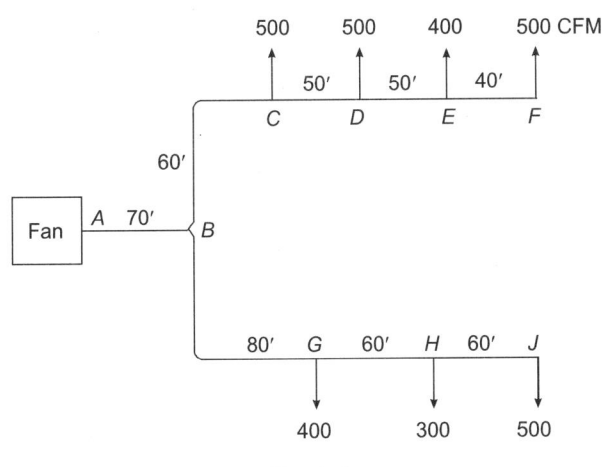

Fig. 6.4

2. Select a design velocity for the fan, using Table 6.1. A velocity of 1400 FPM may be chosen, which should be reasonably quite for the application.
3. Assuming friction loss rate of 0.13 inch water/ 100 ft, read off equivalent round duct diameter in Fig. 6.2 corresponding to CFM for each section. The value is 20.5 inch diameter for 3100 CFM. The friction loss rate of 0.13 inch water gives economical design of duct and is selected for G.I. duct material.
4. The rectangular duct sizes are read from Fig. 6.3 and there can be various combinations as can be seen in the figure. These duct sizes can also be adopted from Table 6.5. In the actual installation, the duct proportion chosen would depend on space available.
5. The pressure loss in the system can be calculated as shown in Table 6.6.

6.2.3 Static Regain Method
This method of duct sizing is often used for high velocity system with long duct runs, especially in large installations.

In this method, an initial velocity in the main duct leaving the fan is selected, in the range of 2500–4000 FPM.

After the initial velocity is chosen, the velocities in each successive section of duct in the main run are reduced, so that the resulting static pressure gain is enough to overcome the frictional losses in the next duct section. The result is that the static pressure is the same at each junction in the main run. There will not be extreme differences in the pressure among the branch outlets, so balancing is simplified.

Example
Determine the duct sizes for the system shown in Fig. 6.5, using the static regain method. Round duct shall be used.

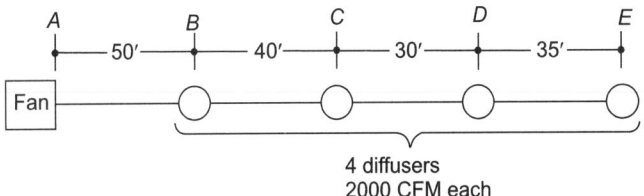

Fig. 6.5

Solution
The results are summarised in Table 6.7.

1. The velocity in the initial section is selected. This being a high velocity system, an initial velocity of 3200 ft/min will be chosen as noise is not the limiting factor here.
2. From Fig. 6.2, the duct size and static pressure loss due to friction in section *AB* is determined. The friction loss per 100 ft is 0.56 inch water (intersection of 8000 CFM and 3200 FPM lines) and therefore the friction loss in this section is 0.56 × 50/100 = 0.28 inch water.
3. The velocity must be reduced in section *BC* so that static pressure gain will be equal to friction loss in *BC*. There will not be complete regain due to

Table 6.6 Equal friction method

Section	CFM	V, ft/min	Friction loss, in. wg. per 100 ft	Length (ft)	Total friction in. wg.	Eq. D,	rect. duct size, in
AB	3100	1240	0.13	70	0.091	20.5	24 × 15
BC	1900	1140	0.13	60	0.078	17	20 × 12
CD	1400	1050	0.13	50	0.065	15	16 × 12
DE	900	900	0.13	50	0.065	12.5	16 × 9
EF	500	889	0.13	40	0.052	10	9 × 9
BG	1200	1029	0.13	80	0.104	14	14 × 12
GH	800	914	0.13	60	0.078	12	14 × 9
HJ	500	889	0.13	60	0.078	10	9 × 9

Table 6.7 Static regain method

Section	CFM	V, ft/min	Eq. D, in	Velocity pressure, in. w.	Friction loss in. w./100 ft	Length ft	Friction loss in. w.	Static pressure regain, in. w.
AB	8000	3200	22	0.64	0.56	50	0.28	
B								0.16
BC	6000	2600	21	0.43	0.40	40	0.16	
C								0.09
CD	4000	2200	18	0.30	0.33	30	0.10	
D								0.09
DE	2000	1700	15	0.18	0.26	35	0.09	

dynamic losses in transition at *B* (Bend losses, etc.). We will assume a 75% regain for the fittings. A trial and error procedure is necessary to balance the regain against the friction loss. Let us try a velocity of 2400 FPM in section *BC*. The friction loss from Fig. 6.2 is 0.32 inch water (intersection of 6000 CFM and 2400 FPM lines)

$$\text{Loss in } BC = \frac{0.32 \text{ inch water} \times 40 \text{ ft}}{100 \text{ ft}}$$

$$= 0.13 \text{ inch water}$$

The static pressure regain available to overcome this loss, using equation of conversion as.

$$H_{v1} - H_{v2} = 0.75 \left[\left(\frac{V_1}{4000} \right)^2 - \left(\frac{V_2}{4000} \right)^2 \right]$$

$$= 0.75 \left[\left(\frac{3200}{4000} \right)^2 - \left(\frac{2400}{4000} \right)^2 \right]$$

$$= 0.21 \text{ inch water}$$

This is too large a regain. Try a velocity of 2600 FPM, and, we get

$$\text{Loss in } BC = \frac{0.40 \text{ inch water}}{100 \text{ ft}} \times 40 \text{ ft}$$

$$= 0.16 \text{ inch water}$$

$$\text{Regain at } B = 0.75 \left[\left(\frac{3200}{4000} \right)^2 - \left(\frac{2600}{4000} \right)^2 \right]$$

$$= 0.16 \text{ inch water}$$
$$H_{v1} - H_{v2} = 0.16 \text{ inch water}$$

This trial is satisfactory. The regain at *B* is approximately enough to overcome the loss in section *BC*, i.e. 0.13 inch water. The duct size of *BC* is 21 inch.

4. Continue the same procedure at transition *C* and with a velocity of 2200 FPM in *CD*, we get regain of 0.09 inch water at *C*. The duct size is 18 inch.

5. The same process at *D* results in a duct size of 15 inch for section *DE*.

One of the disadvantages of static regain method is that some of the duct sections are larger than those found by equal friction method.

The above explanations regarding friction loss, factors affecting aspect ratio of rectangular ducts, fan operating costs lead us to the conclusion that following factors shall be kept in consideration while designing any duct system:

1. Friction loss or pressure loss.
2. Heat gain or loss.
3. Economical consideration.

6.2.4 Concept of Equivalent Length

The theoretical method of determining the friction loss through any duct length is to multiply the friction loss in inch of wg per 100 ft of length of duct with the equivalent length of duct.

Equivalent length of duct

= Straight run of duct + Equivalent length of (elbows + fittings + reduction in diameter).

To calculate equivalent length, proceed as follow:

a. Calculate equivalent area

$$= \frac{\text{Volume of air in CFM}}{\text{Velocity of air in FPM}}$$

b. Using Table 6.5 calculate rectangular duct size.
c. Determine friction rate from chart given in Fig. 6.2 using CFM, equivalent round duct diameter.
d. Calculate all duct sizes of the network.
e. Calculate equivalent length of longest duct with elbows, fittings. This is based on assumption that all the branches from AHU will have equal friction. The elbows with different radius, depth of duct and width of duct have different equivalent lengths. Conversion tables are available which can be used. However, this conversion becomes a complex process.

Instead, a rough approximation can be done as follow.

Equivalent length of elbow/fittings = Add 10 to 20% of total length of the longest loop of duct depending upon the number of elbows, bends, fittings likely to be used.

f. Multiply the equivalent length with the friction rate so calculated to calculate equivalent frictional loss in duct network.

g. Total fan static pressure is calculated as follow.

Fan static pressure
= Total equivalent frictional loss in duct network + Loss in filters + Terminal supply pressure required in room to be air-conditioned.

6.2.5 Effect of Heat Gain/Loss on Duct Design

In any duct system, the loss or gain of heat is bound to take place. This heat transfer takes place because the duct passes through an unconditioned space, where outside temperature is very different from that of the air passing inside the duct. The heat load calculation must take into account this loss or gain, or else the conditions cannot be provided in the space to be air-conditioned.

This compensation can be provided either by reducing the temperature of supply air or by increasing the volume of air supplied. If the temperature of supply air is reduced, there is all likely hoods that condensation may take place due to metal surface temperature falling below the dew point temperature of room or return air temperature. The supply air temperature is always less

than the dew point temperature of room air (summer applications) corresponding to the dry-bulb temperature of air in the room. But, this supply air temperature cannot be allowed to fall beyond a certain limit, i.e. there is a limit. The maximum difference between dew point temperature and supply air temperature beyond which, if supply air temperature falls, condensation may occur on metal surface. This limit for various velocities is given in Table 6.8.

6.3 IMPORTANT CONSIDERATIONS IN DUCT DESIGN

However, there are few guidelines which help in designing the duct system so that minimum heat loss takes place.

1. Aspect ratio, i.e. the ratio of long side to short side of duct should be as small as possible. The lower aspect ratio has lesser surface area and the heat loss/gain lesser, usually aspect ratio of 4:1 is the most ideal.

2. Lesser velocity has more time to come in contact with metal surface, and hence more heat transfer. Lesser air volume at lesser velocity is the worst scenario and maximum condensation take place.

3. Add insulation to reduce heat transfer where it is most likely to cause condensation. Condensation will most likely take place, where return air enters the annular space above false ceiling (non-ducted return) picking up the moisture content from the

Table 6.8 Maximum difference between supply air temperature and room dew point temperature without condensing moisture on ducts

Air-conditions surrounding duct		Air velocity in stratight run of duct (FPM)*											
		Painted	Bright metal	Painted	Bright metal	Painted	Bright metal	Painted	Bright metal	Painted	Bright metal	Painted	Bright metal
DB (F)	RH (%)	400		800		1200		1600		2000		3000	
	45	20	15	15	9	11	8	8	5	7	4	5	3
	50	18	13	13	8	10	7	7	5	6	4	4	3
	55	15	11	11	7	8	6	6	4	5	3	4	2
74–100	60	13	10	10	6	7	5	5	3	4	3	3	2
	70	9	7	7	4	5	4	4	2	3	2	2	2
	80	6	4	4	3	3	2	2	2	2	1	1	1
	85	4	3	3	2	2	2	2	1	2	1	1	1
Value of (f2/U)–1		.90	.66	.66	.42	.49	.31	.37	.24	.31	.20	.23	.15

Equation: $T_{dp} - T_{sa} = (T_{rm} - T_{dp})\left(\dfrac{F_2}{U} - 1\right)$

T_{dp} = Duct surface temperature, (room dew point).
T_{sa} = Supply air DBT in duct.
F_2 = Film heat transfer coefficient on out side of duct. (BTU/hr (sq ft) (°F).

T_{rm} = Room dry bulb temperature.
U = Overall heat transfer coefficient of duct BTU/HR (sq ft) (°F).
(1.65 for painted duct, 1.05 for bright metal duct).

* Reprinted with permission of M/S Carrier Corporation from *Handbook of Air-conditioning System Design*.

room (particularly in monsoon). As the return air from the farthest end travels towards AHU, it comes in contact with supply air duct. Here, the volume is very small and velocity is also very low, the condensation may take place. The volume of air is added from rooms towards the AHU side. More volume has lesser heat transfer and condensation chances reduce as shown in Fig. 6.6.

Thus, insulation shall be provided on the supply air duct. The maximum chances of condensation are near the tail end of SA duct while minimum are nearer the AHU. Thus, as a practice insulation is provided on the SA duct leaving the first 15 m from AHU.

This can be better understood with the help of following formula.

Heat transfer = CFM × Sp. heat × (ΔT).

Thus, for the same amount of heat transfer to take place, more the volume involved in heat transfer, less the ΔT becomes and hence less scope for condensation.

6.4 INSULATING MATERIALS

To limit the heat loss, insulation shall be provided outside the duct. Similarly, to limit the transmission of sound of fan to room, the insulation shall be provided inside the SA duct, particularly near the room. Various insulating materials are available for different kind of applications.

The comparative chart of widely used materials for HVAC applications are described along with their thermal and physical properties in Table 6.9.

6.5 ECONOMICAL CONSIDERATIONS

The duct design shall be so calculated that it strikes a balance between first cost of installation as well as running and operation cost. Since the space comes at a value, hence as less space as possible shall be utilized by air-conditioning ducts. While ducts pass through corridors or rooms above false ceiling, the greater depth will demand more floor height which certainly cannot be provided. Generally for a good air-conditioning system, floor height of 3.6 m shall be demanded from the architects. However, where fan coil units are to be used and all the air is to be discharged to atmosphere, 3 m height shall suffice. However, in some applications due to restriction of floor height, return or supply air ducts can pass through the rooms instead of corridors above false ceiling.

Where nothing works due to space restrictions, the only possibility is to increase the velocity in the system to reduce the cross-sectional area. However, velocity can be increased up to the extent of keeping the vibrations noise in limit. It is also limited from the limitation of increase in fan HP and consequently increased energy consumption throughout the life of air-conditioning system.

As a result, the most economical selection of supply air velocity is between 1200–1400 FPM.

The aspect ratio of duct already described in the previous section shall be taken into consideration while selecting the duct sizes.

The most economical rectangular section, equivalent to round duct diameter for equal friction, must be

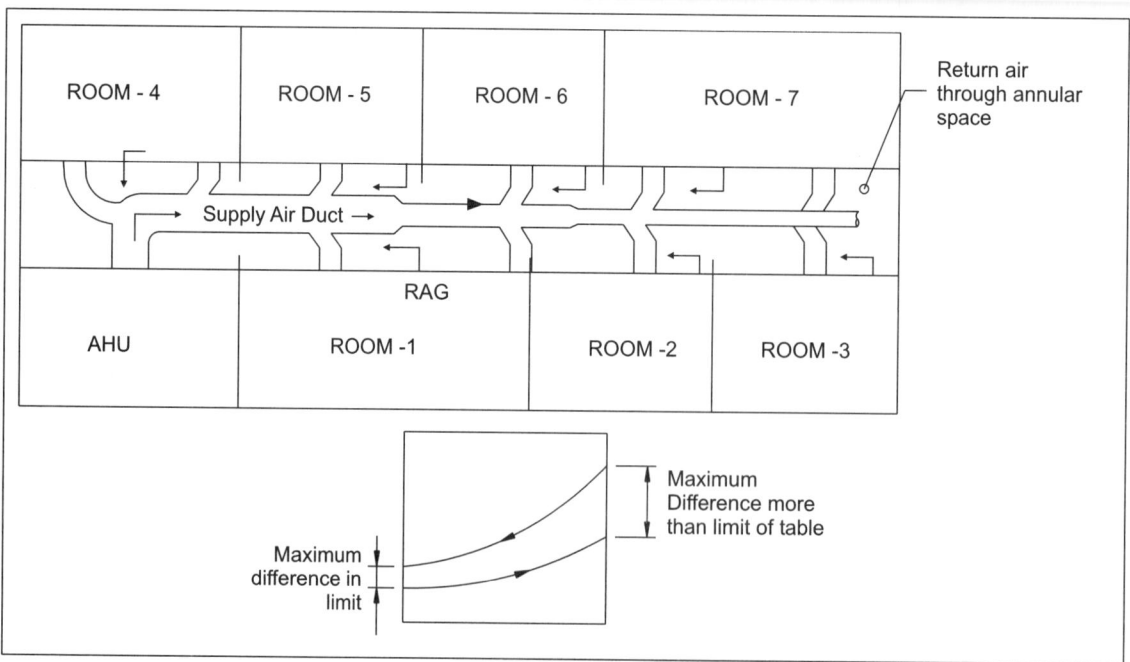

Fig. 6.6 Relation between volume of air and chances of condensation

Table 6.9 Comparison of properties of thermal insulation materials

S. no.	Characteristics	Rigid urethane foam	Expanded polystyrene	Resin bonded glass wool	Resin bonded mineral wool	Phenolic foam
1.	Physical form	Freon-filled closed cellular rigid plastic foam	Air-filled closed cellular rigid thermoplastic foam	Open-celled resilient mat. of resin bonded glass wool	Open-celled resilient mat. of resin bonded fibre drawn form rock or slag	Open celled rigid phenolic foam
2.	Density (Kg/m^3)	32	13–25	12–48	40–144	24–32
3.	Thermal conductivity (Kcal/mhr °C)	0.014 (initial) 0.019 (design)	0.028 (initial) (for 18 kg/m^3 density) 0.032 (design)	0.032 (for 32 kgs/m^3 density)	0.035 (for 48 kg/m^3 density)	0.021 (initial) 0.0266 (design)
4.	Compressive strength at 10% deformation (KN/m^2)	172	70–100	Negligible	Negligible	75–116
5.	Fire performance	Self extinguishing	Normal variety not self extinguishing as per IS 4671.	The glass fibre melt and fuse during fire and the resin burns up.	Although the mineral fibre is incombustible, the resin is affected by fire.	Self-extinguishing (non-flammable)
6.	Water vapour permeability (per-inch)	2.5	2.4	Very high	Very high	10
7.	Water absorption	Negligible	Negligible	Very high	Very high	Very high

adopted. Arbitrarity selecting the width and depth of duct for the same cross-section may not only increase the friction loss, but may also cause more heat loss or gain.

6.6 MODERN TOOLS FOR THE CALCULATION OF DUCT SIZE

Nowadays, various manufacturers have developed softwares to calculate the duct size based on velocity, max friction allowed and CFM to be handled. Manual tools like ductolator are also available. These tools allow the design engineer to select the CFM to be handled, match it with the desired velocity. Thus, he can check the friction that will be present in the system. If this friction rate is within desirable limits, he can select various round duct sizes and equivalent rectangular duct size.

Duct design-friction chart is another tool to calculate duct sizes. This chart is already given in Fig. 6.2.

Chilled Water System Design

Early designs of chilled water system were based on using separate chillers for each building. As energy became more expensive, the owners and designers became more interested in efficient use of electricity. The largest electricity consumption in any building is on account of HVAC system.

Nowadays large sized chillers up to 10,000 TR are very commonly used. The electricity consumption which used to be in the range of 1–2 kW/ton has come down to 0.5 kW/ton. This is all due to efficient designs of cooling towers, fan and pumps and due to development of variable speed drives.

It was also observed that HVAC system is generally designed to operate at the extreme conditions. But the operation of the plant in these extreme conditions is generally limited to very few days or months and even in these few extreme days, the load pattern does not remain the same. This led to a thought process to design the systems which could operate at partial capacity with the decrease in load or atmospheric conditions.

Thus, considering a simple situation in which 15 buildings in a campus could have individual chillers or could have one large chiller in the campus, the saving in energy would be as follow.

a. Individual chillers for 15 buildings, chiller of 150 TR each @ 0.65 kW/TR at full load running, 0.80 kW/TR at part load running, 100 TR actual load on each building (each machine operating at part load).
Energy consumed = 15 × 100 × 0.8 = 1200 kW.

b. Two chillers of 1000 TR each @ 0.55 kW/TR at full load and 0.75 kW/TR at part load, 100 TR actual load on each building (one machine operating at full load and another at part load).
Energy consumed = 1000 × 0.55 = 550 kW
= 500 × 0.75 = 375 kW
Total = 875 kw

The above saving, i.e. (1200 – 875 = 325 kW) is on account of chilling machine whereas, similar saving will be effected in the performance of cooling tower, fan and pumps.

It is necessary to understand the basic operation of a chilled water system as depicted in Fig. 7.1.

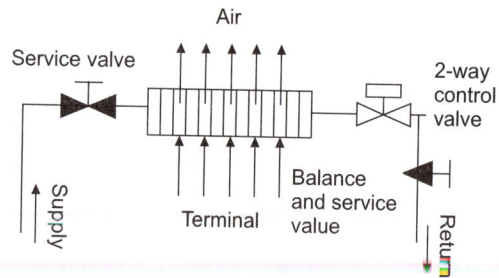

Fig. 7.1 Load system schematic

In this system, it is important to note, that to take advantage of large chilling machine's high efficiency performance, the system must operate with certain control devices, which can regulate the operation of the system at the most efficient level.

7.1 REGULATING DEVICES

1. Variable speed (frequency) drives.
2. Two way/three way regulating valves.
3. Variation of chilled water system schemes.

7.1.1 Variable Speed (Frequency) Drives

These drives are nowadays used in HVAC applications to provide the most energy efficient means of controlling the flow of air and water with changed load patterns. As already described, the HVAC systems are designed for the peak load conditions, the peak existing only for a fractional part of the year. For majority of time, they have excess capacity. Hence to save this energy, variable speed (frequency) drives are used to match the capacity of the systems to the actual requirement of the buildings.

To simplify the concept of saving of energy, let's understand the effect of variation of load on various components of air-conditioning system.

Let's begin our journey from the air-conditioned space. Any variation of air-conditioning load, say, due to reduced occupancy, outside climatic change or reduced outside temperature in the night hours gets reflected in the AHU. The AHU can either.

1. Reduce the volume of air supplied in the air-conditioned space maintaining the same temperature.
2. Reduce the temperature of supplied air by maintaining a constant volume of air supplied to maintain the desired conditions inside the room.
3. Both.

The temperature of supplied air in 2 above can be reduced by closing the valves of chilled water pipe line or bypassing the surplus chilled water. But, the chilled water machine continues to generate the same quantity of water which is getting bypassed. The chilled water pumps are pumping the same quantity of water, which is being bypassed due to reduced load. This causes a sheer wastage of energy.

To reduce the energy consumption, the VFD can be used so that the pumps can run at a lower speed and pump less quantity of water. This will result in reduced quantity of condenser water requirement thereby saving energy consumed in condenser pump. The cooling tower will require less quantity of natural air, and thus speed of CT fan can be reduced to add to energy saving. This can further be optimized by reducing the speed of AHU fan.

7.1.1.1 How VFD Helps in Saving the Energy?

The components of HVAC system include condenser water pump, primary and secondary chilled water pumps which are centrifugal types. The types of fans used in AHU, FCU and cooling tower include forward or backward curved centrifugal fans, plug fans, vane axial fans or variable pitch vane axial fans.

For all these pumps and fans type, without taking system effects into consideration (e.g. pressure drop along pipe work or duct work), the following basic equation apply.

$$\text{Power absorbed} \propto \frac{\text{Flow} \times \text{Pressure}}{\text{Efficiency}} \quad \dots \text{(i)}$$

Affinity Laws

Flow	$\propto$	Speed
Static pressure	$\propto$	Speed^2
Input power	$\propto$	Speed^3

As can be seen from above equations, controlling flow minimizes the power absorbed by the pumps and fans. However, different methods of controlling flow results in different reduction of absorbed power at reduced flow rates. For fans, air flow can be controlled using outlet dampers, inlet guide vanes, variable pitch fans, eddy current couplings and VFD. For pumps, water flow can be controlled using bypass (3 way) valves, throttling (2 way) valves and VFD's.

7.1.1.2 Comparison of Power Absorbed

Figure 7.2 shows that for the same air flow rate (80% of operating/design point) the different flow control methods result in following absorbed motor powers (as a percentage of the power absorbed at the operating/design point).

Damper control	:	absorbs 93 %
Inlet guide vanes	:	absorbs 70 %
Eddy current coupling	:	absorbs 67 %
VFD	:	absorbs 51 %

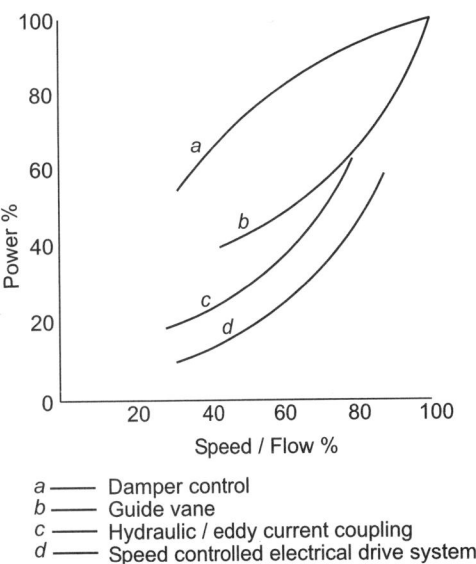

a ——— Damper control
b ——— Guide vane
c ——— Hydraulic / eddy current coupling
d ——— Speed controlled electrical drive system

Fig. 7.2 Comparison of power absorbed for different fan/air flow control methods

Figure 7.3 shows a typical pump curve (plot of pressure/head versus flow). On many projects an oversized primary chilled water or condenser water pump is selected, due to safety margins in the design process and then a two-way throttling valve is used to set the flow to the chiller design point. In the Fig. 7.3, a pump with a design point S1 (pressure P1 and flow F1) has been selected, and then a throttling valve has been used to reduce the flow to system design flow operating point S2 (pressure P2).

Equation (i) above shows that for a constant efficiency, the power absorbed by a pump is proportional to flow (*x*-axis) multiplied by pressure (*y*-axis), (i.e. it can be represented by the area of a rectangle under the curves in Fig. 7.3). It can be seen that the area of the rectangle at the system operating point S2 is slightly less than that at the pump's design point S1. Therefore, the power absorbed by the pump using a throttling valve to control the flow is slightly less than that at the pump's design point.

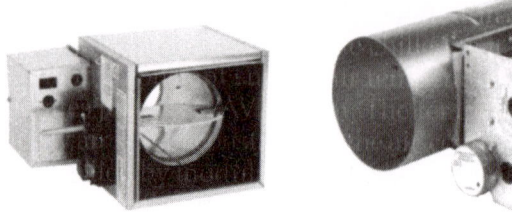

Fig. 7.4 Variable air volume systems

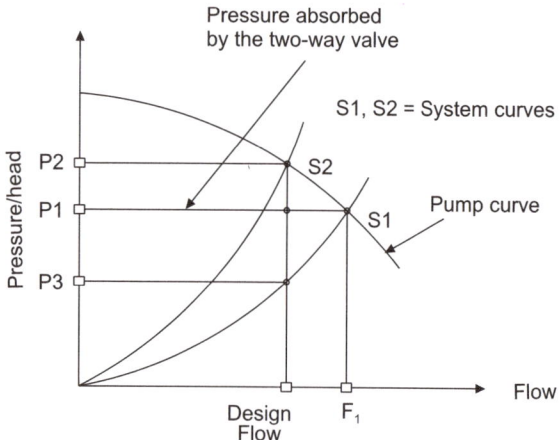

Fig. 7.3 Two-way (throttling) valves and VFD pump/water flow control

However, using a VFD, instead of putting a restriction in the system (i.e. the valve) across which pressure is dropped resulting in a reduced flow, the speed of the pump is reduced resulting in the pump delivering a reduced pressure head P3 to provide the same design flow. It can be easily seen that the area of this rectangle is much smaller than that at point S2. (Note, if a three-way valve instead of two-way is used, the same power is absorbed at all flow rates, because the valve just serves to divert the flow of water from the load but the total flow remains constant).

Figures 7.2 and 7.3 readily shows that using a VFD to control the flow of air or water in an HVAC system is the most energy efficient method resulting in a system with the lowest operational costs.

Another application of VFD is to change the method of maintaining design conditions inside air-conditioned space. This is known as variable air volume system.

7.1.1.3 Variable Air Volume System

Variable air volume (VAV) systems are the most energy efficient method of maintaining buildings environmental conditions. In a VAV system the temperature of the supply air is kept constant by modulating the cooling coil chilled water valves, and the volume of air to each zone is varied by the VAV boxes modulation, based on the temperature of the zone. Some form of flow control (e.g. inlet guide vanes or VFD) is used for the supply fan to maintain the required static pressure in the supply duct as the VAV boxes modulate open and closed (Fig. 7.4).

Depending on the complexity of the design, there may be temperature sensors in the supply air duct, on mixed air area and rooms to control or monitor the temperatures. Digital controller maintains a constant supply air temperature by controlling chilled water (heating if relevant). Relative humidity control may also be provided by a duct humidifier.

Using VFD for VAV applications, a pressure sensor, measuring the supply air static duct pressure, typically two-thirds the way along the duct, is connected directly to the VFD. The VFD in turn through its controller operates in closed loop control to maintain the static pressure at the required set point. As the VAV boxes close on getting a signal from the temperature sensor, the increase in static pressure is detected by the sensor and the VFD reacts to reduce the speed/flow of the supply fan to maintain the pressure at set point.

Thus, in the traditional constant air volume (CAV) AHU's, there are no means of flow control (as implied by its name). The conditioned space receives design air flow at all times and the chilled water valves are modulated to vary the temperature of the supply air based on the room or return air temperature. As such there are very few options to optimise energy savings.

However, where a constant air volume (CAV) AHU supplies a large single zone (e.g. hospitals, airports, hotels, movie theatres/cinemas, shopping malls), there is an opportunity to apply a VFD to provide energy savings by simulating a VAV system. These types of buildings often have a large variable load profile depending on the occupancy of the zone. They therefore provide an opportunity to vary the volume of air supply dependant on the occupancy.

This is achieved by fitting a VFD to control the speed of the fan and therefore the volume of air supplied to the single zone based on the room or return air temperature (just as VAV boxes do in a VAV system). In addition a temperature sensor is fitted in the supply duct and used to modulate the chilled water valves to maintain a constant supply air temperature. This then operates just as a VAV system. The temperature of the supply air is kept constant but the air-conditioned space is controlled by modulating the volume of air supplied.

Typically the VFD should be programmed with a minimum speed of approximately 70% to maintain the air quality, and then its speed varied from the speed/ flow to maximum depending on either temperature (as above) or an air quality sensor (e.g. CO_2) installed in the zone which will fluctuate depending on occupancy.

7.1.1.4 Other Applications

The same type of analysis can be applied to other VFD applications including.

- Filter control in clean room applications where a VFD is used to maintain the static pressure downstream of the pre-filters to maintain the same volume flow rate to the clean room irrespective of the filter condition (clean or clogged). When the filter is clean, the VFD will run the fan at slower than full speed and therefore save energy.

- Condenser water pumps are traditionally throttled using a two-way valve to set the flow to the design flow rate of the chiller, just as for the primary chilled water pump above. By opening the valve and instead using a VFD to slow the pump speed and set the flow, the system will run more efficiently.

- Cooling tower fans which are traditionally not controlled, have a simple on/off control or two speed motors. Using a VFD, a temperature sensor installed in the cooling tower basin or condenser water return lines can be directly connected to the VFD. The VFD will operate the cooling tower fan at whatever speed/flow rate is required to ensure that the return water temperature to the chiller remains at design point. This result in large energy saving, reduced mechanical wear and lower acoustic noise.

- Primary chilled water pumps: These are almost identical to condenser water pumps, hence a VFD can again be used as an "electronic" valve.

- Secondary chilled water pumps, which can be controlled to maintain only the necessary differential pressure across the furthest most point in the system, thereby saving energy when the cooling coil chilled water valves are in any position apart from fully open.

7.1.2 Throttling or Two-way Control Valve

Control valves are a critical part of the variable volume hydronic system. The control valve is responsible for properly varying flow through the water coil at a variety of building load conditions. Because of their critical nature, great care must be taken when selecting control valves so they may perform properly. Undersized valves may provide insufficient capacity and oversized valves provide poor control. Let us examine control valves now in more detail.

7.1.2.1 Control Valve Components

- *Actuator*: Causes valve motion in response to an external signal.
- *Body*: Portion of valve which regulates the flow of fluid.
- *Trim*: All portions of the valve in contact with the fluid (seats, disc, stem, etc.).
- *Disc*: Part which makes contact with the valve seat when the valve is closed.
- *Plug*: Characterizes the flow of fluid. The disc is often an integral part of the plug.

Equal percentage type valves are typically applied to cooling and heating coils because of their favourable flow characteristic. It provides a high degree of control accuracy with wide variations in pressure, flow rates, load changes, and other variables, such as long time lags. When combined with the heat transfer characteristic of a cooling coil, the change in stem position almost provides a linear change in heat transfer if an equal percentage control valve is used. Figure 7.5a depicts the relationship between flow and heat transfer for a cooling coil, while Fig. 7.5b depicts the relationship between lift and flow for an equal percentage control valve. By combining the first two curves Fig. 7.5c, we can depict the relationship between the control valve's

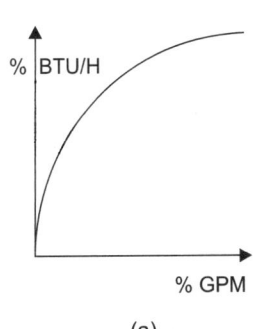

(a)

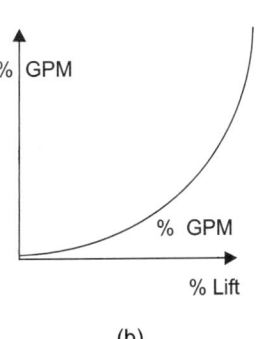

(b)

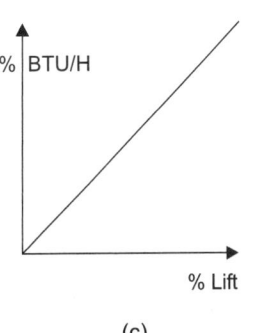

(c)

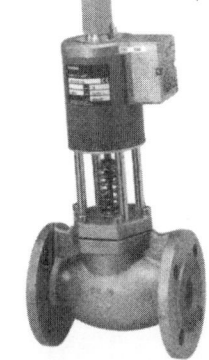

(d) A typical two-way control valve

Fig. 7.5 Relationship between coil

lift and heat transfer characteristic of the cooling coil. Figure 7.5d shows the photo of a typical two-way valve.

In addition to the flow characteristic, the valve's rangeability and authority can affect the linear relationship described above. Valve rangeability is defined as the ratio of maximum controlled flow over minimum controlled flow, i.e.

$$Vr = \text{Flow (max)}/\text{Flow (min)}$$

The greater the rangeability of the valve, the greater the ability of the valve to control accurately during low flow conditions as well as design flow conditions. For example, if a valve is selected with a rangeability of 30, the minimum controlled flow is approximately 3.3%.

$$Vr = 30 = 100 \,\%/\text{Flow (min)}$$
$$\text{Flow (min)} = 100/30 = 3.3\%$$

Valve authority, or beta, is defined as the ratio of minimum differential pressure over maximum differential pressure.

$$\text{Beta} = \Delta P \text{(valve)}/\Delta P \text{(max)}$$

See Fig. 7.6 below for a better indication.

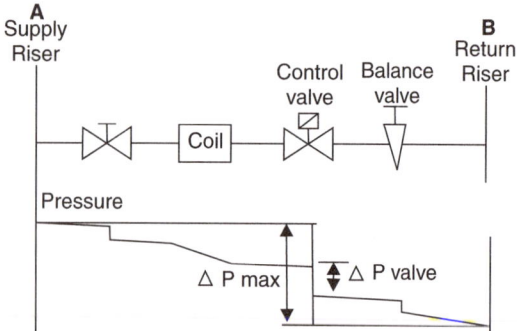

Fig. 7.6 Difference pressure across control valve

In variable volume systems, control valves operate the majority of the time at reduced loads and under higher differential pressures than those at design flow. The following picture Fig. 7.7 depicts the distortion of

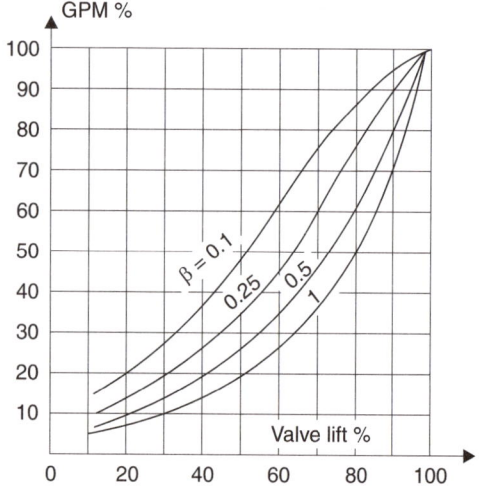

Fig. 7.7 Valve authority of an equal percentage valve

the equal percentage valve's flow characteristic with varying valve authority.

7.1.2.2 Control Valve Sizing

As stated earlier, sizing control valves is extremely important in variable volume hydronic systems because

- Undersized valves provide insufficient capacity.
- Oversized valves provide poor control.

There are basically five factors to be considered when sizing a control valve.

1. Maximum flow capacity.
2. Differential pressure across valve at maximum flow.
3. Fluid specific gravity and viscosity.
4. Inlet pressure and temperature at maximum flow.
5. Maximum differential pressure across valve at any flow.

7.1.2.3 Maximum Flow

Maximum flow across the valve is usually determined by the hydronic system designer's design and coil selection. As discussed earlier, the design flow rate of the coil is a function of the maximum heat load of the space being cooled. The maximum flow rate of the coil is obviously the maximum flow rate used for selecting the coil's control valve, and is the first step towards calculating valve C_v (defined in next section).

7.1.2.4 Valve Pressure Drop

The next step in calculating the valve C_v is to determine the pressure drop across the valve when the valve is in the full open position. A valve's C_v is determined by the construction of the valve itself, and is defined as the amount of flow (in gpm) that will flow through a fully open valve resulting in one psi of pressure drop across the valve. Increasing the pressure drop across the valve may improve controllability, but it also increases the pumping horsepower required to flow the design gpm through the valve and coil circuit as well as increasing noise. For these reasons, a "not to exceed" maximum pressure drop across the valve is often stated. On the other hand, too low a pressure drop across the valve is also undesirable. If the pressure differential allocated to the valve is 10 % or less of the load circuit pressure drop (*A* to *B* in Fig. 7.6), the valve selected will be so large that it cannot throttle effectively until near its fully closed position. The result is poor control, noise and excessive trim wear. In general, 25–50% of pressure drop across the load circuit (*A* to *B* in Fig. 7.6) is a good rule of thumb.

Now that maximum valve flow and maximum allowable pressure drop at design flow are known, valve C_v can be calculated.

$$C_v = \frac{\text{Maximum flow}}{\text{Design } \Delta P}$$

7.1.2.5 Fluid Specific Gravity and Viscosity

For fluids other than water, the C_v formula must be adjusted to accommodate the more viscous nature of the fluid. Thus, the valve C_v should be as shown below.

$$C_v = \frac{\text{Maximum flow}}{\sqrt{\text{Design } \Delta P / \text{Sp Gr}}}$$

where, Sp Gr is the specific gravity of the fluid.

7.1.2.6 Inlet Pressure and Temperature

For long valve life, the operating pressure and temperature should not exceed the valve rating. Although most of the equipments are designed with some margin of safety, it is generally better to select a valve rated for the highest inlet pressure and temperature which the valve is likely to experience.

7.1.2.7 Maximum Differential Pressure Across the Valve at any Flow

This parameter is an important criteria for valve selection for two reasons.
1. Choosing the appropriate actuator.
2. Avoiding valve noise and possible valve cavitation.

In variable volume systems, the valve actuator must be selected to be capable of supplying sufficient force to close the valve against the maximum pump head pressure. In variable volume closed loop systems all loads could be reduced causing the pump to ride back on the pump curve. As the pump rides back on the curve, the head being produced by the pump increases. The control valve actuator must be strong enough to continue modulating the valve closed as this pressure increases. In the worst case, this could be the "shutoff pressure" being produced by the pump.

Finally, excessive valve noise and valve cavitation are caused by high liquid velocities. As water passes through a valve, it is accelerated in such a manner that pressure is decreased below vapour pressure and bubbles form. Then immediately downstream of the valve, velocity decreases and pressure increases so the bubbles collapse causing noise and excessive wear on the valve and piping. Maximum valve differential pressure can be calculated using the following equation.

$$P_{\text{max.}} = K_c (\text{inlet pressure} - \text{vapour pressure}).$$

For example, butterfly valves are used sometimes as modulating control valves for large loads. Let us examine the maximum differential pressure across the valve at 100 psig (114.7 psia) inlet pressure with water at 68 °F (vapour pressure = .339 psia). The butterfly valve's $K_c = 35$.

$$P_{\text{max.}} = K_c (\text{inlet pressure} - \text{vapour pressure}).$$
$$P_{\text{max.}} = 0.35 (114.7 - 0.339).$$
$$P_{\text{max.}} = 40.0 \text{ psid.}$$

The manufacturer's of the two-way control valve provide with ready-reckoner tables or sliding rules or computer software to select the two-way valve with proper valve authority and actuator motor.

7.1.3 Bypass or Three-way Valve

The three-way valve as the sketch shown in Fig. 7.8 is used to bypass the chilled water if the load on the system reduces. This bypassed water again mixes with the supply water to AHU and passes through the cooling coil again and the cycle continues.

The three-way valves are also motorised. The selection of three-way valve is done on the similar principles as for two-way valves except that the actuator is selected based on the difference in pressure between supply line and the by-pass line, because the actuator has to operate against the pressure difference between these two lines.

Certainly the actuator for two way valve will be of higher capacity than that of three-way valve.

Fig. 7.8 A typical three-way control valve

7.2 VARIOUS ARRANGEMENTS OF CHILLED WATER CONNECTIONS

The piping that is used to circulate hot or chilled water for air-conditioning is called a hydronic piping system.

7.2.1 CHILLER PIPING (EVAPORATOR SIDE)

There are many ways to pipe a chiller. There are also many ways to pipe the chilled water distribution system. We will analyze a variety of distribution piping methods here. For now we will look only at the piping within the chiller plant. The designer must consider many factors when choosing a piping scheme. In all cases, the pumps will be pumping into the chiller (Fig. 7.9).

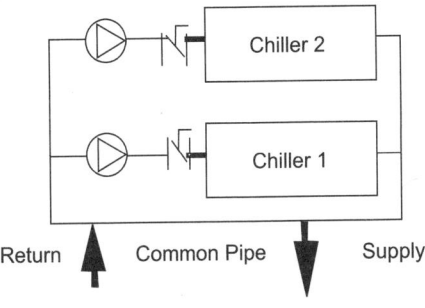

Fig. 7.9 Chiller pump location

Figure 7.10 is the preferred way to pump chillers in a primary-secondary pumped system.

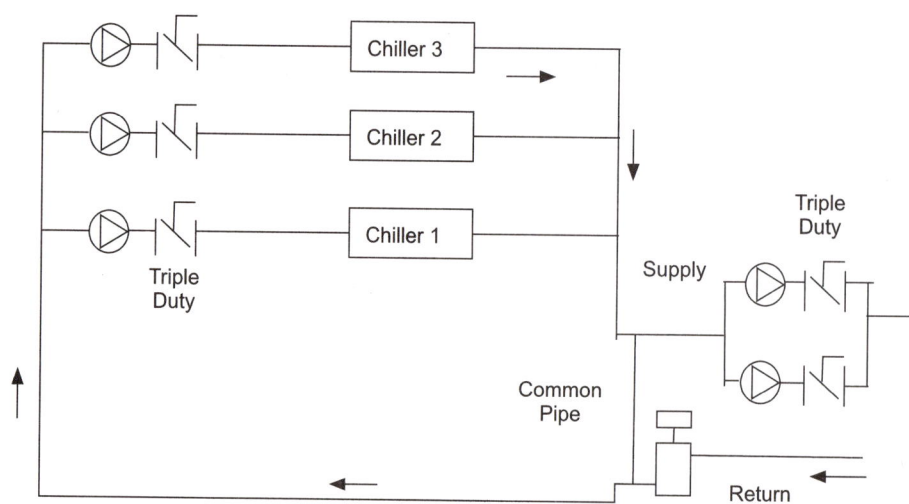

Fig. 7.10 Chiller pump for primary loop

The advantage to this piping method is that the flow can be balanced to the exact requirements of the chiller. This design permits the selection of multiple chillers of varying capacity. The pump is matched to the chiller.

Redundancy can be addressed through additional piping and valving, (Fig. 7.11). This piping arrangement will permit the operator to isolate pumps and or chillers, which need to be serviced and reroute the path of the water to another chiller of equal size. When this method is used, valve position must be verified and be apparent to the operator to ensure that proper valves are open or closed.

A second method involves piping the chiller pumps with a common suction and discharge header, (Fig. 7.12).

The primary reason for piping chillers in this manner is pump redundancy. If all chillers have the same flow rate and pressure drop, all the pumps can also be identical. In this manner, if the headers are valved properly, a single redundant pump can be used as a standby. Any combination of pumps equal to the number of chillers in operation can be selected to provide the proper flow to the chillers. Typically, the number of pumps installed is equal to the number of chillers installed, plus one.

The downside of this configuration is that it is very difficult to control the flow of water to a mixture of pumps and chillers of different sizes.

Care must also be taken when starting and stopping the chillers. When piped in this manner the chillers are hydraulically coupled. If the chiller's control valve is opened prior to the pump starting, flow through the other chiller(s) will drop suddenly, resulting in control instability.

Likewise, if a pump is started prior to the valve opening up, the other chiller(s) will experience an overflow condition. This may also cause instability and or water hammer. A flow limiting valve may also be employed to limit the potential of overflowing the chiller(s).

Some existing systems may be piped as a primary pumped system shown in Fig. 7.13.

In this configuration the pump provides flow into the entire system. Since this is a constant volume system, and the designer wants to follow the chiller manufacturer's recommendations, a modulating valve is

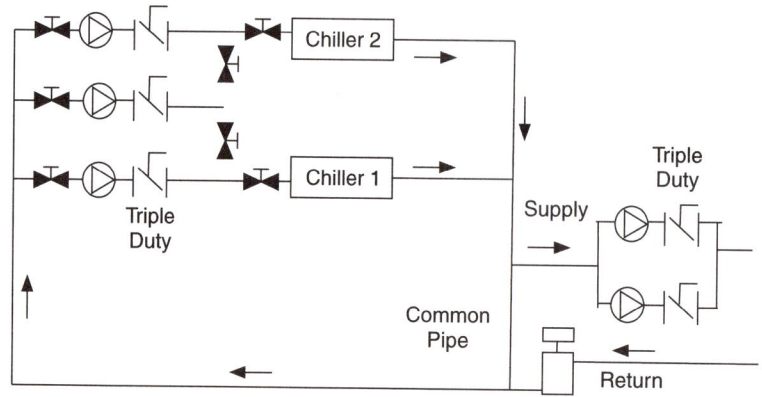

Fig. 7.11 Piping and valving to increase pump flexibility

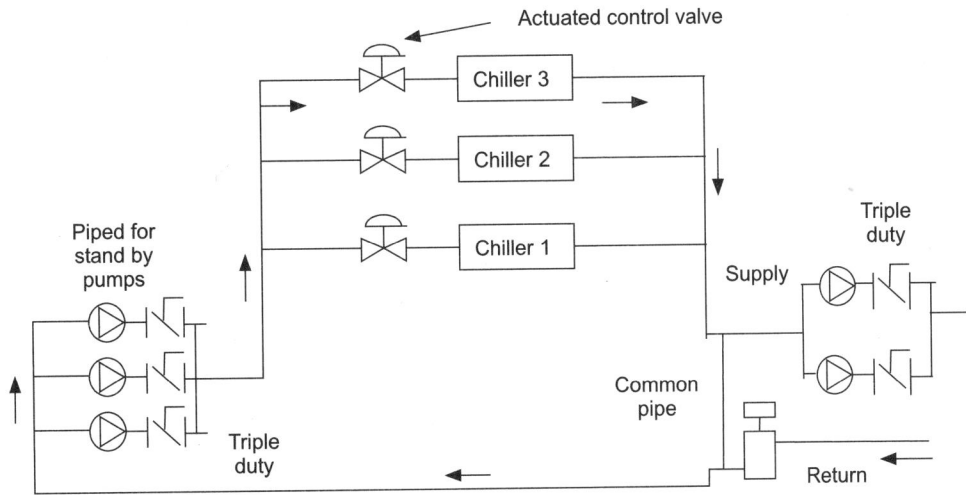

Fig. 7.12 Piping and valving to increase pump flexibility

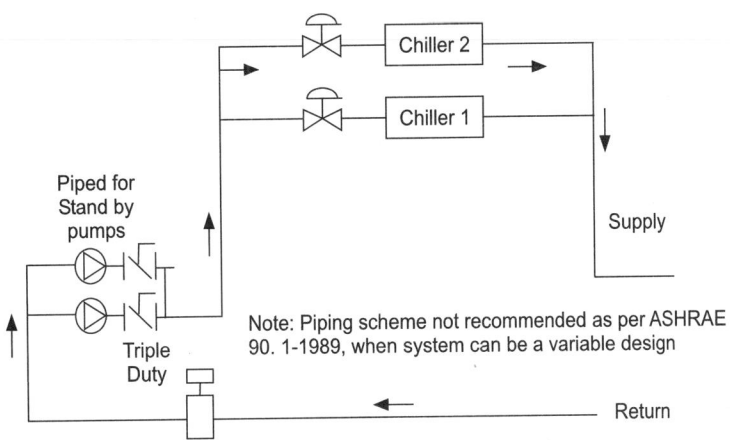

Fig. 7.13 Primary pumped system

provided at the inlet of each chiller to control the flow to each chiller. A triple duty valve is also provided for each pump to balance the flow rate out of each pump.

This is not the recommended method of piping for new designs. ASHRAE states in Standard 90.1–1998, section 9.5.5, all HVAC systems with pumping horsepower greater than 7-1/2 shall be designed for variable flow. Primary pumping is not variable volume, but constant volume.

7.2.2 CHILLER PIPING (CONDENSER SIDE)

There are even more ways to pump condensers and cooling towers. As with the evaporator side, the location of the pumps is important. NPSH is critical on the condenser side because most condenser loops are open to the atmosphere. The elevation of the cooling tower and atmospheric pressure are the only two positive values applied to the pump suction. Pressure drops occur through the piping, valves, strainers, etc. For this reason, the pumps are located prior to the condenser

to limit the large pressure drop of this component. The pumps should be located as low as possible and as close as possible to the tower. Avoid overhead suction piping and provide a slope on horizontal pipe runs to the pump suction.

Condenser pumps can be piped with a header system or each can be pumped directly. The cooling towers can also be piped direct or headered. A third convention is multiple condensers for a single multi-celled tower. The advantages and disadvantages of both methods are similar to those for the evaporator portion of the chiller-first cost and redundancy. We must also consider particular control considerations. Let us look at the advantages and disadvantages of the different piping methods.

Figure 7.14 shows direct pumps, with its own, dedicated cooling tower and condenser.

The distinct advantage in this layout is complete independence. Since the suction lines are separate, no equalization line between the towers is required.

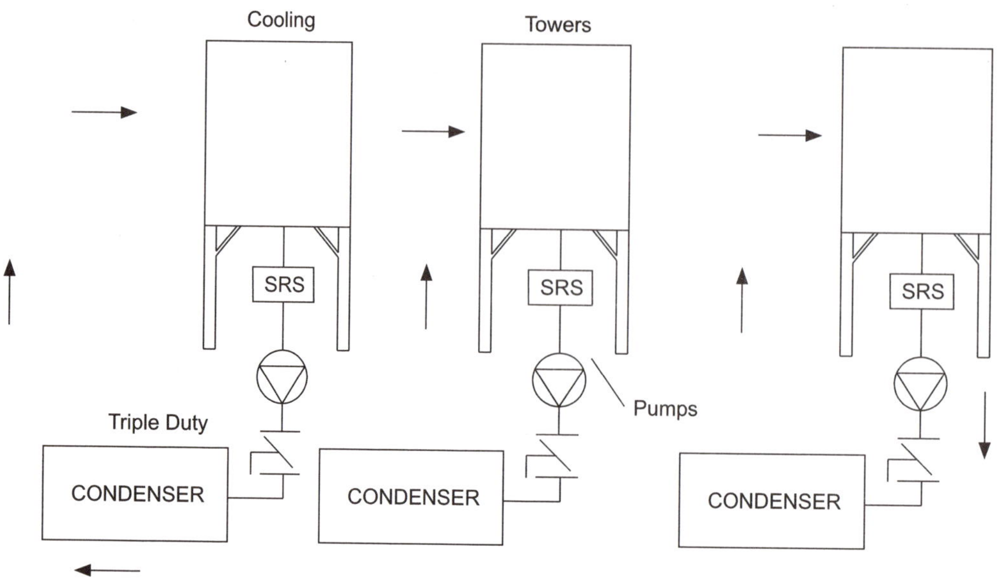

Fig. 7.14 Individually pumped condensers and cooling towers

As discussed previously in the evaporator piping section, a dedicated pump per condenser and tower simplifies balance and controllability, especially if the equipments are of unequal sizes. The disadvantage is the additional first cost of piping and equipment.

Figure 7.15 is an example of pumps headered together, individual condensers, and a multi-cell cooling tower.

As with evaporator piping discussed earlier, the advantage of this type of pumping is the low cost of redundant pumping. The disadvantage is the difficulty in balancing dissimilar tower cells. All must be equal in size (pump flow and head) and piped similarly.

Figure 7.16 utilizes a pump dedicated for each condenser and multiple towers with headered supply and return piping with an equalization line.

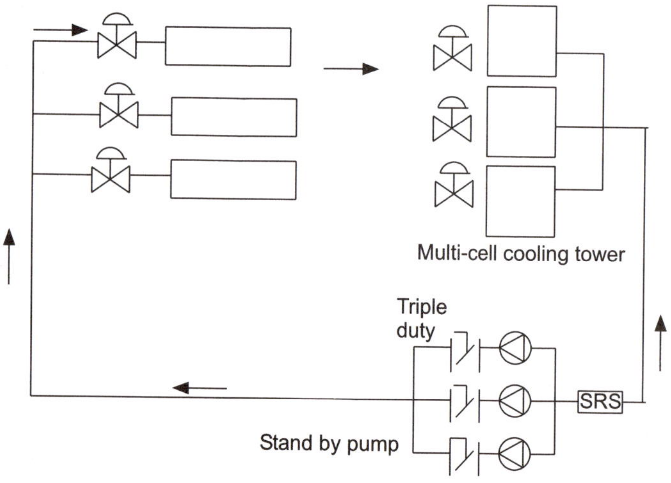

Cooling tower sump water level should remain constant. Too high a level results in overflow. Too low a level can lead to vortexing and the introduction of air. When multiple towers are piped together, an equalization line must be installed so that all the towers operate at the same level. The head available to create the necessary flow is the sump water level difference between the towers. This is a small value often less than one foot. The equalization line should be sized such that the pressure drop does not exceed the maximum allowable sump level change.

Figure 7.17 has a dedicated pump for each tower. The pump discharge is headered together as are the condenser suction and discharge.

By headering the pump discharges, flow can be directed to one or a combination of individual condensers. Care must be taken to ensure proper balancing of each condenser. All pumps and condensers must be equal in size and capacity.

Fig. 7.15 Multi-celled cooling tower

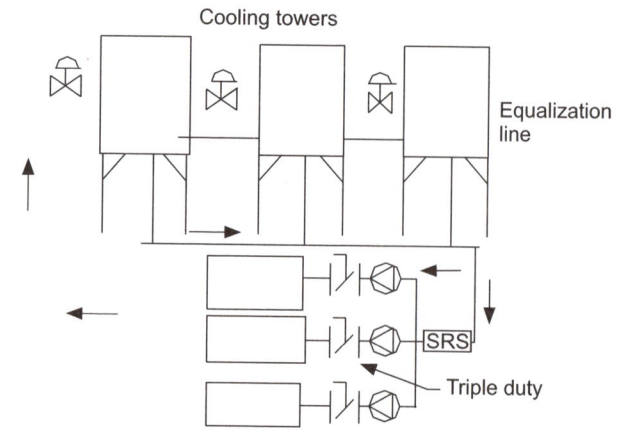

Fig. 7.16 Multiple towers with headered condensers

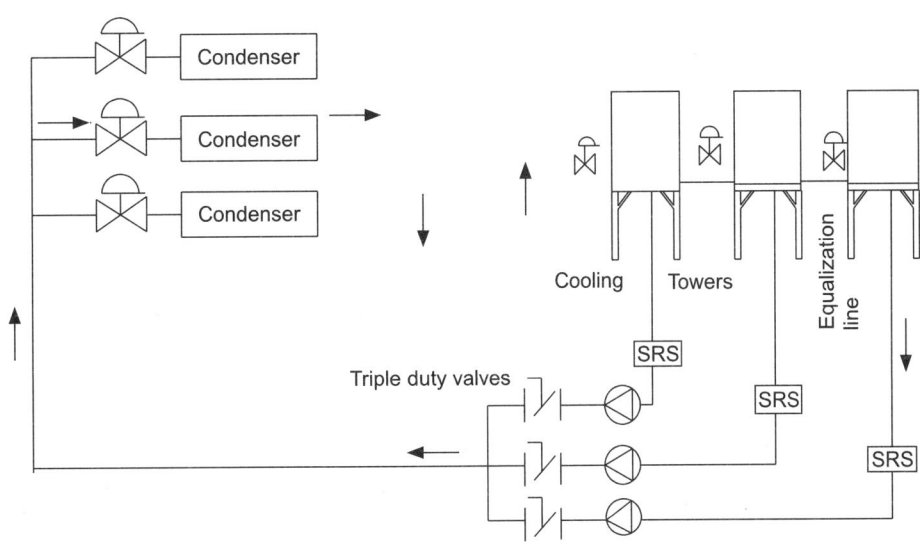

Fig. 7.17 Dedicated pumps for multiple cooling towers

Figure 7.18 has individual supply and return headers for all components, pumps, condensers, and cooling towers.

Once again, by headering the suction and discharge of the pumps, towers and condensers, some flexibility is gained. Should a tower, condenser or pump fail, or be shut-off for the purpose of maintenance, proper valving will permit the operation of a combination of different pumps, towers and/or condensers. As before, all components must be of equal size and capacity.

Figure 7.19 is a piping schematic for three-way valve piping.

Note: With three-way piping, do not bypass to the pump suction, bypass to the sump to avoid potential pumping problems due to entrained air.

Figure 7.20 shows the tower controlled with a two-way valve. However, a less expensive two-way valve control configuration can be formed by placing the valve on the discharge of one of the condensers so that a smaller, less expensive valve is used. The disadvantage of this arrangement is the only condenser that can be used during winter operation.

7.2.3 Water Pipe Connections (Load Side)

This article describes the various ways to pipe the terminal units in the low side system of HVAC application. The low side system consists of air handling units (AHU) or terminal units piped together in various methods explained below as per the application. The connection between the piping and air handling units may be made in any of the following six basic ways.

    a. Series loop.

    b. One pipe main.

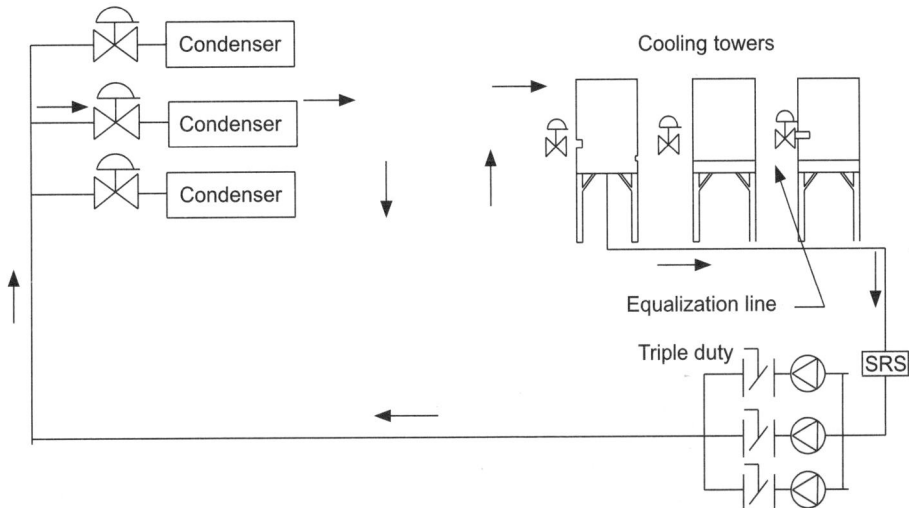

Fig. 7.18 Condenser water system with each individual component headered

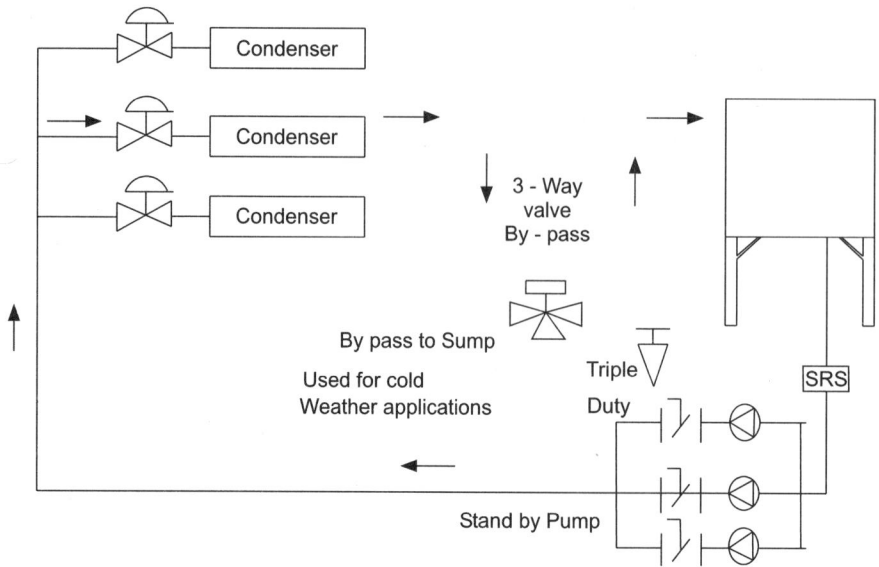

Fig. 7.19 Three-way valve controlled tower

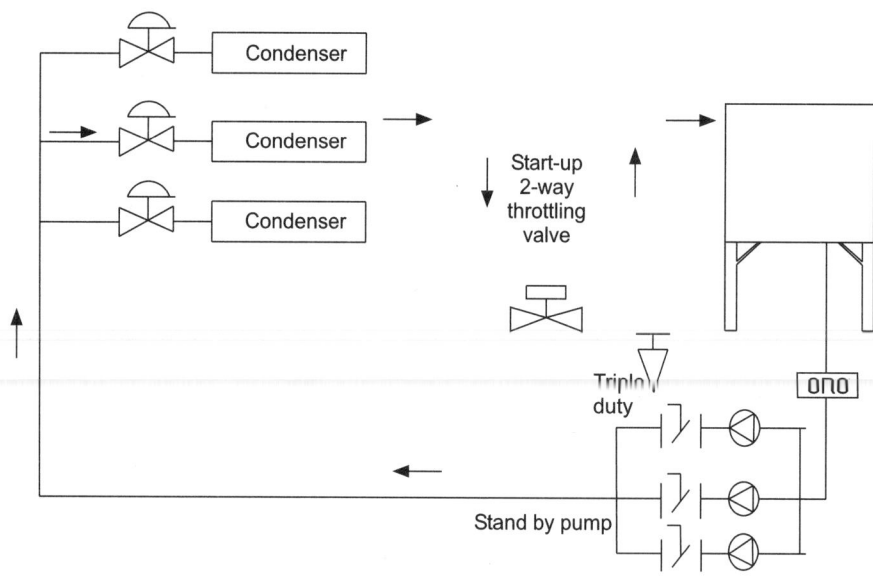

Fig. 7.20 Two-way valve control of tower on supply header

c. Two pipe direct return.
d. Two pipe reverse.
e. Three pipe system.
f. Four pipe system.

7.2.3.1 Series Loop System

In this system shown in Fig. 7.21, all the AHU's are connected in series and one loop is formed. In this system, all the water that enters the circuit will pass through all the units and will return to the generator.

Disadvantage

i. The maintenance or repair of any terminal unit require shutdown of entire system.

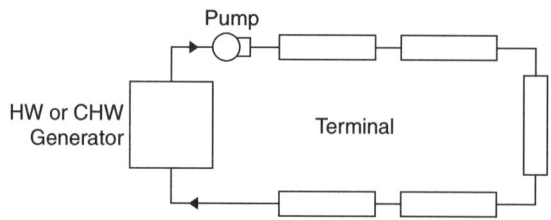

Fig. 7.21 Schematic series loop piping system

ii. Individual capacity control cannot be achieved by changing water flow.

iii. The number of units that can be connected on one circuit is limited by the continuous temperature drop through each successive unit.

7.2.3.2 One Pipe Main

The system shown in Fig. 7.22 is the one pipe main system. In this system, one main pipe feeds the water to the AHU's which are mounted in parallel to the main pipe.

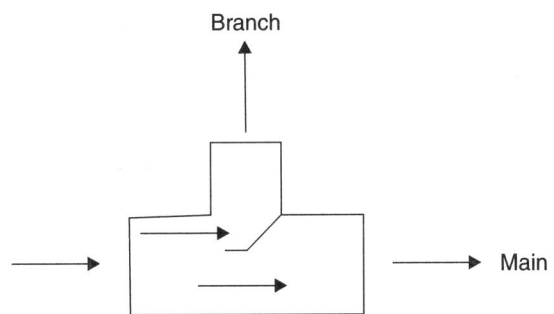

Fig. 7.22 Schematic piping system

The advantage of this type of system is that by locating a control valve, one can easily control the flow of water through AHU and capacity control is achieved. However, the water seeks the path of least resistance and tends to flow in a straight line than flowing through the AHU, thus starving the unit.

7.2.3.3 Two Pipe Direct Return

To overcome the problem explained in one pipe main and to get the water temperature supplied to each terminal unit to be equal, the two pipe system is used. In this system, there are two mains, one for supply water and one for return. Each terminal units is fed by an individual supply branch, a return branch carries the water back to the return main. In this manner, all units receive water directly from the source. This will be clear from Fig. 7.23.

The disadvantage of this system is the unit that is located near to the supply mains gets more water as the water has to travel less, and hence less resistance is encountered. Similar is the case with return piping. The farthest unit gets least water as maximum friction is encountered in both supply and return pipes. Though this problem can be solved by providing balancing valves so that only that much quantity of water is supplied to the unit as is required by it. But, this process increases the cost of the system, as balancing valves are very costly.

7.2.3.4 Two Pipe Reverse Return

The balancing problem is solved to a great extent by installing two pipe reverse return system. The circuit length for each unit is balanced so that water encounters the same resistance for each unit. As shown in Fig. 7.24, the supply pipe of water for first unit is the smallest in length but the return pipe of the same unit is the longest when compared to other units.

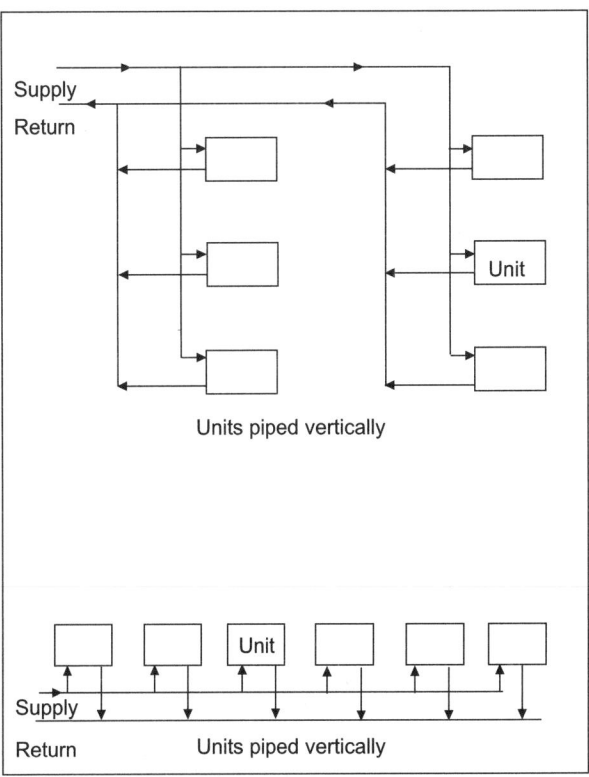

Fig. 7.23 Direct return water piping system

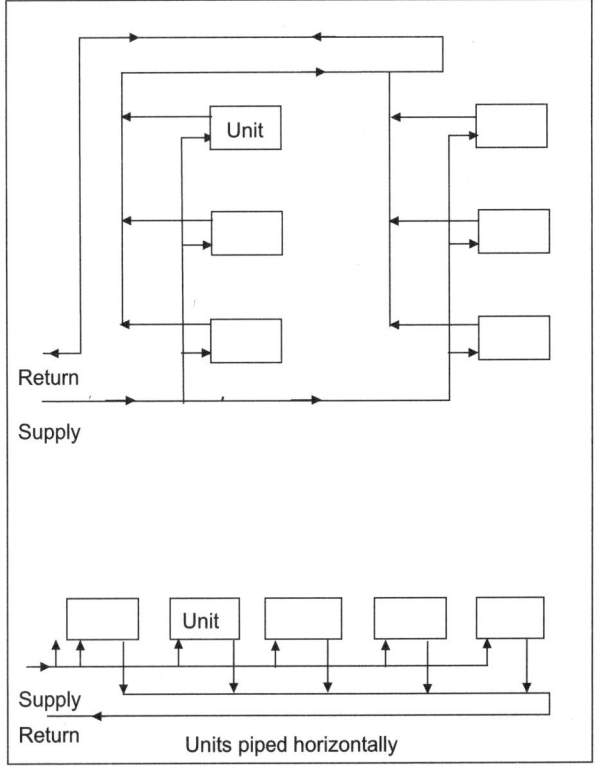

Fig. 7.24 Reverse return piping

Similarly, the supply pipe of last unit is the longest while its return pipe is the smallest. This way total length/total friction encountered by water for each unit is equalized and thus water is supplied in equal quantity to each unit.

7.2.3.5 Combination System

Figure 7.25 shows the combination arrangement which combines the features of all the system described above.

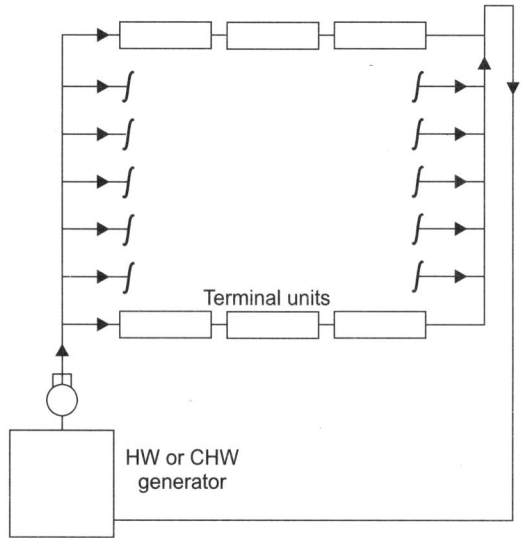

Fig. 7.25 Combination reverse return (riser) and series loop system

This type of system may be chosen for high rise buildings, where separate control of each unit on a floor is not needed, yet flow balance will be simple, and costs reduced.

7.2.3.6 Three Pipe System

The two pipe system can be used for either hot water or chilled water separately. But in modern buildings, heating and cooling both are required in respective seasons. Hence, supply of hot water from generator and chilled water from chiller is required to be supplied to AHU or terminal unit.

In three pipe system, shown in Fig. 7.26, there are two supply mains, one for chilled water and one for hot water. Both the pipes are connected to AHU through a three-way valve that determines whether hot or cold water is supplied to AHU. The return is common and receives water from each unit. The connections to the units can be made either by direct or by reverse return.

In the hot water system, there is problem of scaling and the hot water pipes tend to corrode faster than chilled water pipes. This will be the case with hot water coils transferring heat to air. Hence, it is preferable to use separate pipes and coils for hot and cold water applications. The applications that require supply of hot

and cold water irrespective of the season, mixing of hot and cold will take place in return piping and efficiency will get reduced.

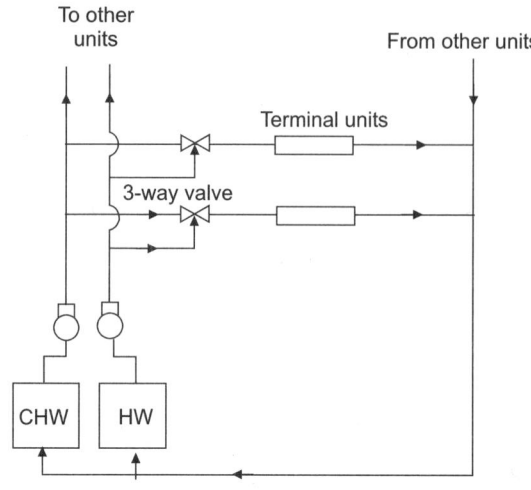

Fig. 7.26 Three-pipe system

7.2.3.7 Four Pipe System

This is basically two separate two pipe systems, one for chilled water and one for hot water, and therefore no mixing occurs. This is an ideal arrangement but of course it is costly and occupies more space.

7.2.4 System Water Temperature

The hydronic systems are classified into temperature categories as follows.

Low temperature hot water (LTW): Temperature below 250 °F.

Medium temperature hot water (MTW): Temperature in the range of 250–350 °F.

High temperature hot water (HTW): Temperature in the range of 350–450 °F.

This kind of categorization is important, as different types of boiler are needed for different applications.

High supply water temperature shall be used so that AHU size can be reduced. But high water temperature requires that the boiler pressure must be increased to prevent the water from evaporating and consequently boiler of greater strength is required.

Similarly, high temperature drop is desirable so that more heat transfer takes place with least quantity of water and consequently smaller pumps and piping may be used.

To achieve a balance between the advantages and disadvantages of low and high temperature applications, it is a usual practice to use LTW system between 180–240 °F and a system temperature drop of 10–40 °F. In HTW system, high temperature drops up to 100 °F are selected to reduce pipe size and power use.

The chilled water system generally uses supply water temperature in the range of 4–50 °F. The system temperature rise usually ranges from 5–15 °F.

7.3 SIZING WATER PIPES

The calculation of size of a hydronic piping system is one of the most important aspect of designing a successful hydronic system. The size of a pipe depends on the following factors.

   i. Water velocity to be used in the system.
   ii. Surface roughness (inside).
   iii. Pipe length.
   iv. Friction loss.

To calculate the size of a pipe, it is necessary to understand its direct linkage to the concept of friction. Friction is a resistance to flow resulting from fluid viscosity and from the walls of the pipe or duct.

The type of flow usually encountered in HVAC application is called *turbulent flow*, and hence, the pressure loss or drop due to friction can be found from the Darcy-Weisbach equation as below.

$$H_f = \frac{FLV^2}{2gD}$$

where

  H_f = Pressure loss (drop) from friction in straight pipe or duct.

  F = A friction factor or roughness of pipe or duct wall.

  L = Length of pipe or duct.

  D = Diameter of pipe or duct.

  V = Velocity of fluid.

The various organizations have experimented with different pipe diameters and materials and calculated H_f. The various parameters like H_f, V, flow, diameter have been shown in a convenient chart shown in Charts 7.1 and 7.2.

Chart 7.1 is for closed piping where water flows continuously in a closed cycle, whereas Chart 7.2 is for open piping system where water is exposed to atmospheric pressure. Thus, if we are able to calculate H_f, select velocity for a particular application, calculate flow requirement for achieving required tonnage, then we can calculate the fourth parameter, i.e. pipe diameter.

7.3.1 Pressure Loss in Pipe Fittings

In addition to pressure loss due to friction encountered in straight run of pipe, the pressure loss also takes place due to bends and fittings in the piping system. These pressure losses in equivalent length of pipe are given in Tables 7.2 to 7.4. The equivalent length of pipe means that the straight run of pipe of equivalent length will have the same pressure drop as that in fittings (bend, elbow, tee, etc.).

7.3.2 Piping System Pressure Drop

Now, we can calculate the total pressure drop in a piping system so that the capacity of the pump in terms of head requirement can be determined.

The system pressure drop is simply the sum of the losses in pressure through each item in one of the paths or circuit from pump discharge to pump suction, including piping, fittings, valves and equipment. Information of pressure drop through equipment is obtained from the manufacture.

To determine the system pressure loss, the pressure losses through only one circuit are considered. This is because the pressure loss is same through every circuit and that is why water from every circuit will flow into the return pipe. If it were not so, then due to pressure differential, water of one circuit will start flowing into another circuit.

This is shown in Fig. 7.27 the pressure loss through circuit *ABD* and *ACD* has to be equal. Pressure drop in terms of equivalent length are valid only for fully open valves.

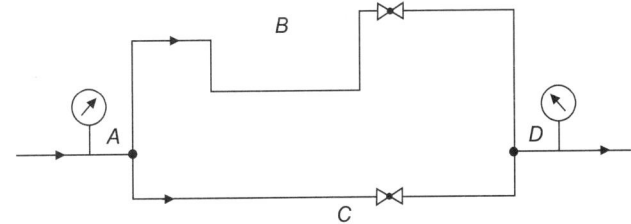

Fig. 7.27 Pressure drop from points *A* to *D* is always the same, regardless of path length

In actual practice, the balancing valves are used to regulate the flow, hence an additional pressure drop is created in that circuit. For this reason, it is customary to select the longest circuit in a system to calculate the pressure drop and to assume that the valves in this circuit are fully open.

Therefore, to find the total system pressure drop in a multi-circuited system, proceed as follows.

   i. Prepare a proportionate sketch of the complete system.
   ii. Calculate equivalent length of each circuit.
   iii. Select the circuit with the largest equivalent length.
   iv. Calculate the pressure drop through this circuit.

Example 7.1

For the steel piping system shown in Fig. 7.28, calculate the required pump head.

Solution

From the energy equation, the required pump head rise F-A is equal to pressure drop due to friction from A-F through the system. But to find this, only the circuit with the highest pressure drop is chosen. This is

ABCC'DEF. Using the friction chart, pressure loss for each item in circuit *ABCC'DEF* is found and summed up. This information is presented in Table. 7.1.

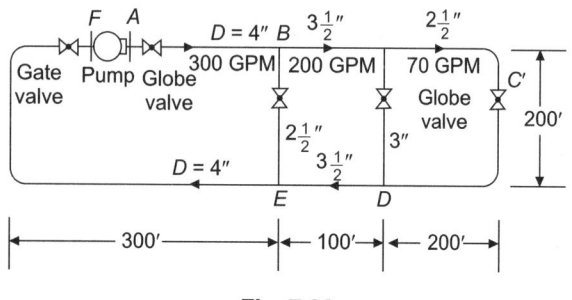

Fig. 7.28

Thus, the total pump head or system pressure drop is calculated and the pump can be designed for required flow and head.

7.3.3 Calculation of Water Quantity

Chilled/hot water is required to meet the requirement of total cooling/heating load of the building. Once the TR (tonnage) of each section of the building fed by a particular AHU is known, we can calculate quantity of water required as below.

$$Q = MS \Delta T$$

Q = Heat to be transferred in terms of TR or kW.

M = Flow rate of water GPM.

S = Sp. heat of water = 8.33.

ΔT = Temperature drop/gain at the inlet and outlet terminals of AHU (ΔF).

7.3.4 SYSTEM PIPE SIZING

Having understood the concept of friction loss and equivalent length of pipe, we now proceed to calculate the size in diameter of different sections of pipe as follow.

  i. Prepare a sketch of the piping system including each AHU/terminal unit.

 ii. Calculate flow rate through each section.

    In a two pipe system, the flow progressively decreases in each supply main section down stream from the pump. Since some of the water branches off at each unit, in the return main, the flow increases in each section. The simplest way would be to calculate from the last AHU and proceed backward adding the water flows to each preceding unit.

iii. Choose a value of friction loss rate to be used for the system piping based on following recommendations.

   a. The friction loss rate should be between 1–5 ft of water/100 ft equivalent length of pipe. Within these limits, values in the higher end are usually used for pipes of larger systems. This is because, in large chilled water piping installations, usually major costing of pipe is due to large sized pipes. Hence, higher fraction rate shall be tolerated in large sized pipes, because it gives pipes of a smaller diameter and hence reduces cost also. Also because the length of large sized pipe is comparatively

Table 7.1 Piping pressure drop calculations for Example 7.1

Section	Item	D, in.	GPM	V FPS	E.L., ft	No. of items	Total length, ft	Friction Loss H_f		
								ft w 100 (ft)		Total ft w
EFAB	Pipe	4	300	7.8			800	5.2		
EFAB	Gate valve				4.5	1	5			
EFAB	Globe valve				110	1	110			
EFAB	90° std ell				11	2	22			
B	Tee				11	1	11			
Subtotal							948	×	5.2/100 =	49.3
BC	Pipe	3½	200	7.0			100	4.8		
C	Tee				9	1	9			
Subtotal							109	×	4.8/100 =	5.2
CD	Pipe	2½	70	4.8			600	3.7		
C'	Glove valve				67	1	67			
CD	90° std ell				6.5	2	13			
D	Tee				6.5	1	7			
Subtotal							687	×	3.7/100 =	25.4
DE		3½	200	7.0			100	4.8		
E	Tee				9	1	9			
Subtotal							109	×	4.8/100 =	5.2
							Pumped head = Total H_f		=	85.1

Chart 7.1 Friction loss for closed piping systems (water) schedule 40 pipe

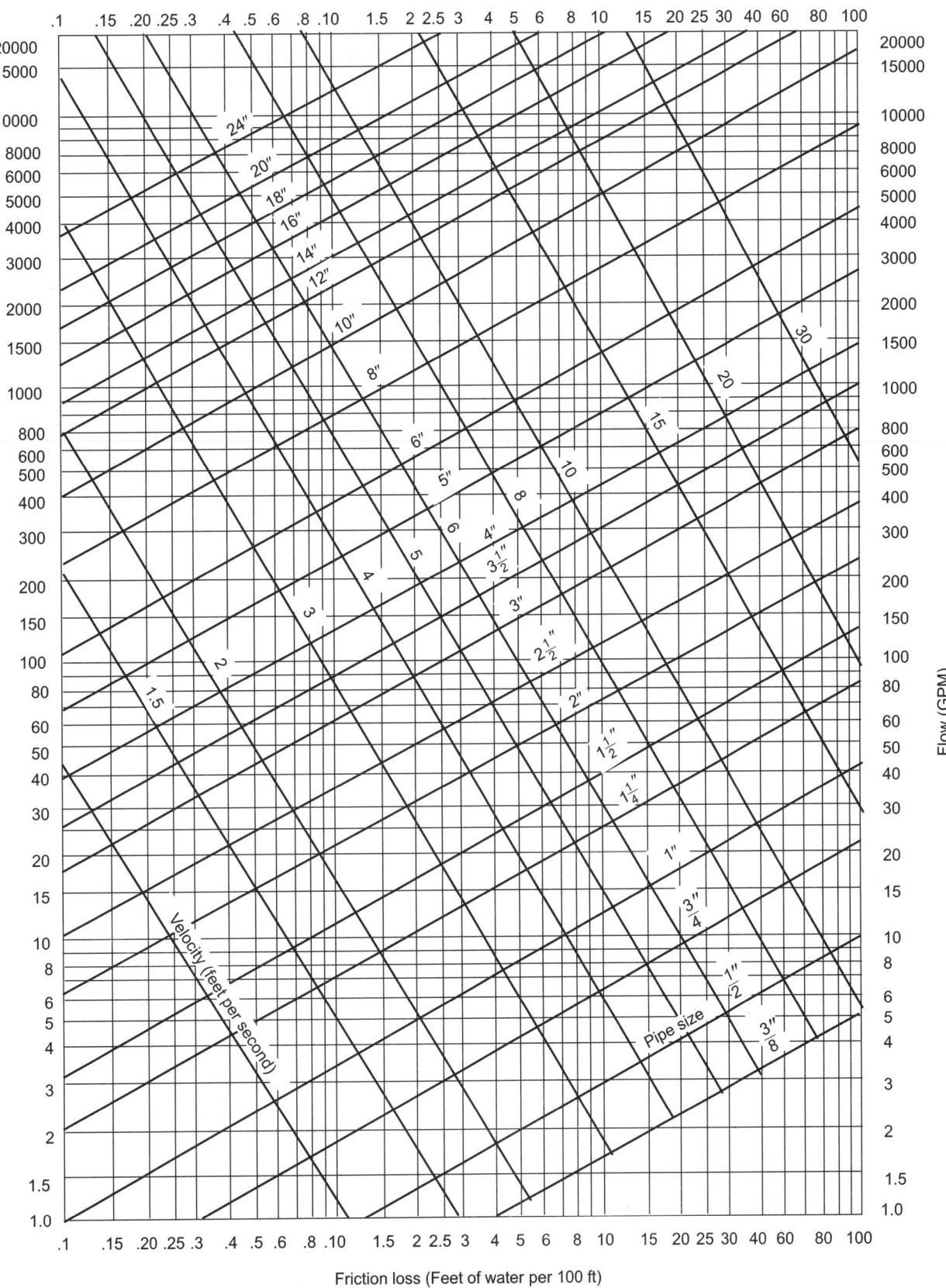

Friction loss (Feet of water per 100 ft)

* Reprinted with permission of M/S Carrier Corporation from *Handbook of Air-conditioning System Design.*

Chart 7.2 Friction loss for open piping systems (water) schedule 40 pipe

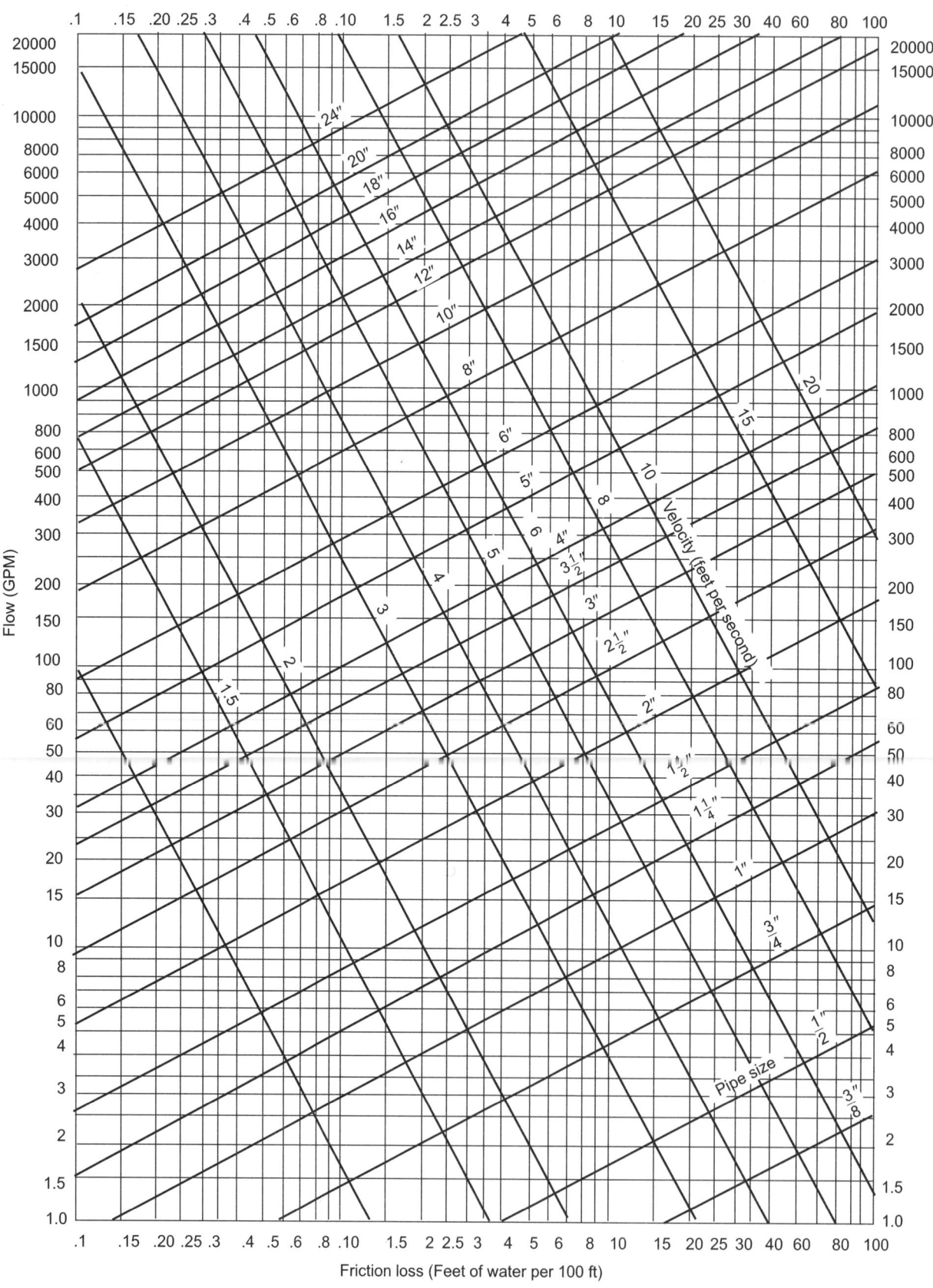

* Reprinted with permission of M/S Carrier Corporation from *Handbook of Air-conditioning System Design.*

small, hence the increase in friction rate will not substantially affect the pump head requirement.

But for small sized pipes, which constitute a major length of total system, lower range of friction rate shall be selected so that overall friction loss may be kept in limit.

b. The velocity in the largest section should not exceed 4-6 FPS in small systems, or 8-10 FPS in large systems. Higher velocities may create noise and vibrations in the system. This aspect shall be particularly taken care of in occupied areas.

c. The velocity in any pipe section should not be below 1.5 FPS. At lower velocities, dirt and air may be trapped in the line blocking water flow.

iv. Based on above guidelines, one can use Chart 7.1 or 7.2 to select the size of the main pipe leaving the pump depending on its flow rate. Often the friction loss rate when applied on chart will result in a size midway between the two standard sizes listed. In such cases, use the size nearest to and preferably of higher diameter than itself.

However, a judgement shall be applied as already described so that higher diameter pipes are not chosen to further higher limits and piping cost is kept in limits.

v. After selecting the new size (nearest to one marked with selected friction rate), a vertical line drawn from this pipe size can see a new friction rate. Always check that this friction loss is between the desired limit, i.e. 1–5 ft/100 ft of pipe. Also check the new velocity that will result by selecting the new pipe size.

vi. This new value of friction rate will now become the basis of selecting the pipe sizes of rest of the system. That is why this procedure is called equal friction method.

vii. When all the piping in the system has been sized, the circuit with greatest total length is determined. The pressure drop in this circuit is then calculated as already described in the example 7.1.

It will be advisable to increase the head of the system so calculated by 15–25% so as to accommodate unforeseen losses in the system or due to deviation from the planned route which may cause additional bends.

Table 7.2 Fitting losses in equivalent metres of pipe (screwed, welded flanged, flared, and brazed connection)

Nominal pipe or tube size, mm	Smooth bend elbows						Smooth bend tees			
	90° Std[a]	90° Long Rad[b]	90° Street[a]	45° Std[a]	45° Street[a]	180° Std[a]	Flow through branch	Straight-through flow		
								No reduction	Reduced 1/4	Reduced 1/2
10	0.4	0.3	0.7	0.2	0.3	0.7	0.8	0.3	0.4	0.4
15	0.5	0.3	0.8	0.2	0.4	0.8	1.9	0.3	0.4	0.5
20	0.6	0.4	1.0	0.3	0.5	1.0	1.2	0.4	0.6	0.6
25	0.8	0.5	1.2	0.4	0.6	1.2	1.5	0.5	0.7	0.8
32	1.0	0.7	1.7	0.5	0.9	1.7	2.1	0.7	0.9	1.0
40	1.2	0.8	1.9	0.6	1.0	1.9	2.4	0.8	1.1	1.2
50	1.5	1.0	2.5	0.8	1.4	2.5	3.0	1.0	1.4	1.5
65	1.8	1.2	3.0	1.0	1.6	3.0	3.7	1.2	1.7	1.8
80	2.3	1.5	3.7	1.2	2.0	3.7	4.6	1.5	2.1	2.3
90	2.7	1.8	4.6	1.4	2.2	4.6	5.5	1.8	2.4	2.7
100	3.0	2.0	5.2	1.6	2.6	5.2	6.4	2.0	2.7	3.0
125	4.0	2.5	6.4	2.0	2.4	6.4	7.6	2.5	3.7	4.0
150	4.9	3.0	7.6	2.4	4.0	7.6	9	3.0	4.3	4.9
200	6.1	4.0	–	3.0	–	10	12	4.0	5.5	6.1
250	7.6	4.9	–	4.0	–	13	15	4.9	7.0	7.6
300	9.1	5.8	–	4.9	–	15	18	5.8	7.9	9.1
350	10	7.0	–	5.5	–	17	21	7.0	9.1	10
400	12	7.9	–	6.1	–	19	24	7.9	11	12
450	13	8.8	–	7.0	–	21	26	8.8	12	13
500	15	10	–	7.9	–	25	30	10	13	15
600	18	12	–	9.1	–	29	35	12	15	18

[a] R/D approximately equal to 1.

[b] R/D approximately equal to 1.5.

Table 7.3 Special fitting losses in equivalent metres of pipe

Nominal pipe or tube size, mm	Sudden enlargement, d/D			Sudden contraction, d/D			Sharp edge		Pipe projection	
	1/4	1/2	3/4	1/4	1/2	3/4	Entrance	Exit	Entrance	Exit
10	0.4	0.2	0.1	0.2	0.2	0.1	0.5	0.2	0.5	0.3
15	0.5	0.3	0.1	0.3	0.3	0.1	0.5	0.3	0.5	0.5
20	0.8	0.5	0.2	0.4	0.3	0.2	0.9	0.4	0.9	0.7
25	1.0	0.6	0.2	0.4	0.4	0.2	1.1	0.5	1.1	0.8
32	1.4	0.9	0.3	0.5	0.5	0.3	1.6	0.8	1.6	1.3
40	1.8	1.1	0.4	0.7	0.7	0.4	2.0	1.0	2.0	1.5
50	2.4	1.5	0.5	0.9	0.9	0.5	2.7	1.3	2.7	2.1
65	3.0	1.9	0.6	1.2	1.2	0.6	3.7	1.7	3.7	2.7
80	4.0	2.4	0.8	1.5	1.5	0.8	4.3	2.2	4.3	3.8
90	4.6	2.8	0.9	1.8	1.8	0.9	5.2	2.6	5.2	4.0
100	5.2	3.4	1.2	2.1	2.1	1.2	6.1	3.0	6.1	4.9
125	7.3	4.6	1.5	2.7	2.7	1.5	8.2	4.3	8.2	6.1
150	8.8	6.7	1.8	3.4	3.4	1.8	10	5.8	10	7.6
200	–	7.6	2.6	4.6	4.6	2.6	14	7.3	14	10
250	–	9.8	3.4	6.1	6.1	3.4	18	8.8	18	14
300	–	12.4	4.0	7.6	7.6	4.0	22	11	22	17
350	–	–	4.9	–	–	4.9	26	14	26	20
400	–	–	5.5	–	–	5.5	29	15	29	23
450	–	–	6.1	–	–	6.1	35	18	35	27
500	–	–	–	–	–	–	43	21	43	33
600	–	–	–	–	–	–	50	25	50	40

Table 7.4 Valve losses in equivalent meters of pipe

Nominal pipe or tube size, mm	Globe[a]	60°-Y	45°-Y	Angle[a]	Gate[b]	Swing check[c]	Lift check
10	5.2	2.4	1.8	1.8	0.2	1.5	
15	5.5	2.7	2.1	2.1	0.2	1.8	Globe
20	6.7	3.4	2.1	2.1	0.3	2.2	and
25	8.8	4.6	3.7	3.7	0.3	3.0	vertical
32	12	6.1	4.6	4.6	0.5	4.3	lift
40	13	7.3	5.5	5.5	0.5	4.9	same
50	17	9.1	7.3	7.3	0.73	6.1	as
65	21	11	8.8	8.8	0.9	7.6	globe
80	26	13	11	11	1.0	9.1	valve
90	30	15	13	13	1.2	10	
100	47	18	14	14	1.4	12	
125	43	22	18	18	1.8	15	
150	52	27	21	21	2.1	18	
200	62	35	26	26	2.7	24	Angle
250	85	44	32	32	3.7	30	lift
300	98	50	40	40	4.0	37	same
350	110	56	47	47	4.6	41	as
400	125	64	55	55	5.2	46	angle
450	140	73	61	61	5.8	50	valve
500	160	84	72	72	6.7	61	
600	186	98	81	81	7.6	73	

8

Chiller Machines and Selection Criteria

The most commonly used refrigeration system is the vapour compression refrigeration system while vapour absorption system is also used nowadays but on a very limited scale.

The vapour absorption system is suitable for places like hotels, laundry, thermal power plant where sufficient heat energy in the form of steam, hot water is available.

But such installations are very limited. In larger number of installations, vapour compression refrigeration is used.

To recapitulate the working principle of vapour compression refrigeration system, the cycle of refrigeration and Carnot cycle is described in the next article.

8.1 REFRIGERATION CYCLE

Refrigeration is the science of producing and maintaining temperature (a product or space) that is lower than that of the surroundings. This involves removing heat from a body at a lower temperature and rejecting it to a body at a higher temperature (usually water or air). As this is against the law of nature, it is necessary to spend some energy in the form of work or heat.

To remove heat from a body and to reject it to the surroundings, it is necessary to have a working medium, which is known as a *refrigerant*. If a space or a product is required to be maintained at a temperature lower than that of the surroundings, it is necessary for the working medium to undergo a few thermodynamic processes in a cyclic manner. Different methods have been evolved over a period of time that are used for producing refrigeration. These methods are:

  i. Vapour compression refrigeration.

  ii. Vapour absorption refrigeration.

  iii. Air refrigeration.

  iv. Steam jet refrigeration.

  v. Non-conventional refrigeration.

Vapour compression and vapour absorption cycles are very commonly used in practice. These cycles are discussed in the next articles.

8.2 UNIT OF REFRIGERATION AND COEFFICIENT OF PERFORMANCE

The unit used for refrigeration is ton and is defined as the heat required to be removed from one ton (200 lb) of water at 32 °F in one day (24 hours) and convert it into one ton of ice at 32 °F.

1 ton of refrigeration

        = 200 Btu/min in British system (BTU)

        = 211 kJ/min in SI system

        = 50.4 Kcal/min in MKS system.

Coefficient of Performance (COP)

COP is a measure of efficiency of refrigeration cycle/system and is defined as

$$COP = \frac{\text{Refrigerating effect}}{\text{Energy spent}}$$

Where, refrigerating effect is the amount of heat removed/absorbed from the substance (air product, etc.) to be cooled.

The energy spent may be in the form of heat in vapour absorption system or in the form of work in vapour compression system. Both must be expressed in the same units.

COP of vapour compression cycle is generally about 4, whereas COP of vapour absorption cycle is between 0.7 to 1.6.

Carnot Cycle

Carnot cycle is an ideal cycle consisting of four reversible processes and it serves as a standard of comparison for other cycles used in practice. In this cycle, heat is transferred from a space or a substance at a lower temperature to another space or a substance at a higher temperature. The schematic of the Carnot cycle is shown in Fig. 8.1, and the cycle is shown on T-S diagram in Fig. 8.2.

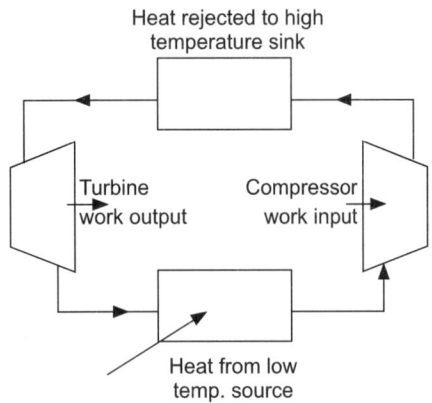

Fig. 8.1 Schematic carnot refrigeration cycle

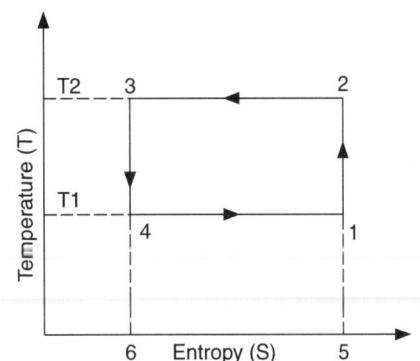

Fig. 8.2 Temperature-entropy diagram for carnot cycle

In this cycle, the processes take place between two given temperatures T_1 and T_2.

where

$T_1 =$ The temperature required to be maintained in the refrigerated space, in °K.

$T_2 =$ The temperature of the surroundings to which heat is rejected, in °K.

The Carnot cycle consists of four reversible processes mentioned below.

1–2 Adiabatic or isentropic compression.

2–3 Isothermal heat rejection.

3–4 Adiabatic or isentropic expansion.

4–1 Isothermal heat absorption or refrigeration.

As all the processes in the cycle are reversible, the cycle is the most efficient when operating between the given temperatures T_1 and T_2.

COP = Refrigerating effect/work done

= area 4-1-5-6/area 1-2-3-4

$$= \frac{T_1 (s_1 - s_4)}{(T_2 - T_1)(s_1 - s_4)}$$

$$= \frac{T_1}{(T_2 - T_1)}$$

It can be easily seen that for higher COP, T_1 should be as high as possible and T_2 should be as low as possible. It can be said in general, that in any refrigeration cycle heat should be removed at the highest possible temperature and heat should be rejected at the lowest possible temperature.

Limitation of Carnot Cycle

i. Reversible isentropic and isothermal processes are impracticable because for heat absorption and rejection, some temperature difference has to be maintained between two substances for heat transfer and both the internal and external friction exist in any process.

ii. Isothermal process requires a very slow speed of operation and it is extremely difficult to design such a mechanism.

iii. If a vapour is used as refrigerant and phase change is allowed during heat absorption and heat rejection processes, wet compression would result, which is difficult to achieve.

8.3 THEORETICAL SINGLE STAGE VAPOUR COMPRESSION CYCLE

The theoretical single stage vapour compression cycle is shown schematically and on T-S and P-H diagrams in Figs. 8.3, 8.4, and 8.5. In this cycle, vapour is used as the working medium, which absorbs heat during evaporation and rejects it to the surroundings (water or air) during condensation.

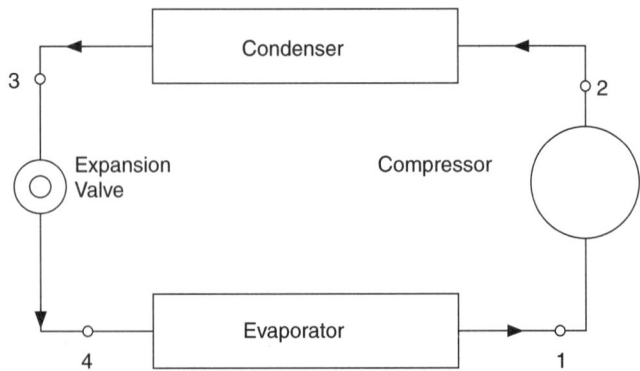

Fig. 8.3 Schematic vapour compression system

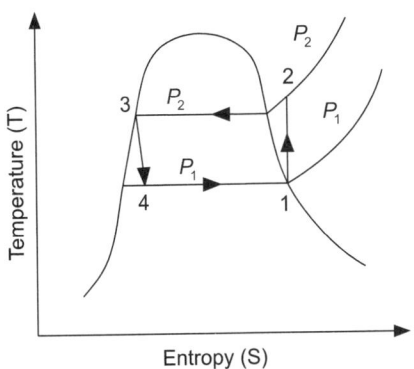

Fig. 8.4 Theoretical vapour compression cycle on T-S diagram

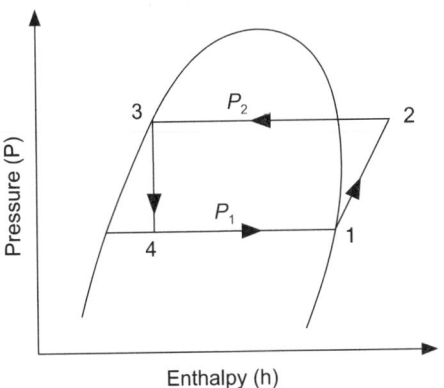

Fig. 8.5 Theoretical vapour compression cycle on P-H diagram

The following processes take place in an ideal vapour compression cycle.

1–2 Reversible Adiabatic (Isentropic) Compression

Dry saturated vapour is received from the outlet of the evaporator at state 1 and is compressed isentropically by the compressor from evaporator pressure p_1 to the condenser pressure p_2 to state 2.

2–3 Reversible Heat Rejection at Constant Pressure (Condensation)

The high pressure, superheated vapour from the compressor outlet enters the condenser at state 2 and rejects heat to the cooling medium (air or water) at constant condensing pressure. The vapour is first desuperheated and then is condensed and leaves the condenser as a saturated liquid at state 3.

3–4 Irreversible Expansion

The saturated liquid at state 3, enters the throttling device and expands to evaporator pressure in an irreversible manner at constant enthalpy to reach state 4, at which point, part of the refrigerant liquid flashes into vapour before it enters the evaporator.

4–1 Reversible Constant Pressure Heat Absorption (Evaporation)

The mixture of refrigerant vapour and liquid entering the evaporator at state 4 absorbs the heat from the surrounding space or product and evaporates at constant pressure and temperature and leaves the evaporator as dry and saturated vapour at state 1.

Therefore

$$COP = \text{Refrigerating effect/work done}$$
$$= (h_1 - h_4)/(h_2 - h_1)$$

Vapour compression cycle would have a COP smaller than that of Carnot cycle, because of the throttling process and the superheating effect at the compressor outlet. The COP can be improved by subcooling the liquid refrigerant and superheating the vapour in the evaporator.

8.3.1 Effect of Variation of Suction (Evaporation) and Discharge (Condensation) Pressures on COP

Decrease in suction pressure at a given discharge pressure increases the work done without causing a significant change in the refrigerating effect and in turn decreases the COP and vice versa. Increase in discharge pressure with a given suction pressure decreases the refrigerating effect and increases the work done, and hence decreases the COP.

8.3.2 Effect of Subcooling Liquid Refrigerant on COP and Refrigerating Capacity

Subcooling the liquid refrigerant increases the refrigerating effect, the work remaining constant, and hence there is an increase in COP.

8.3.3 Effect of Suction Superheat

To avoid wet compression, it is essential to ensure that no liquid refrigerant enters the compressor. This can be achieved if the refrigerant vapour is superheated before it enters the compressor by one of the following methods.

   i. Suction vapour and liquid refrigerant heat exchanger.

   ii. Superheating the suction vapour in the evaporator by using the thermostatic expansion valve.

   iii. Heat exchange with the surroundings.

Superheat of about 5 °C is generally used. The suction superheat has the following effects.

   a. The refrigerating effect increases in the first two cases.

   b. The specific volume increases from v_1 to v_2, which would decrease the mass flow rate in the given compressor.

c. The energy for compression of the refrigerant vapour per unit mass increases because of the diverging nature of the isentropic lines.

8.3.4 Suction-Liquid Heat Exchanger

Figure 8.6 shows the schematic diagram of suction-liquid heat exchanger, in which the refrigerant vapour leaving the evaporator outlet at state 1 and at lower temperature t_1, enters the heat exchanger, absorbs heat from the refrigerant liquid at higher temperature t_3, gets superheated to temperature t_1', and leaves the heat exchanger at state 1'. The process is shown on P-H diagram in Fig. 8.7. During this heat exchange, the condenser liquid gets subcooled from temperature t_3 to t_3' and leaves the heat exchanger at state 3'.

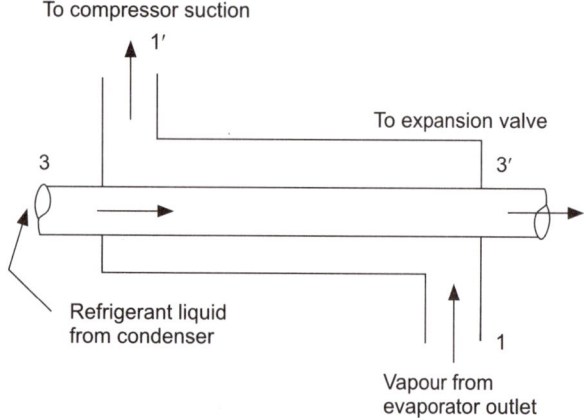

Fig. 8.6 Suction-liquid heat exchanger

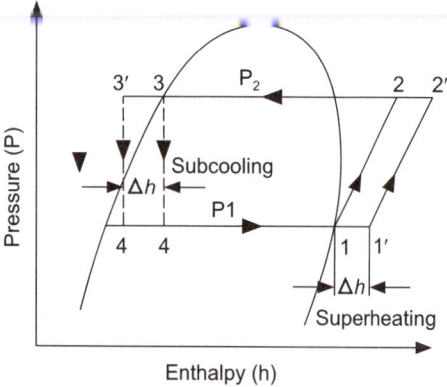

Fig. 8.7 Superheating of vapour and subcooling of liquid on P-H diagram

The heat exchange

= $h_1' - h_1$

= Increase in enthalpy of suction vapour due to superheating

= $h_3 - h_3'$

= Decrease in enthalpy of refrigerant liquid due to subcooling.

8.3.5 Actual Vapour-Compression Cycle

The actual vapour compression cycle differs considerably from the theoretical cycle due to the irreversibilities in various processes.

This irreversibility occurs mainly because of the temperature difference required to be maintained for heat transfer, the pressure drop taking place in the condenser and the evaporator and the irreversibility of the compression process. The theoretical cycle is shown by 1-2-3-4 whereas the actual irreversible cycle is shown by 1'-2'-3'-4' in Fig. 8.8. The COP of actual cycle is significantly lower than that of the theoretical cycle.

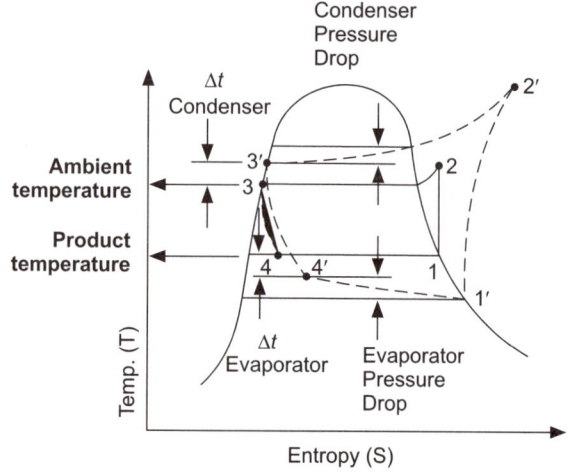

Fig. 8.8 Theoretical vapour compression cycle on P-H diagram

8.3.6 Multistage Vapour Compression Cycle

For low temperature applications, refrigerant vapour is required to be compressed over a large pressure range, giving rise to a high compression ratio and high discharge temperature that produces adverse effects on the performance of the refrigeration system. In such applications, the refrigerating effect per kg decreases, the work done per kg increases and the power/ton increases.

Generally, it can be said that if the compression ratio (r_p) is beyond 6-7 (9 in extreme cases), and if the evaporation temperature is below (–) 40 °C, multistaging is used. These values would vary with the refrigerant used and with the size of the plant. For refrigerants having large values of C_p/C_v, a lower limit of r_p should be used, e.g. for ammonia having C_p/C_v of 1.29, r_p should be less than 8, whereas for R-22, having C_p/C_v of 1.15, r_p should not exceed 9.

8.3.6.1 Advantages of Multistaging

With multistaging and inter cooling, the discharge temperature of the vapour can be decreased and following adverse effects on the plant can be avoided.

1. Low discharge temperature

Single stage compression with high pressure ratio increases the compressor discharge temperature, which may cause the lubricating oil breakdown, valve carbonization and maintenance problem. It will give rise to high temperature stresses and in extreme cases may shorten the life of the compressor. The discharge temperature should not be allowed to exceed 140 °C, as far as possible.

2. Volumetric efficiency

High compression ratio decreases the volumetric efficiency, and hence the capacity per unit displacement of compressor becomes small. With multistaging, the compressor displacement can be reduced, which may bring down the compressor cost.

3. Compression work

Multistaging reduces the work of compression and hence increases the COP. This reduces the operating cost of the plant.

4. De-superheating

De-superheating of compressor discharge vapour by intercooling reduces the displacement of the second stage compressor.

Intercooling will add to the capital cost and one must do a proper economic analysis of the increased capital cost and reduced operating cost of multistaging before taking a final decision.

8.3.6.2 Methods of Multistaging

Multistaging is also referred to as compounding. In compound systems the compressors are interconnected in series and the discharge from the low stage compressor is admitted to the suction side of the high stage compressor. The compressor may have two or more stages of compression contained in a single compressor (internally compounded), and one or more cylinders may be isolated from the other so that they may act as an independent stage of compression.

Interstage pressure

The interstage pressure for optimum plant performance is the geometric mean of the evaporator and the condenser pressures. In actual practice interstage pressure used may be different from this theoretical value based on practical considerations and research data available in this regard.

Number of stages

The number of stages used in compounded system depends primarily upon the temperature of eva-poration. For R-12, two stages are used up to a temperature of (−) 60 °C and three stages are used up to a temperature of (−) 73 °C whereas for R-22, two stages are used up to a temperature of (−) 68 °C and three stages are used up to a temperature of (−) 80 °C.

8.3.6.3 Cascade Refrigeration System

There are many industrial and medical applications where very low temperatures are required to be maintained, e.g. for blood storage, a temperature of (−) 80 °C is required whereas for precipitation, hardening of special alloy steels, a temperature of (−) 90 °C is required. For such applications, it is recommended to use cascade system instead of multi-stage compression system.

Cascade refrigeration system is a system with two independent vapour compression systems combined together in such a way that the evaporator of the high temperature system serves as the condenser of the low temperature system, but the refrigerants in the two systems are different and are not allowed to mix with each other. The intermediate heat exchanger is known as a cascade condenser. A schematic layout and T-S diagram of a typical cascade system are shown in Figs. 8.9 and 8.10 respectively.

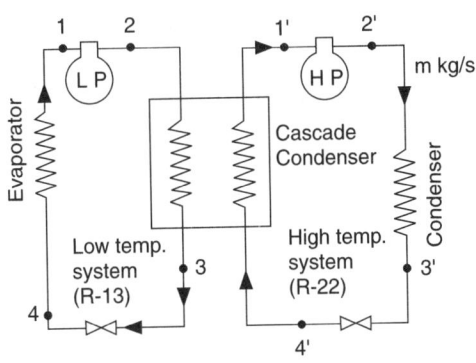

Fig. 8.9 Cascade system

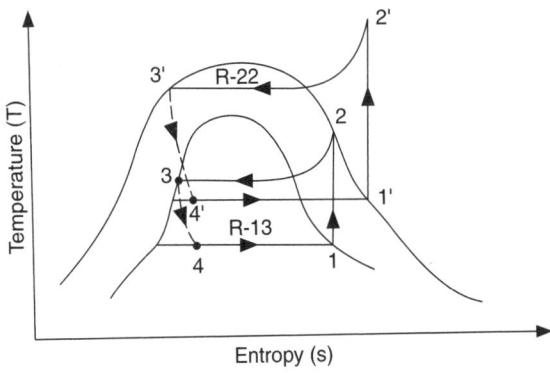

Fig. 8.10 Cascade system on T-S diagram for R-13 and R-22

When a refrigerant has suitable evaporator pressure but a very high condenser pressure, cascade system is used which employs refrigerants having progressively lower boiling points.

One advantage of the cascade system is that the oil from one circuit, does not pass into the other circuit, and hence the problem of balancing the oil in the crankcase between separate stage compressors can be avoided.

An important disadvantage of the cascade system is that an overlap in temperature has always to be provided in the cascade condenser. The condensing temperature of the low temperature system is usually about 5 °C above the evaporating temperature of the high temperature system. This overlap gives rise to higher power consumption.

8.4 REFRIGERANTS AND THEIR EFFECT ON ENVIRONMENT

Refrigerant is the medium through which heat of an air-conditioning system is transferred usually through evaporation and condensation. The properties of good refrigerant are:

1. It should be nontoxic, noninflammable and non-explosive.
2. It should be chemically stable and should not react with lubricant and foodstuff.
3. It should allow consumption of minimum power per ton of refrigeration or high COP.
4. It should have high latent heat, low specific volume of vapour, low compression ratio, low specific heat of liquid, high specific heat of vapour and low freezing point.

CFC and HCFC refrigerants leaking into the environment produce two undesirable effects on the environments, i.e.

1. Ozone depletion
2. Global warming.

The Montreal protocol signed in 1987 by European economic committee and 24 nations banned the use of CFC's globally to be achieved in a phased manner. Subsequently 162 countries have signed the treaty.

CFC-11, CF-114, R-12, R-22 have been found to be effecting the ozone and global warming and have been proposed to be replaced with new refrigerants. Out of these alternatives available, HFC-134 a (R-134 a) has been found to be best suited with respect to capacity and efficiency of the system. Global warming and ozone depletion potential of some refrigerants can be seen in Annexure A.

Ammonia is a natural refrigerant and is readily available at a comparatively low cost and has been used widely in industrial and commercial systems such as ice plants, cold storage, food freezing plants, dairies, breweries, etc. The properties in Table 8.1 shows that ammonia has many properties that make it a very good refrigerant. However, a few properties of ammonia shall be taken into account before deciding the application where it can be used.

a. For the same compression ratio, its discharge temperature is very high, hence water cooling of compressor cylinder may be required. This also puts a limit on the compression ratio up to which it can be used.
b. It has corrosive action on ferrous metals.
c. It is toxic.
d. It does not mix with lubrication oil and hence draining of oil has to be done cautiously using oil separater.
e. It dissolves in water and can affect food stuff.

8.5 COMPRESSOR

Compressor is the 'heart' of vapour compression system. It sucks the superheated vapours from evaporator and compresses it to raise its temperature and pressure above atmospheric, so as to discharge the heat into atmosphere which otherwise would not be feasible. The process of compression is required only to increase the temperature because heat flows from higher temperature to lower temperature (water or air), but in the process pressure also increases as it is the inherent property of any fluid. If we had any process by which we could increase the temperature without requiring to increasing the pressure, the compressor would not be required. But we know this is not possible.

Types of Compressors

With the change of technology, the nature of compression process has also undergone a substantial change. Earlier, it used to be only reciprocating compressors that gave way to centrifugal compressors and the latest in this change is scroll and screw compressors. Chart 8.1 gives classification of various kinds of compressors.

8.5.1 Reciprocating Compressors

These are also called positive displacement compressors. The name comes from the fact that whatever gases (vapours) are sucked in the suction process, all of those are positively driven out in the discharge process except for those in the clearance volume. The smallest has one cylinder and the largest has normally 16 cylinders.

The capacity (TR) depends on various factors but the most important being the swept volume. The smallest capacity can be for a fraction of a TR, while the largest being for about 300 TR or more.

The reciprocating compressors are normally designed for compression ratios of 8 to 10. Compression ratio r_p is the ratio of discharge pressure in absolute units to suction pressure in absolute units. Some of the heavy-duty types of reciprocating compressors are designed for ratios of 10 to 12. But at such high compression ratios, the volumetric efficiency suffers.

Table 8.1 Comparative refrigeration performance per kW of refrigeration for ammonia

Refrigerant / Properties	Ammonia	R-22	R-134 a
Evaporator pressure, MPa	0.236	0.296	0.16
Condenser pressure, MPa	1.164	1.194	0.77
Compression ratio	4.94	4.03	4.81
Refrigerating effect $\frac{kJ}{kg}$	1102.23	162.46	150.71
Refrigerating flow rate $\frac{kg}{s}$	0.00091	0.00616	0.00664
Sp vol. of suction gas m^3/kg	0.5106	0.0774	0.1224
Compressor displacement L/S	0.463	0.476	0.812
Compressor power kW	0.207	0.210	0.226
COP	4.84	4.75	4.42
Compressor discharge temperature k	371	326	316

The very word reciprocating means to and fro. It is a compressor where to and fro motion of a piston is converted into rotary motion by using a crankshaft.

Single acting compressors are most commonly used. Double acting compressors were used in a few cases in very old days and they are not in vogue any more for ACR applications.

Staging

The concept of multistaging has already been described in the previous article. Single stage com-pressors can be used either for water-cooled or air-cooled applications. This is because of the limitation on compression ratios. Single stage compressors are normally suitable for evaporating temperatures of around (–) 30 °C with condensing temperatures of + 40 to + 45 °C in case of halocarbons such as R-22, whereas they are limited to about 15 to (–) 20 °C evaporating and + 40 °C condensing for ammonia refrigerant. The BHP/TR for such compressors varies from 0.95 to 1 for standard air-conditioning applications.

The low-stage vapours have high specific volume, and therefore the number of low-stage cylinders is always higher than the high-stage cylinders. The typical combinations will be 21, 31, 42, 51, 63, 72, 93, 102, etc. where the last digit shows the high-stage cylinders. The combinations of low-stage and high-stage have to be balanced for the various flows at different pressures and temperatures.

8.5.2 Rotary Compressors

8.5.2.1 Vane Type

These are of fixed vane type or rotating vane (eccentric) type and are generally available in the range of 3 to 50 HP.

8.5.2.2 Rolling Piston Type

These are for small capacities of about 3 HP.

8.5.2.3 Trochoidal

These are small Wankel rotary type positive displacement compressors, which can run up to 9000 rpm and are used up to 2 TR capacities.

8.5.2.4 Scroll

Scroll compressors are orbital motion, positive displacement machines that compress with two interfitting, spiral-shaped scroll members. The motor stator is rigidly attached to the shell. The rotor is shrink fit onto the eccentric shaft.

Chart 8.1 Classification of compressors

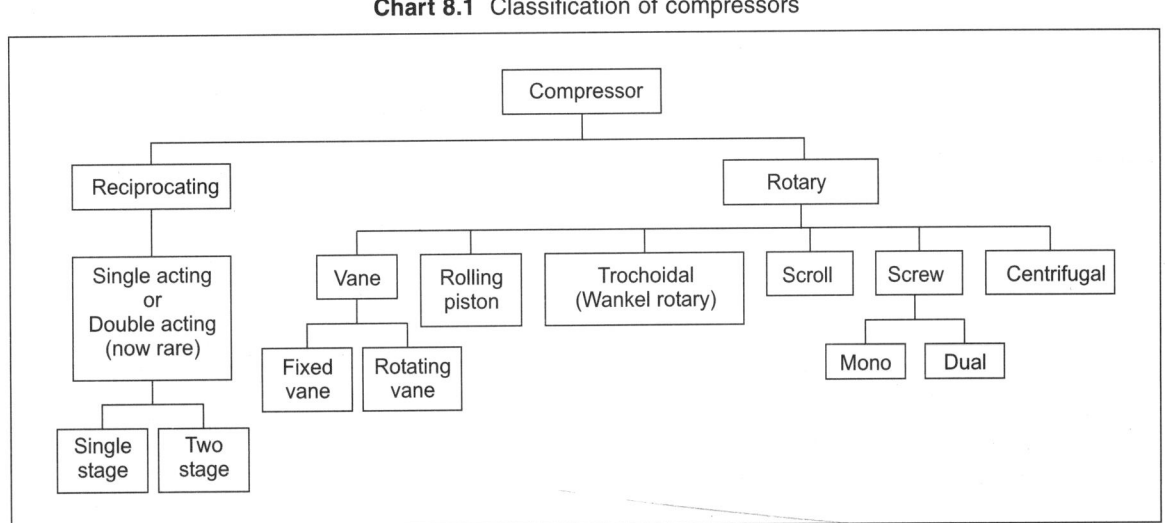

How the Scroll Compressor Works?

The scroll compressor is designed around two identical spirals or scrolls that, when inserted together, form crescent-shaped pockets. During a compression cycle, one scroll remains stationary while the other orbits round the first. As this motion occurs, gas is drawn into the scrolls and moved in increasingly smaller pockets toward the center. At this point, the gas, now compressed to a high pressure, is discharged from a port in the center of the fixed scroll [*see* Fig. 8.11a].

During each orbit, several pockets of gas are compressed simultaneously, creating smoother, nearly continuous compression.

They are currently used in residential and commercial air-conditioning, refrigeration and heat pump applications as well as in automotive air-conditioning and are developed up to about 100 TR applications.

They have better efficiency (COP) and can withstand liquid slop-overs compared to reciprocating type in that range but are somewhat costlier in the first cost although power consumption is lower. These compressors operate in the range of –32 to 12.8 °C for evaporator and 27 to 68 °C for condenser.

Since, the scroll requires no mechanical valves, it eliminates the valve losses associated with reciprocating type compressors. Also the absence of suction and discharge valves makes the operation extremely quite. A cut section of the scroll compressor is shown in Fig. 8.11b.

8.5.2.5 Screw Compressors

These are the main types of rotary compressors used for refrigeration application. There are two types namely.

  i. Mono-screw.

  ii. Dual-screw.

These compressors can be used for quite low temperatures of (–) 40 °C or lower with standard condensing temperatures or for very high condensing temperatures encountered in air-cooled applications. They can be used for high compression ratios of 20 to 25 in single stage. Although a number of applications also use two-stage versions, mainly when there is substantial load at the intermediate level. These compressors are available up to 190 TR applications.

Mono-screw compressors are available both in hermetic and open versions. Hermetics are preferred

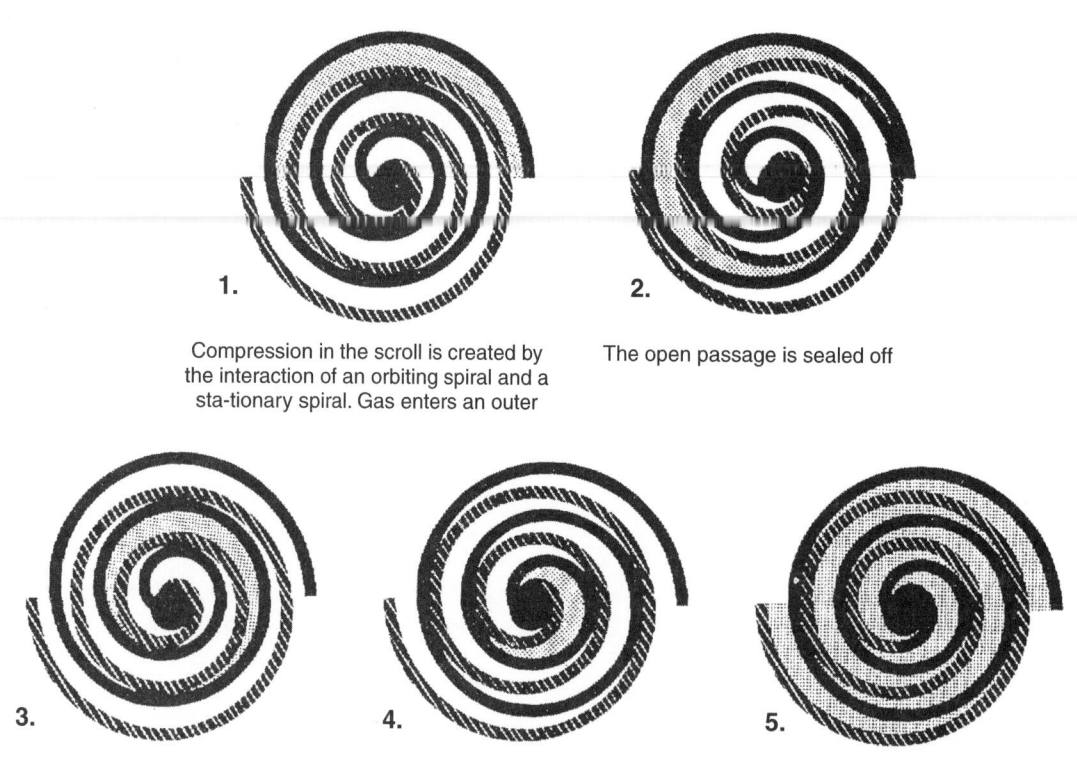

1. Compression in the scroll is created by the interaction of an orbiting spiral and a sta-tionary spiral. Gas enters an outer

2. The open passage is sealed off

3. As the spiral continues to orbit, the gas is compressed into an increasingly smaller pocket.

4. By the time the gas arrives at the center port, discharge pressure has been reached.

5. Actually, during operation, all six passages are in various stages of compression at all times, resulting in nearly continuous suction and discharge

Fig. 8.11a Operation of scroll compressor

Key Components

1. Discharge plenum
2. Thermal valve
3. Fixed scroll
4. Orbiting scroll
5. Crankcase
6. Counterweight
7. Eccentric shaft
8. Lower bearing ring
9. Lower bearing
10. Thrust washer
11. Magnet
12. Oil tube
13. Shell
14. Rotor
15. Stator
16. Suction tube
17. Electric terminal
18. Terminal cover
19. Suction baffle
20. Slider block
21. Internal pressure relief valve
22. Discharge tube
23. Check valve

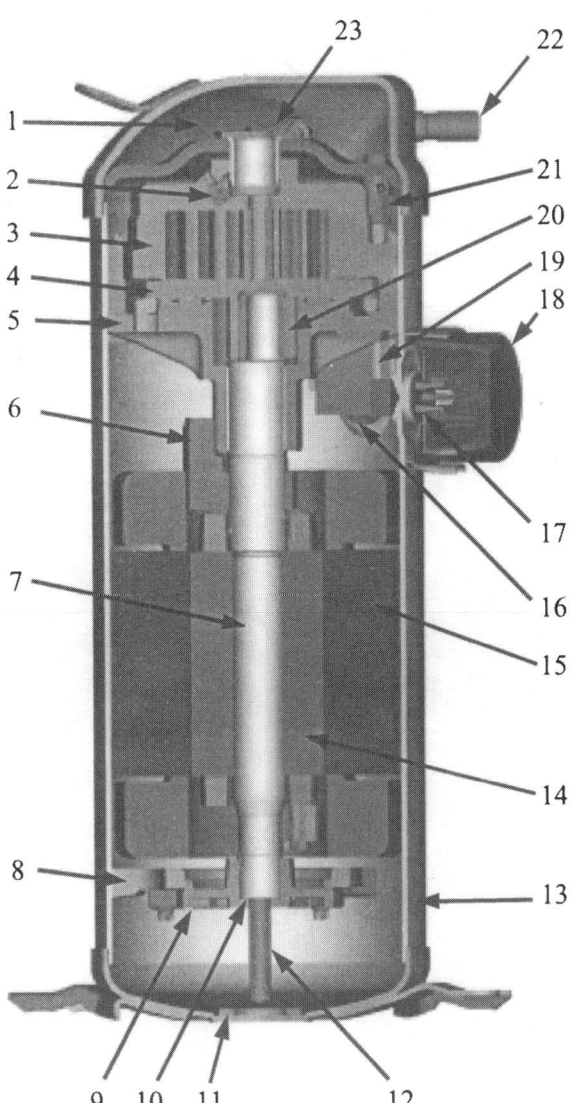

Fig. 8.11b Scroll compressor function description

for air-conditioning whereas open are used for refrigeration applications mostly with ammonia as refrigerant. Multiple mono-screw compressors are used in standard packaged chillers both for water-cooled and air-cooled applications. Various refrigerants such as R-22, R-404A, R-407C, R-410A are being used in line with the requirements of phasing out of HCFCs. Section of a typical mono-screw compressor is shown in the Fig. 8.11c.

Reliable Bearing Design

As the main screw rotor shaft and gate-rotor shafts cross at right angles in the single-screw compressor, sufficient space to locate each bearing is obtained. Because of this unique characteristic for the single-screw compressor, bearing lives can be designed much higher compared to twin-screws.

How Mono-screw Works?

The combination of six-screw chambers and eleven rotor teeth smooths out discharge flow by developing more and smaller pulsations. Meshing sequence is staggered to eliminate pure tones. This unique geometry combined with the absence of oil, results in significant noise reduction and less vibration transmitted through the piping system. An extremely short leakage path, compared to twin-screw compressors, greatly improves compressor efficiency.

Suction

Gas flows through the suction port into open ends of available grooves. As the screw rotates, gate-rotor teeth enter and seal grooves in sequence, trapping gas in the chambers defined by the three sides of the groove, the compressor casing and the gate-rotor tooth surface.

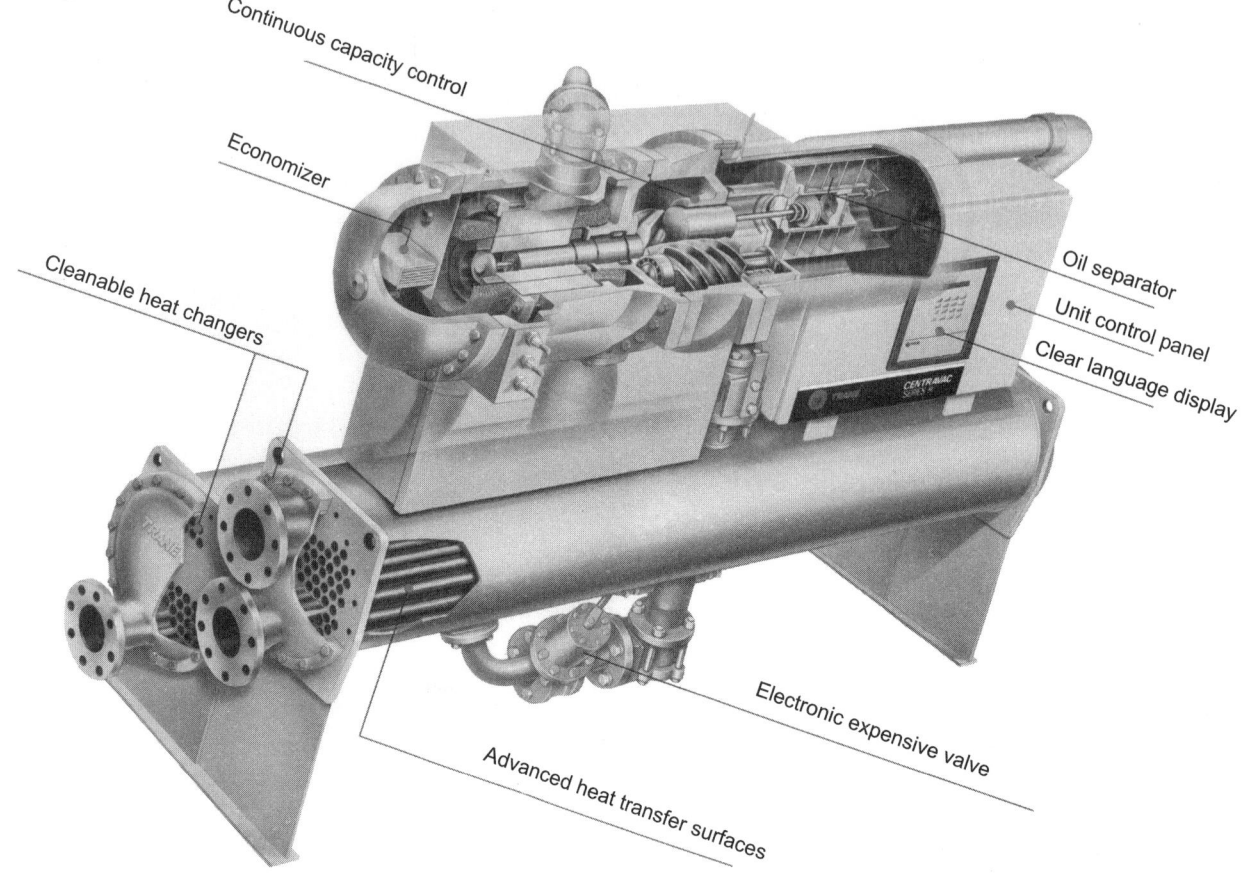

Fig. 8.11c Section of a rotary screw chiller

Compression

Volume within each tooth chamber decreases as the screw rotates. Gas is continuously compressed until the groove leading edge presses the discharge port.

Discharge

Release of gas into the discharge chamber ends the compression cycle. Gas continues to be discharged until the gate-rotor reduces groove volume to zero. With two gate-rotors on opposite sides of the screw, this process occurs twice with each screw revolution, doubling the compression capacity of the unit.

Twin Screw Compressors (Double Screw or Dual Screw)

These compressors consist of two mating helically grooved rotors-male (lobes) and female (flutes or gullies or anti lobes) in a stationary housing with inlet and outlet gas ports as shown in Fig. 8.12.

The refrigerant gas mainly flows in an axial direction. Most commonly used combinations of male + female rotors are 4 + 6, 5 + 6, 5 + 7. Normally male rotor is driven by a prime mover and female follows it, but in

some designs female rotors are driven through timing gears.

Twin helical screws are preferred for industrial refrigeration applications although they are also sometimes used for air-conditioning and heat pump applications. Commercially available compressors are suitable for common refrigerants like R-22, R-134a, R-404A, R-407C, R-410A and R-717, etc.

Compression Process

Compression process is obtained by direct volume reduction with pure rotary motion. For clarity, the following description of the three basic compression phases is limited to one male rotor lobe and one female rotor interlobe space Fig. 8.13.

Suction

As the rotors begin to unmesh, a void is created on both the male side (male thread) and the female side (female thread). The gas is drawn in through the inlet port. As the rotors continue to turn, interlobe space increases in size, and gas flows continuously into the compressor. Just prior to the point at which the interlobe space leaves the inlet port, the entire length of the interlobe space is completely filled with gas.

Compression

Further rotation starts the meshing of another male lobe with another female interlobe space on the suction end and progressively compresses the gas in the direction of the discharge port. Thus, the occupied volume of the trapped gas within the interlobe space is decreased and the gas pressure consequently increased.

Discharge

At a point determined by the designed built-in volume ratio, the discharge port is uncovered and the compressed gas is discharged by further meshing of the lobe and interlobe space.

During the re-meshing period of compression and discharge, a fresh charge is drawn through the inlet port on the opposite side of the meshing point. With four male lobes rotating at 3600 rpm, four interlobe volumes are filled and give 14,400 discharges per minute. Since the intake and discharge cycles overlap effectively, a smooth continuous flow of gas results.

8.5.3 Centrifugal Compressors

Centrifugal compressors sometimes referred to as turbo compressors belong to the same class of other turbo machines such as fans, propellers and turbines. These machines are high-speed machines and the volute casing converts kinetic energy into pressure energy. They can handle high volumetric flows as compared to reciprocating machines.

Centrifugal compressors are most commonly used for chilled water applications in air-conditioning. With multi-staging they can be adapted to low temperatures and high condensing temperatures. High head special single-stage impeller can develop higher-pressure ratios required for ice-skating rinks.

Both hermetic (actually semi-hermetic) and open designs are used widely.

Centrifugal compressors whether for air-conditioning or low temperature and brine applications used to be imported prior to 1969 in India.

The user must consider the following factors before selecting any machines.

1. The dead weight and operating weight of machines vis-à-vis the strength of floor on which this machine is to be mounted.

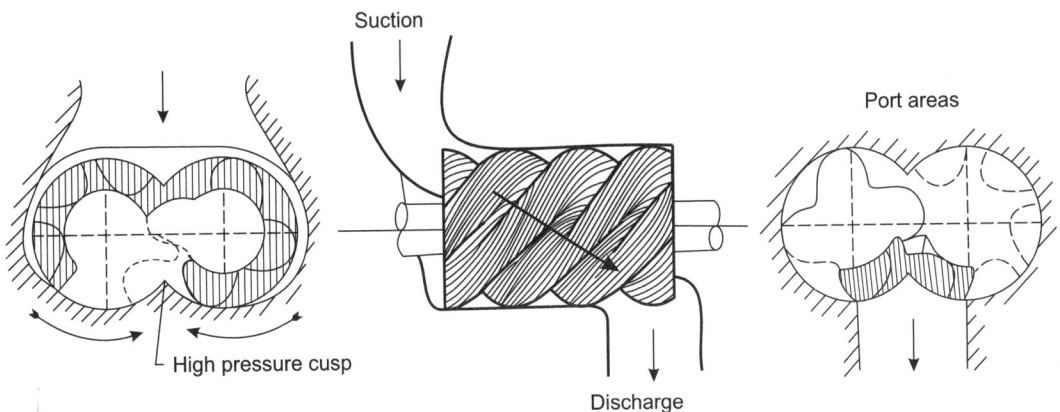

Fig. 8.12 Twin screw composser

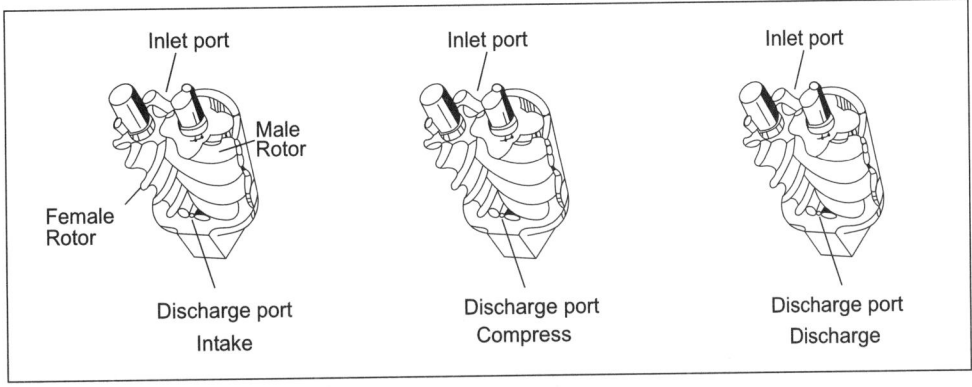

Fig. 8.13 Compression process

2. The dimensions of the machine, i.e. length and width.

3. Power consumption called I_{kW}/ton or input kW per ton. This is a very important factor as increasing the I_{kW} from 0.65 to 0.68 can make a difference of Rs. 4.00 lakh in terms of electricity consumption per year for a 600 ton machine as shown below.

   Increase in power consumption = 600 ton × (0.68 − 0.65) × 12 hours × 30 days/month × 12 month × Rs. 5/- per unit = Rs. 3,88,800/-.

Nowadays machines with I_{kW} of 0.55 are available.

Dual Compressor Centrifugal

The latest technology in centrifugal chillers is dual compressor centrifugal chillers Fig. 8.14. The advantage of dual compressor chiller are:

a. Less floor space requirement than multiple single compressor chiller.

b. At least 35% saving in installed cost compared to installing two separate chillers.

c. Eliminates piping to the second chiller and to additional pumps.

d. Fewer valves and controls.

e. Less rigging cost.

f. Less control wiring.

g. Twenty percent energy saving compared to two single units.

Fig. 8.14 Dual compressor chiller

The advantage of using dual compressor chiller will be clearer from the example given below.

Part Load Efficient

Most of the chillers usually spend 99% of their operating hours under part load conditions, and most of this time at less than 60% of design capacity. One compressor of a dual chiller operates with the full heat transfer surface of the entire unit. Thus, one 500 ton compressor on a 1000 ton dual chiller utilizes 1000 tons of evaporator and condenser surface. This increases the compressor's capacity and also results in very high efficiency. Typical efficiencies for a dual compressor chiller is like this.

A. Full load 0.55 kW/TR (6.5 COP).

B. 60% load, one compressor 0.364 kW/TR (9.6 COP).

Lower Installed Cost

Let us take an example where job requirement is 1200 TR with 50% backup.

Single Compressor Method

In this method, one has to use 2 × 600 TR + 1 × 600 TR, i.e. 3 × 600 TR machines = 1800 TR total installed capacity.

Dual Compressor Method

In this method, one will use.

   i. 1 × 750 TR dual compressor chiller at full load = 750 TR.

   ii. 1 × 750 TR dual compressor chiller at 60% load = 450 TR.

      Total = 1200 TR.

Thus, only 3 compressor out of 4 will be running 99% of the time when the load is 60%. However, redundancy is provided by 4th compressor.

The dual compressor centrifugal chiller are available from 160 to 2500 TR.

8.5.4 Limitation of Various Types of Compressors

Hermetic

Fully hermetic compressor which are also called fully welded compressors should be used for compact systems where the refrigerant lines (suction and discharge) are not too long. Proper oil recovery back to the compressor can become a problem and compressor may get damaged or burnt if lub oil does not return back in time due to long lines. Other types of open or semi-hermetic compressors should be considered where systems are not compact and/or extensive refrigerant piping with several bends and accessories is involved.

Semi-hermetic

These are not fully hermetic. This means to say that the compressors can be opened in the field without cutting the casing. These are of bolted construction so that in case of any problem, the compressor interiors can be opened and inspected. These types of compressors are cooled by refrigerant itself, and hence do not reject any heat in the surroundings and there is no need for special external mechanical ventilation. These compressors are compact and quiet. Large capacity centrifugal or screw packaged chillers are invariably termed as "hermetic chillers". The word "hermetic packaged chiller" is really a 'semi-hermetic' compressor system. This is a misnomer, but so long as everyone understands it properly, it is Okay.

Open

These types of compressors are used right from fractional to large tonnage systems. The driver (usually electric motor) is outside the compressor casing. Some type of coupling or drive set is required for connecting it to the driver. A good mechanical seal is used on the shaft to prevent leakage of any refrigerant vapours. Compressors can be direct driven at 1450 rpm on 50 Hz, 4 pole motors of 1750 rpm on 60 Hz. These speeds become 2900 rpm and 3500 rpm respectively for 2 pole motors.

The speed can be increased through various gear trains. Many-a-time, reciprocating open compressors are driven by using large flywheels and pulleys with a number of V belts.

Open compressors are universally used for ammonia refrigerant, as hermetic motors usually use copper for windings. Ammonia reacts with copper. Special purpose ammonia hermetic motors with aluminium windings are built and hermetic ammonia compressors are used to a very small limited extent but are not commercially available yet.

8.6 VAPOUR ABSORPTION REFRIGERATION

The high power consumption of vapour compression refrigeration is due to the presence of compressor. However, in vapour absorption, the compressor is eliminated. The machine works on the principle of good affinity of lithium bromide for water. The water is used as the refrigerant.

At normal pressure, water boils at 100 °C. The boiling point can be lowered by lowering the pressure. In a vapour absorption machine, the water that is being used as a refrigerant is introduced in an evaporator under a vacuum of 6 mm of mercury. At this pressure, water boils at 3.9 °C so the liquid refrigerant water instantly turns to water vapour as it enters the evaporator. This change of state requires energy, which is obtained by taking the heat from the surroundings and in process causing refrigeration.

Once the refrigerant water has evaporated in the evaporator, it must then be condensed back to a liquid. This is achieved by absorbing it into lithium bromide (LiBr) which is hygroscopic, i.e. it has ability to absorb water. The view of a single stage as well as a double stage vapour absorption machine is shown in Fig. 8.15a, b.

The single stage absorption cycle is explained in Fig. 8.16. The cycle works as follow.

1. Solution Pump

A dilute lithium bromide solution is collected in the bottom of the absorber shell. From here, a hermetic solution pump moves the solution through a shell and tube heat exchanger for pre-heating Fig. 8.17a.

Fig. 8.15a Single stage vapour absorption machine

Fig. 8.15b Double stage vapour absorption machine

2. Generator

After exiting the heat exchanger, the dilute solution moves into the upper shell. The solution surrounds a bundle of tubes which carries either steam or hot water. The steam or hot water transfers heat into the pool of dilute lithium bromide solution. The solution boils, sending refrigerant vapour upward into the condenser and leaving behind concentrated lithium bromide. The concentrated lithium bromide solution moves down to the heat exchanger, where it is cooled by the weak solution being pumped up to the generator Fig. 8.17b.

3. Condenser

The refrigerant vapour migrates through mist eliminators to the condenser tube bundle. The refrigerant

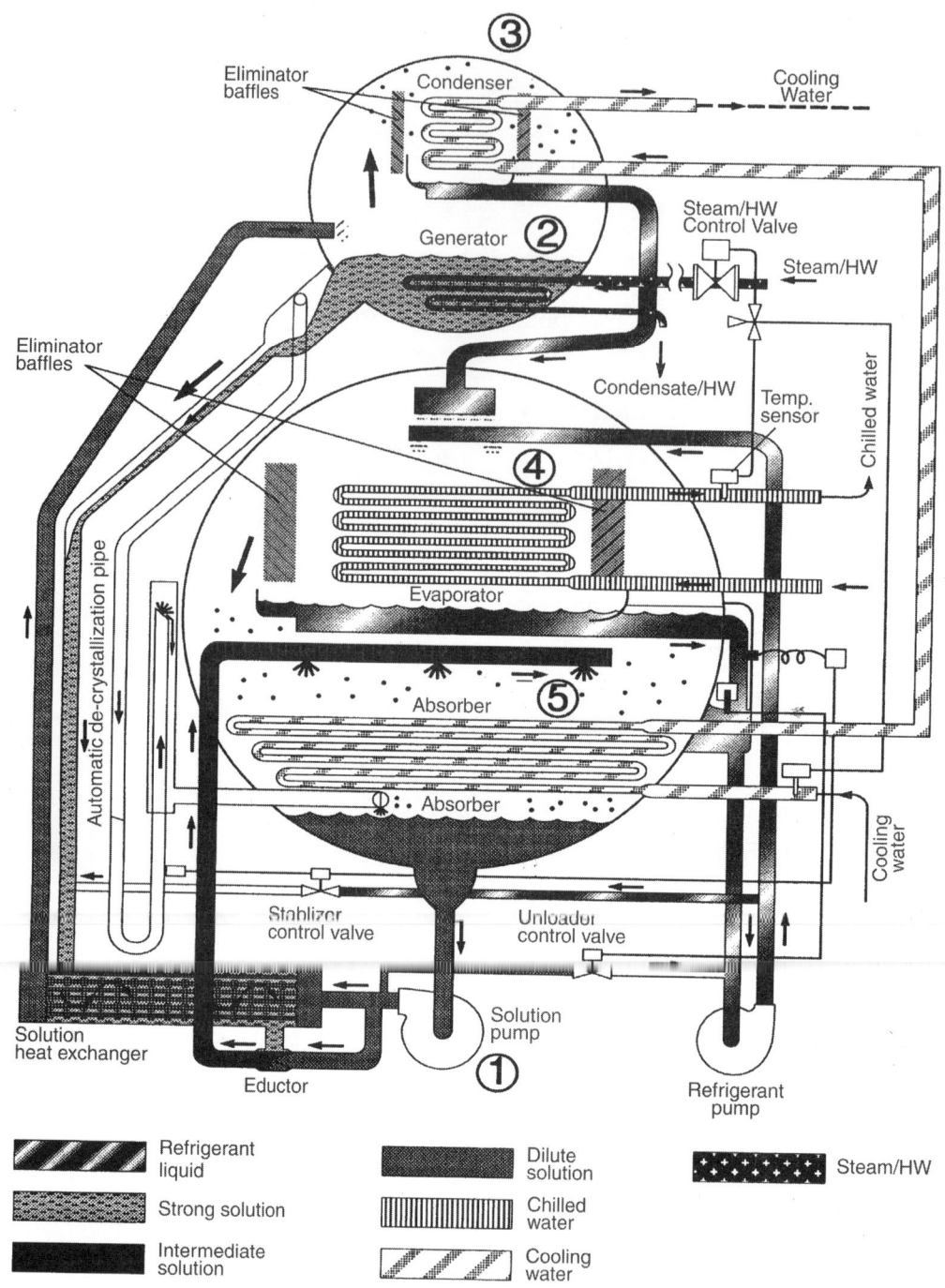

Fig. 8.16 Single stage absorption cycle

vapour condenses on the tubes. The heat is removed by the cooling water that moves through the inside of the tubes. As the refrigerant condenses, it collects in a trough at the bottom of the condenser Fig. 8.17c.

4. Evaporator

The refrigerant liquid moves from the condenser in the upper shell down to the evaporator in the lower shell and is sprayed over the evaporator tube

bundle. Due to the extreme vacuum of the lower shell (6 mm Hg vacuum), the refrigerant liquid boils at approx 3.9 °C, creating the refrigerating effect. This vacuum is created by hygroscopic action, i.e. the strong affinity LiBr has for water in the absorber (Fig. 8.17d).

5. Absorber

As the refrigerant vapour migrates to the absorber from the evaporator, the strong LiBr solution from the

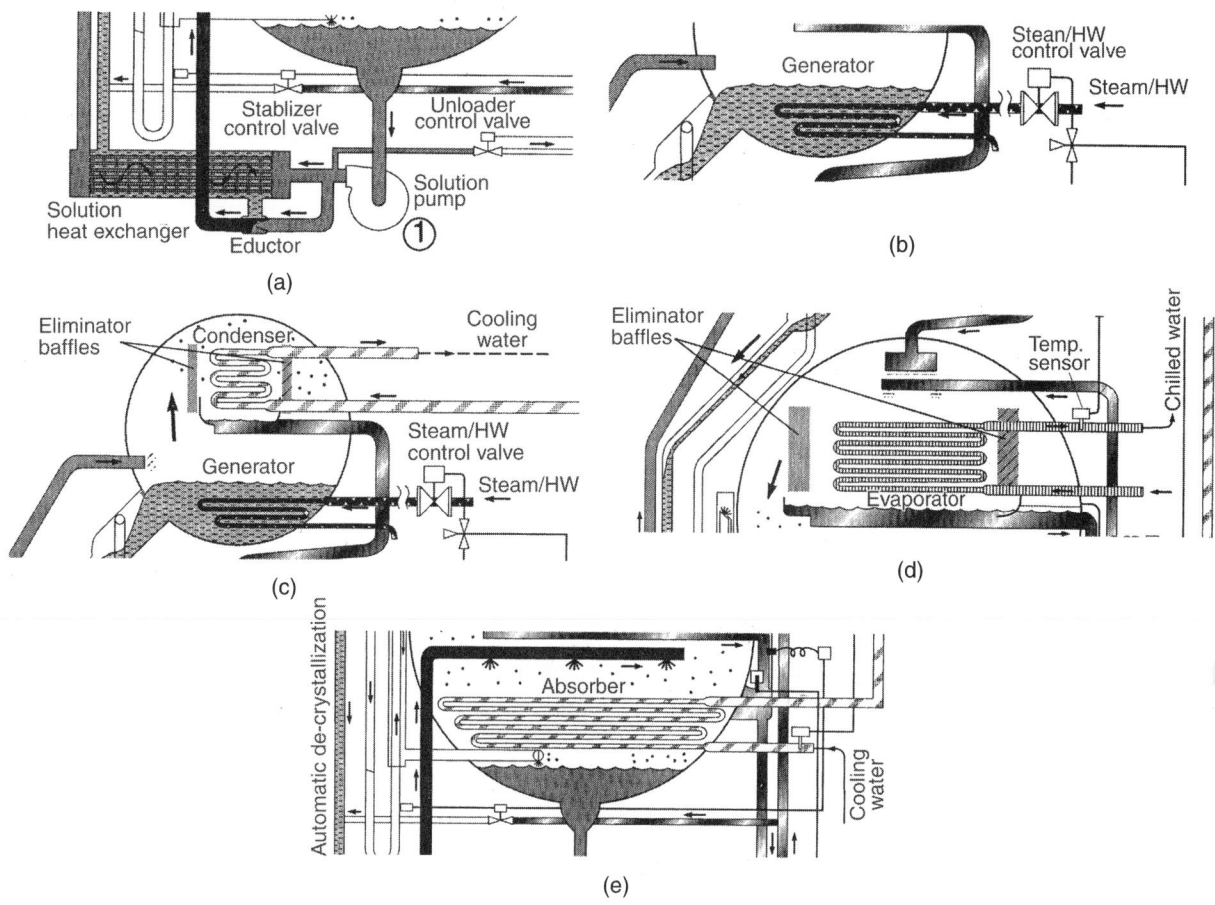

Fig. 8.17 Various stages of vapour absorption cycle

generator is sprayed over the top of the absorber tube bundle. The strong LiBr solution actually pulls the refrigerant vapour into solution, creating the extreme vacuum in the evaporator. The absorption of the refrigerant vapour into the LiBr solution also generates heat which is removed by the cooling water. Now the dilute LiBr solution collects in the bottom of the lower shell, where it flows down to the solution pump. The chilling cycle is now completed and the process begins once again (Fig. 8.17e).

This was about basic principle of vapour absorption refrigeration which we have read many times before. The advantages include practically NIL electricity consumption, no problem of Freon gases leaking and causing ozone depletion or global warming. Modern electronics and microprocessors control the performance of machines to provide much higher efficiencies and variation of capacities from 10% to 100%. The maintenance is much simpler as there are no moving parts. The capacity of single stage chillers is from 120 – 1377 TR.

Basic Requirements

The decision in favour or against the selection of absorption chiller for a particular application shall be guided by the availability of surplus steam at a desired pressure as given below.

Parameter	Single stage absorption	Double stage absorption
Quantity of steam required TR	14.5 Kg/Hr/TR	6.5 Kg/Hr/TR
Working pressure of steam	1.5 Kg/cm^2 saturated	8 Kg/cm^2 saturated
Maximum working pressure in water system	8 Kg/cm^2	8 Kg/cm^2

This much quantity of steam is easily available in hotel industries, hospitals, process and printing industry as the steam after being utilized for laundry, processing purposes goes waste and has to be otherwise thrown in the atmosphere. This steam is available at the desired pressure and can be effectively used for the purpose of air-conditioning.

Sometimes due to lean season or otherwise, when the steam generation is not sufficient or is not available at the desired pressure, then the steam has to be generated by burning fuel. The burner is available in the absorption chiller itself. However, the generation of steam by burning the fuel is not an economical option and the cost of providing air-conditioning per TR is even costlier than by centrifugal machine.

Hence, a proper cost-utilization analysis shall be done before a absorption machine is selected.

8.7 COMPARISON OF CHILLER MACHINES

All the chiller types described in this chapter have relative advantages and disadvantages. The availability of chillers up to a certain rating put a restriction of their use for high capacity installations. The efficiency of compressor is measured in terms of COP, i.e. coefficient of performance. The energy consumed vis-à-vis refrigerating effect produced is different for different kind of machines. Table 8.2 lists this parameter for various machines along with the maximum tonnage up to which various machines are available.

Thus, while selecting a particular type of chiller, designer shall specifically look at the limitations imposed by various compressors, the energy consumption, availability in terms of tonnage and refrigerant to be used. All these parameters have an important bearing on the proper selection of machine, not only in terms of capital cost but the operating cost also.

Table 8.2 Comparison of energy consumption of various chillers

S. No.	Type of chiller	Size range-tons	Full load operating cost kW/ton
1.	Reciprocating-water cooled.	5–240	0.86
2.	Reciprocating-air cooled.	5–450	1.2
3.	Screw-water cooled	125–675	0.6–0.7
4.	Screw-air cooled	100–300	1.1
5.	Centrifugal-packaged	150–2000	0.5–0.7
6.	Centrifugal-variable speed	150–1200	0.3–0.6
7.	Absorption-single stage	200–1500	9 litre/ton-hour
8.	Absorption-two stage	200–1500	4.5 litre/ton-hour

9

Mechanical Ventilation Fan Design

The most common application of mechanical ventilation is to supply fresh air to maintain an acceptably non-odorous atmosphere by diluting suspended contaminants, carbon monoxide and carbon dioxide concentration in the air. Another application of mechanical ventilation is in controlling the fire by exhausting the smoke and is generally provided where smoke cannot escape the promise on its own by means of natural ventilation, e.g. in basements of a building.

The mechanical ventilation is a detailed engineering involving the kind of fans to be used, static pressure to be maintained during normal conditions of occupation and during fire operation, the material and type of fans, their control, automatic operation etc.

9.1 MANDATORY REQUIREMENT OF NBC

National Building Code 2005 Part 8 has tabulated the minimum outside fresh air per person for various applications (Ref. Table 9.1) for a mechanically ventilated or air conditioned space. As per Annexure-C of NBC Part-8, it is required that staircases have to be ventilated at each landing by means of vent, opening to atmosphere, not less than 0.5 sq mt in the external wall and the top. If the staircase is in the center of the building, i.e. cannot be ventilated to atmosphere, then a positive pressure of 50 Pa shall be maintained inside. Such arrangement for pressurization shall operate automatically with the fire alarm.

NBC-2005 also specifies that shafts shall be pressurized to a positive pressure of 50 Pa and lift lobby to a pressure of 25–30 Pa if the lifts are in the core of the building. The mechanical ventilation system shall be able to operate automatically with the fire alarm. Same requirement has been specified for lift lobby pressurization when lifts are communicating with the basement.

The ventilation of basements, as per this code, shall be with vents with cross-sectional area not less than 2.5 percent of floor area spread evenly around the parameter of the basement; to be provided in the form of grilles, shafts. It specifies that alternatively a system of air inlets shall be provided at basement floor levels and smoke outlets at basement ceiling level. In multi-storeyed basements, the supply duct can be common to all floors but exhaust duct of each basement compartment shall be separate. ASHRAE standard 62–200 amended has specified minimum ventilation rates (Table 9.2a). The air circulation rate has been specified in the code (Ref. Table 9.2b) and with this, CO level can be maintained within 29 mg/m^3 with peak levels not to exceed 137 mg/m^3.

The system of mechanical supply and exhaust fans with a network of supply/exhaust ducts is a requirement imposed by the complicated modern buildings with multi-storeyed basements. The designer of a mechanical ventilation system has to be thus aware of all parameters and calculations required to calculate volume of air to be handled and the static pressure, type of fan, computerized flow dynamics (CFD) techniques to be used so as to design an efficient system.

9.2 FAN ENGINEERING BASICS

The important element in designing a mechanical ventilation system is the calculation of volume of air (CFM) and the static pressure under which this volume is to be handled. While CFM requirement is calculated from Table 9.2, it is necessary to calculate the static pressure so that the system works efficiently and the entire area is supplied with fresh air and exhausted of foul air or smoke in the minimum time.

Air has to move in a network of ducts and for propelling the air forward, fan must apply certain

Table 9.1 Outdoor air requirements for ventilation[1] of air conditioned areas and commercial facilities (Abstract from NBC–2005)

Sl. No.	Application	Estimated maximum[2] Occupancy Persons/100 m²	Outdoor air Requirement l/s/Person	(l/s/m²)	Remarks
(1)	(2)	(3)	(4)	(5)	(6)
1.	**Food and beverage service**				
	Dining rooms	70	10		
	Cafeteria, fast food	100	10		
	Kitchen (cooking)	20	8		Make up air for food exhaust may require more ventilating air. The sum of the outdoor air and transfer air of acceptable quality from adjacent spaces shall be sufficient to provide an exhaust rate of not less than 27.5 m³/h.m² (7.5 l/s²). Independent of room size.
2.	**Hotels, motels, resorts, dormitories**				
	Bedrooms	15			
	Living rooms			15	
	Baths			18	Installed capacity for intermittent use.
	Lobbies	30	8		
	Conference rooms	50	10		
	Assembly rooms	120	8		
	Dormitory sleeping areas	20	8		
	Office space	7	10		
	Reception areas	60	8		
	Telecommunication centers and data entry areas	60	10		
	Conference rooms	50	10		
3.	**Public spaces**				
	Corridors and utilities			0.25	
	Public restrooms, l/s/wc or urinal		25		Normally supplied transfer air.
	Locker and dressing rooms			2.5	Local mechanical exhaust with no re-circulation recommended
	Elevators			5.0	Normally supplied by transfer air.
	Retail stores, sales floors and show room floors				
	Basement and street	30		1.50	
	Upper floors	20		1.00	
	Storage rooms	15		0.75	
	Dressing rooms			1.00	
	Malls and arcades	20		1.00	
	Shipping and receiving	10		0.75	
	Warehouses	5		0.25	
	Smoking lounge	70	30		Normally supplied by transfer air, local mechanical exhaust; exhaust with no re-circulation recommended.
4.	**Specialty shops**				
	Barber shop	25	8		
	Beauty parlour	25	13		
	Supermarkets	8	8		
5.	**Sports and amusement**				
	Spectator areas	150	8		When internal combustion engines are operated for maintenance of playing surfaces, increased ventilation rates may be required.
	Game rooms	70	13	2.50	
	Swimming pools (pool and deck area)			2.50	Higher values may be required for humidity control.
	Playing floors (gymnasium)	30	10		
	Ballrooms and discos	100	13		
	Bowling alleys (seating area)	70	13		

(Contd...)

Table 9.1 (*Contd...*)

SI No. (1)	Application (2)	Estimated maximum[2] Occupancy Persons/100 m^2 (3)	Outdoor air Requirement l/s/Person (4)	(l/s/m^2) (5)	Remarks (6)
6.	**Theatre**				
	Ticket booths	60	10		
	Lobbies	150	10		Special ventilation will be needed to eliminate
	Auditorium	150	8		special stage effects (for example, dry ice
	Stages, studios	70	8		vapours, mists, etc.)
7.	**Transportation**				
	Waiting rooms	100	8		
	Platforms	100	8		Ventilation within vehicles may require special
	Vehicles	150	8		consideration.
8.	**Education**				
	Classrooms	50	8		
	Laboratories	30	10		Special contaminant control systems may be
	Training shop	30	10		required for processes or functions including
					laboratory animal occupancy.
	Music rooms	50	8		
	Libraries	20	8		
	Locker rooms			2.50	
	Corridors			0.50	
	Auditoriums	150	8		
9.	**Hospital, nurses and convalescent homes**				
	Patient rooms	10	13		Special requirements or codes provides and
	Medical procedure	20	8		pressure relationships may determine minimum
	Operating rooms	20	15		ventilation rates and filter efficiency.
	Procedure recovery and ICU		20	8	Generating contaminants may require higher rates.
	Autopsy			2.50	Air shall not be re-circulated into other spaces.
	Physical therapy	20	8		
	Correctional cells	20	10		
	Dining halls	100	8		
	Guard stations	40	8		

[1] This table prescribes supply rates of acceptable outdoor air required for acceptable indoor air quality. These values have been chosen to dilute human bioeffluents and other contaminants with an adequate margin of safety and to account for health variations among people and varied activity levels.

[2] Net occupiable space.

amount of pressure to it. This pressure can be categorized as:

a. Static pressure

b. Velocity pressure

9.2.1 Static Pressure

Static pressure is the difference between absolute pressure at a point and the atmospheric pressure. This is also known as gauge pressure as it is the pressure indicated by pressure gauge. Consider Fig. 9.1 in which the static pressure has been clearly depicted as being zero, positive or negative. In Fig. 9.1a, the pressure inside the system exactly balances the atmospheric pressure so that level of liquid in two tubes of monometer is exactly the same. In Fig. 9.1b, the inside

pressure is more than atmospheric pressure (open end of tube) and thus liquid level in the open end of tube is higher than the other end and the system is said to be under positive pressure.

In Fig. 9.1c, the liquid levels are reverse of Fig. 9.1b and thus the system is said to be under negative pressure as compared to atmospheric pressure.

The atmospheric pressure is 10333 mm of water column while in normal HVAC applications, the pressures encountered are in the range of 200–500 mm of water column (Higher ranges are encountered in HVAC application for operation theaters which require HEPA filters causing more drop in pressure). When a ventilation fan sucks air to deliver it into the duct (Fig. 9.2), the pressure on the inlet side becomes negative while on the outlet of the fan, it becomes positive.

Table 9.2a Minimum ventilation rates (Abstract from ASHRAE standard 62–2001)

Occupancy category CFM/ Person	People outdoor air rate CFM/ Person	Area outdoor air rate (CFM/sq.ft.)
Lecture hall (Fixed seats)	7.5	0.06
Art class room	10	0.18
Computer lab	10	0.12
Music/ Theatre/ Dance	10	0.06
Multi use assembly	7.5	0.06
Restaurant dining room	7.5	0.18
Cafeteria/ Fast food dining	7.5	0.18
Conference/ Meeting	5	0.06
Corridors		0.06
Bedroom/ living room	5	0.06
Office room	5	0.06
Reception area	5	0.06
Auditorium seating	5	0.06
Courtroom	5	0.06
Libraries	5	0.12
Museums/ galleries	7.5	0.06
Mall common areas	7.5	0.06
Supermarket	7.5	0.06
Gym/ stadium (play area)		0.3
Spectator area	7.5	0.06
Stages/ studios	10	0.06

Table 9.2b Recommended rate of air circulation for different areas (Abstract from NBC-2005)

Sl. No. (1)	Application (2)	Air change per hour (3)
1.	Assembly rooms	4–8
2.	Banks/building societies	4–8
3.	Bathrooms	6–10
4.	Bedrooms	2–4
5.	Billiard rooms	6–8
6.	Cafes and coffee bars	10–12
7.	Canteens	8–12
8.	Churches	1–3
9.	Cinemas and theatres	10–15
10.	Club rooms	12, Min
11.	Conference rooms	8–12
12.	Dairies	8–12
13.	Dance halls	12, Min
14.	Entrance halls	3–5
15.	Factories and workshops	8–10
16.	Garages	6–8
17.	Glass houses	25–60
18.	Gymnasium	6, Min
19.	Hospitals-sterlising	15–25
20.	Hospital-wards	6–8
21.	Hospital domestic	15–20
22.	Laboratories	6–15
23.	Laundries	10–30
24.	Lavatories	6–15
25.	Lecture theatres	5–8
26.	Libraries	3–5
27.	Living rooms	3–6
28.	Offices	6–10
29.	Recording studios	10–12
30.	Restaurants	8–12
31.	Schoolrooms	5–7
32.	Shops and supermarkets	8–15
33.	Stores and warehouses	3–6
34.	Squash courts	4, Min
35.	Swimming baths	10–15
36.	Toilets	6–10
37.	Utility rooms	15–20

Note: The ventilation rates may be increased by 50% where heavy smoking occurs or if the room is below ground.

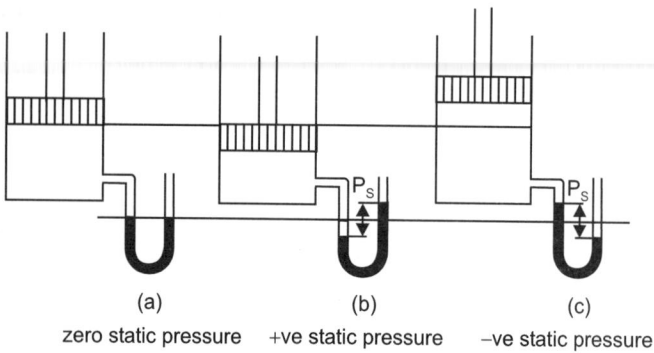

(a) zero static pressure   (b) +ve static pressure   (c) −ve static pressure

Fig. 9.1 Static pressure

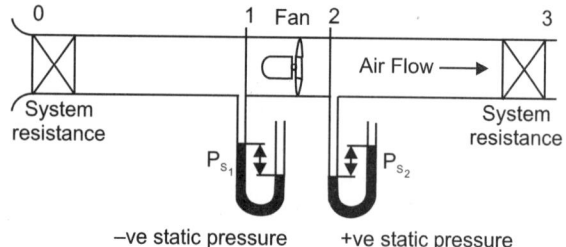

−ve static pressure   +ve static pressure

Fig. 9.2 A fan supplying fresh air to an enclosed space (ducted inlet, ducted outlet installation)

9.2.2 Velocity pressure

Also known as dynamic pressure, it is the component of pressure associated with velocity imparted to air by the fan. Air moving at velocity 'v' and having density 'ρ' has a velocity pressure (P_v):

$$P_v = 1/2 \, \rho v^2 \text{ Pascals}$$

The velocity pressure is always a positive quantity and in fact it is due to this reason that air moves forward. The velocity 'V' is the average velocity of air flowing in a circular duct, zero at surface and maximum at the center, with recordings taken at various points and deriving average. The quantity (Q) of air flow is thus calculated with average velocity and cross sectional area 'A' of the duct:

$$Q = V \times A$$

where Q = air flow quantity in m^3/s

V = Average velocity in m/s

A = Cross sectional area of duct in m^2

9.2.3 Total Pressure

The total pressure with which air is flowing is defined as the sum of static pressure and velocity pressure:

$$P_t = P_s + P_v$$

Now, P_s = Static pressure (can be –ve or +ve)

P_v = Velocity pressure (can be +ve only)

Refer Fig. 9.2 again, as the fan sucks the air, the pressure becomes negative here and the air is drawn inside with velocity V, the net pressure here is

$$P_{t_1} = -P_{s_1} + P_v$$

where P_{s_1} is the static pressure at suction

and P_v is the velocity pressure

Similarly as the fan discharge air, the static pressure (P_{s_1}) is positive along with velocity pressure P_v, the net pressure is

$$P_{t_2} = P_{s_2} + P_v$$

Example

Let air is passing through a round duct with diameter of 250 mm. The average velocity of air in the duct is 10 m/s. Find velocity pressure and total pressure at suction and discharge of fan.

Also assume $P_{s_1} = -20$ mm, $P_{s_2} = +5$ mm

Duct area $= \dfrac{\pi}{4}d^2 = \dfrac{\pi}{4} \times (.25)^2 = 0.05 \ m^2$

Quantity of flow $= V \times A = 10 \times 0.05 = 0.5 \ m^3/sec$

$$= 10.62 \ CFM$$

Velocity Pressure $\left(P_{v_1} = P_{v_2}\right) = \dfrac{1}{2}\rho V^2 = \dfrac{1}{2} \times 1.2 \times 10^2$

$$= 60 \ Pa = 6mm$$

$$P_{t_1} = -P_{s_1} + P_v = -20 + 6 = -14 \ mm$$

$$P_{t_2} = +P_{s_2} + P_v = 5 + 6 = 11 \ mm$$

Let us extend this example to fan. The total pressure created by the fan moves the air through the system and this total pressure at equilibrium is equal to the total losses in pressure in the system so as to give the air desired velocity to move through the system.

In other words, energy imparted by fan to air is consumed in overcoming the losses like frictional losses, filters, coil, dampers etc. as well as in providing energy to the air to make forward motion and reach destination.

Fan static pressure = Fan total pressure – Fan velocity pressure

$$= \left(P_{t_2} - P_{t_1}\right) - P_{v_2}$$

$$= \left(P_{s_2} + P_{v_2}\right) - \left(P_{s_1} + P_{v_1}\right) - P_{v_2}$$

$$= P_{s_2} - P_{s_1} - P_{v_1}$$

Thus fan static pressure P_{S_F} is not always equal to differece of static pressure before and after fan.

In the above example:

Fan total pressure $= P_{t_2} - P_{t_1} = 11 - (-14)$

$$= 25 \ mm$$

Fan static pressure $= P_{s_2} - P_{s_1} - P_{v_1}$

$$= 5 - (-20) - 6$$

$$= 19 \ mm$$

Increase in static pressure from inlet to outlet

$$= P_{s_2} - P_{s_1} = 5 - (-20) = 25 \ mm$$

Thus it can be seen that fan static pressure need not necessarily be equal to rise in static pressure from inlet to outlet.

9.2.4 Analysis of Pressure Changes

Let us now examine the changes taking place in static velocity and total pressure in a system with the help of illustration (Refer Fig. 9.3).

As can be seen, the suction mouth of the duct is bell shaped and suction and discharge both are ducted.

Level 0, the static, velocity and total pressures are zero just outside the duct. As air enters the duct, velocity pressure increases from 0 to 6 mm (air accelerates from 0 to 10 m/s) while static pressure drops by equal margin, i.e. 6 mm or it can be said that static pressure converts into velocity pressure making total pressure constant (i.e. no loss at entry).

Level 1, the air has traveled certain distance from level 0 to level 1 and in the process has suffered some frictional losses in duct, filters and coils. This drop in total pressure is 14 mm. As air velocity remains constant (due to uniform duct dia), the velocity pressure also remains constant and thus the static pressure drops by 14 mm. At plane 1, various pressures are:

$$P_v = 6 \ mm$$

$$P_s = -6 - 14 = -20 \ mm$$

$$P_t = -20 + 6 = -14 \ mm$$

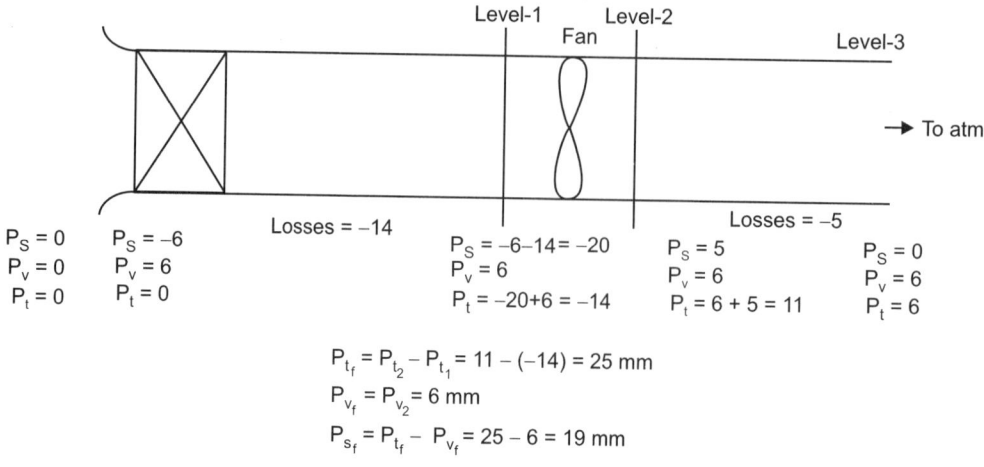

$$P_{t_f} = P_{t_2} - P_{t_1} = 11 - (-14) = 25 \text{ mm}$$

$$P_{v_f} = P_{v_2} = 6 \text{ mm}$$

$$P_{s_f} = P_{t_f} - P_{v_f} = 25 - 6 = 19 \text{ mm}$$

Fig. 9.3 Ducted inlet, ducted outlet

Level 2, On entering the fan, the fan has raised the pressure of air and various values become

$$P_{v_2} = 6 \text{ mm (constant)}$$

$$P_{s_2} = 5 \text{ mm (given)}$$

$$P_{t_2} = 6 + 5 = 11 \text{ mm}$$

Level 3, (Outlet) Here again the velocity pressure remains constant, but static pressure drops due to duct friction, losses in grilles and dampers by 5 mm so that the various pressures at level 3 becomes.

$$P_v = 6 \text{ mm}$$

$$P_s = 0 \text{ mm (atmospheric discharge)}$$

$$P_t = 0 + 6 = 6 \text{ mm}$$

This velocity pressure of 6 mm is lost as air is discharged into atmosphere at velocity pressure of 6 mm.

A simple analysis shows that fan has done work on air by raising the total pressure from −14 mm to 11 mm, i.e. 25 mm, out of which 19 mm was lost in friction encountered on the way and balance 6 mm was used to accelerate the air from 0 to 10 m/s. This velocity pressure was ultimately lost due to discharge in atmosphere.

Similar exercise (Figs. 9.4 to 9.6), show three other situations having different type of inlets/outlets and fan position.

Figure 9.6 is an example of ductless system like an exhaust fan ventilating directly to atmosphere without any resistance in between.

The velocity pressure that is going waste into the atmosphere in the form of high discharge velocity is an unnecessary burden on fan and energy is being wasted. This velocity pressure can be recovered by providing a diffuser (a divergent duct piece) as shown in Fig. 9.7. This diffuser causes velocity to reduce as the air reaches

the outlet of diffuser and in the process converts into static pressure increase, keeping total pressure constant. This will enable designer to select a fan of low static pressure as system pressure loss is taken care, to some extent, by this increase in static pressure.

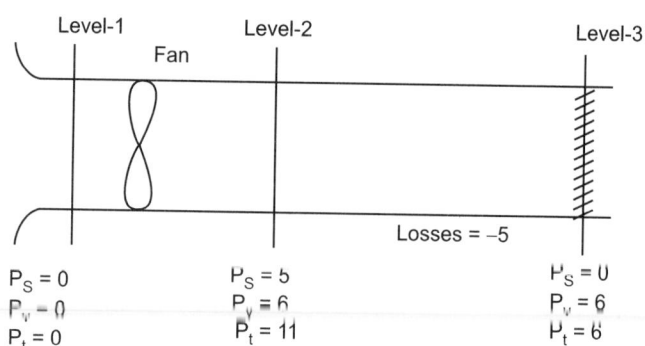

$$P_{t_f} = P_{t_2} - P_{t_1} = 11 - 0 = 11 \text{ mm}$$

$$P_{v_f} = P_{v_2} = 6 \text{ mm}$$

$$P_{s_f} = P_{t_f} - P_{v_f} = 11 - 6 = 5 \text{ mm}$$

Fig. 9.4 Free inlet, ducted outlet

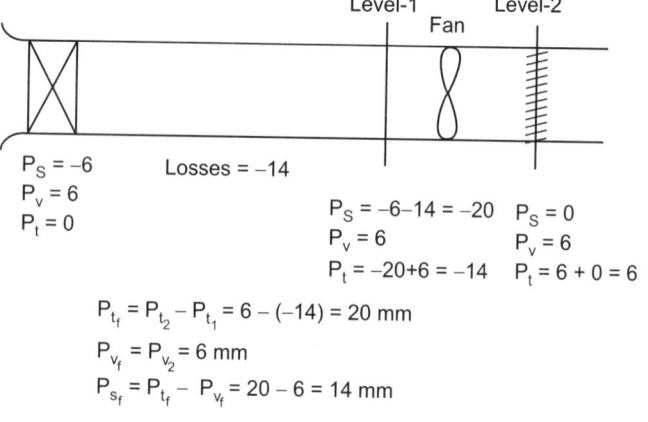

$$P_{t_f} = P_{t_2} - P_{t_1} = 6 - (-14) = 20 \text{ mm}$$

$$P_{v_f} = P_{v_2} = 6 \text{ mm}$$

$$P_{s_f} = P_{t_f} - P_{v_f} = 20 - 6 = 14 \text{ mm}$$

Fig. 9.5 Ducted inlet, free outlet

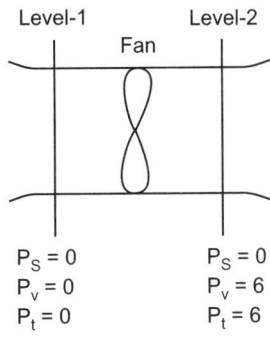

$$P_{t_f} = P_{t_2} - P_{t_1} = 6 - 0 = 6 \text{ mm}$$
$$P_{v_f} = 6 \text{ mm}$$
$$P_{s} = 6 - 6 = 0 \text{ mm}$$

Fig. 9.6 Free inlet, free outlet

Let us calculate the benefit in terms of saving of fan power by providing a diffuser at the discharge end of duct.

Let

Velocity in duct and fan outlet = 10 m/s

Velocity pressure, $P_v = 60 \text{ Pa} = 6 \text{ mm}$

Pressure loss in duct = 14 mm

Total fan pressure $\left(P_{t_2} - P_{t_1}\right) = 6 - (-14) = 20 \text{ mm}$
(as velocity pressure is lost at discharge)

Let us provide a diffuser in the outlet so as to regain half of velocity pressure, otherwise being lost at discharge.

Gain in static pressure = 3 mm (half of velocity at discharge pressure)
Pressure loss in duct = 14 mm

Total fan pressure $\left(P_{t_2} - P_{t_1}\right) = 3 - (-14) = 17 \text{ mm}$

Thus the fan with less static pressure will be selected and thus less initial cost of installation.

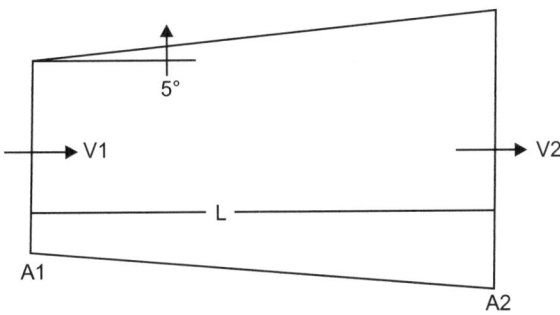

Fig. 9.7 A diffuser

9.3 SELECTION OF FANS

Fans are necessary to enable the air flow over heat exchanger like AHU in HVAC, duct network, provide fresh air in the building to maintain required oxygen level, replenish fresh air, exhaust smoke and other gases from basement.

The fans are classified into two types:

i. Centrifugal fan
ii. Axial fan

The difference in two categories of fans is in the direction of air flow through the fan.

9.3.1 Centrifugal Fans

In this category of fans, the air is pulled along the fan shaft and then blown radially away from the shaft. The centrifugal fans are further subdivided into following categories according to shape of blades (Ref. Fig 9.8).

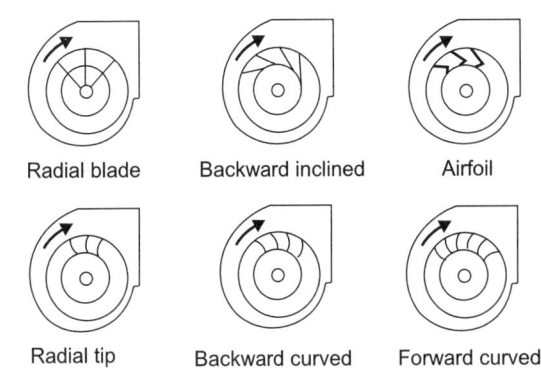

Fig. 9.8 Various types of centrifugal fan impeller blades

i. Forward curved fans
ii. Backward curved fans
iii. Radial fans

The performance characteristic of a fan is a graph between volume flow rate of air and the pressure developed. Other performance characteristic of importance is the relation between efficiency and brake horse power. The performance characteristic of forward and backward curved fans is shown in Fig. 9.9 and 9.10. It can be seen that:

i. For both type of fans, the pressure developed has a slight peak in the middle range of flow, then the pressure drops off as the flow increases.

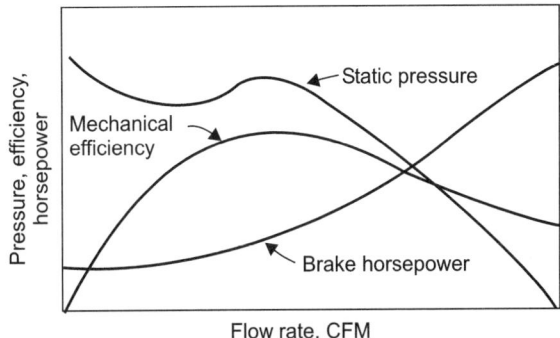

Fig. 9.9 Typical performance characteristics of a forward curved blade centrifugal fan

ii. The BHP required for forward curved fan increases sharply with flow, but with backward curved fans, the BHP increases only gradually, peaks at a maximum and then falls off.

iii. Efficiency is highest in the middle ranges of flow.

iv. A higher maximum efficiency can be achieved with a backward curved blade fan.

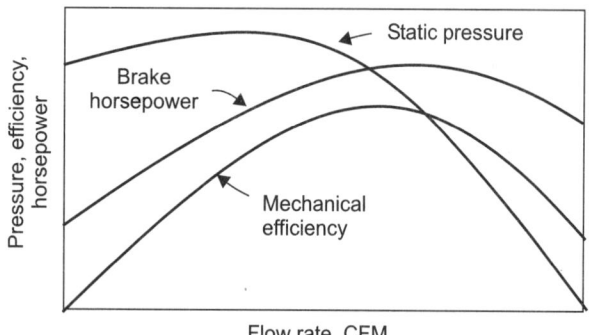

Fig. 9.10 Typical performance characteristics of a backward curved blade centrifugal fan

The choice of best type of fan to be used for a given application depends on fan performance characteristics and other features. Centrifugal fans are most commonly used in ducted air conditioning. Forward curved are usually lower in cost than backward types for the same performance. The operating cost is however higher due to lower efficiency. The rising BHP curve could result in overloading of the motor if operated at a condition beyond the selected CFM. Backward curved blade fans are generally more expensive than forward types, but have lower operating costs due to higher efficiency. The limiting horsepower characteristics reduces the possibility of overloading the motor, if the fan is delivering more air than it was designed for.

9.3.1.1 Aerofoil Blade Fans (AF)

This is another category of centrifugal fans (backward curved) with the difference that impeller in backward curved fans is made of single thickness sheet while in aerofoil fans the impeller is made hollow with thin sheet. The efficiency of aerofoil blade fans is the highest, approaching 92%, but these fans are limited only by the cost of manufacturing. They are used for clean air applications as they are more prone to corrosion due to thin sheet of metal used for making impeller.

Single inlet vs double inlet fans

A single inlet centrifugal fan sucks air from one side only while the handling capacity of fan is almost doubled if a clockwise and an anticlockwise fans are combined into a single unit. This unit is thus called double inlet double width (DIDW). This unit can handle 90% more air compared to single unit wheel, but power consumption is almost doubled. This kind of unit saves

space for handling the same volume of air compared to two single inlet units.

9.3.1.2 Efficiency Comparison

The variation in efficiency of various categories of centrifugal fans is given below (Table 9.3):

Table 9.3 Efficiency comparison of various categories of centrifugal fans

Parameter		Backward curved		Forward curved
	Aerofoil	Backward curved	Backward inclined	
Blades	5–15	5–15	5–15	25–65
Max efficiency(%)	92	84	77	71
Cost	High	Medium	Medium	Medium to low
Static pressure	Very high (75 cm)	Very high	High	Low (12–15 cm)
Power curve	Non overloading	Non overloading	Non overloading	Overloading

9.3.2 Axial Flow Fans

In axial flow fans, the direction of flow of air is parallel to axis of the fan. The air at the outlet of fan is highly swirling and vortex like and thus stabilizing arrangement has to be provided, otherwise heavy pressure loss will take place. The various categories of axial flow fans are:

9.3.2.1 Propeller Fan

The simplest application fans like exhaust fans is the example of propeller fans. These fans can handle high volume but cannot create high pressure and are thus used in applications without duct. The efficiency is low usually 55–60%

9.3.2.2 Tube Axial Fans

The utilization of this category of fan is in applications where velocity profile downstream is not very important, e.g. exhausting air with inlet duct. Blades

Fig. 9.11 A tube axial fan

are similar to aerofoil construction and the clearance between blade tip and housing is very small enabling little more efficiency than propeller fans.

9.3.2.3 Vane Axial Fans

This is an improvement over tube axial fan as guide vanes are provided on downstream side to guide the air flow and improve air distribution. The pressure loss is thus avoided to a great extent. The blades are of aerofoil design. The above features enable vane axial fans to deliver with highest efficiency and pressure of all categories of axial flow fans.

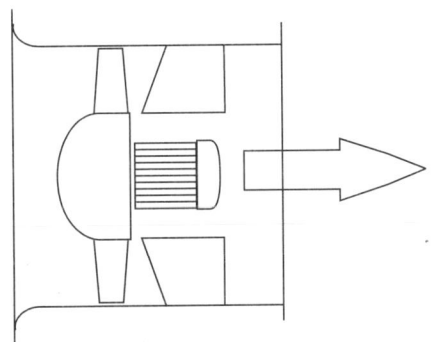

Fig. 9.12 A vane axial fan

Table 9.4 compares the efficiency and other parameters of different categories of axial flow fans:

Table 9.4 Efficiency comparison of various categories of axial flow fans

Parameter	Vane axial	Tube axial	Propellers
Blades	5–18	4–10	3–8
Efficiency	80–85%	70–75%	55–60%
Cost	High	Medium	Low
Static pressure	High (upto 20 cm)	Medium	Low (upto 2 cm)
Power curve	Non over-loading	Non over-loading	Non over-loading
Housing	Cylindrical with guide vane	Cylindrical	Annular ring

9.3.3 Special Application Fans

There are specific applications where centrifugal and axial flow fans are modified suitably to serve the specific purpose as mentioned below:

9.3.3.1 Roof Extractors

Normally used for shop floor applications to exhaust the air from the roof. The wheel and blades are directly mounted on motors. This is another category of propeller fan with low pressure and high volume handling, though backward curved blades are also used. The impeller-motor assembly is placed inside an aluminium casing to protect the fan from rain and for safety reasons.

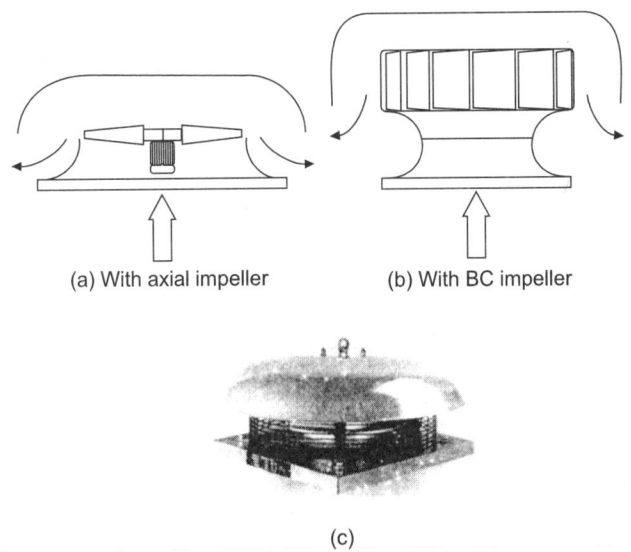

(a) With axial impeller   (b) With BC impeller

(c)

Fig. 9.13 Roof ventilator

9.3.3.2 In-line Fans

These fans have impeller in a tubular casing and the flow is inline with wheel axis and duct deriving its name "Inline fans". The impeller blades can be backward curved or aerofoil type. The scroll housing is not provided and impeller is mounted directly on motor. The assembly is placed in an aluminium casing. The pressure capability is lower than backward curved fans but more than axial fans. They are increasingly used due to their compact design, low noise and for HVAC applications involving low pressure return air and exhaust applications.

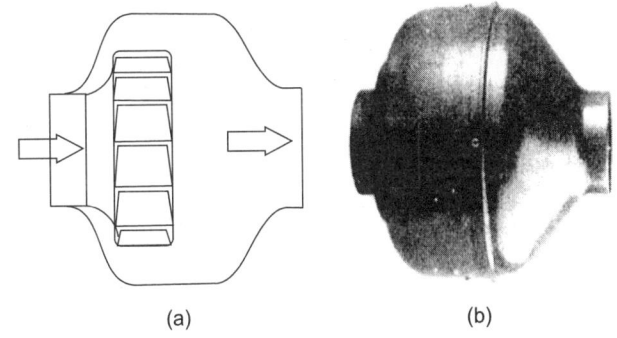

(a)   (b)

Fig. 9.14 A tubular centrifugal or inline fan

9.3.3.3 Crossflow Fans

In this design, forward curved blades are used and the air passes twice through the impeller. The only difference with forward curved centrifugal impeller is that air enters the impeller radially inward. These fans find applications in air curtains, wall mounted indoor air conditioning unit. These are low efficiency fans with maximum efficiency reaching upto 45%.

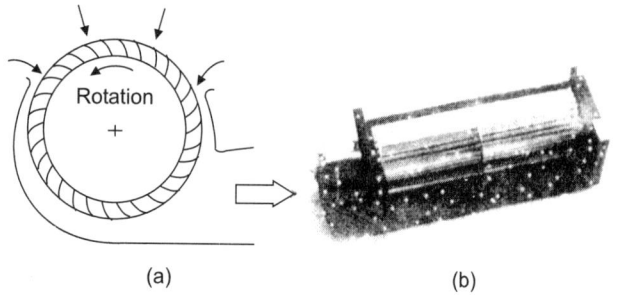

(a)   (b)

Fig, 9.15 A crossflow fan

9.4 FAN OPERATING POINT

The system curve of a fan is parabola

$P_s = kQ^2$ where k is a constant and Q is the flow

In Fig. 9.16 below, the point of intersection of fan curve and system curve is the fan operating point

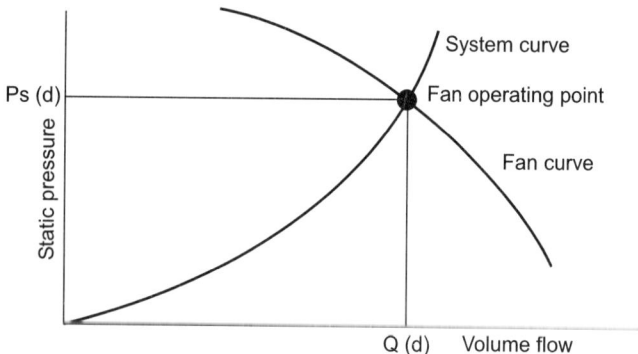

Fig. 9.16 Fan operating point

If the system pressure drop $P_{s(d)}$ at design flow $Q(d)$ is estimated correctly, the fan operating point will be as shown in figure. However, if the system pressure drop is estimated wrongly and is actually lower than estimated, the fan operating point will lie on another system curve (Fig. 9.17) which is again a parabola. The

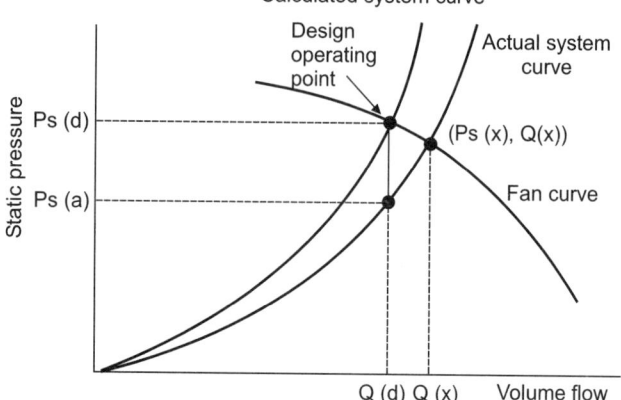

Fig. 9.17 Dynamics of fan system combinations

discharge will increase to $Q(x)$ and the fan operating point will shift to right on fan curve giving lower efficiency. The design flow can still be achieved by reducing the speed of fan till the fan curve passes through the point $P_{s(a)}$, $Q(d)$.

If the pressure drop is more than estimated, the flow will be less than designed flow. The actual system curve will be to the left of design curve. Sometimes, it so happens that clogging of filters causes more pressure drop, creating the same condition. In such cases, fan speed will have to be increased.

9.5 FAN PERFORMANCE PARAMETERS

The various parameters and formulaes requiring considerations and calculation on daily basis by a designer are given below:

1. Outlet velocity (m/s) $= \dfrac{\text{Flow volume}(\text{m}^3/\text{s})}{\text{Fan outlet area}(\text{m}^2)} = \dfrac{Q}{A}$

2. Fan velocity pressure $(P_{v_f}) = \dfrac{1}{2}\rho v^2$

    where ρ is density of air in kg/m^3

3. Fan total pressure $(P_{t_f}) = \dfrac{1}{2}\rho v^2$

    $= $ Fan velocity pressure + Fan static pressure

    $= P_{v_f} + P_{s_f}$

4. Fan output power $= Q\left(\text{m}^3/\text{s}\right) \times P_{t_f}\,(Pa)$

5. Fan total efficiency

    $(\eta_t) = \dfrac{\text{Fan output power}\,(w_0)}{\text{Fan shaft power}\,(w_f)} \times 100\%$

6. Fan static efficiency $(\eta_s) = \dfrac{Q \times P_s}{w_f} \times 100\%$

In any fan selection sheet shown in Table 9.5, the manufacturer specify total efficiency and static efficiency separately. Let's understand why this static efficiency is important to us.

Static efficiency is the power required to develop static pressure. If in an exhaust application, air is being discharged at high velocity without regaining it in terms of useful static pressure, the power corresponding to P_{v_f} will be lost. Thus η_s is a measure to define the effectiveness of the system being used to regain velocity pressure into static pressure.

This concept will be more clear with the help of following example in which two fans are compared, one exhausting at a higher velocity into the atmosphere than another.

Let $Q = 4000$ CFM $= 2.11$ m^3/s

$P_s = 50$ mm $= 490$ Pa

Outlet velocity of fan 1 (v_1) $= 10$ m/s

Outlet velocity of fan 2 (v_2) $= 14$ m/s

Total efficiency of fan 1 as well as 2 $= 65\%$

Fan-1

$$P_{v_f} = \frac{1}{2}\rho v_1^2 = \frac{1}{2} \times 1.2 \times 10^2 = 60 \text{ Pa}$$

Fan shaft power (w_1) $= \dfrac{2.11 \times (490 + 60)}{0.65} = 1785$ watts

Static efficiency (η_s) $= \dfrac{2.11 \times 490 \times 100}{1785} = 57.9\%$

Fan-2

$$P_{v_f} = \frac{1}{2}\rho V_2^2 = \frac{1}{2} \times 1.2 \times 14^2 = 118 \text{ Pa}$$

Fan shaft power (w_2) $= \dfrac{2.11 \times (490 + 118)}{0.65} = 1974$ Pa

Static efficiency (η_s) $= \dfrac{2.11 \times 490 \times 100}{1974} = 52.37\%$

Thus, even with the same total efficiency, the fan with lower static efficiency will require more shaft power.

Thus while selecting fans from different manufacturers, always select on the basis of higher static efficiency even if they have equal total efficiency.

9.6 FAN LAWS AND THEIR IMPORTANCE

Fan laws are the laws that govern the performance of a fan and help us to derive the performance parameters of another geometrically similar fan. The fans are geometrically similar if:

1. The number of blades and vanes are same
2. All angular dimensions are same
3. All linear dimensions change in proportion to the wheel diameter.

In addition, the fans must be dynamically similar, i.e. Mach number, Reynolds number, surface roughness and gap size should vary only significantly.

The fan laws are categorized in three category:

Fan law-1 shows the effect of changing size, speed, density on volume airflow rate, pressure and power level.

Fan law-2 shows the effect of changing size, pressure or density on volume airflow rate, speed and power level.

Fan law-3 shows the effect of changing size, volume airflow rate or density on speed, pressure and power.

The fan laws as shown in Table 9.6 can be used to predict the performance of any fan when test data is available for any fan of same series.

Table 9.6 Fan laws

Law No.	Dependent variables			Independent variables
1a	Q_1	$=$	Q_2	$\times$ $(D_1/D_2)^3 (N_1/N_2)$
1b	P_1	$=$	P_2	$\times$ $(D_1/D_2)^2 (N_1/N_2)^2 \rho_1/\rho_2$
1c	W_1	$=$	W_2	$\times$ $(D_1/D_2)^5 (N_1/N_2)^3 \rho_1/\rho_2$
2a	Q_1	$=$	Q_2	$\times$ $(D_1/D_2)^2 (P_1/P_2)^{1/2} (\rho_2/\rho_1)^{1/2}$
2b	N_1	$=$	N_2	$\times$ $(D_2/D_1) (P_1/P_2)^{1/2} (\rho_2/\rho_1)^{1/2}$
2c	W_1	$=$	W_2	$\times$ $(D_1/D_2)^2 (P_1/P_2)^{3/2} (\rho_2/\rho_1)^{1/2}$
3a	N_1	$=$	N_2	$\times$ $(D_2/D_1)^3 (Q_1/Q_2)$
3b	P_1	$=$	P_2	$\times$ $(D_2/D_1)^4 (Q_1/Q_2)^2 \rho_1/\rho_2$
3c	W_1	$=$	W_2	$\times$ $(D_2/D_1)^4 (Q_1/Q_2)^3 \rho_1/\rho_2$

Notes:

1. Subscript 1 denotes fan under consideration. Subscript 2 denotes tested fan.
2. For all fans laws $(\eta_t)_1 = (\eta_t)_2$ and (Point of rating)$_1$ = (Point of rating)$_2$.
3. P equals either P_{t_f} or P_{sf}.

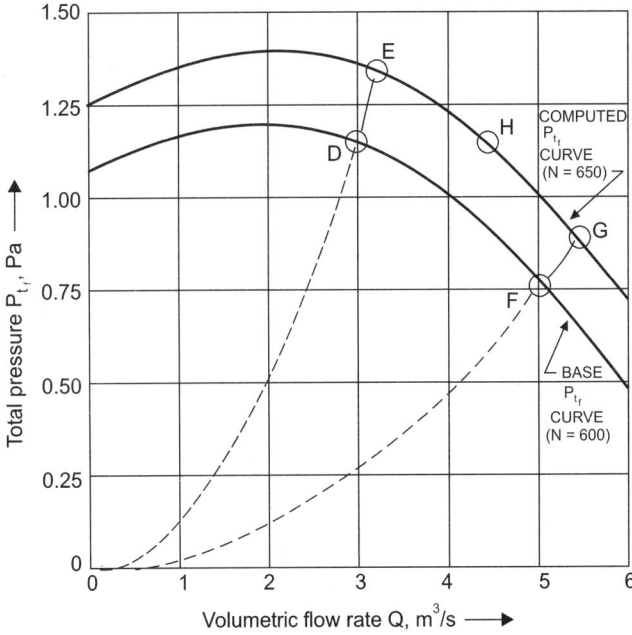

Fig. 9.18 Example application of fan laws

Figure 9.18 (example application of fan laws) shows the effect of change in fan speed for an equal diameter fan. In the figure, base curve is already given. But for generating P_t curve for a different fan speed ($N = 650$), the following calculation can be performed.

At point D

$$Q_2 = 3 \text{ m}^3/\text{s and } P_{t_{f2}} = 228 \text{ Pa}$$

Using fan law 1a at point E

$$Q_1 = 3\left(\frac{650}{600}\right) = 3.25 \text{ m}^3/\text{s} \ (D_1 = D_2)$$

Using fan law 1b

$$P_{t_{f_1}} = 228\left(\frac{650}{600}\right)^2 = 268 \text{ Pa} \ (\rho_1 = \rho_2)$$

This way, total pressure curve $P_{t_{f_1}}$ at $N = 650$ rpm may be generated by computing additional points from data on the base curve, such as point G from point F.

If equivalent points are joined, as shown by dashed lines, the resulting curve is a parabola which is defined by equation $\Delta P = kQ^2$.

Refer Fig. 9.17 once again where actual pressure drop is less than design pressure drop. It can be seen that fan speed should be reduced by a factor $Q(d)/Q(x)$ so as to obtain the flow $Q(d)$. If fan speed is not reduced, the actual flow and pressure drop will be $Q(x)$ and $P_s(x)$. This reduction in speed will also reduce pressure from $Ps(x)$ to $Ps(a)$. The fan equation $P = kQ^2$ states that the pressure is proportional to square of speed and therefore square of flow. This indicates that when fan speed is increased or decreased, the operating point moves along the system curve and hence magnitude of speed change can be calculated directly from required change in flow volume.

The capacity of fan delivery can also be controlled by inlet vane control or by use of dampers (refer Fig. 9.17). Dampers or vanes placed at inlet to the fan or away from fan outlet are better way to control as compared to vanes closer to fan outlet. This is due to the reason vanes placed near fan outlet result in high pressure losses because of turbulence at high speed.

9.7 CALCULATIONS OF CAPACITY OF FANS

National Building Code 2005 specifies in Annexure-C of Part-4 (Fire and Life Safety) that mechanical extractors shall be designed to permit 30 air changes per hour in case of fire. However, for normal operations, NBC Part-8 Section-3 specifies the various rates of air circulation for different areas and 12 air changes per hour during normal operations in basement. This circulation rate helps in maintaining the CO and CO_2 levels as specified earlier. The same rate of circulation has been specified wherever smoke venting facility has been installed for the purpose of exit safety. The applications specified in clause 3.4.12 of NBC are window less buildings, underground structures, large area factories, hotels and assembly buildings (including cinema halls).

The exhaust fans while exhausting smoke create negative pressure and the air is drawn in to dilute the smoke. It is necessary that provision of this make up air (usually 90% of exhaust capacity) is made either through ramps (upper basement only) or through inlet shafts opening on the ground floor. For other basements, inlet shafts have to be provided invariably. The inlet shafts shall be so located so as not to cause short circuit of air path with exhaust fan locations. The duct network for supply and exhaust air shall be so planned that both type of ducts drop to floor level. Since the car fumes in car parkings tends to be heavier than normal air, they can be picked up more efficiently by planning the ducts this way (Fig. 9.19).

Thus, during normal operations (non-fire conditions), set of fans (inlet and exhaust) and exhaust ducts designed for 12 air changes will operate. However, during fire conditions, the additional set of fans designed for additional 18 air changes per hour will

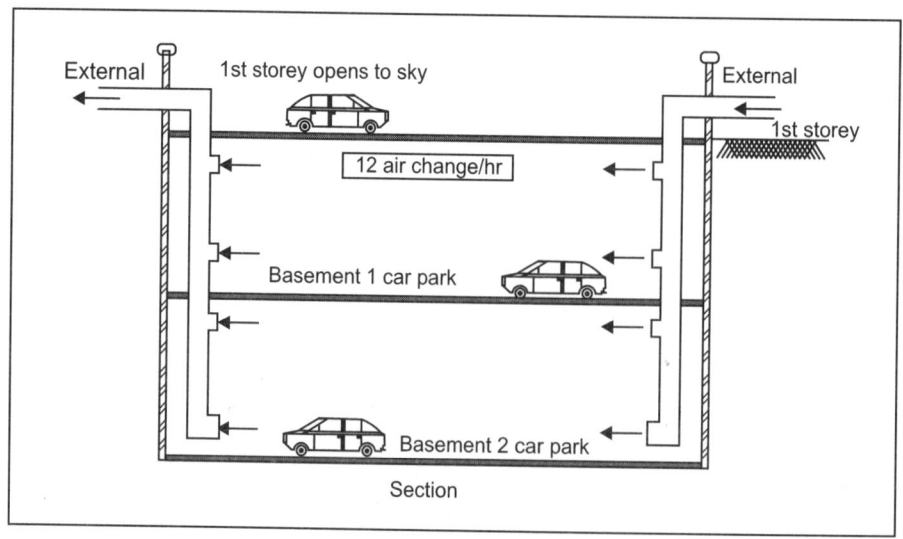

Fig. 9.19 Smoke purging system

operate so as to effectively produce 30 air changes per hour as per NBC-2005. Ventilation system can be designed as supply only, exhaust only or combination of both. The combined system is more efficient as it creates negative pressure in the parking areas which eliminates the possibility of toxic fumes going into any occupied areas around them.

The following case study is an example of volume of air to be handled in case of normal/fire conditions, shaft size calculations, fan capacities for a typical installation (Fig. 9.20, 9.21).

1. Area details
   1.1 Parking area of lower basement = 125000 sq. ft.
   1.2 Basement height = 10.8 ft.
   1.3 Volume = 125000 × 10.8
       A = 1350000 Cu. ft.
   1.4 Parking area of upper basement = 106000 sq. ft.
   1.5 Height = 10.8 ft.
   1.6 Volume = 106000 × 10.8
       B = 1144800 Cu. ft.
   1.7 Total volume A + B = 2494800 Cu. ft.

2. Design basis
   2.1 Parking area
       Normal Fresh air : 12 Air change per hour (ACPH)
       Normal exhaust air : 12 ACPH
       Emergency fresh air : 18 ACPH
       Emergency exhaust air : 18 ACPH
       Total emergency fresh air : 12 + 18 = 30 ACPH
       Total emergency exhaust air : 12 + 18 = ACPH
   2.2 Pump room
       Total Fresh air : 15 ACPH
       Total Exhaust air : 20 ACPH
   2.3 Electrical room
       Total fresh air : 20 ACPH
       Total exhaust air : 20 ACPH

3. Basement CFM and shaft size calculations
   3.1 Lower basement (L.B.)

   Normal fresh air (12 ACPH) $= \dfrac{1350000 \times 12}{60}$

   $= 270000$ CFM

   Normal exhaust air (12 ACPH) $= \dfrac{1350000 \times 12}{60}$

   $= 270000$ CFM

   Emergency fresh air (18 ACPH) $= \dfrac{1350000 \times 18}{60}$

   $= 405000$ CFM

   Emergency exhaust air (18 ACPH) $= \dfrac{1350000 \times 18}{60}$

   $= 405000$ CFM

   3.2 Lower basement shaft size requirement

   Normal fresh air (limited to 1000 FPM) $= \dfrac{270000}{1000}$

   $= 270$ sq ft

   Emergency fresh air (limited to 1800 FPM) $= \dfrac{405000}{1800} \times 0.85$

   $= 190$ sq ft

Note:
   a. During emergency 85% of fresh air quantity can be adopted so as to create vacuum condition enabling more effectiveness of smoke exhaust.
   b. During emergency, sound level is not the criteria, hence velocity upto 1800 FPM can be adopted while in normal ventilation velocity limitation is 1000 FPM to keep sound in limits.

   Total fresh air shaft size: 270 + 190 = 460 sq ft

   Normal exhaust air (limited to 1400 FPM) $= \dfrac{270000}{1400}$

   $= 193$ sq ft

   Emergency exhaust air (limited to 2200 FPM):

   $\dfrac{405000}{2200} \times 0.85 = 156$ sq ft

   Total exhaust air shaft size : 193 + 156 = 349 sq ft

Note:
   a. Keeping exhaust velocity more than supply velocity creates negative pressure and draws more fresh air from outside.
   b. In emergency, velocity upto 2200 FPM can be allowed as noise is not the criteria but quickness of smoke removal is.

   3.3 Upper basement (U.B.)

   Normal fresh air (12 ACPH) $= \dfrac{1144800 \times 12}{60}$

   $= 228960$ CFM

   Normal exhaust air (12 ACPH) $= \dfrac{1144800 \times 12}{60}$

   $= 228960$ CFM

   Emergency fresh air (18 ACPH) $= \dfrac{1144800 \times 18}{60}$

   $= 343440$ CFM

   Emergency exhaust air (18 ACPH) $= \dfrac{1144800 \times 18}{60}$

   $= 343440$ CFM

   3.4 Upper basement shaft size requirement

   Normal fresh air (limited to 1000 FPM) $= \dfrac{228960}{1000}$

   $= 229$ sq ft

   Emergency fresh air (limited to 1800 FPM)

   $= \dfrac{343440}{1800} \times 0.85 = 162$ sq ft

   Total fresh air shaft size = 229 + 162 = 391 sq ft

   Normal exhaust air (limited to 1400 FPM) =

   $\dfrac{228960}{1400} = 164$ sq ft

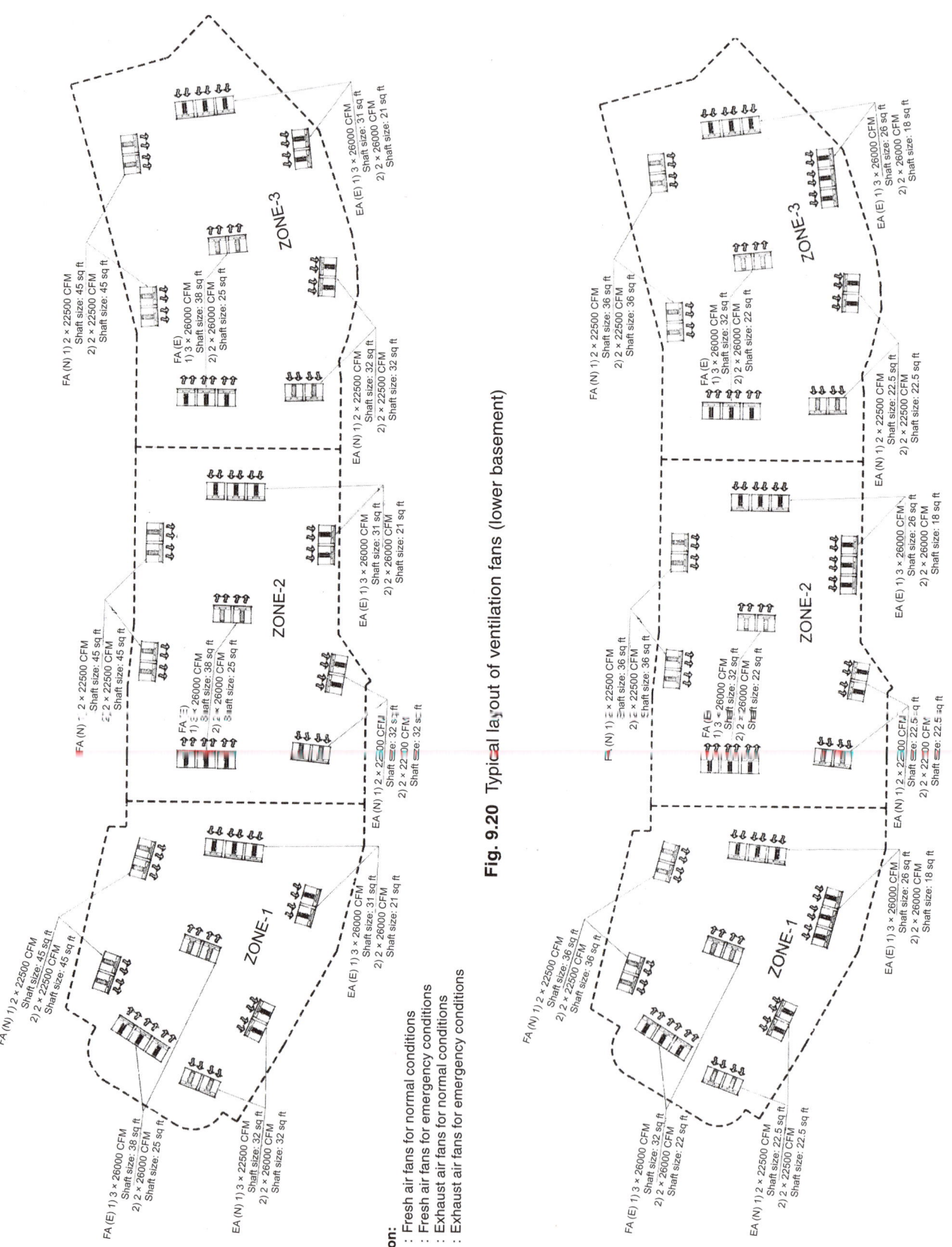

Fig. 9.20 Typical layout of ventilation fans (lower basement)

Fig. 9.21 Typical layout of ventilation fans (upper basement)

Notation:
FA (N) : Fresh air fans for normal conditions
FA (E) : Fresh air fans for emergency conditions
EA (N) : Exhaust air fans for normal conditions
EA (E) : Exhaust air fans for emergency conditions

Emergency exhaust air

(limited to 2200 FPM) $= \dfrac{343440}{2200} \times 0.85 = 133$ sq ft

Total exhaust air shaft size $= 164 + 133 = 297$ sq ft

As a normal practice, the fresh air (both normal and emergency) is drawn from the ramps by the negative pressure created by exhaust fans. A design should tally the cut out/opening calculated as above with the ramp openings available. If ramp openings are insufficient, then provision of additional shaft and fresh air supply fans shall be invariably made.

4.0 Electrical room ventilation requirement

Sub station (93000 ft³) Fresh air = 31000 CFM
@20 ACPH

Exhaust air = 31000 CFM

5.0 Pump room

Pump room (51648 ft³) Fresh air = 12912 CFM
@15 ACPH

Exhaust air @20 ACPH = 17200 CFM

6.0 Lift well pressurization (Lower basement to terrace)

Air leakage at lift door of size 2.1×1.0 m with a gap of 1 mm all around.

Leakage area $= 2.1 \times 1 \times .004 = .0084$ sq m

(2 sides near wall and double in the center makes 4 mm or 0.004 m)

Air leakage (As per ASHRAE) $= 0.839\, A\sqrt{P}$

(ASHRAE Handbook – HVAC applications)

ΔP to be maintained 50 Pa as per NBC-2005

$Q_1 = 0.839 \times .0084\sqrt{50}$

$Q_1 = .049$ m³/sec

Leakage volume at ground floor (open at all times during fire) with ingress velocity of 1.0 m/s

$Q_2 = 2.1 \times 1.0 \times 1.0 = 2.1$ m³/s

Total pressurization air quantity (for tower having LB + UB + G + 38)

$Q = Q_1 \times$ (no. of landings – 1) $+ Q_2$

(GF already considered above)

$= 0.049 \times (41 - 1) + 2.1$

$= 4.06$ m³/sec

Accounting 15% lift shaft leakages

$Q = 4.06 \times 1.15 = 4.67$ m³/sec

7.0 Staircase pressurization

Assuming door size as 2.1×1.5 m, leakage area with a 1 mm gap all around

$A = 2.1 \times 1.5 \times .004 = 0.0126$ m²

$Q_1 = 0.839 \times 0.0126 \times \sqrt{50} = 0.07475$ m³/s

Assuming that door at ground floor and one floor at terrace remains always open, as during emergency, public rushes either to ground floor or rushes upward to assemble at terrace.

$Q_2 = 2 \times A \times V$

$= 2 \times (2.1 \times 1.5) \times 1.0 = 6.3$ m³/s

Total pressurization air quantity (tower with LB + UB + G + 38)

$Q = Q_1$ (No. of landing – 2) $+ Q_2$

$= 0.07475\,(41 - 2) + 6.3$

$= 9.22$ m³/s

Accounting for 5% duct losses $= 9.22 \times 1.05$

$= 9.68$ m³/s

The lower and upper basements are too large to be served by one set of fans. Accordingly, in terms of availability of fan capacity with the manufacturers and also to effectively handle the fresh air and smoke, the basements are divided into strategic zones. Each zone has its own sets of supply and exhaust fans, its own electrical panel and wiring that is independent of other zones. It is also important to note that every zonal electrical panel shall be receiving electric supply through an independent cable directly from the substation.

The duct network has to be designed for normal applications. But for emergency conditions, the axial flow fans shall be feeding and exhausting air without ducts. This is basically to avoid melting of ducts that handle high temperature exhaust.

Now-a-days, trend is to design totally ductless system. This is for the reason that ducts used under normal applications have to handle high temperature smoke during fire conditions. The ducts/hangers may not sustain such a high temperature and may give away. Instead jet fans are used to suck the air/smoke from one point and transmit it to the next jet fan which further transmits it to next jet fan till it is finally exhausted through axial flow fans. The numbers, handling capacity and strategic location of jet fans is a matter of computational flow dynamics (CFD) analysis which is to be performed by the manufacturer and is out of preview of this book.

The Tables 9.7 and 9.8 below gives the number of fresh and exhaust air fans along with approximate number of jet fans for each zone of the basement and the approximate power requirement for each fan so that circuit breakers of electrical panels can be designed. It is also important that different fans are not looped on one circuit so as to have reliability of operation in emergency even if one circuit fails.

9.8 MECHANICAL VENTILATION AND SMOKE CONTROL

Wherever ducts are used for extraction of smoke from the building, the system should be so designed that it withstands the temperature of hot gases and does not crumble under temperature and pressure. In an air conditioned building, usually supply of cold air is through ducts but return is from the annular space created between ceiling and false ceiling.

Table 9.7 Quantities of ventilation fans (Lower basement with three zones)

Description Zone-1	Zonal CFM = Total LB CFM/no. of zones	CFM of each fan selected	No. of fans	kW rating	Total shaft size (sq ft)	Zonal shaft size (sq ft)
FA (normal)	90000	22500	4	5.5	270	90
Exhaust (normal)	90000	22500	4	5.5	193	64
FA (emergency)	135000	26000	5	7.5	190	63
Exhaust (emergency)	135000	26000	5	7.5	156	52

Table 9.8 Quantities of ventilation fans (Upper basement with three zones)

Description Zone-1	Zonal CFM = Total UB CFM/ no. of zones	CFM of each fan selected	No. of fans	kW rating	Total shaft size (sq ft)	Zonal shaft size (sq ft)
Fa (normal)	76320	22500	4	5.5	229	76
Exhaust (normal)	76320	22500	4	5.5	164	55
FA (emergency)	114480	26000	5	7.5	162	54
Exhaust (emergency)	114480	26000	5	7.5	133	44

The hot gases from the room under fire travel from this annular space, the hangers supporting the asbestos or metallic false ceiling should be of non combustible material.

Similarly, when supply or return duct is crossing one zone and enters the other, precautions must be taken that smoke from one zone does not spread to the other zone (that is not under fire) or does not spread to common lobbies where public gather for safe evaluation (Fig. 9.22)

The following precautions and engineering practices need to be followed to prevent above incidents from happening:

9.8.1 Duct Material and Fire Barrier

Ducts for air conditioning and mechanical ventilation should be made of aluminium, steel and should be covered with non combustible covering or lining. The material of lining should be such that it generates minimum amount of smoke or toxic gases. The Fig. 9.23 shows such an arrangement. As can be seen here, the material providing lining over the duct, where it crosses a fire wall, should be of some fire rating as the wall itself. Similar care should be exercised in case of air conditioning pipes crossing a fire wall. The purpose is to prevent ingress of smoke from one fire zone to another due to leakage from areas around duct or pipes.

9.8.2 Prevent Spread of Fire Between Zones

As far as possible, ventilation duct should not pass through refuge lobby or exit passage ways, but where it cannot be avoided, this part of duct should be encased in a fire resistant structure having fire rating equal to main structure itself. The other method is to provide fire damper of approved fire rating, wherever duct penetrates such sensitive areas. It is important that the space around fire dampers is sealed with non combustible material of equivalent fire rating as shown in Fig. 9.22, 9.23.

9.8.3 Location of Fire Damper

Fire dampers are meant to prevent flow of smoke from the zone under fire to non fire zone. These dampers operate on receiving a signal from fire alarm panel. Though fire dampers provide safety in preventing the spread of fire, but they can be counter productive, if installed in places where they should not be, like:

a. Return air duct which also serves as a smoke extract duct.

b. Openings in walls of protected shaft when these openings have kitchen exhaust duct passing through it.

c. Any air pressurizing system like exit staircase, lift lobby or emergency lobby.

d. Duct network in basement or car park ventilation which requires no restriction in path of smoke flow like closed dampers, say, due to human error or malfunctioning.

9.8.4 False Ceiling Material

The false ceiling which contains ducting above it shall be made of fire rated board. The Singapore code requires that ducts are not to be located above fire rated floor, ceiling or roof ceiling assembly unless such ducts were part of prototype tested for required fire resistance rating.

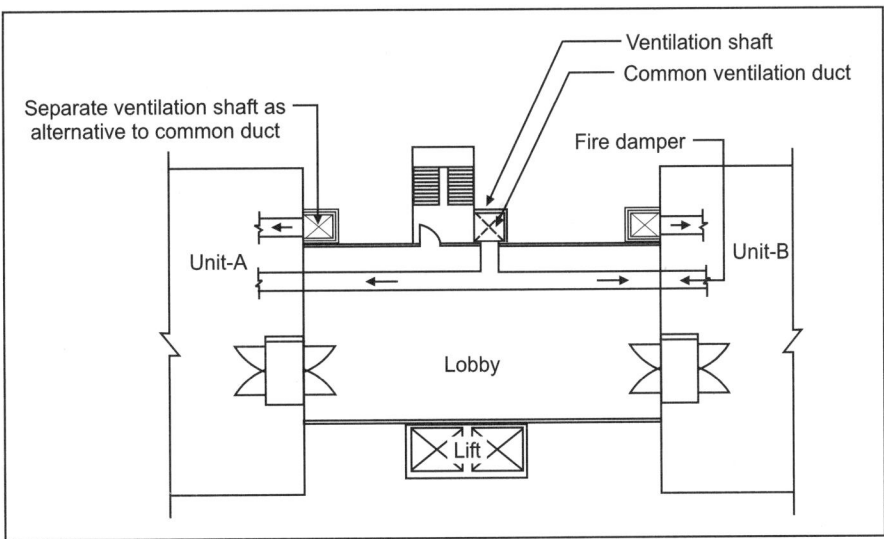

Fig. 9.22 Segregation of fire zone

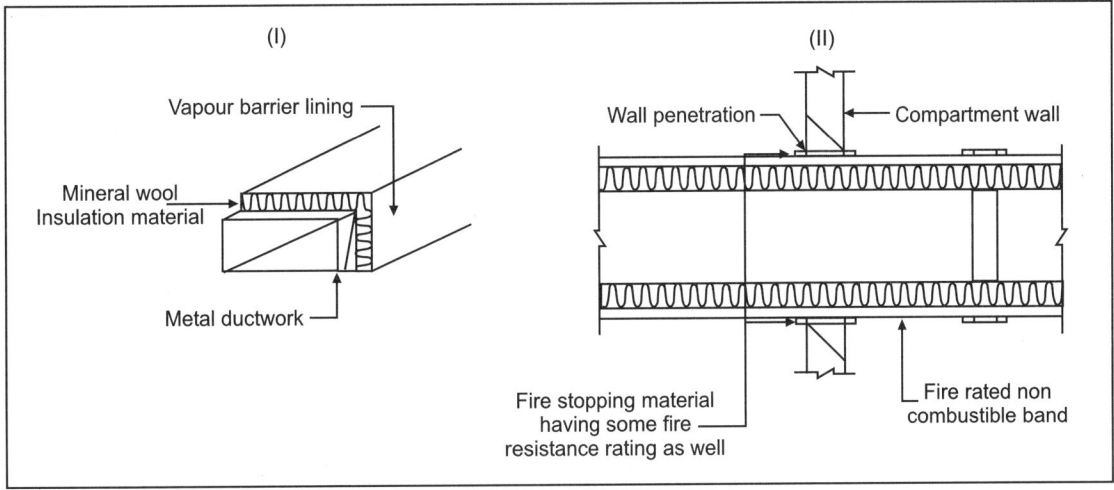

Fig. 9.23 Use of fire lining in ventilation ducts

9.8.5 Exhaust Fan in AHU

Provide an exhaust axial fan in the AHU room connected to exhaust shaft or open to atmosphere. The interlock shall be so made that as soon as AHU stops running on receiving a fire signal, the exhaust fan shall start creating negative pressure and the smoke is sucked through return air path from the rooms under fire and exhausted into atmosphere.

9.8.6 Lift/Staircase/Emergency Lobby Ventilation

The mechanical ventilation system for each staircase shall be independent and at least 5 m away from the nearest exhaust shaft to prevent smoke being sucked again. In the buildings with 4 or more storeys, the staircase pressurization shall be through a vertical duct running throughout the vertical height and pressurized air shall be supplied at alternate landing through grilles (Fig. 9.24).

Mechanical ventilation system for emergency/escape lobbies shall be exclusive to them and provide 10 air change per hour.

The mechanical ventilation system for pressurization of staircase shall maintain an air flow with a velocity not less than 1.0 m/s. This velocity shall be achievable with two successive floor doors and main discharge door in fully open position.

To avoid pressure built up in occupancy area or staircase, adequate measures shall be taken to allow air leakage as pressure build up will create difficulty in opening the doors to occupancy area. The escape of pressurizing air can be achieved through window, vents, opening below and above doors. However, the

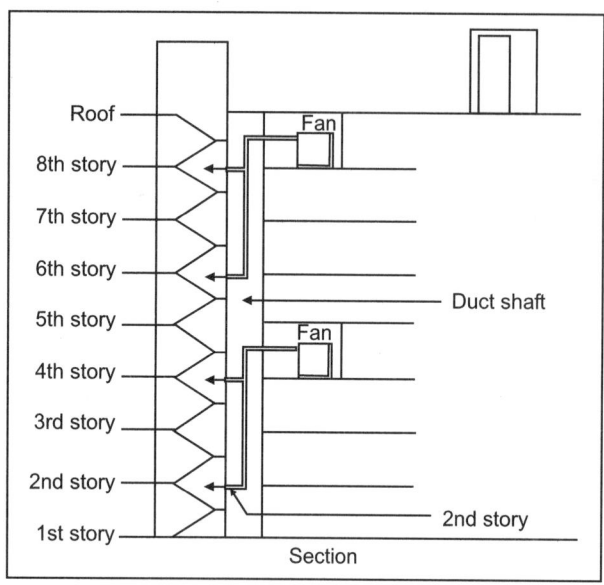

Fig. 9.24 Mechanical ventilation of staircases

rate of supply of pressurized air to pressurized area shall be sufficient to make up for the loss through leakages.

The calculation of air leakage through lift and staircase doors has already been explained in the preceding example.

9.8.7 Basement Ventilation

The exhaust and supply air fans should be electrically interlocked so that failure of one exhaust fan automatically shuts down the corresponding section of supply air fan so as to prevent fresh air (oxygen) being bumped into the basement.

9.9 TEMPERATURE RATING OF FAN, MOTORS

Smoke exhaust fans shall be tested according to ASHRAE standard 149–200 "Laboratory method of Testing Fans Used to Exhaust Smoke in Smoke Management Systems". There are different applications where smoke exhaust fans are used, e.g. office/

residental/commercial buildings, chemical process industries, power house, tunnels, etc. Each application require different fire rating of fan and motor material, and accordingly, the material and design of fan has to be selected. So what makes one fan more capable of sustaining higher temperature than another fan? Each model of fan has a recommended maximum operating temperature which is limited by the weakest component, say a bearing or a fan impeller which determine the maximum temperature beyond which this component will fail causing total failure of fan.

In general, construction material of fan impeller blades is the most obvious element to consider when deciding the temperature withstand capability of a fan, e.g. Aluminium withstands maximum temperature upto 250° F, standard carbon steel upto 750° F and 316 stainless steel upto 1000° F. Other construction considerations include bearing type, drive component selection, means of ventilation and cooling of drive components and insulation options.

The Singapore manual of mechanical ventilation adopts 250° C rated smoke exhaust fans, though NBC-2005 does not specify the temperature rating of fans to be adopted. As there is enough combustible material in any commercial building, the fire rating temperature of 250° C is in general adopted everywhere.

The motors are classified in various categories by standard NEMA (National Electrical Manufacturer's Association) as per maximum allowable operating temperature and are rated as below in Table 9.9.

Operation temperature = reference temp + allowable temp rise + allowance for hot spot winding

For class F motor:

Operation temperature = 40° C + 105° C + 10° C

= 155° C

In general, a motor should not operate with temperatures above the maximum. Each 10° C rise above the rating may reduce the motor lifetime by one half. It is important to note that insulation classes are directly related to motor life.

Table 9.9 Motors classification on temperature rating

Temperature tolerance class	Maximum operative temperature allowed		Allowable temp rise at full load (1.0 service factor)	Allowable temp rise (1.15 service factor)
	°C	°F	°C	°C
A	105	221	60	70
B	130	266	80	90
F	155	311	105	115
H	180	356	125	–

Note: Allowable temperature rises are based upon a reference ambient temperature of 40° C.

A motor operating at 180° C will have an estimated life of

– only 300 hours with class A
– 1800 hours with class B insulation
– 8500 hours with class F insulation

It is thus very clear that the motor insulation class shall be selected very carefully, commensurate with the operating temperature of fan. But certainly, a fan rated for handling 250° C smoke gases, must have H-class motor only.

9.10 BASIC FAN SELECTION

The fan type and its characteristics affect the selection of a fan. For any one point on performance curve, there may be many different fans which may satisfy the parameters required. But based on any set of properties, e.g. fan size, efficiency or motor size, etc. there is only one best fan that will be suitable for that application. Let us understand this situation with an example. To deliver 60,000 CFM at 7.0 inches total pressure at a density of 0.075 lb/ft³, several different fans will satisfy this rating as shown in Table 9.10.

It can be seen from the above table that for economy in capital cost, 48" axial fan should be selected. Similarly, for economy in operating cost, 48" axial or 60" backwardly inclined fan can be selected. The optimum fan to satisfy future increase in load would be 48" axial utilizing a blade angle change. Centrifugal fans would require an increase in speed by 16% with a corresponding horsepower increase of 58%, while motor for 48" axial would simply require better insulation so as to handle the increased power. As already explained, the performance of the fans can be changed by various methods so as to achieve the desired flow or pressure.

Example

Let us calculate the CFM and static pressure requirement of a ventilation fan installed in basement to exhaust the air.

In the diagram shown in Fig. 9.25 the air is sucked from two ducts and in the plenum AB, both the streams merge and are thrown out by the fan to the atmosphere.

Let us assume duct velocity as 2000 FPM (basement ventilation fire condition). This discharge velocity from fan is also 2000 FPM. The volume of air sucked from

Table 9.10 Comparison of various fans

Item	Axial	Radial			Backward		
Diameter	48"	66"	60"	54"	66"	60"	54"
Speed (RPM)	1750	603	685	816	630	729	870
Power (BHP)	81.4	91.0	95.0	103	77.1	77.1	80.0
Efficiency(%)	81.0	72.5	69.5	64.0	85.5	85.5	82.5
Cost factor	1.0	2.5	2.1	2.0	2.6	2.2	2.1
Motor (HP)	100	100	125	125	100	100	100

Courtesy: M/s Greenheck : Product application guide

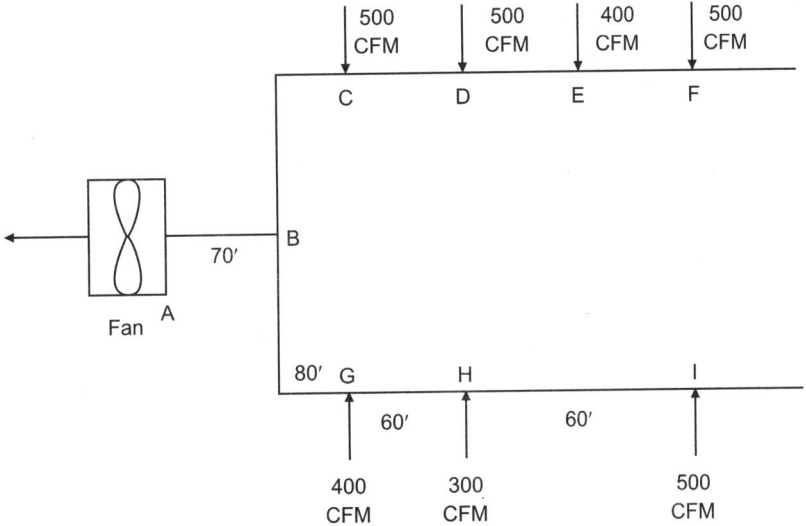

Fig. 9.25 Typical ducting arrangement

grilles are shown in the diagram. The Table 9.11 below shows the total frictional loss encountered, calculated with equal friction method.

a. From Fig. 6.2 in Chapter 6, read friction loss as 0.3 in Wg. per 100 ft. at the intersection of 2000 FPM line and 3100 CFM line

b. The equivalent round duct diameter is 17 inches. Similarly calculate equivalent round duct diameter for each section but keeping friction constant at 0.3 in Wg. per 100 ft.

c. The rectangular duct sizes are read from Fig. 6.3.

d. Calculate total pressure loss in the system as given in the table.

$$= 1.41 \text{ inch water}$$
$$= 1.41 \times 251 \text{ Pa}$$
$$= 354 \text{ Pa}$$

e. In addition fan is generating a velocity pressure corresponding to 2000 FPM or 10 MPS.

$$= \frac{1}{2} \times \rho \times v^2 \text{ Pa}$$
$$= \frac{1}{2} \times 1.2 \times (10)^2$$
$$= 60 \text{ Pa}$$

f. In the discharge section of the fan, the velocity pressure will be converted into static pressure by using a divergent piece called the plenum either before or after the fan. Thus total fan pressure for which fan is to be selected = 354 − 60

$$= 294 \text{ Pa}$$
$$= 1.17'' \text{ static pressure}$$

g. From data table of a leading manufacturer, relevant row of which is reproduced below in Table 9.12, the selected fan shall be for 1.25" SP, having 16.5" diameter running at 1475 RPM, consuming 1.06 HP and delivering 3122 CFM (manufacturer's data table may be referred).

Table 9.11 Duct size calculation

Section	CFM	Vol. FPM	Friction loss inch Wg per 100 ft.	Length ft.	Total frictional loss inch Wg	Eq. round dia inch	Rect. duct size inch
AB	3100	2000	0.3	70	0.21	17.5″	12″ × 22″
BC	1900	1750	0.3	60	0.18	14.0″	8″ × 22″
CD	1400	1600	0.3	50	0.15	12.0″	10″ × 12″
DE	900	1500	0.3	50	0.15	11.0″	10″ × 11″
EF	500	1300	0.3	40	0.12	8.5″	10″ × 6″
BG	1200	1550	0.3	80	0.24	11.5″	10″ × 11″
GH	800	1450	0.3	60	0.18	10.25″	10″ × 9″
I II	500	1300	0.3	60	0.18	8.5″	10″ × 6″
Total					1.41		

Table 9.12 A sample fan selection table[1]

Vol.	Capacity	3/4″ SP		7/8″ SP		1″ SP		1 1/4″ SP		1 1/2″ SP		15″ SP	
FPM	CFM	RPM	HP	RPM	HP	RPM	HP	RPM	HP	RPM	HP	RPM	HP
2000	3122	1329	0.795	1366	0.86	1405	0.929	1475	1.06	1539	1.21	3693	12.00

1. Selection table for wheel diameter 16.5″, overall width 7 9/16″, outlet area 1.56 sq ft.

The chapters forming part of this section are.

1. Internal Wiring and Electrical
   Distribution System   Chapter 10
2. Substation   Chapter 11
3. Electric Generator   Chapter 12
4. Pumping System   Chapter 13
5. Illumination Engineering   Chapter 14

Chapter 10

The design of internal wiring is specifically meant to frame the electrical distribution system of a building in a set of rules. The framing of rules for wiring is important because undersized wires or distribution boards have the effect of producing sparking or melting of wires and other components. The modern building is supposed to have wiring with a life of at least 15 years. This is because the interior finishes with plaster of Paris, wooden architecture or other materials does not allow the wiring to be replaced because of any sparking or faults at a later stage. This will involve digging the path of wire in the walls and hence not only expenditure on repairs but also disturbance to the client and ongoing business. Hence proper planning in the design and distribution system of wiring is essential that has been explained as a set of rules in this chapter.

Chapter 11

This chapter explains in detail the designing of a substation and the various components that are part of it. An over designed substation means wastage of money while an under designed means constant power failures and breakdown of costly components and heavy repair expenses. The protection equipment shall be accurate to act in time to protect against the desired

fault level. The method of calculating the rating of various components like current transformer, potential transformer, ACB, bus bar sizes, transformer, cable size, voltage drop, etc. has been described in very easy to understand language and with practical examples.

Chapter 12

Nowadays, the reliability of continuous supply of electricity from state electricity departments is becoming poor due to large gap between supply and demand of power. The high rise office buildings, IT, shopping malls, residential societies need DG set of suitable capacity as much as they need air-conditioning or lifts and escalators. It is not sufficient to simply specify the kVa rating in the purchase order. Designer/consultant has to be sure about the various methods of excitation, output voltage control to get the best performing DG set. He must also know the calculation and distribution of resistive and inductive loads of the building, which will have a large impact on the selection of kVa rating of DG set. This chapter describes in detail all of the above aspects with diagrams and simple numericals.

Chapter 13

The pumps are the backbone of any water supply system. The calculation of water demand and head of pumping call for detailed calculation starting from population expected in a building or a town, number of pumping hours, piping material and associated frictional losses. The pumps are not only used for domestic purpose but in air conditioning and fire fighting applications as well. The power consumed by these pumps in high rise buildings is on an increase and efforts must be made to design and operate these pumps in the most efficient way. This chapter explains all of these calculation in easy step by step approach along with various methods with which energy efficiency can be achieved.

Chapter 14

With the development of various software programmes, the in door and out door illumination calculations, number and type of fixtures to be used have become very easy. But behind every computing technique, there is real engineering which must be known so that intent and purpose of design can be translated in output results. The chapter starts from basic optics to detailed polar curve diagrams and lux calculation for various applications by using various fixtures. Emphasis has been given to design this chapter from the point of view of imparting detailed knowledge to engineers with easy to solve examples.

10

Internal Wiring and Electrical Distribution System

The design and understanding of wiring system inside a building is considered to be the most simple job and is generally left to the discretion of a wire man who has gained certain expertise over a period of time.

But the fact remains that only through a proper selection and design consideration given to select wiring, conduit, rating of various accessories and specifications given in various IS codes, the electrical installation can be considered as safe.

The main cause of fire due to overloading and short circuit is due to improper planning by a designer and leaving the system in the hands of unskilled hands.

This chapter describes the various principles to be followed in designing electrical wiring system and distribution boards inside a building. Simultaneously, the method of measurement and payment of wiring system is also described because sometimes the low quality of work done gets overpaid due to poor knowledge of the architect, builder or an engineer.

10.1 POINT WIRING

10.1.1 Definition

A point (other than socket outlet point) shall include all works from the controlling MCB or switch to:

   i. Connector/ceiling rose in case of ceiling/exhaust fan, light fitting and call bell.
   ii. Lamp holder.

10.1.2 Scope

The following items are covered in the scope of point wiring.

   i. *Conduit.* Steel/PVC and accessories like bends, tee, etc. required for continuity and its fixing.
   ii. Wire.

   iii. Switch boxes (Steel/PVC) and related accessories, switches regulators, sockets, phenolic laminated sheet, etc.
   iv. Control switch or MCB if specified.
   v. 3 pin or 6 pin socket, ceiling rose/connector in case of 5 amp point.
   vi. connection/interconnection inside the switch board or between points on the same circuit or looping from the nearest available point, looping of neutral, etc.
   vii. Earth conductor from one metallic switch box to another and also to socket outlets.
   viii. In the case of modular switch socket, the modular box, switch, socket and regulator, etc.

10.1.3 Measurement of Point Wiring

The point wiring shall be measured on the basis of counting of exhaust fan, ceiling fan, call bells and fittings.

No separate payment shall be made for interconnection, looping from the nearest point or for loop earthing conductor between metal switch boxes. The counting of points is a vague term as the length of wiring can vary from a 1–2 room flat to a large conference hall. Hence, the rates of point shall be fixed accordingly before floating tenders, e.g.

   i. Group A wiring for type I, II and III residential accommodation as defined in Government sector or up to 100 sq m of covered area.
   ii. Group B wiring for type IV and above residential accommodation or from 100 to 200 sq m of covered area.
   iii. Group C for all non residential buildings such as officer, hospitals, departmental stores, educational institutions, libraries, etc.

10.1.4 Point Wiring for Socket Outlet Points

i. The light plug (5 A/6 A) and power (15 A/16 A) point wiring shall be measured on linear basis from the respective tapping point to socket outlet point.

ii. The metal/PVC switch box with switch/MCB, socket outlet shall be measured and paid as a separate item.

10.1.5 Group Control Point Wiring

i. If a number of points are controlled by one switch, then first point shall be counted from switch to first nearest outlet. The distance from first outlet to next outlet and so on shall be treated as separate points each. Classification shall remain same for type of building as group A, B, C.

10.1.6 Twin Control Light Point

i. A light point controlled by two numbers of two-way switches shall be measured as two points from fitting to switches on either side.

ii. Here again no recovery shall be made for not providing any accessories normally required for two independent points.

10.1.7 Multiple Controlled Call Bell Points Wiring

i. In the case of single call bell controlled from more than one place, the point shall be measured from call bell to connector/ceiling rose as one point. Similarly from this connector/ceiling rose to next one and so on treated as separate points.

ii. Again no recovery shall be made for non-provision of accessories normally required for two independent points.

10.2 CIRCUIT AND SUBMAIN WIRING

i. **Circuit wiring** is the wiring from the distribution board of a floor to the switch board controlling various points or switch board housing 5 A/15 A socket outlets.

ii. **Submain wiring** is the wiring from one main distribution board to the another and so on.

Measurement

i. The circuit and submain wiring shall be measured on liner basis for the actual length and number of wires and conduit used.

Mostly circuit/submain item is paid as a single item both for wire and conduit combined together on the basis of pre-settled rate.

ii. The length of circuit shall be measured from distribution board to the first nearest switch board irrespective of whether neutral is taken to switch box or not.

Generally neutral is looped on the roof from point to point and taken directly to distri-

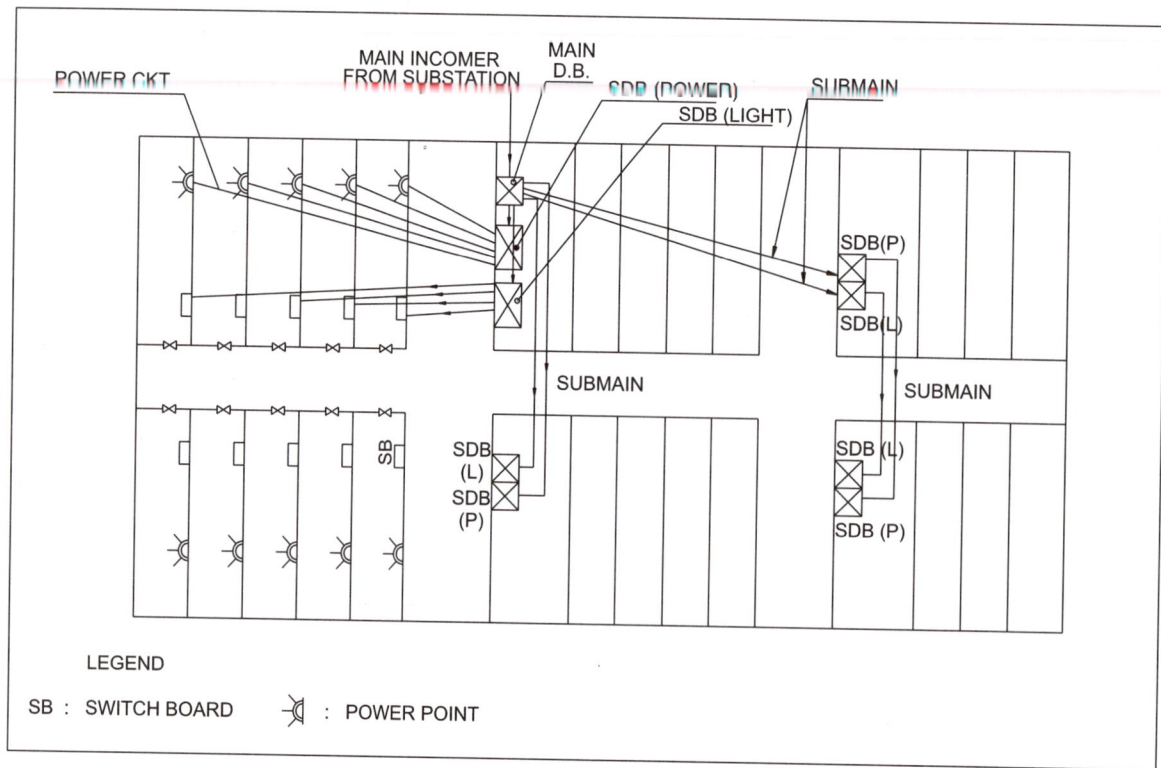

Fig. 10.1 Distribution system in a typical building

bution board unless there is on board 5 A socket outlet in which case neutral is brought in switch board.

iii. The conduit of different sizes can carry different number of wires as listed in Table 10.1. As the conduits run in walls and roof, it is a practice that these circuit and point wiring may be taken in the same conduit which are to be taken in the same direction. This will save the number of conduits that are to be laid in the roof. (For more strength of roof, the number of conduits shall be as minimum as possible).

Again the circuit wiring is measured on linear basis and if clubbed with point wiring, circuit shall be measured on linear basis and point in the normal way. But here the rate of circuit shall be without conduit (as the conduit in this case has been paid with point). It shall be only for drawing the required number of wires in the exiting conduit (otherwise you are paying for conduit two times).

iv. Protective conductor (loop earthing) which run along the circuit wiring and submain wiring shall be measured on linear basis and paid separately.

10.3 SYSTEM OF DISTRIBUTION

10.3.1 Control at the Point of Entry of Supply

There shall be a linked switch with fuse or circuit breaker on each live conductor of the supply mains at the point of entry.

10.3.2 Distribution

i. The wiring shall be done on a distribution system through main and/or branch distribution boards.

ii. Main distribution board shall be controlled by a circuit breaker. Each outgoing circuit shall be controlled by a circuit breaker or only a fuse on the phase or live conductor (as in the case of a TPDB).

iii. The branch distribution board shall be controlled by a circuit breaker. Each outgoing circuit shall be provided with a fuse or miniature circuit breaker (MCB) of specified rating on the phase or live conductor.

iv. Triple pole distribution boards shall not be used for final circuit distribution, but if used then in such special cases, the triple pole distribution boards shall be of HRC fuse type or MCB type only.

Table 10.1 Maximum number of PVC insulated 650/1100 V grade aluminium/copper conductor cables in a conduit conforming to IS 694–1990

Nominal cross-sectional area of conductor in sq mm	20 mm		25 mm		32 mm		38 mm		51 mm		64 mm	
	S	B	S	B	S	B	S	B	S	B	S	B
1	2	3	4	5	6	7	8	9	10	11	12	13
1.50	5	4	10	8	18	12	–	–	–	–	–	–
2.50	5	3	8	6	12	10	–	–	–	–	–	–
4	3	2	6	5	10	8	–	–	–	–	–	–
6	2	–	5	4	8	7	–	–	–	–	–	–
10	2	–	4	3	6	5	8	6	–	–	–	–
16	–	–	2	2	3	3	6	5	10	7	12	8
25	–	–	–	–	3	2	5	3	8	6	9	7
35	–	–	–	–	–	–	3	2	6	5	8	6
50	–	–	–	–	–	–	–	–	5	3	6	5
70	–	–	–	–	–	–	–	–	4	3	5	4

Note:

1. The above table show the maximum capacity of conduits for simultaneous drawing of cables.
2. The columns headed S apply to runs of conduits which have distance not exceeding 4.25 m between draw in boxes and which do not deflect from the straight by an angle of more than 15 degrees. The columns headed B apply to runs of conduit which deflect from the straight by an angle of more than 15 degrees.
3. Conduit sizes are the nominal external diameters.

v. The loads of the circuits shall be divided, as far as possible, evenly between the number of ways of distribution boards, leaving at least one spare circuit for future extension.

vi. The neutral conductors (incoming and outgoing) shall be connected to a common link (multi way connector) in the distribution board and be capable of being disconnected individually for testing purposes.

vii. 'Power' wiring shall be kept separate and distinct from 'lighting' wiring, from the level of circuit, i.e. beyond the branch distribution boards.

viii. Wiring shall be separate for essential loads (i.e. those fed through standby supply) and non-essential loads throughout.

ix. Circuit on opposite poles of a three wire DC system shall be kept apart unless they are enclosed in earthed metal casing, suitably marked to indicate the risk of shock due to the voltage between the conductors contained.

10.3.3 Wiring System

i. Wiring shall be done only by the "looping system". Phase or live conductors shall be looped at switch boxes and neutral conductors at the point outlets.

ii. No joint shall be made in the wiring along the run. If at all a joint is required it shall be in a metal box and with approved mechanical connectors.

iii. Lights, fans and call bells shall be wired in the 'lighting' circuits. 15 A/16 A socket outlets and other power outlets shall be wired in the 'power' circuits. 5 A/6 A socket outlet shall also be wired in the 'power' circuit both in residential as well as non-residential buildings.

iv. The wiring throughout the installation shall be such that there is no break in the neutral wire except in the form of linked switchgear.

10.3.4 Joints in Wiring

i. No bare conductor in phase and/or neutral or twisted joints in phase, neutral, and/or protective conductors in wiring shall be permitted.

ii. There shall be no joints in the through-runs of cables. If the length of final circuit or submain is more than the length of a standard coil, thus necessitating a through joint, such joints shall be made by means of approved mechanical connectors in suitable junction boxes.

iii. Termination of multistranded conductors shall be done using suitable crimping type thimbles.

10.4 RATINGS OF OUTLETS (TO BE ADOPTED FOR DESIGN)

i. Incandescent lamps in residential and non-residential buildings shall be rated at 60 W and 100 W respectively.

ii. Ceiling fans shall be rated at 60 W. Exhaust fan, fluorescent tubes, compact fluorescent tubes, HPMV lamps, HPSV lamp, etc. shall be rated according to their capacity. Control gear losses shall be also considered as applicable.

iii. 5 A/6 A and 15 A/16 A socket outlet points shall be rated at 100 W and 1000 W respectively, unless the actual values of loads are specified.

10.5 CAPACITY OF CIRCUITS

i. Lighting circuit shall not have more than a total of 10 points of light, fan and socket outlets, or a total connected load of 800 W, whichever is less.

ii. Power circuit shall be designed with only one outlet per circuit in non-residential buildings. The circuit shall be designed based on the load. Where not specified, the load shall be taken as 1 kW per outlet. (In the non-residential building, it is assumed that all the power outlets may have some connected load all the time, hence any looping with another outlet may increase the burden on the power circuit).

iii. Power circuit in residential buildings shall be designed for not more than two outlets (15 A/16 A and 5 A/6 A) per circuit. The ratings for load calculation purposes shall however be taken as per the type of outlets.

iv. Load more than 1 kW shall be controlled by an isolator or miniature circuit breaker.

10.6 CONFORMITY TO IE ACT, IE RULES, AND STANDARDS

i. All electrical works shall be carried out in accordance with the provision of Indian electricity act, 1910 and electricity rules, 1956 amended up to date.

ii. The works shall also conform to relevant Indian standard codes of practice (COP) for the type of work involved (*see* Appendix A).

iii. All components shall conform to relevant Indian standard specification, wherever existing. Materials with ISI certification mark shall be preferred. However, for conduits, wiring cables, piano/modular switches and socket outlets, ISI marked materials shall only be permitted wherever available.

10.7 RATING OF COMPONENTS

i. All components in a wiring installation shall be of appropriate ratings of voltage, current, and frequency, as required at the respective sections of the electrical installation in which they are used.

ii. All conductors, switches and accessories shall be of such size as to be capable of carrying the maximum current which will normally flow through them, without their respective ratings being exceeded.

10.8 WIRING / CABLES

i. Conductors of wiring cables (other than flexible cables) shall be of aluminium or copper, as specified. However, wiring for socket outlets in all residential and non-residential buildings shall be done by using copper conductor cables only.

ii. The smallest size of conductor for 'lighting' circuits shall have a nominal cross-sectional area of not less than 1.5 sq mm. The minimum size of conductor for 'power' wiring shall be 4 sq mm. The 5 A /6 A socket outlet shall also be wired with 4 sq mm conductor as it is seen that the common man has the tendency to use appliance of any rating by changing the plug or to draw connection for various equipments by using multi-pin sockets. This puts the load on the system and subsequent burnings, etc.

iii. Stranded aluminum conductor shall not be used in wiring cables up to and including 6 sq size.

10.9 ACCESSORIES

i. Power (15 A /16 A) outlets shall be controlled by single pole piano/modular switches or by MCB's, where specified. Only MCB's shall be used for controlling industrial type socket outlets, and power outlets above 1 kW.

ii. Control switch shall be placed only in the live conductor of the circuit. No single pole switch or fuse shall be inserted in the protective (earth) conductor, or earthed neutral conductor of the circuit.

iii. In an earthed system of supply, socket outlets and plugs shall only be of 3 pin type, the third pin shall be connected to earth through protective (loop earthing) conductor. 2 pin or 5 pin sockets shall not be permitted to be used.

iv. Preferably shutter type (interlocking type) of sockets shall be used.

v. 5 A/6 A and 15 A/16 A socket outlets shall be installed at the following positions, unless otherwise specified.

a. Non-residential buildings 23 cm above floor level.
b. Kitchen 23 cm above working platform and away from the likely positions of stove and sink.
c. Bathroom-socket outlet for shaver/hair drier is permitted and at least 1 m away from shower.
d. Rooms in residences 23 cm above floor level.

vi. Unless and otherwise specified, the control switches for the 5 A/6 A and 15 A/16 A socket outlets shall be kept along with the socket outlets.

10.10 SWITCH BOX COVERS

Phenolic laminated sheets of approved shade shall be used for switch box covers. These shall be of 3 mm thick synthetic phenolic resin bonded laminated sheet as base material and conforming to grade P-I of IS 2036 – 1974.

10.11 FITTINGS

i. Wires used within pre-wired fittings shall be flexible with PVC insulation and 14/0.193 mm (minimum) copper conductors. The leads shall be terminated on built-in terminal block, ceiling rose or connector, as required.

ii. Fittings using discharge lamps shall be complete with power factor correction capacitors, either integrally or externally. An earth terminal with suitable marking shall be provided for each fitting for discharge lamp.

iii. Fittings shall be installed such that the lamp is at a height of 2.5 m above floor level or as per the actual requirement.

10.12 SWITCHGEAR AND CONTROLGEAR GENERAL ASPECTS

i. All items of switchgear and distribution boards (DBs) shall be metal clad type, except those forming part of cubicle type switch boards, in which case the board design shall be such as not to permit direct contact.

ii. The type, ratings and/or categories of switchgear and protective gear shall be specified.

iii. RCDs (ELCBs) where specified, shall conform to the requirements of current rating, fault rating, single phase or three phase configuration and sensitivity.

iv. Independent earth terminal block. Every distribution board (single phase as well as 3 phase) shall have an earth terminal block identical to, but independent from neutral terminal block, to enable termination of protective (loop earthing) conductors (incoming as well as outgoing) individually by screwed connection and without twisting.

v. Earthing terminal (1 for single phase and 2 for 3 phase) shall be provided on the metal cladding of switches and DBs for body earthing. These shall be suitable marked.

vi. Each distribution board shall be provided with a circuit list giving details of each circuit which it controls and the current rating of the circuit, and size of the fuse element.

10.13 COMPLETION PLAN AND COMPLETION CERTIFICATE

i. For all major works completion certificate after completion of work shall be submitted to the engineer-in-charge. This shall include the insulation resistance test results of the entire installation, results of the earthing resistance, electrodes and their description and all other relevant details as per the proforma in Appendix B.

ii. Completion plan drawn to a suitable scale in tracing cloth with ink indicating the following, along with three blue print copies of the same shall also the submitted.
   a. General layout of the building.
   b. Location of main switch board and distribution boards, indicating the circuit numbers controlled by them.
   c. Position of all points, their controls and position of junction boxes and tee hidden inside the plaster on wall/roof. (This is essentially required as any repair work in future require chipping of plaster at the place of tee or junction box).
   d. Types of fittings, viz. florescent, pendants, brackets, bulk head, fans and exhaust fans, etc. and their location.

10.14 EARTHING

The essential requirements of earthing are given in IS code of practice on earthing (IS 3043–1987).

The electrical distribution system in the building is with earthed neutral (i.e. neutral earthed at the transformer/generator end), earthing of metallic body of equipment and non-current carrying metallic components in the substation, lightning protection, computer installations and hospital operation theaters.

Earthing requirements are laid down in Indian electricity rules 1956.

10.14.1 Statutory Requirement

i. All medium voltage equipment shall be earthed by two separate and distinct connections with earth. In the case of high and extra high voltages, the neutral points shall be earthed by at least two separate and distinct earth electrodes. Each of the earth electrodes can be provided at the generating station or substation, and may be earthed at any other point, provided such earthing causes no interference. If necessary, the neutral may be earthed through suitable impedance.

ii. Necessary protective device shall be provided against earth leakage.

10.14.2 Supply System Requirement

" System earthing" is provided to preserve the security of the supply system. This is done by limiting the potential of live conductors with reference to earth, to such values as consistent with the level of insulation applied. Earthing the neutral point of the transformer ensures reasonable potential to earth, including at the time when the HV supply is impressed on the transformer. Earthing also ensures efficient operation of protective gear in the case of earth faults.

10.14.3 Special Earthing Requirements

i. *Static earthing* is provided to prevent building up of static charges, by connections to earth at appropriate locations for example, operation theaters in hospitals. (For details, please refer to IS 7689–1974 and the national electrical code).

ii. *Clean earth* may be needed for some of the data processing equipment. These are to be independent of any other earthing in the building. (For details, please refer to IS 10422–1982 and IS 3043–1987).

iii. Earthing is essentially required in protection of buildings against lightning. (this has been explained in the articles ahead).

10.14.4 Material of Earth Equipments

i. Copper.
ii. GI.

10.14.5 Type of Earthing Systems

i. Pipe earth electrode: GI pipe.
ii. Plate earth electrode: GI/copper plate.
iii. Strip or conductor: GI/copper strip earth electrode.

The selection of right type of electrode for right installation is very essential as otherwise the whole process is defeated. The selection is given in Table 10.2.

The method of earthing is shown in Figs 10.2 and 10.3 for pipe earth and plate earth respectively.

The minimum size of earth electrode shall be as per Table 10.3.

Table 10.2 Selection of type of electrode for different kind of installations

Type of electrode	Application
GI pipe	Internal electrical installation, with incoming switch gear up to 200 A.
GI plate	i. For internal electrical installations with incoming switch gear larger than 200 A.
	ii. Neutral earthing of transformers, generating sets up to 500 kVa.
	iii. Lightning conductors.
Copper plate	Neutral earthing of transformers/ generating sets above 500 kVa.
Strip/conductor	Locations where it is not possible to use other types.

Table 10.3 Selection of the size of earthing conductor

Type of electrode	Material		Size
Pipe	GI Medium class		40 mm dia. 4.5 m long (without any joint).
Plate	i.	GI	60 cm × 60 cm × 6 mm thick.
	ii.	Copper	60 cm × 60 cm × 3 mm thick.
Strip	i.	GI	100 sq mm section.
	ii.	Copper	40 sq mm section.
Conductor	i.	GI	5 mm dia. (6 SWG).
	ii.	Copper	4 mm dia. (8 SWG).
Strip	i.	GI	25 mm × 4 mm.
	ii.	Copper	20 mm × 3 mm.

Note: Galvanisation of GI items shall conform to class IV of IS 4736–1986.

The dimensions and method of connection as shown in respective figures are essential to achieve the desired results of earthing.

The length of buried strip in strip type earthing above shall not be less than 15 meters. This length shall be adequately increased, if necessary on the basis of the information available about soil resistance so that the required earth resistance is obtained.

10.14.6 Earth Bus

i. Two copper strips, each of size 50 × 5 mm shall be provided as earth bus in a 11 kV substation and/or DG set room irrespective of the capacity of the transformer or the DG set. Each of these strips shall be connected to an independent earth electrode. The two earth leads from the body of each transformer/panel/generating set, etc. shall be connected to these two strips of earth bus. The two strips of earth bus shall be bonded together.

ii. The neutral earth leads of the transformer and or generator alternator shall not be connected to this earth bus. They shall be connected directly to individual earth electrodes.

10.14.7 Location of Earth Electrode

i. Normally an earth electrode shall not be located closer than 1.5 m from any building. The excavation for earth electrode shall not affect the foundation of the building in which case the electrode shall be located further away from the building.

ii. The location shall be such that soil has a reasonable chance of remaining moist.

iii. The distance between two or more electrodes (plate/pipe) shall not be less than 2 m.

iv. The strip or conductor electrode shall be buried in a trench not less than 0.5 m deep.

v. The artificial treatment of soil shall be done using charcoal/coke and salt.

vi. The earthing conductor from the electrode up to the building shall be protected from mechanical injury by a medium class 15 mm dia GI pipe in case of wire and by 40 mm dia. pipe in case of strip which shall be buried 30 cm deep in earth.

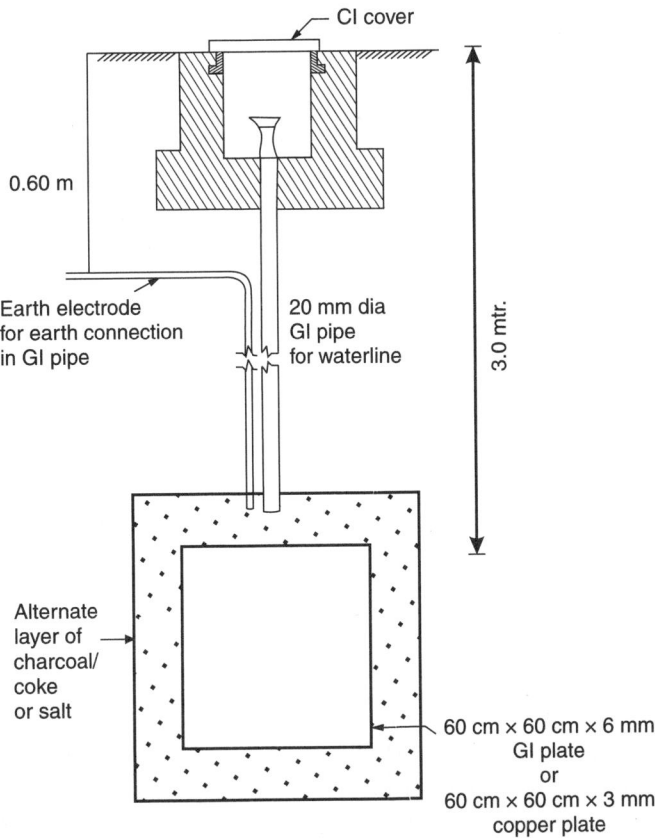

Fig. 10.2 Plate earthing

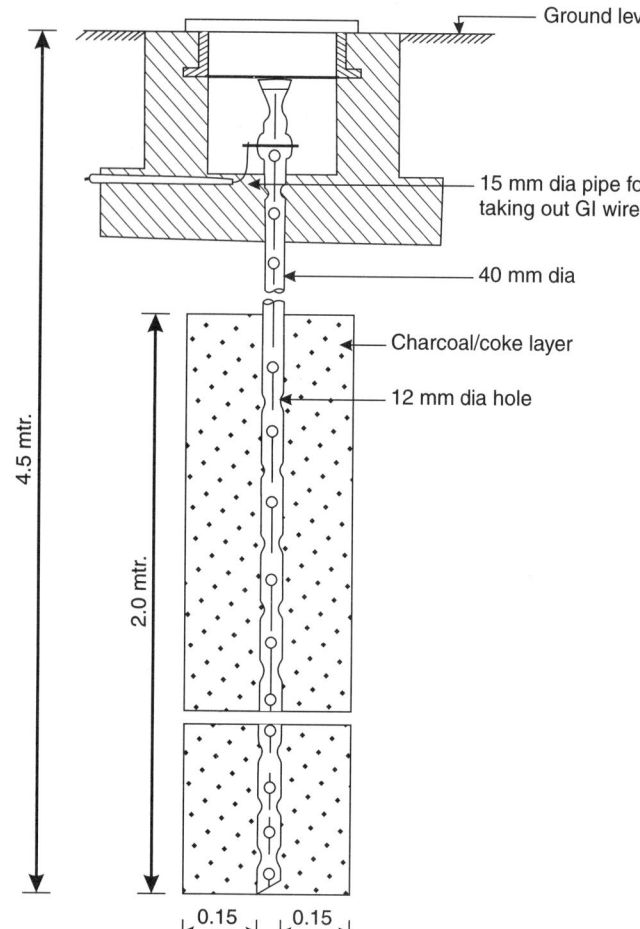

Fig. 10.3 Pipe earthing

Labels in figure:
- Ground level
- 15 mm dia pipe for taking out GI wire
- 40 mm dia
- Charcoal/coke layer
- 12 mm dia hole
- 4.5 mtr.
- 2.0 mtr.
- 0.15
- 0.15

10.14.8 Earth Resistance

i. The earth resistance at each electrode shall be measured. No earth electrode shall have resistance greater than 5 ohms as measured by an approved earth testing apparatus. In rocky soil the resistance may be up to 8 ohms.

ii. Where the above stated earth resistance is not achieved, necessary improvement shall be made by additional provisions, such as additional electrode(s), different type of electrode, or artificial chemical treatment of soil, etc.

10.14.9 Number of Earth Electrodes

The Indian electricity rules 1956 vide rules 33, 61, 67 throws light on this matter and should be complied with. The non-current carrying metal parts of all apparatus utilising power supply at voltage exceeding 250 volts shall be earthed by two separate and distinct connections to earth bus or to two separate and distinct earthing sets.

The number of earthing electrodes for substations and generating sets shall be as under.

i. For neutral earthing of each transformer: 2 sets.

ii. For body earthing of all the transformers, HT/LT panels and other electrical equipment, in the substation/power house: 2 sets.

iii. For neutral earthing of each generating sets: 2 sets.

iv. For body earthing of all the generating sets, LT panel and other electrical equipment in the generator room: 2 sets.
    Note: Where generator and substation equipment are located together in the same building, the body earthing can be common for all the electrical equipment in the building.

v. Separate earth electrode shall be provided for lightning arrester/lightning conductor.

10.15 PROTECTION OF BUILDING AGAINST LIGHTNING

The IS 2309–1989 gives the detailed requirement of providing protection to building against lightning. This article covers the protection of building which are not explosive in nature, i.e. public buildings are the scope of this article.

10.15.1 Principle and Zone of Protection

A. Principle of Protection

i. The principle for protection of buildings against lightning is to provide a conducting path between earth and the atmosphere above the building through which the lightning discharge may enter the earth without causing damage to the building. If adequately earthed metal parts of proper proportions are provided and spread properly on and around the building, damage can be largely prevented.

ii. The required conditions of protection are generally met by placing all the air-terminals, whether in the form of vertical finials or horizontal conductors, on the upper most part of the buildings or its projections, with lightning conductors connecting the air terminals with each other and to the earth.

B. Zone of Protection

i. General
   The zone of protection is the volume within which a lightning conductor gives protection against a direct lightning stroke by directing the stroke to itself. For a vertical conductor rising from ground level, the zone is described as a cone with its apex at the tip of the conductor and its base on the ground. For a horizontal conductor, the zone is defined as the volume

generated by a cone with its apex on the horizontal conductor from and to end.

ii. Protective angle

1. This cannot be precisely stated, since it depends upon the severity of the stroke and the presence within the protective zone of conducting objects providing independent paths to the earth. All that can be stated is that the protection afforded by a lightning conductor increases as the assumed protective angle decreases.

2. a. However, for the practical purpose of providing an *acceptable degree* of protection for an ordinary structure, the protective angle of any single component part of an air termination network, namely, either one vertical, or one horizontal conductor is considered to be 45 degrees.

   b. Between three or more vertical conductors, spaced at a distance not exceeding twice their height, the equivalent protective angle may, as an exception, be taken as 60 degrees to the vertical.

3. Protective angles of zones of protection for some forms of air termination are illustrated in IS 2309–1989.

C. Components

i. Air terminations

1. Air termination networks may consist of vertical or horizontal conductors, or combinations of both.

2. For the purpose of lightning protection, the vertical and horizontal conductors are considered equivalent and the use of pointed air terminations, or vertical finial is, therefore, not regarded as essential.

ii. Down conductors

1. **General.** The function of a down conductor is to provide a low impedance path from the air termination to the earth electrode so that lightning current can be safely conducted to the earth. In practice, depending upon the form of a building, it is often necessary to have many down conductors in parallel, some or all of which may be a part of the building structure itself.

2. **Recommended number.** The position and spacing of down conductors on large structures are often governed by architectural convenience. However, recommendations for their number are given below.

i. Structure having a base area not exceeding 100 sq m need have only one down conductor, except when built on a bare rock, or where access for testing is difficult.

ii. For a structure having an area exceeding 100 sq m, the number of down conductors should be at least the smaller of the following.

   a. One plus an additional one for each 300 sq m, or a part thereof, in excess of the first 100 sq m

   b. One for each 30 m of the perimeter of the structure protected.

iii. Tall structures presenting inspection difficulties.

D. Illustrations

The IS 2309–1989 provides several arrangements for air terminations, down conductors, voltage gradient along ground surface near to masts, towers, columns and single down conductors, re-entrant loops, typical joints, earth termination, bonding to building services, fixing of lightning conductors, test points, typical forms of vertical air terminations, etc. which may be referred to.

10.15.2 Material of Protecting Components

The material of air termination, down conductor, earth termination shall be reliably resistant to corrosion.

   i. Copper.

   ii. Copper clad steel.

   iii. Galvanised steel.

   iv. Aluminium.

The size of conductors and its shape are given in Tables 10.4 and 10.5.

Table 10.4 Shapes and minimum sizes of conductors for use above ground

S. no.	Material and shape	Minimum size
1.	Round copper wire or copper clad steel wire	6 mm diameter
2.	Stranded copper wire	50 sq mm or (7/3.00 mm dia.)
3.	Copper strip	20 × 3 mm
4.	Round galvanized iron wire	8 mm diameter
5.	Galavanize iron strip	20 × 3 mm
6.	Round aluminium wire	8 mm diameter
7.	Aluminium strip	25 × 3 mm

Table 10.5   Shapes and minimum sizes of conductors for use below ground

S. no.	Material and shape	Minimum size
1.	Round copper wire or copper clad steel wire	8 mm diameter
2.	Copper strip	32 × 6 mm
3.	Round galvanized iron wire	10 × 6 mm
4.	Galavanize iron strip	32 × 6 mm

10.15.3 Earth Termination Network

i. An earth station comprising one or more earth electrodes as required, should be connected to each down conductor. Number of down conductors are already calculated vide 10.15.1c above.

ii. Each of the earth stations should have a resistance not exceeding the product given by 10 ohms multiplied by the number of earth electrodes to be provided therein. The whole of the lightning protective system, including any ring earth, should have a combined resistance to earth not exceeding 10 ohms without taking account of any bonding.

iii. If the value obtained for the whole of the lightning protection system exceeds 10 ohms, a reduction can be achieved by extending or adding to the electrodes, or by interconnecting the individual earth terminations of the down conductors by a conductor installed below ground, sometimes referred to as a ring conductor.

Buried ring conductors laid in this manner are considered to be an integral part of the earth termination network, and should be taken into account when assessing the overall value of resistance to earth of the installation.

iv. A reduction of the resistance to the earth to a value below 10 ohms has the advantage of further reducing the potential gradient around the earth electrode when discharging lightning current. It also further reduces the risk of side flashing to metal in, or of structure.

v. Earth electrodes should be capable of being isolated and a reference earth point should be provided for testing purposes.

10.16 ILLUMINATING THE BUILDING

10.16.1 Selection of Number of Fittings

The efficiency of the people working in well lit building increases many folds and also reduces the biological stress on them. The selection of proper lux level on the table, corridors, street, etc. is a very important design consideration and needs years of practice and experience on the part of a designer. Table 10.6 gives the lux level required for various activities in a building (Simplified from NBC 2005).

Table 10.6   Lux level requirement for different applications

S. no.	Location	Lux level in Office building	Hospitals
1.	Entrance	150	150
2.	School/gen. office/ conf. room/operation theater/dispensary/lab	300	300
3.	Drawing office	450	—
4.	Corridor	70	70
5.	Stairs	100	100
6.	X-Ray/Bathroom	100	100
7.	Kitchen/Laundry	200	200
8.	OPD/Ward	—	150

The calculation of number of fittings in a room or hall or corridor can be made with simple mathematics as below (detail in Chapter 14).

Number of fittings

$$= \frac{\text{Lux requirement} \times \text{Area in sq m}}{\text{Lumen output of bulb} \times \text{Transmission efficiency}}$$

Transmission efficiency will vary with the kind of reflector used, distance of fitting from the surface where a pre-determined lux level is required or on the age of fitting or bulb being used and is a matter of experience and manufacturer's data, e.g. in a simple 40 watt tube light twin mirror optic fitting, the lumen output is 2400 lumen. The transmission efficiency in mirror optic fitting can be considered to be 0.7 to 0.8.

Then in an area of 100 sq m, the number of times fittings required will be

$$\text{Number of fittings} = \frac{300 \times 100}{2 \times 2400 \times 0.7} = 9 \text{ fittings.}$$

Depending on symmetry requirements this number can be reduced or increased to 8 or 10 respectively.

10.16.2 Selection of Type of Fitting

The heat generated by the present day bulbs has a great effect on the air-conditioning requirements. Today, we are not using simply incandescent lamps, but metal halide, sodium vapour lamps are commonly used.

The selection of a proper kind of fitting and bulb is governed by the lumen it generates, heat produced, number of hours a bulb is rated to operate, the colour rendering effect, i.e. the burning temperature of the filament or gas inside the bulb has a significant effect on the colour output of the light generated. The most ideal colour effect is produced at 4200 °K. Table 10.7 illustrates the data related to various kind of lamps used these days.

Table 10.7 Data related to various luminaires

Lamp type	Range	Luminous flux	Efficiency Lm/W	Life hours	Colour rendering
GLS (Incandescent)	25 – 1000 W	230 – 18000	9 – 18	1000	Excellent
Halogen	300 – 1000 W	5100 – 22000	17 – 22	2000	Excellent
Flourescent moderate	18 – 40 W	1015 – 2450	49 – 77	5000	Good to
HPMV	80 – 400 W	3500 – 22000	44 – 58	5000	Moderate
HPSV	70 – 1000 W	6000 – 130000	83 – 119	12000 – 15000	Fair
LPSV	18 – 35 W	1800 – 4500	100 – 129	10000	Poor
Metal halide					
HPI-T	250 W	17000 W	70 – 90	10000	Good
HPI-T	400 W	31500 W	70 – 90	10000	Good
HPI-T	1000 W	81000 W	70 – 90	10000	Good
HPI-T	2000 W	189000 W	70 – 90	10000	Good
HPI-BU	250 W	17500 W	70	10000	
HPI-BU	400 W	27600 W	70	10000	
MHNTD	70 W	5500 W	75 – 80	6000	Excellent
MHNTD	150 W	11250 W	75 – 80	6000	Excellent

11 Substation

Electric substation is the heart of electrical and electromechanical services provided in any type of residential, commercial or office complex. With the increase in the use of electrical gadgets, reliability of power supply including its flexibility and economy in capital cost as well as maintainability are of utmost importance. Before proceeding to designing a sub-station, a realistic estimation of load including maximum demand is necessary to decide transformer capacity, layout and distribution system. The next step would involve fault level calculation, selection of equipment, finalization of specification, etc.

11.1 ELECTRICAL LOAD ESTIMATION

Any design of electrical substation starts with the estimation of the connected load and maximum demand. Each type of load has its characteristics depending on the place, as well as type of installation and has its own utilization factor. In addition, maximum demands of different types of loads do not occur at the same time and the maximum demand on the substation will be fraction of sum of maximum demands of different types of loads. This fraction may be called the diversity factor.

The substation capacity may be decided based on the sum total of maximum demand of the loads mentioned above plus a margin of 20 to 30%. The maximum can be arrived by taking a diversity factor of 0.8 on the sum total of the maximum demands of each service arrived above. However, in respect of residential colonies, a diversity factor of 0.5 may suffice.

Let us work out an example of a small residential cum commercial/institutional development (or called mixed use development) comprising of high rise towers in a clusters (Fig. 11.1). The design of electrical

distribution system is explained with the help of an example comprising of residential, commercial and institutional buildings.

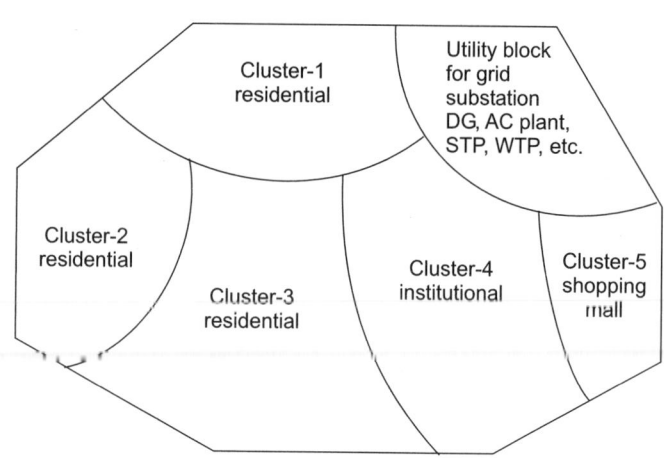

Fig. 11.1 A typical township-mixed use

In the figure, respective areas have been marked. The residential and commercial sectors are medium to high rise towers while institutional is low rise. The residential pockets comprise of 2B + G + 16 stories while commercial are 2B + G + 5 story high. The residential towers are a mix of 2/3/4 BHK units.

At this stage, the calculation of electrical load for a substation is based on certain assumptions related to electric power consumption likely in these premises. These assumptions are given in Table 11.1 and the load has been calculated in Table 11.2.

The example given here is a very typical example of the townships coming up in various parts of the country now. Though more details may be required to work out total load of a building including electrical load of basement, local shopping centres, staff quarters, dispensaries and other amenities, but the same have been omitted here for the sake of convenience.

Table 11.1 Load consumption pattern in various utilities

Type of utility	Connected load (KW)	Diversity	Remarks
A. Residential			
i. 4 BHK	7	75%	
ii. 3 BHK	6	75%	
iii. 2 BHK	5	75%	
iv. 1 BHK	4	75%	
B. Non Residential			
i. Commercial/Shopping mall	86 w/sq mtr	70%	Includes air-conditioning
ii. University teaching facility	59 w/sq mtr	60%	
iii. University residential	59 w/sq mtr	60%	load
iv. University – other facilities	86 w/sq mtr	60%	
C. Overall diversity for grid substation		50%	
D. Overall diversity for centralized DG power			
i. Residential		40%	
ii. Non-residential		100%	

Table 11.2 Calculation of demand load and equipment capacities (Refer Fig. 11.1)

Cluster No.	Type of Utility	Numbers/ Built up area (sq mt)	Connected load (KW/unit or sq mt)	Sanctioned load (KW)	Diversity	Demand load (KW)	Load at 0.95 PF and 0.85 LF	Transformer Capacity 11 KV/415V	Transformer capacity 132/11 KV (MVA)
		Number of units							
1.	4 BHK	980	7	6860	75%	5145	8.69	3 × 2500 KVA + 1 × 1500 KVA	
	3 BHK	300	6	1800	75%	1350			
	2 BHK	140	5	700	75%	525			
						7020			
2.	4 BHK	280	7	1960	75%	1470	5.18	2 × 2500 KVA	1 × 30
	3 BHK	300	6	1800	75%	1350			
	2 BHK	364	5	1820	75%	1365			
						4185			
3.	4 BHK	864	7	6048	75%	4536	15.7	6 × 2500 KVA + 1 × 1000 KVA	
	3 BHK	900	6	5400	75%	4050			
	2 BHK	1092	5	5460	75%	4095			
						12681	29.57		
		Area (sq. mt.)							
4.	Institutional						13.1	5 × 2500 KVA + 1 × 1000 KVA	1 × 30
	i. University	110000	86	9460	60%	5676			
	ii. Hospital	80000	86	6880	60%	4128			
	iii. Student's hostel	22000	59	1298	60%	779			
						10583			
5.	Shopping mall	200000	86	17200	70%	12040	14.91	6 × 2500 KVA	
						12040	30.61		
A.	Total demand load					46509			
B.	Grid demand load 50% diversity					23255	28798 30000 (approx.)		
C.	Centralized DG demand load								
	Residential – 40% diversity					9554			
	Non-residential – 100% diversity					22623			
D.	Total DG demand load					32177			

The transmission and power connection to townships is available depending on grid demand load. The power level at which this connection is available also varies from state to state in India. In the state of Uttar Pradesh, the permitted voltage levels are given in Table 11.3.

Table 11.3 Supply voltage levels for grid connections in U.P. (India)

Type of system	Load (KVA)	Distribution voltage level
LT	Less than 6.25 KVA	230 V, 1 phase
	6.25 to 63 KVA	400 V, 3 phase, 4 wire
HT	63 to 3000 KVA	6.6/11 KV, 3 phase, 4 wire
	3000 to 100,000 KVA	33 KV, 3 phase
EHT	More than 100,000 KVA	132/220 KV

The demand load worked out in the case study is 46.5 MVA. Thus, the connection from 132 KV grid shall be required for meeting the demand of township being planned.

The next phase of planning is to decide the voltage level at which this power is to be transmitted to various clusters. This decision purely depends on the compactness or vastness of the township, distance from substation to switching station of each cluster and the number of substations to be planned and the land value.

The above town can be assumed to be spread on 100 acres (400000 sq mtr) of land considering the volume of construction and built up area (BUA) involved. It can be fairly estimated that the distance from substation to switching station will be considerably long and LT network may not be possible due to heavy voltage drop taking place. Also from the point of view of redundancy and maintenance of the system, it is better to have separate switching stations for each cluster receiving supply at 11 KV.

The transmission of power in very large townships can be at 33 KV level. This is a matter of weighing cost of converting:

a. 132/66 KV to 11 KV and then transmitting at 11 KV upto cluster level, stepping down to 415 volts including the cost of equipments, land required for installing grid station at 132/66 KV and further converting it to 11 KV.

b. 132/66 to 33 KV, transmitting at 33 KV upto cluster level and converting 33 KV to 11 KV or directly to 415 V at cluster level or intermediate levels.

In the present case, transmission of power at 11 KV from grid substation to cluster level switching station has been considered. In the single line diagram (Fig. 11.2), distribution of power is shown. The fault level of equipment to be selected and the fault withstand capacity of cable have also been considered before

selecting the amperage and short circuit withstand capacity of switchgears.

Consider the calculation for cluster – 1:

Demand load (KW)   :   7000 KW

$$\text{Current drawn (Amps) at 11 KV} = \frac{KW}{\sqrt{3}\,V}$$

$$= \frac{7000}{\sqrt{3} \times 11}$$

$$= 367 \text{ Amps}$$

Cable size selection can be performed by taking some standard manufacturer's cable and performing voltage drop calculations. But be aware that in HT system, voltage drop in rarely a problem but short circuit withstand capacity is. Thus cables in HT system are designed for fault levels as shown in Table 11.4.

Table 11.4 Selection of cable size at 11 KV

Cable size (sq mm)	Current carrying capacity in ground (amp)	Deration @ 20% (amp)	No. of runs to carry 367 amp load	Rated Short circuit withstand capacity for 1 sec (KA)	Optimal selection (sq mm)
3c × 400	400	320	2	37.60	
3c × 300	355	284	2	28.20	
3c × 240	315	252	2	22.56	3c × 240
3c × 185	275	220	2	17.39	
3c × 150	245	196	2	14.10	

In Table 11.4, if one was to select cable just to carry current of 367 Amp, then 2 runs of 3c × 150 sq mm were sufficient. But let's check this selection for voltage drop as in Table 11.5. The length of cable has been assumed to be 1000 meters. From this calculation, it is seen that voltage drop is not even one percent of system voltage. Thus, this selection is acceptable as far as voltage drop is concerned.

Let's further examine this selection for system's short circuit fault expected and the fault withstand capacity of this cable.

In HT and EHT distribution, fault levels are high and the fault level of 350 MVA and 1000 MVA are considered for calculation of short circuit current in 11 KV and 33 KV systems respectively. In this case at 11 KV, the fault level current for 1 second will be.

Short circuit current KA (1 sec duration):

$$= \frac{\text{Fault level (KVA)}}{\sqrt{3} \times \text{Voltage (KV)}}$$

$$= \frac{350 \times 1000}{\sqrt{3} \times 11} \times \frac{1}{1000}$$

$$= 18.37 \text{ KA}$$

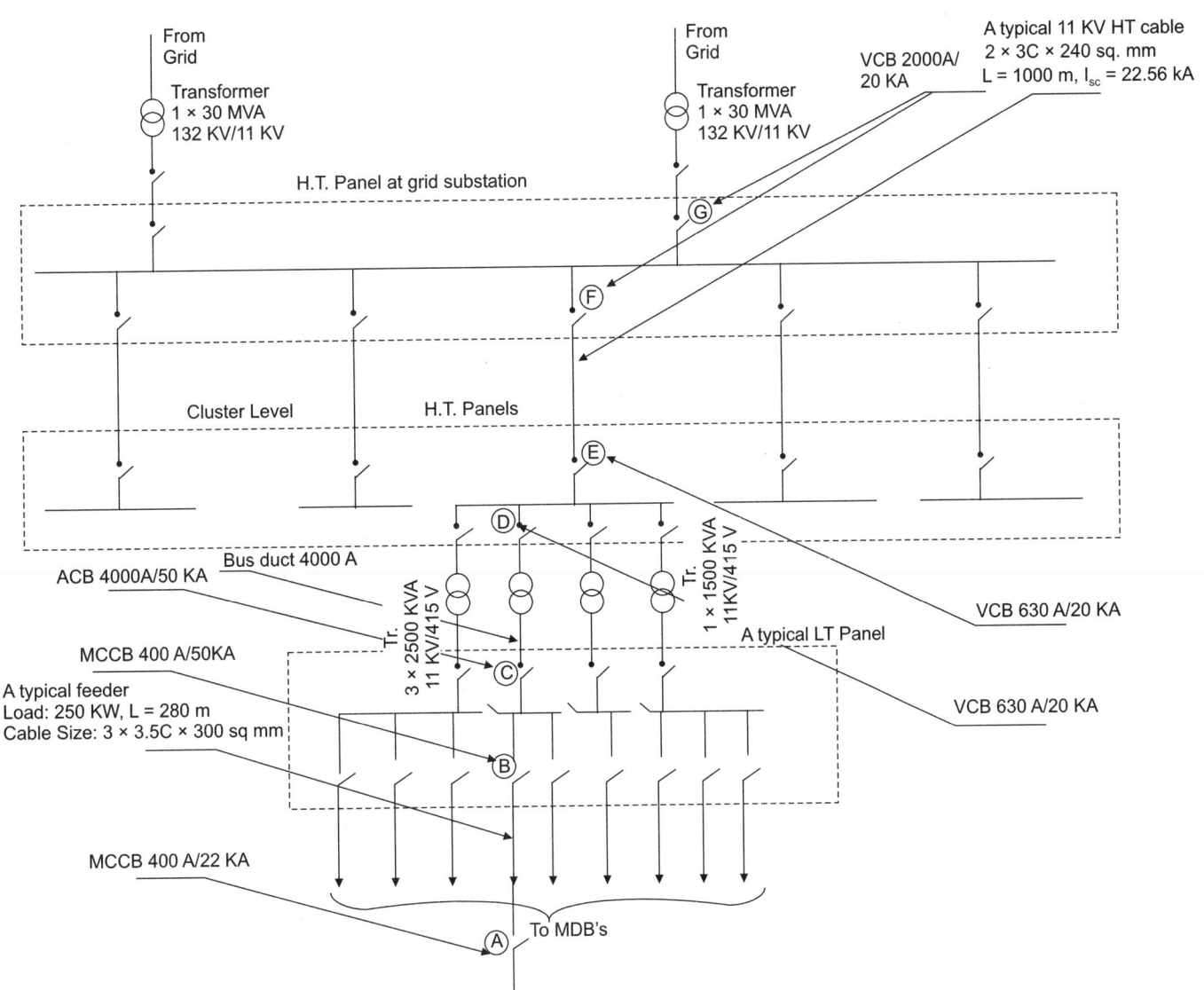

Fig. 11.2 Single line diagram of power distribution

From Table 11.4, it can be seen that 3c × 150 sq. mm. cable will not be able to sustain the fault level likely here. The minimum level which can sustain fault of 18.37 KA is 3c × 240 sq. mm. Hence, two runs of 3c × 240 sq mm cable will be utilized here.

The cable sizes for different feeders from LT panel to end use point can now be calculated and the calculations for the same can be seen in Table 11.5.

Determining Switchgear Sizes

The next step is to calculate the ampere and short circuit withstand capacity for which every switchgear or breaker shall be selected. The farther the switchgear is from the source of fault, less will be the breaking capacity required. This is due to resistant and impedance of electrical cables which reduce the value of fault, down the circuit.

A fault generated in HT line or transformer will be highest due to high voltage levels. This fault will travel down the circuits through cables. The impedance of transformer and resistance/impedance of cables will be calculated to determine the fault level likely to be present at various points till end use point.

a. Transformer (132/11 KV):

$$\text{Full load current } (I_L) = \frac{\text{KVA rating of transformer}}{\sqrt{3} \times \text{voltage on secondary}}$$

$$= \frac{30000}{\sqrt{3} \times 11}$$

$$I_L = 1575 \text{ Amp}$$

In star connection $I_L = I_P$

Therefore $I_P = 1575$ Amp.

Percentage impedance = 12.5% (As per IS code for 30 MVA transformer)

$$\text{Impedance (ohms)} = \frac{\Delta V_P}{I_P}$$

Table 11.5 Voltage drop calculation (Refer Fig. 11.2)

Cable type	Load (KW)	Load current (Amp)	Length (m)	Cross section cable (sq mm)	Current rating (Amp)	No. of core	Power factor	De-rating factor	De-rated current carrying capacity (Amp)	No. of runs selected	Resistance /km (ohm) (r)	Reactance /km (ohm) (x)	Imped-ance /km $\sqrt{r^2+x^2}$	Voltage drop $\dfrac{\sqrt{3}mbc}{1000\times j}$	Percent voltage drop
	a	b	c	d	e	f	g	h	i	j	k	l	m	n	o
From grid to cluster transformer															
11 KV Al Cond XLPE cable	7000.00	367.00	1000.00	400.00	315.00	3.00	1.00	0.80	252.00	2.00	0.10	0.10	0.14	44.43	0.40
	7000.00	367.00	1000.00	300.00	355.00	3.00	1.00	0.80	284.00	2.00	0.13	0.10	0.16	52.12	0.47
	7000.00	367.00	1000.00	240.00	315.00	3.00	1.00	0.80	252.00	2.00	0.16	0.10	0.19	60.71	0.55
	7000.00	367.00	1000.00	185.00	275.00	3.00	1.00	0.80	220.00	2.00	0.21	0.11	0.24	75.01	0.68
	7000.00	367.00	1000.00	150.00	245.00	3.00	1.00	0.80	196.00	2.00	0.27	0.11	0.29	90.90	0.83
From cluster transformer to LT panel															
1.1 KV Al Cond XLPE cable	2500.00	3478.00	100.00	300.00	315.00	3.50	1.00	0.80	252.00	14.00	0.12	0.07	0.14	6.11	0.06
From LT panel to main dist'n board															
1.1 KV Al Cond XLPE cable	250.00	434.75	280.00	300.00	455.00	3.50	0.80	0.75	341.00	3.00	0.12	0.10	0.14	9.92	2.39

where ΔV_P = Voltage drop on full load

= Percent impedance of transformer × Phase voltage

$$= \frac{12.5}{100} \times \frac{11000}{\sqrt{3}} \left(V_P = \frac{V_L}{\sqrt{3}} \right)$$

$$= 794 \text{ volt}$$

Impedance (ohms) $= \dfrac{794}{1575}$

$$= 0.5042 \text{ ohm}$$

Maximum fault current $(I_f) = \dfrac{V_L}{\sqrt{3} \times \text{Impedance}}$

$$= \frac{11000}{\sqrt{3} \times 0.5} = 12590 \text{ A}$$

$$= 12.59 \text{ KA} \qquad \dots (1)$$

b. System fault level

For HT at 11 KV, the system must be designed for 350 MVA fault level Calculating the maximum fault current at 350 MVA.

$$I_f = \frac{350}{\sqrt{3} \times 11}$$

$$= 18.37 \text{ KA} \qquad \dots (2)$$

c. Calculated fault on cable

Let us assume the size and length of cable connecting 132/11 KV transformer and incoming breaker of HT panel be 4 runs of $3c \times 400$ Sq. mm and 150 meter length so as to calculate the impedance of cable as well. As calculated in Table 11.6.

Impedance of cable (ohm) $= \sqrt{R^2 + X^2}$

where, R = resistance (ohm) per km length of cable

X = reactance (ohm) per km length of cable

Impedance of cable (ohm) $= \sqrt{.0038^2 + .0159^2}$

$$= 0.0164 \text{ ohm}$$

Total impedance = impedance of transformer + impedance of cable

$$= 0.5042 + 0.0164$$

$$= 0.5206 \text{ ohm}$$

Fault current $= \dfrac{11}{\sqrt{3} \times 0.5206}$ ohm

$$= 12.2 \text{ KA} \qquad \dots (3)$$

d. Fault with stand capacity of cable

Also consider the short circuit withstand capacity of cable selected *i.e.* 3c × 400 sq. mm.

From manufacturer's table

$$I_{Sc} \text{ of cable} = 37.60 \text{ KA} \qquad \dots (4)$$

From the four different KA values calculated, it is certain that switchgear with I_{Sc} value less than 37.6 KA

shall be selected so that in the event of a fault, it can wait for 1 second for the fault to be cleared and then trip to protect the cable equipments and life down the circuit. The ratings of different switchgears for a typical circuit of single line diagram (Fig. 11.2) are calculated in Table 11.6.

The power factor (P.F.) on LT side plays a crucial role. The low P.F. attracts penalty from the power supply company. The correction to P.F. is carried out using suitable capacitor banks in substation. The upstream side of LT panel is thus taken care of, but downstream the cable is carrying high reactive power. As explained in the sections ahead, the P.F. is best improved when correction is applied near the heavy inductive load, say motors, chillers, etc.

11.2 COMPONENTS OF A SUBSTATION

The substation comprises of various equipments like the one which handle HT power called HT panels, those which convert high voltage to low voltage called transformers and those which handle low voltage to be supplied for the use of the consumer called LT panel. These equipments are provided with the following accessories for the efficient and safe use of the system.

a. HT panel comprises current transformer, potential transformer, HT circuit breaker and protection relays like over current, earth fault relays.

b. Transformer to convert HT supply to LT supply with protection devices like bucholze relay.

c. LT panel comprising circuit breakers, current transformers, outgoing switches or MCCB's and protection relays like over current and earth fault relays.

The method of calculating the capacity of various devices as listed above is described in the following articles.

11.2.1 Capacity and Number of Transformers

The capacity and number of transformers in a substation determines the reliability of the system. It is essential that substation is not only designed with spare capacity to take care of the future loads, but it should also be so designed that in case of any break down, the power supply is not adversely affected. Installation with single transformers should be avoided and minimum of two transformers be provided. The loading on transformers should be restricted not to exceed 85% to take into account future loading.

It would be advisable to restrict the number of transformers in a substation to only 2–3 transformers to avoid having a complicated LT system and nearest standard size should be chosen.

Table 11.6 Switchgear/breaker amperes and short circuit capacity (Refer Fig. 11.2)

Breaker No.	Transformer capacity (KVA)	Percent Impedance of transformer	Circuit voltage level (KV)	Load on circuit (KW)	Load current (Amp)	Size of cable sq mm	Length of cable (KM)	No. of runs of cables	Resistance of cable selected (Ohm/KM)	Reactance of cable selected (Ohm/KM)	Impedance of cable selected (Ohm)	Impedance of cable up in the circuit	Total cable impedance (Ohm)	Impedance of transformer (Ohm)	Total impedance	Maximum fault current (I_f) KA	Max fault value in MVA	Check value against I_{sc} cable (KA)	Check value against 1000 MVA/ 350MVA fault level (KA)	Selectable I_{sc} of breaker (KA)	Breaker capacity calculated (Amp)	Breaker capacity selected (Amp)	
G	30000	12.50	11.00	–	–	400.00	0.15	4.00	0.00	0.02	0.02	0.00	0.02	0.50	0.52	12.20	232.61	37.60	18.37	20	1574.59	2000	Minimum design fault level 350 MVA or 18 KA at 11 KV.
F	30000	12.50	11.00	–	–	0.00	0.00	0.00	0.00	0.00	0.00	0.02	0.02	0.50	0.52	12.20	232.61	37.60	18.37	20	1574.59	2000	
E	30000	12.50	11.00	7000	367.40	240.00	1.00	2.00	0.16	0.10	0.15	0.02	0.17	0.50	0.67	9.42	179.64	22.56	18.37	20	367.40	630	Minimum breaker size of 11 KV is 630 Amp
D	30000	12.50	11.00	7000	367.40	240.00	1.00	2.00	0.16	0.10	0.15	0.17	0.32	0.50	0.82	7.68	146.32	22.56	18.37	20	367.40	630	
C	2500	6.25	0.41	2500	3478.01	0.00	0.00	0.00	0.00	0.00	0.00	0.32	0.32	0.00	0.32	0.73	0.52	0.00	48.69	50	3478.01	4000	Minimum design fault level 35 MVA or 50 KA at 415 volt.
B	9000	6.25	0.41	250	347.80	0.00	0.00	0.00	0.00	0.00	0.00	0.32	0.32	0.00	0.32	0.74	0.53	0.00	48.69	50	347.80	400	The common bus is fed by 4 transformers of 3 × 2500 + 1 × 1500 KVA. Hence total fault level and impedance calculated with 9000 KVA
A			0.41	250	347.80	300.00	0.28	3.00	0.12	0.10	0.12	0.32	0.44	–	0.44	0.54	0.39	28.20	–	22	347.80	400	

11.2.2 Limitation on Transformer Size due to LT Switchgear

Requirement of immediate resumption and restoration of supply after short-circuit/fault makes application of circuit breaker control on the LT side of transformer mandatory. The LT circuit breakers manufactured are having a short-circuit withstand capacity of 31 MVA/35 MVA and hence, fault level on the LT bus bar should be restricted to less than 43/50 kA. As a thumb rule percentage impedance of a transformer can be considered as 5 percent (1/20) and hence, a symmetrical short-circuit on LT side would result in a maximum fault level equivalent to 20 times the rating of the transformer.

Thus, in respect of 1600 kVa transformer the fault level works out to 32 MVA (1600 × 20) and similarly for 2000 kVa and 2500 kVa transformers the fault level would work out to 40 MVA and 50 MVA respectively. Hence in case of transformers of 2000/2500 kVa capacity the fault withstand capacity of LT breaker and bus bar system should be properly decided and designed.

11.2.3 Incoming Supply and Fault Level

As a practice, most licensee's insist HT supply for demands exceeding 100 kVa. Wherever HT supply is given, the fault level of the licensee's system should be taken into consideration for deciding the specification of HT breakers to be used. The fault levels of 11 kV HT system are generally 150 MVA/250 MVA/350 MVA. The fault level of HT grid is being taken as 350 MVA. Similarly the fault level for 33 kV distribution is taken as 1000 MVA.

11.3 DEVELOPING SINGLE LINE DIAGRAM

Having finalised the substation capacity, the number and capacity of the transformer; the single line diagram of the substation can be developed. The number of incoming supply can be decided based on the importance of the building. Mostly duplicate supply (one as a stand by) should be taken from the supply company. In important installations having two independent incoming supplies, a bus-coupler may also be provided and outgoing feeders to transformers may be distributed on both the sides so as to ensure that in case of fault in one section of the bus-bar, the other section can be kept in use. Typical single line diagram for HT layout is at Fig. 11.3.

The number and capacity of transformers decides the number and rating of incoming feeders on the LT panel board. Wherever two or more incoming is necessary, bus-coupler should be provided. The number of outgoing feeders can be decided based on the design of LT distribution system. While deciding the number and capacity of outgoing feeders, extra care should be taken to decide capacity of feeder connecting the emergency supply through generating set.

Apart from providing outgoing feeders for connecting different existing loads, provision should be made on each section of the LT panel board to connect power factor control panel. In addition, some spare outgoing feeders should be kept for future loads.

11.4 FAULT LEVEL CALCULATION

The aim of the electrical design is to select equipment to withstand prospective fault and also clear the fault

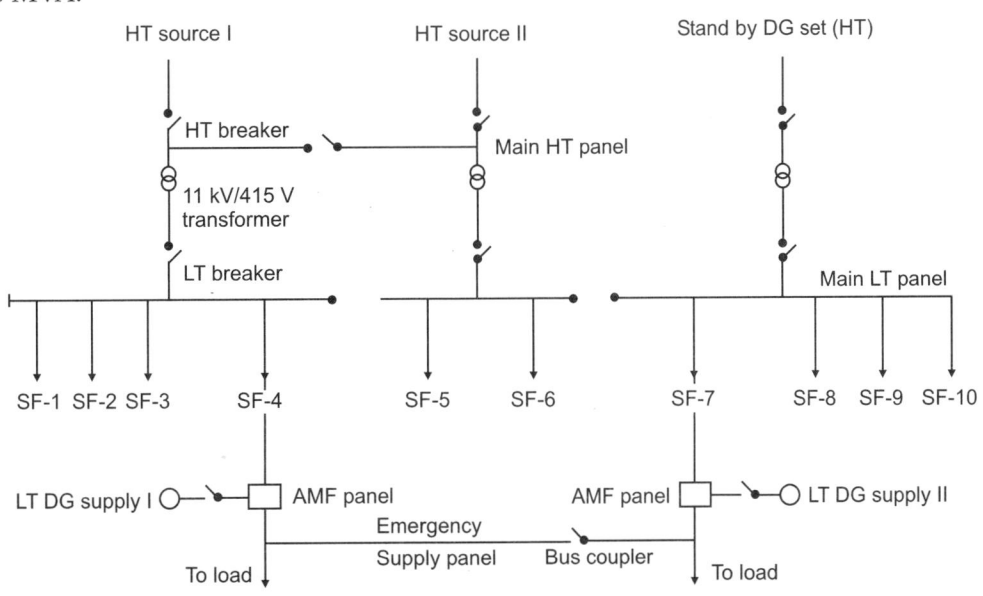

SF = Supply feeder to load

Fig. 11.3 Typical single line diagram of substation

before it can cause damage to the equipment. For this purpose, calculation of fault level is very essential. In case of HT system, the fault level should match with the fault level of the licensee's system. BIS has specified the short-circuit apparent power of the system for various system voltages in IS 2026 (Part I) 1977 according to which the short-circuit apparent power shall be 500 MVA when highest system voltages are 7.2, 12, 17.5 and 24 kV.

Though it does not specify the short-circuit apparent power for LT networks, the local supply companies requires the LT equipments to withstand short-circuit fault level of 35 MVA.

Percentage impedances for various rated power transformer is listed in Table 11.7.

Table 11.7 Percentage impedances of transformers of various ratings

Rated power kVa	Impendance voltage percent
Up to 630	4.0
631–1250	5.0
1251–3150	6.25
3151–6300	7.15
6301–12500	8.35
12501–25000	10.00
25001–200000	12.5

Example

A transformer of 1600 kVa, the voltages at the primary and secondary winding being 11 kV and 415 V respectively. Find the short-circuit symmetrical current at the terminals of LT winding.

Percentage impedance indicates the voltage drop that takes place when full load current is being drawn.

Impedance in ohms can be worked out as a reference to the HT side or the LT side.

Fault current is the value of the current that flows when there is a short-circuit at the terminals and so it will be the current that flows when the full rated voltage is consumed by the impedance.

Full load currents is $= 1600/\left(\sqrt{3} \times 415\right) = 2225$ A on LT side.

While it is $= 1600 \times 1000/\left(\sqrt{3} \times 11000\right) = 84$ A on HT.

Refer Table 11.7 for the values of impedances and select percentage impedance as 6.25% in the case of transformers of 1600 kVa ratings.

Impedance of 6.25% means the voltage drop on full load will be 6.25% of 11000 V (referred to HT) or 6.25% of 415 V (referred to LT).

Impedance value will be

$$V_{\text{drop on full load}} = \frac{6.25 \times 11000}{100 \times \sqrt{3}} \text{ or } = \frac{6.25 \times 415}{100 \times \sqrt{3}} \text{ as}$$

referred to HT and LT respectively.

$$= 397 \text{ V} \qquad \text{or} \quad = 15 \text{ V}$$

$$Z = \frac{397}{84} \qquad \text{or} \quad = \frac{15}{2225}$$

$$= 4.726 \text{ ohm} \quad \text{or} \quad = 0.006742 \text{ ohm}$$

$$I_{\text{fault}} = \frac{11000}{\sqrt{3} \times 4.726} \quad \text{or} \quad = \frac{215}{\sqrt{3} \times 0.006742}$$

$$= 1343 \text{ A} \qquad \text{or} \quad = 35539 \text{ A}$$

$$= 1.34 \text{ kA} \qquad \text{or} \quad = 35.53 \text{ kA}$$

While, $V_{\text{fault (HT)}} = \dfrac{\sqrt{3} \times 11000 \times 1343}{10^{6}} = 25.58$ MVA.

While, $V_{\text{fault (LT)}} = \dfrac{\sqrt{3} \times 415 \times 35539}{10^{6}} = 25.54$ MVA.

Thus, fault at both the windings will be same in terms of its strength though fault current will be different for the respective windings.

11.5   FAULT WITHSTAND CAPACITY

The fault withstand capacity of an equipment has two parameters, i.e. fault level and duration for which the equipment can withstand fault.

Example

An ACB having 1 sec. fault withstand capability of 50 kA can withstand mechanical and electrical shock of 50 kA for 1 sec. If duration of fault persists beyond 1 sec. the fault withstand capacity would be calculated as below.

Fault withstand capacity for T second

$$= \frac{\text{Fault withstand capacity for 1 sec.}}{T^{1/2}}$$

Thus, an equipment having 1 second fault withstand capacity of 50 kA will have 3 second rating of $50 \text{ kA}/3^{1/2} = 28.9$ kA.

Manufactures generally declare fault withstand capacity for 1 sec. or 3 sec. Fault withstand capacity of HT panel, air circuit breaker, fuse switch/switch fuse, transformers is available from manufacturer's tables.

In respect of LT panels, bus-ducts and rising mains; the fault withstand capacity should be specifically considered. The bus-bar and bus-ducts should not have a fault withstand capacity less than that of the main incoming breakers. As a practice 50 kA fault level is considered for LT panels which corresponds to 35 MVA (refer example above).

In number of cases fault withstand capacity of 50 kA is mentioned even for rising main which however may not always be necessary. This also results in a very high cost. For deciding the fault withstand capacity of the rising mains fault level near the rising mains should be calculated taking into consideration the transformer impedance and impedance of cable as described earlier.

11.6 PROTECTION SYSTEM AND FAULT DISCRIMINATION

Having selected equipment to withstand the prospective fault level, protection system is to be designed to effect fast clearance of fault to ensure least damage to load as well as the equipment. It is also to be ensured that the fault is cleared well with in the time specified in the fault withstand capacity. Normally following protections are provided.

HT Panel. Inverse definite minimum time (IDMT) relay with two overload (50–200% in steps of 25%) and one earth fault (10–40% in steps of 5%) relay.

Transformer. The transformer is protected from external faults by protections provided in HT and LT panels. Protection against inter-winding fault in transformer is done by providing Buchholz relay for all transformers of capacity 800 kVa and above.

LT ACB. Built in overload and instantaneous release.

SW fuse/fuse switch. HRC fuses.

LT ACB. These days microprocessor based air circuit breakers are manufactured. These breakers have in-built over-current trip device, under-voltage trip and thermal overload magnetic trip device. This results in a small size panel, lesser wiring hassles. The rated breaking capacity and short-circuit time are improved to a great extent due to microprocessor control.

Table 11.8 Parameters of a typical microprocessor controlled ACB

Parameters	Voltage	630 A	1600 A
Rated breaking capacity (I_{cs} RMS)/ Rated making capacity (kA peak)	415 V	45/94.5	50/105
Short time current I_{cw} kA for 1 sec (3 sec RMS)	415 V	45 (25)	50 (45)
Total breaking time (m sec)/ total closing time	415 V	30/40	30/40

Typical making/breaking capacity, short-circuit time and total breaking time are shown in Table 11.8.

To ensure that upstream breakers do not trip in case of fault in the down stream network, discrimination of fault is necessary. Since down stream breakers will normally have a lower overload and short-circuit setting the down stream relay will act faster. Even then there is possibility of upstream breaker tripping affecting other feeders.

The total time (T_t) for breaker to trip from the time the fault occurs is.

$$T_t = t_{\text{relay}} + t_c + t_b.$$

T_t = Total tripping time.

T_{relay} = Time taken by relay to act.

T_c = Time taken by tripping coil to actuate.

T_b = Time taken by breaker to open.

For any fault in down stream network, the total tripping time should be less than the time taken by relay of the next upstream relay to act. This can be achieved by adjusting the time setting of each relay.

11.7 CAPACITORS RATING AND SELECTION

In the earlier days of electricity development, the electric loads were linear. These loads were due to induction motors, heaters, bulbs. The voltage was followed by current and there were only a few problems. But with the scientific developments, many equipments were invented which produced non linear loads, e.g. computer, motor control, drives, electronic ballasts, etc. These non linear loads cause.

1. Low power factor by producing reactive power.
2. Harmonics.
3. Transients.
4. Voltage variations.

11.7.1 Reactive Power

Every electric load that works with magnetic fields generated by above equipments, including transformers and welding machines, produces a varying degree of electrical lag between voltage and current that is called inductance. This lag maintains the current sense positive for certain times even though, negative going voltage tends to reverse it. During this time, negative power or energy is produced and fed back into the network. When the current and voltage have the same sign, the same amount of energy is again needed to build-up the magnetic field in inductive loads. This magnetic reversal energy is called reactive power (Fig. 11.4). In alternating voltage networks (50/60 Hz), such a process repeats 50 or 60 times a second. So an obvious solution is to briefly store the magnetic reversal energy in capacitors and relieve the network of this reactive energy. For this reason, automatic reactive power consumption system (detuned/conventional) are installed.

Such systems consist of a group of capacitor units that can be cut in and cut out and which are driven by a power factor controller as determined by value taken from current transformer.

Every form of electrical energy has thus two components, namely active power and reactive power. The active power contributes for actual work done and energy meters record the active energy and consumer

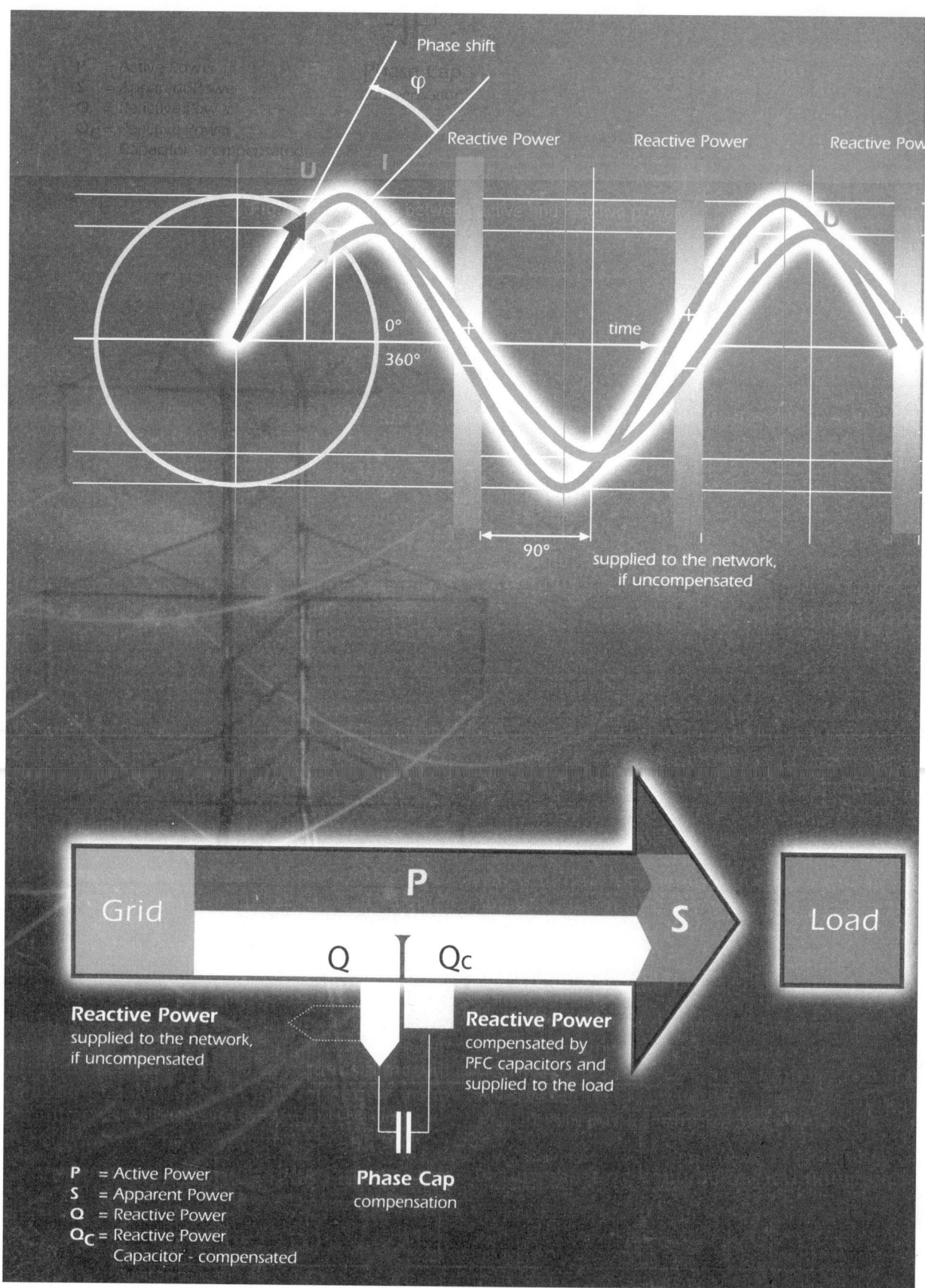

Fig. 11.4 Relation between active and reactive power

is billed for active component only. The reactive component results in higher load current, voltage drop and system losses.

Figure 11.5 illustrates the concept of losses in the system due to reactive component, due to which apparent power is more than the reactive power.

The ratio of active and apparent power is termed as power factor.

Thus, power factor (PF)

$$= \frac{\text{Active power}}{\text{Apparent power}} = \cos \phi$$

As per Fig. 11.5, the power factor should theoretically be as near to unity as possible, by installation of capacitors, but it is not a common practice to produce a unity power factor since it may result in over compensation due to load changes and may result in leading power factor which will causes problems of high voltage, equipment burning, etc.

Mathematically active and reactive components can be defined as

$$P_a = \sqrt{3} \ VI \cos \phi$$

Where P_a is the amount of input power (apparent power) converted into active power.

$$P_r = \sqrt{3} \ VI \sin \phi$$

Where P_r the amount of input power converted into reactive component.

Whereas, the total input power is represented by

$$P = \sqrt{3} \ VI = \sqrt{P_a^2 + P_r^2}$$

It is this reactive power component which has been fed in the network without compensation that can be stored in the capacitor and then used for the next change of magnetism. To calculate capacitor reactive power Q_c in var (volt ampere reactive), the following equation is used.

$$Q_c = P_a (\tan \phi_1 - \tan \phi_2)$$

The value of (Tan ϕ_1 – tan ϕ_2) is given in Table 11.3.

$$Q_c = P_a \times (\tan \phi_1 - \tan \phi_2)$$

Example

Actual motor power

$$P = 100 \ \text{kW}$$

Q_c (kvar) $= P_a \times f =$ active power (kW) × factor f

Actual $\cos \phi = 0.61$

Target $\cos \phi = 0.96$

Capacitor reactive power Q_c

$$P_a = S \times \cos \phi = \text{apparent power} \times \cos \phi$$
$$Q_c = 100 \times 1.01 = 101.0 \ \text{kvar}$$

Select value for factor 'f' from Table 11.9 at intersection of actual $\cos \phi = 0.61$ and target $\cos \phi = 0.96$, which is 1.01.

The expression for kvar rating of capacitor is

$$Q_c = V_c I_c$$

$$Q_c = \frac{V_c \times V_c}{X_c} = \frac{(V_c)^2}{X_c}$$

$$X_c = \frac{1}{\omega.c} = \frac{1}{2\pi fc}$$

$$Q_c = V_c^2 \times 2\pi fc$$

The above expression is for single phase PFC application when the capacitor is connected across the phase and neural and is subjected to phase neutral voltage.

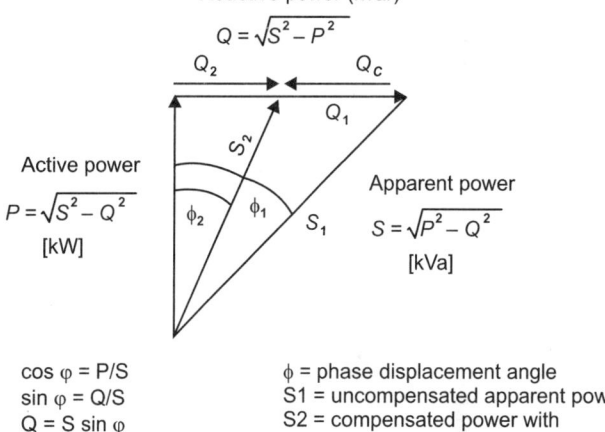

cos φ = P/S   φ = phase displacement angle
sin φ = Q/S   S1 = uncompensated apparent power
Q = S sin φ   S2 = compensated power with
Q = P tan φ   capacitors for compensation

Fig. 11.5 Losses due to reactive power

But when the capacitor is connected in three phase application, the expression is as below.

i. Capacitor connected in star.
    The capacitor is subject to a voltage of $V_L / \sqrt{3}$.
    Thus, total kvar compensation is calculated as

$$Q_{\text{Tot}} = \frac{3(V_L)^2}{(\sqrt{3})^2} \times \omega \times c$$

$$C_{\text{star}} = \frac{Q_{\text{Tot}}}{(V_L)^2 \times 2\pi f}$$

ii. Delta connection
    The capacitor is subjected to the voltage V_L, phase to phase, thus total kvar compensation is

$$Q_{\text{Tot}} = 3(V_L)^2 \omega c$$

$$C_{\text{delta}} = \frac{Q_{\text{Tot}}}{3 V_L^2 \omega} = \frac{Q_{\text{Tot}}}{3 V_L^2 \ 2\pi f}$$

Table 11.9 Table for definition of capacitor reactive power Q_c

Current (actual) tan φ	cos φ	Achievable (target) cos φ									
		0.80	0.82	0.85	0.88	0.90	0.92	0.94	0.96	0.98	1.00
		Multiplication factor of									
3.18	0.30	2.43	2.48	2.56	2.64	2.70	2.75	2.82	2.89	2.98	3.18
2.96	0.32	2.21	2.26	2.34	2.42	2.48	2.53	2.60	2.67	2.76	2.96
2.77	0.34	2.02	2.07	2.15	2.23	2.28	2.34	2.41	2.48	2.56	2.77
2.59	0.36	1.84	1.89	1.97	2.05	2.10	2.17	2.23	2.30	2.39	2.59
2.43	0.38	1.68	1.73	1.81	1.89	1.95	2.01	2.07	2.14	2.23	2.43
2.29	0.40	1.54	1.59	1.67	1.75	1.81	1.87	1.93	2.00	2.09	2.29
2.16	0.42	1.41	1.46	1.54	1.62	1.68	1.73	1.80	1.87	1.96	2.16
2.04	0.44	1.29	1.34	1.42	1.50	1.56	1.61	1.68	1.75	1.84	2.04
1.93	0.46	1.18	1.23	1.31	1.39	1.45	1.50	1.57	1.64	1.73	1.93
1.83	0.48	1.08	1.13	1.21	1.29	1.34	1.40	1.47	1.54	1.62	1.83
1.73	0.50	0.98	1.03	1.11	1.19	1.25	1.31	1.37	1.45	1.63	1.73
1.64	0.52	0.89	0.94	1.02	1.10	1.16	1.22	1.28	1.35	1.44	1.64
1.56	0.54	0.81	0.86	0.94	1.02	1.07	1.13	1.20	1.27	1.36	1.56
1.48	0.56	0.73	0.78	0.86	0.94	1.00	1.05	1.12	1.19	1.28	1.48
1.40	0.58	0.65	0.70	0.78	0.86	0.92	0.98	1.04	1.11	1.20	1.40
1.33	0.60	0.58	0.63	0.71	0.79	0.85	0.91	0.97	1.04	1.13	1.33
1.30	0.61	0.55	0.60	0.68	0.76	0.81	0.87	0.94	1.01	1.10	1.30
1.27	0.62	0.52	0.57	0.65	0.73	0.78	0.84	0.91	0.99	1.06	1.27
1.23	0.63	0.48	0.53	0.61	0.69	0.75	0.81	0.87	0.94	1.03	1.23
1.20	0.64	0.45	0.50	0.58	0.66	0.72	0.77	0.84	0.91	1.00	1.20
1.17	0.65	0.42	0.47	0.55	0.63	0.68	0.74	0.81	0.88	0.97	1.17
1.14	0.66	0.39	0.44	0.52	0.60	0.65	0.71	0.78	0.85	0.94	1.14
1.11	0.67	0.36	0.41	0.49	0.57	0.63	0.68	0.75	0.82	0.90	1.11
1.08	0.68	0.33	0.38	0.46	0.54	0.59	0.65	0.72	0.79	0.88	1.08
1.05	0.69	0.30	0.35	0.43	0.51	0.56	0.62	0.69	0.76	0.85	1.05
1.02	0.70	0.27	0.32	0.40	0.48	0.54	0.59	0.66	0.73	0.82	1.02
0.99	0.71	0.24	0.29	0.37	0.45	0.51	0.57	0.63	0.70	0.79	0.99
0.96	0.72	0.21	0.26	0.34	0.42	0.48	0.54	0.60	0.67	0.76	0.96
0.94	0.73	0.19	0.24	0.32	0.40	0.45	0.51	0.58	0.65	0.73	0.94
0.91	0.74	0.16	0.21	0.29	0.37	0.42	0.48	0.55	0.62	0.71	0.91
0.88	0.75	0.13	0.18	0.26	0.34	0.40	0.46	0.52	0.59	0.68	0.88
0.86	0.76	0.11	0.16	0.24	0.32	0.37	0.43	0.50	0.57	0.65	0.86
0.83	0.77	0.08	0.13	0.21	0.29	0.34	0.40	0.47	0.54	0.63	0.83
0.80	0.78	0.05	0.10	0.18	0.26	0.32	0.38	0.44	0.51	0.60	0.80
0.78	0.79	0.03	0.08	0.16	0.24	0.29	0.35	0.42	0.49	0.57	0.78
0.75	0.80		0.05	0.13	0.21	0.27	0.32	0.39	0.46	0.55	0.75
0.72	0.81			0.10	0.18	0.24	0.30	0.36	0.43	0.52	0.72
0.70	0.82			0.08	0.16	0.21	0.27	0.34	0.41	0.49	0.70
0.67	0.83			0.05	0.13	0.19	0.25	0.31	0.38	0.47	0.67
0.65	0.84			0.03	0.11	0.16	0.22	0.29	0.36	0.44	0.65
0.62	0.85				0.08	0.14	0.19	0.26	0.33	0.42	0.62
0.59	0.86				0.05	0.11	0.17	0.23	0.30	0.39	0.59
0.57	0.87					0.08	0.14	0.21	0.28	0.36	0.57
0.54	0.88					0.06	0.11	0.18	0.25	0.34	0.54
0.51	0.89					0.03	0.09	0.15	0.22	0.31	0.51
0.48	0.90						0.06	0.12	0.19	0.28	0.48
0.46	0.91						0.03	0.10	0.17	0.25	0.46
0.43	0.92							0.07	0.14	0.22	0.43
0.40	0.93							0.04	0.11	0.19	0.40
0.36	0.94								0.07	0.16	0.36
0.33	0.95									0.13	0.33

As a conclusion, we can say

$$C_{\text{delta}} = \frac{C_{\text{star}}}{3}$$

So PFC configurations are usually delta connected, because star connection requires three times the capacitance of a delta connection.

Example

Calculate capacitor ratings for
i. Star connection [Fig. 11.6a]
ii. Delta connection [Fig. 11.6b]
When the given parameters are:
Induction motor 220 kW
Network 440 Vac, 3 phase
Frequency 50 Hz
Power factor
Current $\cos \phi_1 = 0.7$
Target $\cos \phi_1 = 0.9$
Since $\cos \phi_1 = 0.7$ hence $\tan \phi_1 = 1.02$
and $\cos \phi_2 = 0.9$ hence $\tan \phi_2 = 0.48$

$$\begin{aligned} Q_c &= P_a (\tan \phi_1 - \tan \phi_2) \\ &= 220 (1.02 - 0.48) \\ &= 118.8 \text{ kvar} \end{aligned}$$

$$V_c = \frac{V_L}{\sqrt{3}} = \frac{440}{\sqrt{3}} = 254 \text{ V}$$

i. Star connection

$$\begin{aligned} C_{\text{star}} &= \frac{Q_{\text{Tot}}}{(V_L)^2 \times 2\pi f} = \frac{118.8 \times 1000}{(440)^2 \times 2\pi \times 50} \\ &= 1954 \,\mu\text{F/line (phase)} \end{aligned}$$

$$C_{\text{Tot}} = 5862 \,\mu\text{F}$$

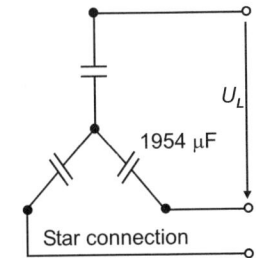

Fig. 11.6a

ii. Delta connection

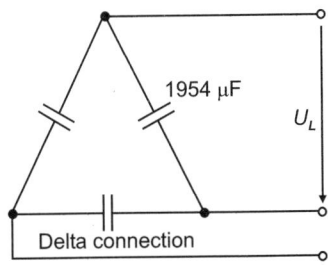

Fig. 11.6b
Fig. 11.6 Star/Delta connections

$$V_C = V_L = 440 \text{ V}$$

$$C_{\text{delta}} = \frac{Q_{\text{Tot}}}{3 \, V_L^2 \times 2\pi f} = \frac{118.8 \times 1000}{3 \times 440^2 \times 2\pi \times 50}$$

$$= 651 \,\mu\text{F/line (phase)}$$

$$C_{\text{Tot}} = 1953 \,\mu\text{F}$$

Thus, we can select 25 kvar × 4 = 100 kvar for compensation from the selection given in Table 11.10a, b. Three capacitors each of 137 µF rating can be selected, i.e. $C_{\text{Tot}} = 4 \times 3 \times 137 \,\mu\text{F} = 1644 \,\mu\text{F/phase}$ or we can select 18.8 kvar × 6 = 112.8 kvar.

Table 11.10a Three-phase capacitors rated voltage 230 Vac, 50/60 Hz, delta connection

	50 Hz	
Output kvar	$I_N A$	$C_N \mu F$
2.5	6.3	3*50
5.0	13.1	3*104
7.5	18.8	3*150
10.4	26.1	3*209
12.5	31.4	3*250

Table 11.10b Three-phase capacitors rated voltage 440 Vac, 50/60 Hz, delta connection

	50 Hz	
Output kvar	$L_N A$	$C_N \mu F$
5.0	6.6	3*27
7.5	9.9	3*41
10.4	13.7	3*57
11.2	14.7	3*61
12.5	16.4	3*69
14.2	18.7	3*78
15.0	19.7	3*82
16.7	21.9	3*92
18.8	24.7	3*103
20.8	27.3	3*114
25.0	32.8	3*137
28.15	37.0	3*154

From the selection table, three capacitors each of 103 µF rating can be selected.

Thus, $C_{\text{Tot}} = 6 \times 3 \times 103 = 1854 \,\mu\text{F/phase}$.

Reduction in apparent power due to PF correction is

$$S_1 - S_2 = \frac{P_1}{\cos \phi_1} - \frac{P_2}{\cos \phi_2} = \frac{220}{0.7} - \frac{220}{0.9}$$

$$= 314 - 244 = 70 \text{ kVa}$$

or 70 × 0.9 = 63 kW of additional power can be supplied and transferred via the network.

The power losses are square of current, thus reduction in loss of power

$$= I_1^2 - I_2^2 = \text{Approx.} \ S_1^2 - S_2^2 = (314)^2 - (244)^2$$

Percentage reduction in losses

$$= \frac{(314)^2 - (244)^2}{(314)^2} \times 100 = 39.6\%.$$

Cable current drawn by motor (uncompensated load)

$$I_1 = \frac{220 \times 1000}{\sqrt{3} \times 440 \times 0.7} = 412 \ \text{A.}$$

Line current drawn by motor (after compensation)

$$= \frac{220 \times 1000}{\sqrt{3} \times 440 \times 0.9} = 320 \ \text{A.}$$

Thus, cable can carry an additional load of 92 A, or designer can reduce cable cross-section.

11.7.2 Methods of Power Factor Correction

So far have studied the method of calculation of reactive power, capacitance, etc. Now we will deal with methods of improvement of power factor. These methods are:

1. Compensation of reactive power with capacitors.
2. Active compensation using semiconductors.
3. Over excited synchronous machines (motor/generator).

The compensation with capacitors is the most common method and we will confine our discussion to various types of controls exercised by capacitor compensation.

The various types of power factor controller are:

1. **Individual or fixed compensation type.** In this method, each reactive power producer is individually compensated. Thus, the capacitor is installed at each load point say with every motor, ballast, etc.
2. **Group compensation.** Here all the reactive power producers are connected as a group and compensated as a whole. Thus, there may be one capacitor bank at one load point of say ten motors, another capacitor bank at another load point of say five welding machines and so on.
3. **Central or automatic compensation.** In this method, when the individual loads are spread over a large distance, then a central control of power factor is desirable.
4. **Mixed compensation.** As the name suggests, this method is a mix of individual or group compensation as well as central compensation.

How Capacitor Improves Power Factor?

In Fig. 11.4, the lagging of current behind voltage at node points produces reactive power. The purpose of power factor correction is to compensate the generated lagging reactive power by leading reactive power at defined nodes.

In this way impermissibly high voltage drops and additional ohmic losses are also avoided. The necessary leading power is produced by capacitors parallel to the supply network, as close as possible to the inductive consumer. Static compensation device reduce the lagging component. But if the conditions of network alter, the leading reactive power can be matched in steps by adding or taking out single power factor capacitors. A typical capacitor control circuit is shown in Fig. 11.7 for delta connected capacitor bank connections along with filters (filters are described in section to follow).

11.7.3 Classification of Capacitors

As all of the equipments used in electrical and mechanical design have standardized nomenclature to define their specification, so is the case with capacitors. The capacitors are classified according to:

1. **Temperature class.** Capacitors are represented by a number followed by a letter, e.g. 40/D. The number is the lowest ambient temperature at which a capacitor may operate. The upper limit temperature is indicated by a letter as shown in Table 11.11. The useful life of a capacitor always depends very much on temperature. Proper cooling of capacitor is a must to ensure that the maximum case temperature is not exceeded. For this purpose, forced cooling may be adopted.

Table 11.11 Capacitor rating based on temperature limit

Temperature class to IEC 831-1	Temperature classes		
	Maximum	Temperature of surrounding air maximum mean for 24 hour	Maximum mean for 1 year
B	45 °C	35 °C	25 °C
C	50 °C	40 °C	30 °C
D	55 °C	45 °C	35 °C

2. **Enclosure of capacitor.** For different models, there are different enclosures. The type of enclosure is indicated by a designation consisting of two letters IP followed by two digits. Table 11.12 indicates the meaning of various digits used.
3. **Maximum admissible over current.** The nominal current is the current resulting for nominal voltage (V_N) and frequency (H_Z) excluding transients. The maximum admissible over current (I_{max}) of 1.5 times I_N is to be maintained by all capacitors.

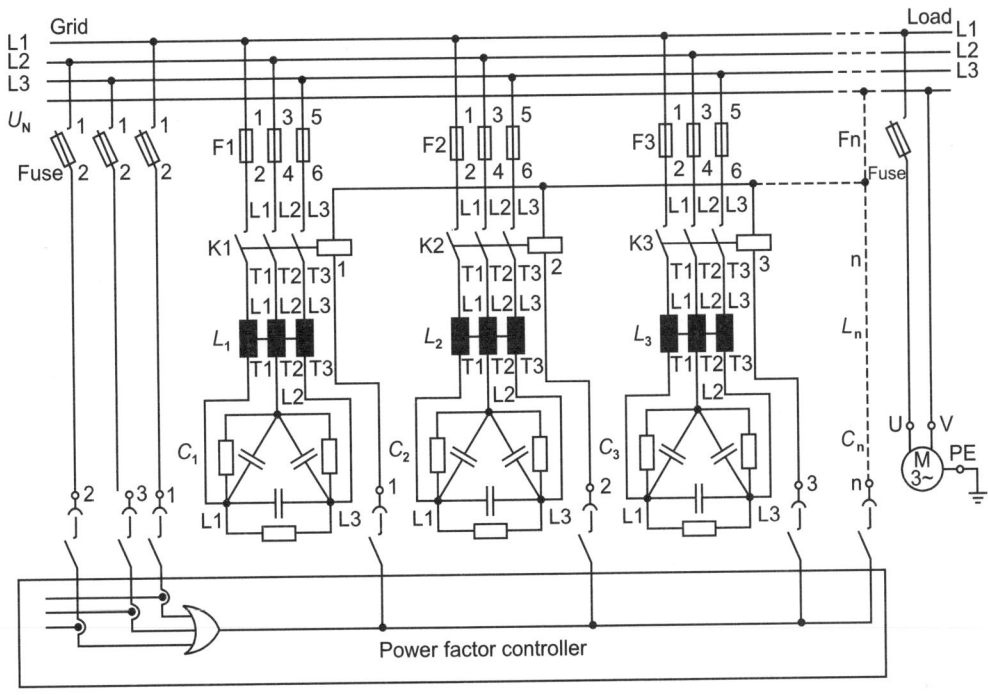

Fig. 11.7 Capacitor control circuit

Table 11.12 Type of capacitor enclosure

Enclosure	First digit	Second digit
IP 00	No protection against touch and ingress of solid foreign bodies.	No protection against ingress of water
IP 20	Protection against finger touch and solid foreign bodies ≥ 12.5 mm diameter.	No protection against ingress of water
IP 41	Protection against tool touch and solid foreign bodies ≥ 1 mm diameter.	Drip-water protection
IP 54	Protection against tool touch and solid foreign bodies ≥ 1 mm diameter, protection against dust deposit.	Splash water protection

4. **Maximum admissible over voltage.** Capacitors shall be suitable for over voltages specified in Table 11.13.

11.7.4 Harmonics and Transients

We studied in the beginning that the modern inductive equipments, e.g. UPS, computers, ballast or switching devices, etc. produce non-linear loads which result in harmonics, transients and voltage variations. All these cause line current pollution. The power factor correction or capacitance of the power capacitor forms a resonant circuit in conjunction with the feeding transformer. The self resonant frequency of this circuit lies typically between 250 and 500 Hz, i.e. in the region of 5th and 10th harmonics (5 × 50 and 10 × 50). Resonance can lead to following undesirable effects.

1. Overloading of capacitors.
2. Overloading of transformers and transmission equipments.
3. Interference with metering and control systems, computers, etc.
4. Amplification of harmonics.
5. Voltage distortion.

Table. 11.13 Maximum admissible over voltage

Frequency 50/60 Hz	Maximum voltage (V_{rms})	Maximum duration	Remarks
Line frequency	1.00* V_N	Continuous duty	Highest mean during entire operating time of capacitor; exceptions (see below) are admissible for times of < 24 hour
Line frequency	1.10* V_N	8 hour daily	Line voltage fluctuations
Line frequency	1.15* V_N	30 minute daily	Line voltage fluctuations
Line frequency	1.20* V_N	5 minute daily	Line voltage fluctuations
Line frequency	1.30* V_N	1 minute daily	Line voltage fluctuations
Line frequency	Such that current does not exceed maximum admissible figure ($I_{max} = 1.5* I_N$) with harmonics		

6. Under balanced load conditions, phase current cancel each other in neutral. In a four wire system, with single phase non-liner loads, odd numbered multiples of the third harmonics (3rd , 9th, 15th) do not cancel, rather add together in neutral conductor. In systems with substantial amount of non-linear single phase loads, the neutral current may rise to a dangerously high level.

These phenomenons are explained in Fig. 11.8.

Voltage Fluctuation

These are mostly caused by power consumption of large loads like welding machines and arc furnaces or by heavy power fluctuations. These random voltage fluctuations are often also referred to as flickers, because they are noticeable as visible alteration in lighting installations. Dynamic voltage compensation is a possible remedy.

Limit for Harmonics

Limit for harmonics in the form of total harmonic voltage distortion (VTHD) in percentage is as Table 11.14.

Table 11.14 Values for total harmonic voltage distortion

Supply system voltage kV	VTHD V (%)	Individual harmonic voltage	
		Odd	Even
0.415	5	4	2
6.6 and 11	4	3	1.75
33 and 66	3	2	1

An example of resonance with capacitor bank in function will make the concept clear. In a typical case with non-linear loads and capacitors in parallel to load

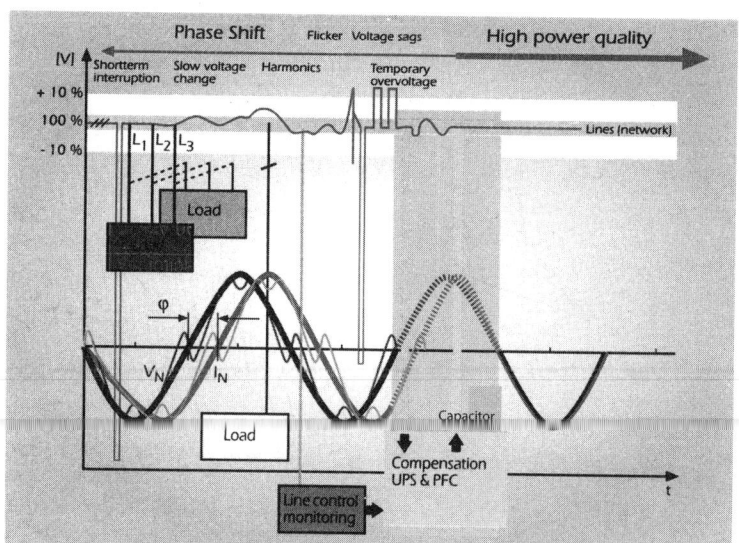

Possibilities of power quality change and forms used

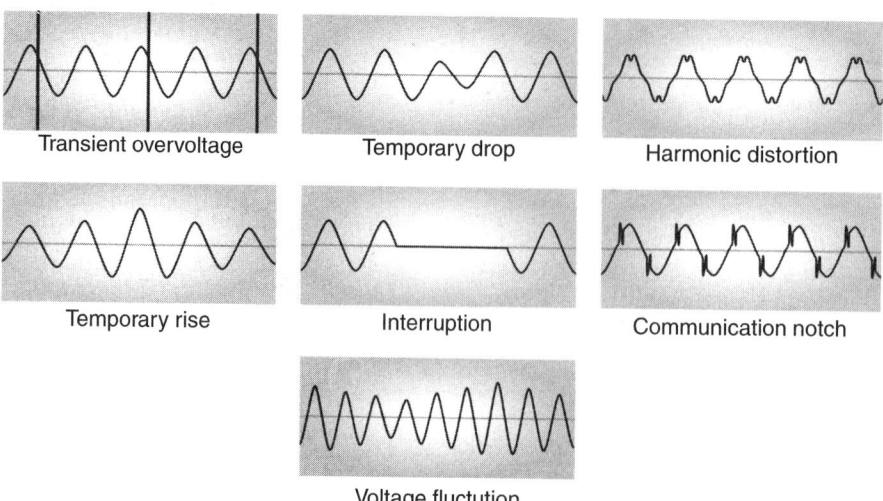

Transient overvoltage   Temporary drop   Harmonic distortion

Temporary rise   Interruption   Communication notch

Voltage fluctution

Fig. 11.8 Power Quality—typical cases of line voltage distribution

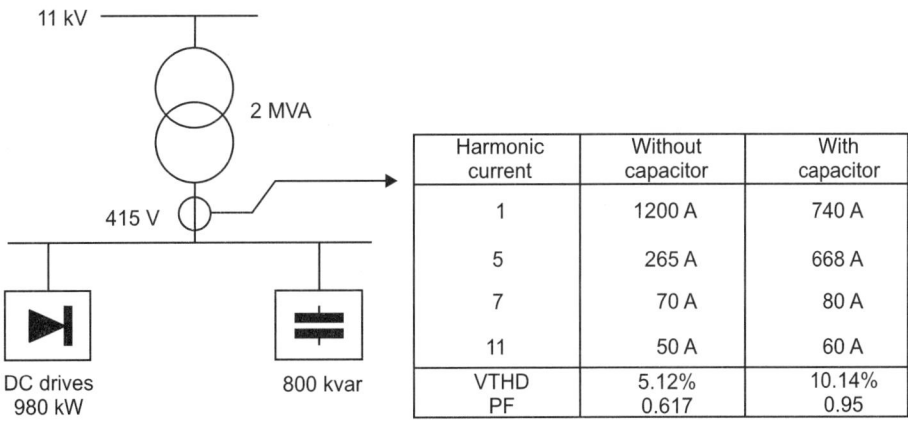

Harmonic current	Without capacitor	With capacitor
1	1200 A	740 A
5	265 A	668 A
7	70 A	80 A
11	50 A	60 A
VTHD	5.12%	10.14%
PF	0.617	0.95

Fig. 11.9 Resonance with capacitor bank

network increased the VTHD to a high level as shown in Fig. 11.9, while power factor certainly improved.

11.7.5 Capacitor Inductor Combination

The capacitor can be installed in a circuit either in series or in parallel to an inductive load. Each one of them produces resonance.

Series circuit. In Fig. 11.10, the capacitor is connected in series with inductor. At the resonant frequency, the impedance reduces to minimal value which is very low and resistive in nature, this results into multiple fold increase in current. The result is.

1. Harmonics are present on the primary side (HT) of transformer.

2. Transformer together with PFC capacitor on LV side act as a series resonant circuit.

3. If this resonant circuit combines with harmonic frequency, resonance occurs.

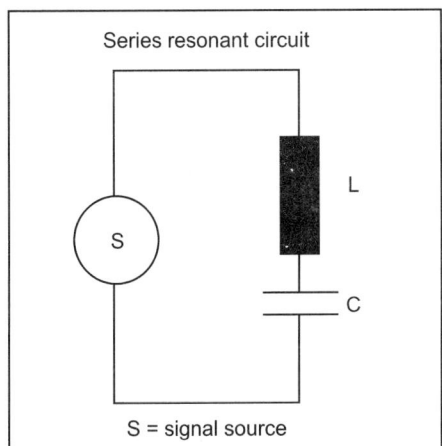

Fig. 11.10 Series resonance

4. Though PFC capacitor will absorb some harmonics from HT side, the amount of absorption

will depend on the relative position of resonant frequency from harmonic frequency.

5. This harmonic current imposes an additional load on PFC capacitor. The voltage on low voltage network is distorted as a result of the resonance.

Parallel circuit. Figure 11.11 explains the connection of capacitor in parallel with inductance. At resonant frequency, inductive reactance is equal to capacitive reactance. The resultant impedance of the circuit increases to a very high value at resonant frequency. Excitation of a parallel resonant circuit results into a high voltage across the impedances and very high circulating current inside the loop. Capacitors and transformers are additionally loaded.

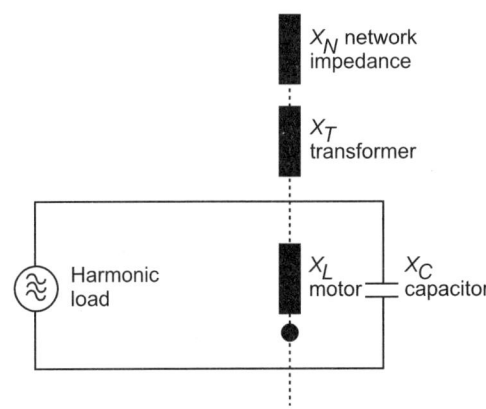

Fig. 11.11 Parallel resonance

If these are the circumstances, then what to do? A designer is required to know about the filter circuits for sucking the harmonics from the power circuit.

11.7.6 Harmonic Filter Circuit

Filter circuits are reactors and capacitors connected in series to form a series resonance circuit. Design and dimensioning of the components have to be done in one of the following ways.

1. **Tuned filter circuit.** Here tuning has to be done for each harmonic frequency which means each harmonic frequency requires its own filter circuit. The harmonic current will be reduced by approximately 90% (Figs. 11.12 and 11.13).

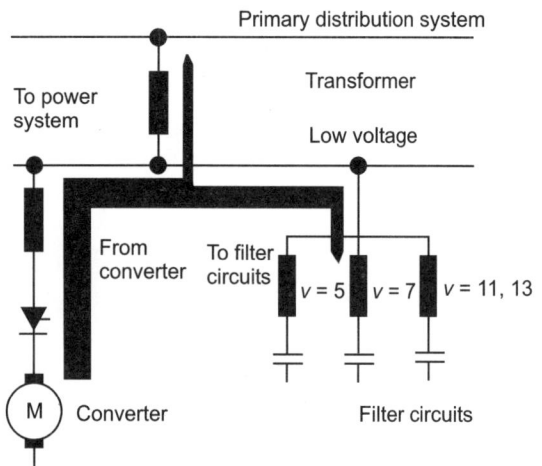

Fig. 11.12   Tuned harmonic filter

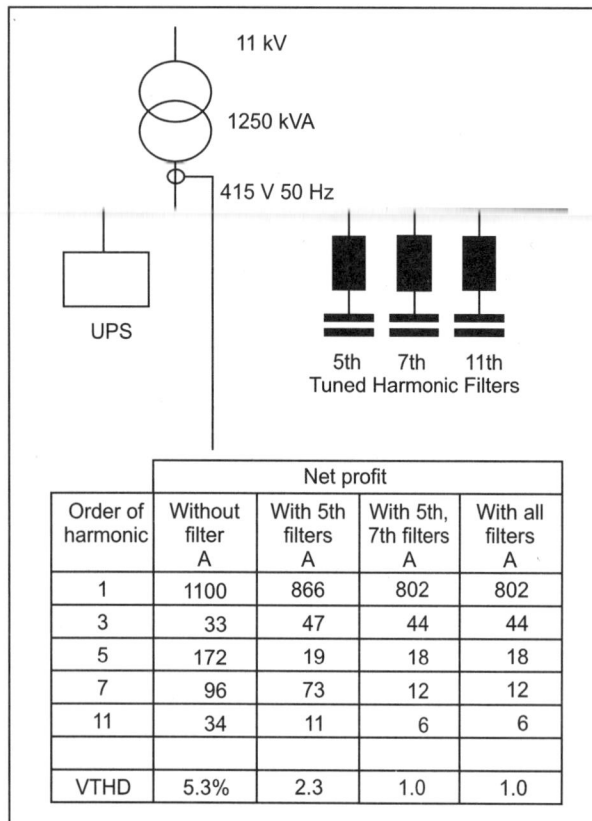

Fig. 11.13   Tuned harmonic filter results

2. **De-tuned filter circuit.** The main purpose of de-tuned filter is to avoid resonance condition of the

capacitor with the transformer inductance. Depending on the de-tuning frequency, more or less all harmonic currents will be sucked from the grid. The circuit remains the same as in Fig. 11.12 but de-tuned capacitors are used. De-tuned filter circuits are common and most effective these days.

Since capacitor along with reactor forms a series resonant circuit. Reactance value can be calculated as follow.

Reactor X_L $= 2\pi FL$

Capacitor $X_c = \dfrac{1}{2\pi fc}$

Network of reactor and capacitor $Z = X_L - X_c$

The reactors connected in series with capacitors result into an increased voltage across the capacitor. Capacitors used for de-tuned filters are therefore required to have voltage ratings higher than the line voltage.

$$V_c = V_N \frac{100}{100 - P}$$

Where

V_c = Capacitor voltage rating required

V_N = Normal line voltage

P = De-tuning required

$$= \left(\frac{f}{f_{res}}\right)^2 \times 100 = 100 \times \frac{X_L}{X_c}$$

$$Q_c = \left(1 - \frac{P}{100}\right)\left(\frac{V_c}{V_N}\right)^2 N_c$$

But, $Q_c = \dfrac{V_c^2}{X_c}$ ..... already shown earlier.

$$\frac{V_c^2}{X_c} = \left(1 - \frac{P}{100}\right)\left(\frac{V_c}{V_N}\right)^2 \times N_c$$

$$\frac{V_c^2}{1/2\pi fc} = \left(1 - \frac{P}{100}\right)\left(\frac{V_c}{V_N}\right)^2 \times N_c$$

or $c = \dfrac{N_c\left(1 - \dfrac{P}{100}\right)}{V_N^2\, 2\pi f}$

Also $P = \dfrac{X_L}{X_c} \times 100 = \dfrac{2\pi fL}{1} \times 2\pi fc \times 100$

So $L = \dfrac{P}{100 \times 4 \times \pi^2 f^2 c}$

	Net profit			
Order of harmonic	Without filter A	With 5th filters A	With 5th, 7th filters A	With all filters A
1	1100	866	802	802
3	33	47	44	44
5	172	19	18	18
7	96	73	12	12
11	34	11	6	6
VTHD	5.3%	2.3	1.0	1.0

The resonant frequency is

$$\omega_r = \frac{1}{\sqrt{L_N \times c}}$$

But it is difficult to calculate the value of L_N as it depends on the load connected to network, the resonant frequency can be approximated by the following formula.

To avoid the resonance condition, the capacitor power output should be less than the critical capacitor output.

$$f_r = 50 \sqrt{\frac{S_T \times 100}{Q_c \times \mu_k}}$$

Where

$\quad S_T$ = Load in kVa.
$\quad \mu_k$ = Percentage impedance of load.
$\quad Q_c$ = Capacitor output kvar.

Example

Find out if the following system configuration causes a risk of resonance.

Transformer S_T = 630 kVa, μ_k = 5%.

Capacitor output Q_c = 250 kVa.

According to formula

$$f_r = 50 \sqrt{\frac{S_T \times 100}{Q_c \times \mu_K}}$$

$$= 50 \sqrt{\frac{630 \times 100}{250 \times 5}} = 355 \text{ Hz.}$$

Thus, it is seen that resonant frequency is critical for the 7th harmonic and the capacitor has to be designed for rating below 250 kvar or even better, a detuned capacitor bank has to be used.

Having done this, a designer should be able to select de-tuned capacitors or reactors or filters. These capacitors are available in various P values by the manufacturer selection table. The tables are given in Tables 11.15, 11.16, 11.17 for grid V = 400 and capacitors of 440 V with 5.67%, 7% and 14% de-tuning. The designer can get the selection tables for different grid voltages and different capacitor voltages from the manufacturer.

11.7.7 Components of a Capacitor Installation

Components of a capacitor installation are described below.

1. **PFC controller.** Modern PFC controller are micro-processor based. The microprocessor analyses the signal from a current transformer and produces switching commands to control the contactors that add or remove capacitor stages. Intelligent control ensures an even utilization of capacitor steps, minimized number of switching operations and optimized life cycle. It displays V, I, f, Q, P_a, S, cos ϕ as well as all odd harmonics, capacitor current, etc.

2. **Fuse.** An HRC fuse of MCCB acts as a safety device for short-circuit protection. The fuse does not protect the overloading of capacitor. Hence, the fuse rating shall be selected accordingly. The HRC fuse rating should be 1.6 to 1.8 times the nominal capacitor current. Table 11.18 gives the selection chart for fuse corresponding to various kvar requirement of capacitor.

3. **Capacitor contactor.** Contactors are electro-mechanical switching elements used to switch capacitors or reactors in standard or de-tuned PFC systems. The switching operation can be performed by electronic switch also called semi-conductors. The latter solution is preferable if fast switching is required for sensitive loads for example. Devices specially designed for capacitor purpose should be used as switching contactors (Fig. 11.14). These contactors should be designed to avoid contact bounce and the resulting high transients. The inrush current must be limited to less than the nominal current of the contactor and the capacitor.

The inrush current spikes would lead to welding of the contactors main contacts. Leading contacts

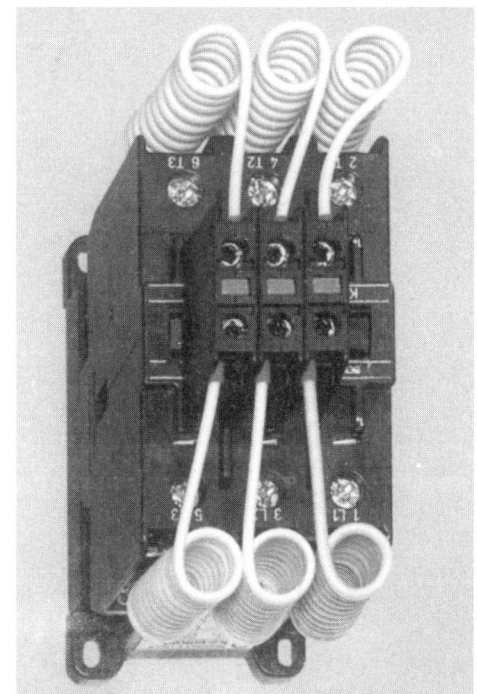

Fig. 11.14 A typical capacitor contactor

Table 11.15 V = 400 V, f = 50 Hz, p = 5.67% (*f$_r$* = 210 Hz)

Power (effective filter output) kvar	Y Capacitance μF	Inductance μH	I_1 A	I_{rms} A	I_{tin} A	Loss W
5	94	6.12	7.65	8.77	15.9	53
6.2	116	4.94	9.49	10.9	19.8	62
7.5	141	4.08	11.5	13.2	23.9	62
10	188	3.06	15.3	17.6	31.9	64
12.5	235	2.45	19.1	21.9	39.8	89
15	281	2.04	23.0	26.3	47.8	89
20	375	1.53	30.6	35.1	63.7	100
25	469	1.22	38.3	43.9	79.7	130
30	563	1.02	45.9	52.6	95.6	164
40	750	0.765	61.2	70.2	127	220
50	938	0.612	76.5	87.7	159	290
60	1126	0.510	91.8	105	191	290
100	1876	0.306	153	175	319	390

Table 11.16 V = 400 V, f = 50 Hz, p = 7 % (*f$_r$* = 189 Hz)

Power kvar	Y capacitance μF	Inductance μH	I_1 A	I_{rms} A	I_{tin} A	Loss W
5	92.5	7.66	7.65	8.03	13.4	52
6.2	115	6.18	9.49	9.96	16.6	52
7.5	139	5.11	11.5	12.1	20.0	61
10	185	3.83	15.3	16.1	26.7	73
12.5	231	3.07	19.1	20.1	33.4	87
15	277	2.56	23.0	24.1	40.1	87
20	370	1.92	30.6	32.1	53.4	100
25	462	1.53	38.3	40.2	66.8	120
30	555	1.28	45.9	48.2	80.1	120
40	740	0.958	61.2	64.3	107	210
50	925	0.766	76.5	80.3	133	210
60	1110	0.639	91.8	96.4	160	270
100	1850	0.383	153	161	267	370

Table 11.17 V = 400 V, f = 50 Hz, p = 14 % (*f$_r$* = 135 Hz)

Power kvar	Y Capacitance μF	Inductance μH	I_1 A	I_{rms} A	I_{tin} A	Loss W
5	85.5	16.6	7.65	7.69	10.8	61
6.2	106	13.4	9.49	9.54	13.4	72
7.5	128	11.1	11.5	11.5	16.2	87
10	171	8.23	15.3	15.4	21.6	87
12.5	214	6.63	19.1	19.2	27.0	100
15	257	5.53	23.0	23.1	32.4	120
20	342	4.14	30.6	30.8	43.2	120
25	428	3.32	38.3	38.5	54.0	210
30	513	2.76	45.9	46.2	64.8	210
40	684	2.07	61.2	61.6	86.4	220
50	855	1.66	76.5	76.9	108	340
60	1026	1.38	91.8	92.3	130	370
100	1710	0.829	153	154	216	450

Table 11.18 Recommended fuse selection (cross-reference)

Fuse (A)	2.5	3	4	5	6	7.5	10	12.5	15	16.7	20	25	30	35	40	50	60	70	80	90	100	Cu-cable mm²
2																						125
10	●	●	●	●																		1.5
16				●	●																	2.5
20						●																2.5
25							●															4
35								●														6
50									●	●	●											10
63												●										16
80													●									25
100														●	●							35
125																●						50
160																	●	●				70
200																			●			95
250																				●	●	120
315																					●	2 × 95

with wiper function are used in these capacitor contactors, i.e. each leading contact is linked to the contactor yoke by a permanent magnet. The leading contact closes before the main contact and open when the main contacts are closed with certainty. The capacitor contactors are suitable for direct switching of low inductance and low loss capacitor banks.

4. **Reactor filters.** These are the devices used in series with capacitor to de-tune the series resonant frequency and helps to prevent capacitor damage. Critical frequencies are 5th and 7th harmonics (250 and 350 Hz). The selection of these reactors has already been described under the de-tune capacitor section.

5. **Capacitor.** The broad field of application for capacitors combined with physical and economic consideration create the need for different dielectric technology. When it comes to low voltage power factor correction, metalized plastic film or polypropylene has demonstrated that it is currently the most suitable and most economic technology. The use of paper as dielectric medium is not preferred these days due to superior properties of polypropylene.

As polypropylene type capacitors seldom produce a pronounced short-circuit, fuses do not offer reliable protection. The fuses are mounted in the circuit for short-circuit protection and not for overload protection. Since no short-circuit is produced in these capacitors, the capacitors are now fitted with disconnector that responds to overpressure. If numerous electric breakdowns occur, the formation of gas produces a fast rise in pressure inside capacitor case (Fig. 11.15). Expansion of cylindrical case beyond a certain degree will separate internal wires and disconnect from the line.

The capacitor produce a leading reactive power to compensate the lagging reactive power. Capacitors should be capable of with standing high inrush currents caused by switching operations (100 times I_N). The calculation and selection of capacitance rating has already been described in proceeding sections.

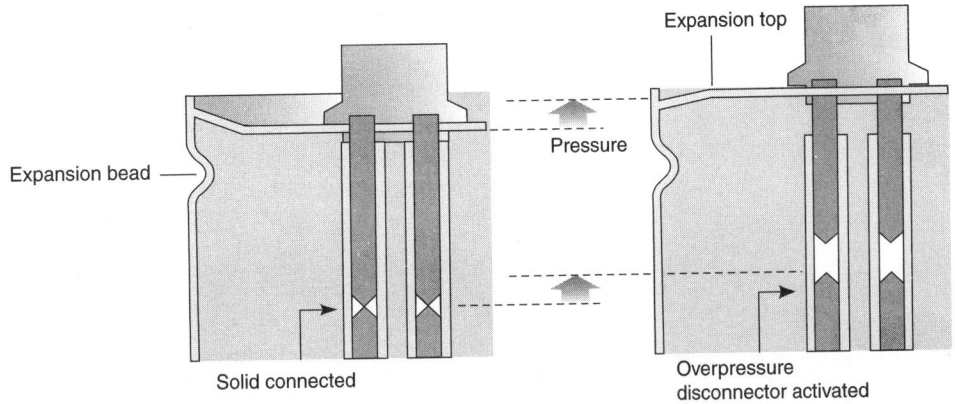

Fig. 11.15 Illustration of pressure inside capacitor

Example

Calculate capacitor bank required for an installation having a maximum demand of 1200 kW being fed from a substation having 2 No. 1000 kVa transformers.

Assume load power factor = 0.80

Generally the load factor in a building is 60% of peak load.

Thus, 60% of maximum demand = 1200 × 0.6 = 720 kW.

Multiplication factor for improving power factor from 0.8 to 0.9 = 0.2680 (from Table 11.9).

Capacitor bank required to improve load power factor = 720 × 0.268 = 192 kvar.

Since most of the substations catering to loads of office building are loaded only for 40 hours in a week and for the balance period, the load is less than 20% of the maximum demand, it is not necessary to consider the magnetizing kvar of the transformer separately. In the above example 2 No. 100 kvar (4 × 25 kvar) capacitor bank may be connected on the two sections of the LT panel being fed by the two transformers. Since, 1000 kVa transformer have magnetizing current of nearly 50 kvar, control circuit of the capacitor bank may be designed to ensure that 50 kvar capacitors remains connected whenever the transformer is ON. This can be achieved by interlocking the switching of 2 No. 25 kvar capacitors in each bank through the NO auxiliary contact of the incoming LT breaker.

Financial Benefits

The power factor correction and control of harmonics gives financial benefits resulting in a pay back period of 12–24 months due to saving in power costs.

11.8 BUS BAR, RISING MAIN AND BUS TRUNKING

In the earlier days of power transmission, only cables were used for carrying the power from substation to various floors of the building or to various zones of an industrial establishment.

Sooner, these were replaced by rising main where air gap between bus bars kept the R-Y-B phases apart. But this resulted in very bulky and wide trunking which would occupy huge space of the shaft in a building or other places. These bus trunking being enclosed by air gaps in a chamber would result in de-rating of bus and thus higher size bus bars would be selected.

The current carrying capacity of a 4 bus, 50 Hz AC. Aluminium bus bar is indicated in Table 11.19.

The above current carrying capacities are based on 50 °C rise in temperature when ambient temperature is 35 °C in still and open atmosphere. But, if the bus bars are to be enclosed in a chamber or temperature rise is to be limited to less than 50 °C , then current carrying capacity will reduce or inversely bus bar size will increase.

Thus de-rating factor if temperature rise is limited to.

1. 40 °C above ambient temperature of 35 °C = 0.88.
2. 30 °C above ambient temperature of 35 °C = 0.75.

Similarly, de-rating factor if bus bars are bound in a closed enclosure is given in Table 11.20.

Example

Select bus bar size for a 1000 amp bus bar with allowable temperature rise of 30 °C above ambient temperature of 35 °C and size of bus bar chamber is 300 × 200 mm located in a well ventilated room.

Assuming ratio of bus bar cross section to area of chamber as < 5%, de-rating factor is 0.75 (from Table 11.20).

Similarly, de-rating factor for lower temperature rise allowed = 0.75.

The rating as per table should be

$$= \frac{1000}{0.75 \times 0.75} = 1777 \text{ amp.}$$

The bus bar as per table above is to be selected for 1940 amp with cross-section of 50.8 × 6.35 mm².

Table 11.19 Current carrying capacity of Al bus bar in standard sections

Size mm²	25.4 × 6.35	38.1 × 6.35	50.8 × 6.35	63.5 × 6.35	76.2 × 6.35	101.2 × 6.35	127.0 × 6.35	152.4 × 6.35	50.8 × 9.53	76.2 × 9.53
Amp	1100	1535	1940	2260	2620	3200	3700	4240	2260	3030
Size mm²	101.6 × 9.53	127.0 × 9.53	152.4 × 9.53	203.2 × 9.53	76.2 × 12.7	101.6 × 12.7	127.0 × 12.7	152.4 × 12.7	203.2 × 12.7	254.0 × 12.7
Amp	3560	4200	4680	5740	3240	3900	4550	5100	6150	6850

Table 11.20 De-rating factor for bus bar

Enclosure	Cross-sectional area of bus bar as % of total cross-sectional area	De-rating factor
a. Outdoors	< 1 %	.95
	< 5 %	.90
	< 10 %	.85
b. Indoors where enclosure itself is in a well ventilated room	< 1 %	.85
	< 5 %	.75
	< 10 %	.65
c. Indoors where and the room temperature is high enclosure itself is poorly ventilated	< 1 %	.65
	< 5 %	.60
	< 10 %	.50

BIS defines the marking arrangement for bus bars and main connections as in Table 11.21.

Table 11.21 Colour coding of bus bars

Bus bar arrangement	Colour	Letter
Three phase	Red, Yellow, Blue	RYB
Two phase	Red, Blue	RB
Single phase	Red	R
Neutral	Black	N
Connection to earth	Green	-

It shall be noted that when the run is horizontal, red should be at the top, left or furthest as viewed from front.

When run is vertical, red should be at the left or furthest when viewed form the front.

Unless the neutral can be readily distinguished, the order shall be RYB, and black.

In DC system with three wires, positive and negative and neutral, neutral should be at the middle.

11.8.1 Latest Trends: Air Insulated Bus Trunking

Further designs over a period of time led to the development of compact bus bar trunking system. These systems have the flexibility of change in layout as the entire trunking is made in modules attached together. These trunkings can be used to supply low voltage electric power to industrial installations and high rise buildings. They are available from 160–1250 amp in aluminium conductor and 125–2000 amp in copper conductor.

This arrangement has possibility of feeding loads up to 400 amp with plug in boxes. The fixing of plug in box is very simple as no screws are required and is simply friction fit in the special notch provided in the compact bus bar enclosure and the jaws go deep inside the enclosure to make contact with bus bars (Fig. 11.16).

The notches for plug in boxes are provided at suitable interval along the height of trunking, thereby providing flexibility of tap off point for power.

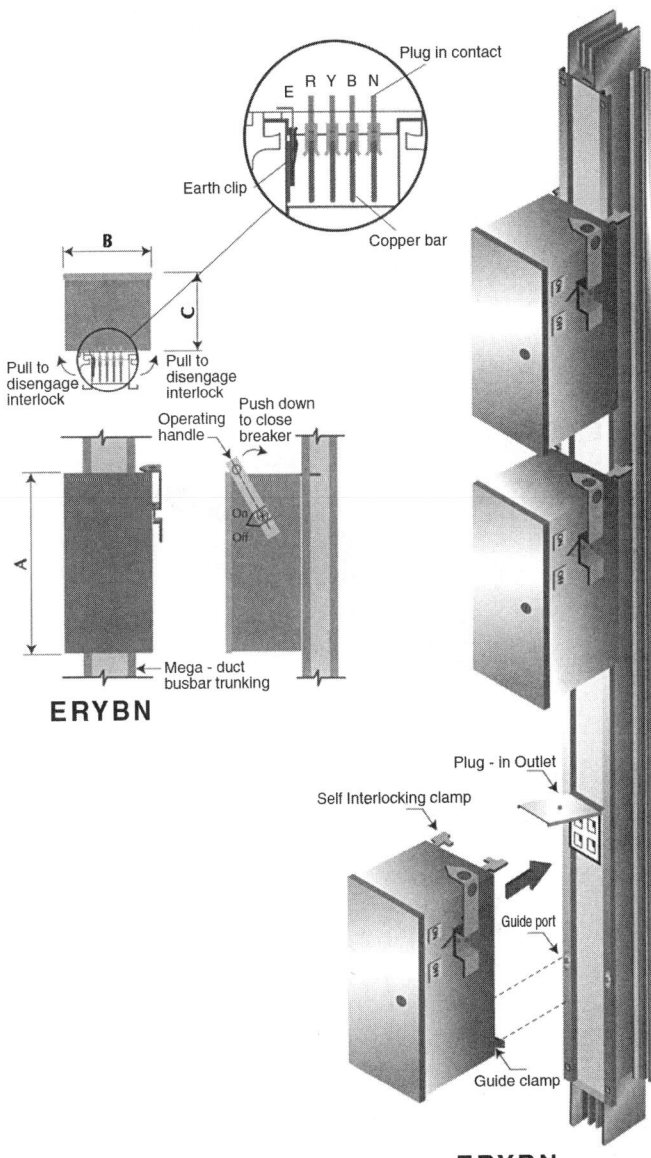

Fig. 11.16 Plug-in-box fixing arrangement

The major advantage of this type of trunking is that two sections can be electrically isolated for maintenance and testing in installed condition without dismantling or removal of any section.

The bus bars are insulated from each other by the provision of paper insulation wrapped around bus bars. But the latest development is high impact, flame retardant, non-hygroscopic insulators made of F class material. Bus bars are insulated with sufficient clearance.

The temperature rise above ambient of 40 °C is as high 55 °C in case of copper bus trunking and 40 °C in case of aluminium bus trunking.

The overall dimension of bus trunking is reduced very much as compared to rising main described earlier. The dimensions of the complete bus trunking is reduced very much as given in Tables 11.22 and 11.23 but conductor cross-section is increased.

The bus trunking may be running horizontally as in an industrial application or vertically as in high rise building. The space required is to be considered carefully beforehand so that architect has a very clear knowledge of size of shaft to be provided. The shaft dimensions have already been given in Chapter 2 earlier.

The sandwich trunking are manufactured from 800–5000 amp with copper bus bars. In aluminium, the trunking is available from 500–3600 amp.

The dimension of bus trunking as well as conductor size reduces to a great extent as compared to air insulated bus trunking as shown in Tables 11.24 and 11.25.

The dimensions of trunking and copper bus bar cross-section are much lesser than aluminium bus trunking. The cross-section of sandwich bus trunking is shown in Fig. 11.17 and the mounting arrangement in Fig. 11.18.

Table 11.22 Dimension of aluminium bus trunking

Current rating amp	160	250	400	500	630	800	1000	1250
Overall dimension mm	147 × 50	147 × 75	147 × 95	147 × 115	147 × 135	147 × 190	147 × 230	147 × 270
Phase cross-section mm^2	90	180	300	420	540	600	840	1080

Table 11.23 Dimensions of copper bus trunking

Current rating amp	125	250	315	400	500	630	800	1000	1250	1500	1750	2000
Overall dimension mm	147 × 60	147 × 60	147 × 75	147 × 75	147 × 95	147 × 95	147 × 115	147 × 135	147 × 190	147 × 230	147 × 230	147 × 270
Phase cross-section mm^2	28	90	120	180	240	300	420	540	600	720	840	1080

However, it must be noted that 50 mm clearance must be provided between two plug in box/units installed on two parallel running bus trunking. Space required for door opening should also be considered. Another 50 mm clearance between trunking and beam/column/pillars should be provided.

11.8.2 Sandwich Insulated Bus Trunking

This is the most modern factory built electrical distribution system and is not only alternative to cables but have many advantages over conventional cable distribution system. Each equipment or installation is protected immediately by a protective device and can be maintained individually without disturbing other distribution networks. Load can be fed directly from plug-in-boxes.

In sandwich bus trunking, the bus bars are in close proximity to each other. The close proximity between phases does not allow mutual inductance between phases yielding low reactance, low impedance, low voltage drop and low power losses. As no air gap is maintained, thus there is no chimney effect and no fire barriers are required. However, the buses are separated at the point where plug-in-box are to be fitted or at joints.

Individual bus bars are covered with multilayer of F class insulation and the bus trunking are manufactured with IP 54 protection.

11.9 INSTRUMENT TRANSFORMERS

Protective relays and meters are actuated by current and voltage supplied by current and voltage transformers. These transformers provide insulation against high voltage of power circuits and also supply the relays and meters with quantities proportional to those of power circuit, but sufficiently reduced in magnitude so that the relays/meters can be made relatively small and inexpensive.

The accuracy requirement of CTs and PTs depend on the type of application. Technically, an entirely safe rule would be to use the most accurate equipment but the same would not always be economically justifiable.

11.9.1 Current Transformers

Accuracy class. It is the designation assigned to a CT whose error remain with in specified limit under prescribed condition of use.

Burden. Usually expressed as apparent power in volt-amp absorbed at specified power factor.

Rated burden. The value of burden on which accuracy requirement are based.

Rated short time current. The value of primary current which the current transformer will withstand

Table 11.24 Dimensions of aluminium sandwich bus trunking

Current rating amp	500	630	800	1000	1250	1350	1600	1750	2000	2500	2750	3000	3600
Overall dimension mm	147	147	147	147	147	147	147	147	147	147	147	147	147
	×	×	×	×	×	×	×	×	×	×	×	×	×
	95	115	145	170	195	220	245	275	340	390	440	490	550
Phase cross-section mm	6	6	6	6	6	6	6	6	6	6	6	6	6
	×	×	×	×	×	×	×	×	×	×	×	×	×
	50	70	100	125	150	175	200	230	125	150	175	200	230
									×	×	×	×	×
									2	2	2	2	2

Table 11.25 Dimensions of copper sandwich bus trunking

Current rating amp	800	1000	1250	1500	1750	2000	2250	2500	3000	3600	4000	4500	5000
Overall dimension mm	147	147	147	147	147	147	147	147	147	147	147	147	147
	×	×	×	×	×	×	×	×	×	×	×	×	×
	95	115	145	170	195	220	245	275	340	390	440	490	550
Phase cross-section mm	6	6	6	6	6	6	6	6	6	6	6	6	6
	×	×	×	×	×	×	×	×	×	×	×	×	×
	50	70	100	125	150	175	200	230	125	150	175	200	230
									×	×	×	×	×
									2	2	2	2	2

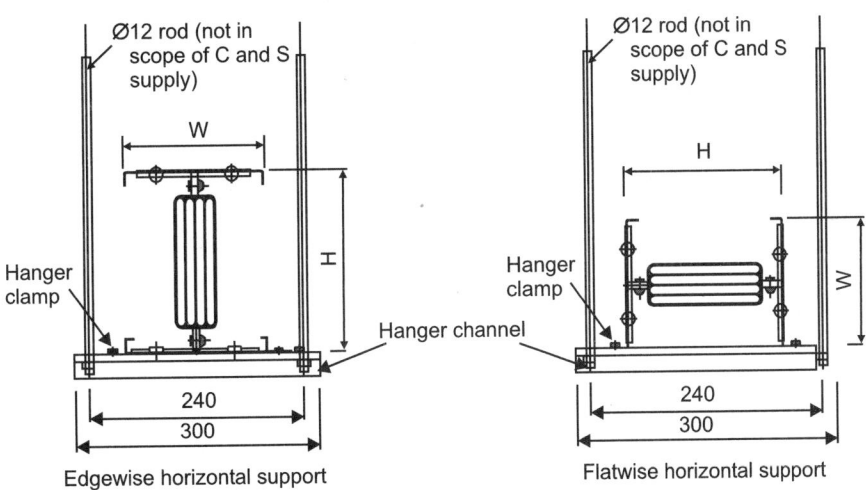

Edgewise horizontal support   Flatwise horizontal support

Fig. 11.17 Cross-section of a horizontal sandwich bus trunking

for a rated time with their secondary circuit short-circuited with out suffering harmful effect.

Accuracy limit factor. The ratio of the rated accuracy limit primary current to rated primary current.

Standard value of rated secondary current. 5A and 1 A.

Rated output. 2, 5, 7.5, 10, 15 and 30 VA.

Limit of temperature. CTs not immersed in oil rise or bituminous compound. Table 11.26 may be referred for class of insulation versus temperature rise.

a. Metering CTs

Accuracy class 0.1, 0.2, 0.5, 1, 3, and 5.

Table 11.26 Temperature vs class of insulation

Class of insulation	Temp. rise (°C)
Y	40
A	55
E	70
B	80
F	105
H	130

For standard accuracy class 0.1, 0.2, 0.5 and 1.0, the current error and phase displacement at rated frequency shall not exceed value given at Table 11.27, when secondary burden is 25 to 100% of rated burden.

Table 11.27 Current errors and phase displacement

Accuracy class	% error at % of rated current				Phase displacement in minutes at % of rated current			
	5	20	100	120	5	20	100	120
0.1	0.4	0.2	0.1	0.1	15	8	5	5
0.2	0.75	0.35	0.2	0.2	30	15	10	10
0.5	1.5	0.75	0.5	0.5	90	45	30	30
1.0	3	1.5	1.0	1.0	180	90	60	60

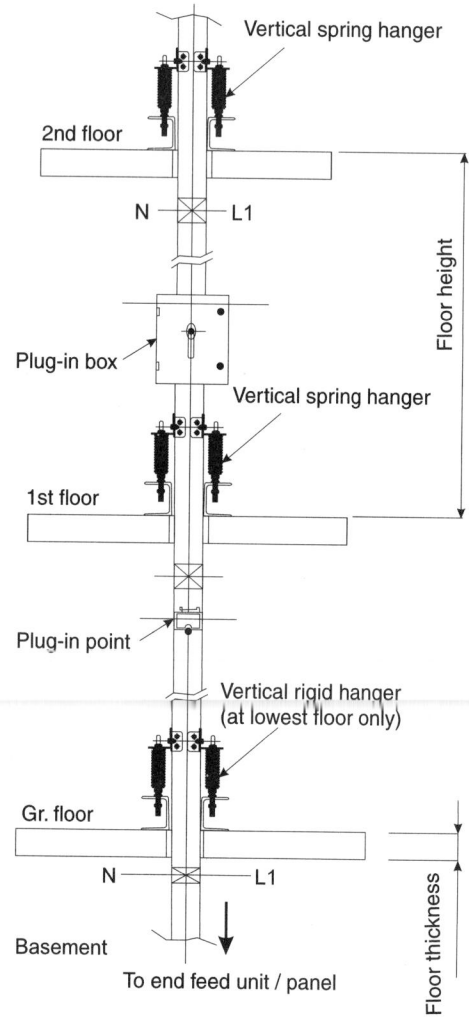

Fig. 11.18 Mounting arrangement of vertical bus trunking

For standard accuracy class 3 and 5 the current error is as given in Table 11.28.

Marking. The measuring CTs are marked by indicating rated output followed by accuracy class and is written as 15 VA class 0.5 or 15/0.5.

b. Protection CTs

The accuracy class shall be designated by the highest permissible percentage composite error at the rated accuracy limit primary current prescribed for the accuracy class concerned followed by *P*. The standard accuracy class are 5 P, l0 P, 15 P.

Protection CTs are designated by accuracy class followed by accuracy limit factor. The standard accuracy limit factor are 5, 10, 15, 20 and, 30 A CT designated as 5 P 10 indicates accuracy class of 5 P and accuracy limit factor 10.

The limits of error is given in Table 11.29.

Marking. Rated accuracy limit factor shall be indicated following the corresponding output and accuracy class.

For example. 30 VA class 5 P/10 or 30/5 P 10.

11.9.2 Potential Transformers

There are 3 types of potential transformers. (i) Measuring, (ii) Protective, (iii) Dual purpose.

Accuracy class. A designation assigned to a potential transformer the error of which remains under specified limits under prescribed conditions of use.

Rated burden. Burden at which rated accuracy class is met.

Rated output 10, 15, 25, 30, 50, 75, **100,** 150, **200,** 300, 400, **500** VA. (Those highlighted are preferred values).

Percentage voltage ratio error and phase displacement of measuring voltage transformer are as shown in Table 11.29.

Table 11.28 Current errors and phase displacement

Accuracy class	% error at 50% of rated current	% of error at 100% of rated current
3	3	3
5	5	5

Table 11.29 Voltage ratio error and phase displacement for PT

Accuracy class	% Voltage ratio error	Phase displacement in minutes
0.1	0.1	5
0.2	0.2	10
0.5	0.5	20
1.0	1.0	40
3.0	3.0	–

11.10 CABLES

HT cables are required to connect HT panel board to transformer and LT cables from transformer onwards. In case of HT cables, the design is based on the fault withstand capacity of cable rather than the current carrying capacity. The cable chosen should have fault withstand capacity matching the HT system fault level. Manufacturer's data may be consulted.

Example

To determine size of HT cable for connecting 11 kV panel board to a 1600 kVa transformer having system fault level of 350 MVA.

$$I_{sh}(kA) = \text{System fault level (MVA)}/1.73 \times 11\,kV$$
$$= 350/1.73 \times 11 = 18.39\,kA.$$

Based on short-circuit current the size of cable can be decided as per formulae below.

$A = I_{sh}(kA) \times t^{1/2}/K$

A = Cross-section of cable in sq mm.

t = Duration of fault (generally taken as 1 sec.).

K = Constant (0.094 for 'Al' and 0.143 for copper).

$A = 18.39 \times 1/\,0.094 = 195$ sq mm 'Al'.

$A = 18.39 \times 1/0.143 = 128$ sq mm 'Cu'.

Next nearest size may be selected.

For determining the size of the LT cables for different feeders, the connected load of the feeders and their approximate lengths may be determined and size of the cable may be decided by using manufacturer's tables keeping in view that voltage drop of should not exceed 8 V (2.5% for LT system) between sending end and receiving end. The cable sizes so worked out are given in Annexure C, Table C.7.

11.11 BATTERY BANK FOR TRIP SUPPLY

Wherever shunt trip release is provided it is necessary to decide the trip supply voltage. In case of HT panel boards 110 V AC trip supply can be taken from the 11 kV/110 V potential transformer provided on the incoming panel. However, heavy fault on HT bus-bar may give rise to a large drop in voltage and situations may be encountered, when trip supply available is very low and same may result in shunt trip release not picking up and there by failing to trip the circuit breaker. To counter such eventuality, it is necessary to provide 24/30 V DC trip supply through battery banks having its own charger. Whenever, DC trip supply is provided, it may also be worthwhile to connect all the indicating lamps on the HT panel board to the DC supply.

11.12 INGRESS PROTECTION (IP CLASSIFICATION)

The degree of protection of equipment like panel boards, rising mains and other equipment like luminaires, etc. are expressed by a two digit number where first digit defines protection against dust and second against moisture/water (IS 10322 part I, 1982).

Example

If ingress protection of an equipment is defined as IP 42, it would mean that the equipment is protected against ingress of solid bodies of size greater than l mm and protected against water falling at an angle of 15° maximum.

Panel boards (HT and LT) are generally manufactured with ingress protection classification of IP 42. Rising mains having ingress protection classification of IP 42, IP 54, IP 55, IP 56 are being manufactured. The higher IP classification results in higher cost of equipment. In most applications rising mains having IP 42 classification would suffice.

11.13 PANELS

After a designer designs the technical specifications of ACB, relays and protection devices, it is time to design and write down the panel specifications. The most intelligently designed panel may have poor appearance or may fail if certain informations are not given to the manufacturer. These are:

1. Classification of panel into various categories depending on usage.
2. Forms of separation between various sections inside the panel.
3. Type test criteria.
4. Protection requirements.

11.13.1 Classification of Panel

The panels are to be classified according to the purpose for which they will be used. This classification helps the manufacturer in segregating the load of one type to one section and the other type to other section or to design separate panels for different applications. This classification is given below.

1. **Power control centre (PCC).** This is the panel used for distribution of power to different applications, e.g. one feeder may supply power to control motors or to a different section of building through a subordinate panel.
2. **Motor control center (MCC).** It is a panel located near the load of motors say in pump room. This panel is specifically designed with capacitors, etc.
3. **Switch distribution board (SDB).** This board is the main distribution board for power supply entry in a building. From here, further distribution is made to subordinate switch boards for various zones of the building.

4. **Light distribution board (LDB).** This is the lowest ranking board in the chain of power supply boards. From here, the supply is fed to various rooms of the building.

5. **Power cum motor control center (PMCC).** This board is common for power supply to various loads and for motor loads. These panels are provided where the building is small and feeding to light, power and motor loads can be done from the same place.

6. **Automatic power factor control panel (APFC).** As already described, capacitor banks are required to improve power factor. The panel housing capacitors, contactors, reactors is called APFC panel. This may be a part of PCC or MCC or may be stand alone type.

7. **Diesel generator synchronizing panel.** This panel is provided at the diesel generator room and houses microprocessor based controls to control the operation of DG sets, synchronizing two DG sets in parallel and controlling the power output.

8. **Programmable logic control (PLC).** This panel is intelligent panel and houses the programmable logic controller which is microprocessor based and it controls and records the various parameters of power generation, supply, breakdowns as well as sequencing of various operations.

9. **Double bus bar panels.** In this panel, both emergency and non-emergency power supplies are run in the same panel using double bus bar arrangement.

11.13.2 Forms of Separation

The panel designing is complete only if various compartments are sealed and insulated from each other so that a person working in one section does not have electric hazard from the other section or in case of fire/sparking in one compartment, others are protected from spread of fire. The various forms are listed in Table 11.30 and shown graphically in Fig. 11.19.

11.13.3 Type Test Criteria

IS 8623 (Part 2) 1993, defines the criteria of type test as partial type test (PTT). In PTT, one sample panel out of a total lot to be manufactured is tested at a central lab authorised for this purpose (in India CPRI, i.e. central power research institute). Based on this design, the manufacturer keeps on manufacturing the panels with the same specifications. But the drawback in this system is that 100% type test is not possible in all the panels as a result of which buyer is not sure if all the panels he has purchased will stand to faults at any time of operation of panels.

The latest trend developing is the fully type test assembly (FTTA). In this system, the manufacturer designs hundreds and thousands of various combinations of possible user requirements, standardize the components and design on a drawing. As and when, requirement from a user is received, the drawing matching to that requirement is taken out and panel is manufactured with only those components. All the panels thus manufactured are fully reliable as they are manufactured with the same components so that the customer is sure about the reliability of panel he is buying. But in this system, the customer is not left with the choice of making suggestions regarding make/brands of various components housed in his panels. This is because panels will be fabricated according to the drawings already standardized by the manufacturer.

11.13.4 Protection Requirements

The panels are manufactured with various classes of protection depending on the application where they are to be used. These requirements are informed to the manufacturers in the form of IP class of protection required. These classifications have been standardized and are given in Table 11.31. The letters IP followed by two numerals and an additional letter define the protection requirement. The meaning and symbol of each class of protection is given in Table 11.31.

Table 11.30 Forms of separation of compartments of panel

	Forms of Separation		
Main criteria	Sub-criteria	Form	Type of construction
No Separation		Form 1	
Separation of bus bars from the functional units.	Terminals for external conductors not separated from bus bars.	Form 2	**Type 1.** Bus bar separation is achieved by insulated coverings, e.g. sleeving, wrapping or coating.
	Terminals for external conductors separated from bus bars.		**Type 2.** Bus bar separation by metallic or non-metallic barriers or partition.
Separation of bus bars from the functional units and separation of all functional units from one another. Separation of the terminals for external conductors from the functional units, but not from each other.	Terminals for external conductors not separated from bus bars.	Form 3 (a)	**Type 1.** Bus bar separation is achieved by insulated coverings, e.g. sleeving, wrapping or coating.
	Terminals for external conductors separated from bus bars.	Form 3 (b)	**Type 2.** Bus bar separation is by metallic or non-metallic rigid barriers or partition.
Separation of bus bars from the functional units and separation of all functional units from one another, including the terminals for external conductors which are an integral part of the functional units.	Terminals for external conductors in same compartment as associated functional unit.		**Type 1.** Bus bar separation is achieved by insulated coverings, e.g. sleeving, wrapping or coating. Cables may be glanded elsewhere.
			Type 2. Bus bar separation is by metallic or non-metallic rigid barriers or partitions. Cables may be glanded elsewhere.
			Type 3. Bus bar separation requirements are by metallic or non-metallic rigid barriers or partitions. The termination for each functional unit has its own integral glanding facility.
		Form 4	**Type 4.** Bus bar separation is achieved by insulated coverings, e.g. sleeving, wrapping or coating. Cables may be glanded elsewhere.
	Terminals for external conductors not in the same compartment as the associated functional unit, but in individual, separate, enclosed protected spaces or compartments.		**Type 5.** Bus bar separation is by metallic or non-metallic rigid barriers or partitions. Terminals may be separated by insulated coverings and glanded in common cabling chambers.
			Type 6. All separation requirements are by metallic or non-metallic rigid barriers or partitions. Cables are glanded in common cabling chamber(s).
			Type 7. All separation requirements are by metallic or non-metallic rigid barriers or partitions. The terminations for each functional units has its own integral glanding facility.

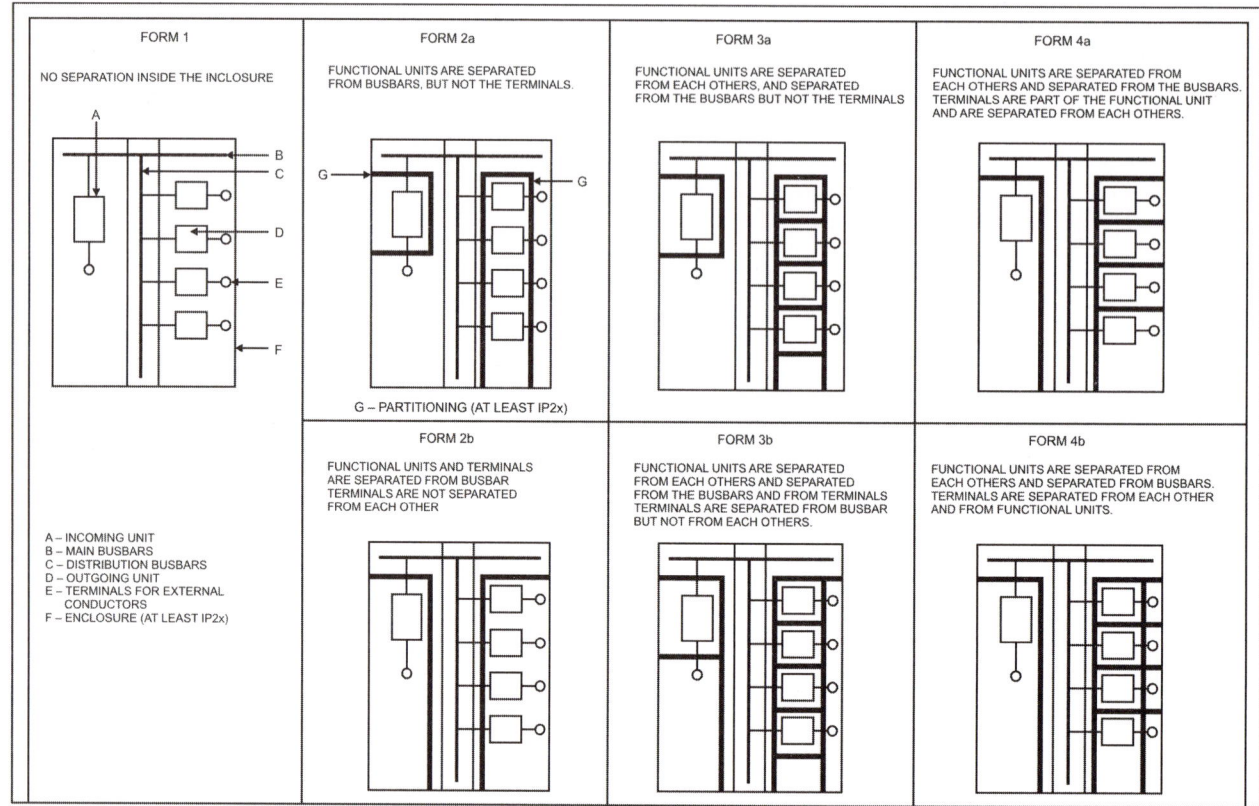

Fig. 11.19 Various forms of separation

Table 11.31 Selection of IP class protection criteria

First numeral		
Protection against ingress of solid objects		*Protection of persons against access to hazardous parts with*
IP	**Requirements**	**Example**
0	No protection	Non-protected
1	Full penetration of 50 mm diameter sphere not allowed. Contact with hazardous parts not permitted	Back of hand
2	Full penetration of 12.5 mm diameter sphere not allowed. The jointed test finger shall have adequate clearance from hazardous parts	Finger
3	The access probe of 2.5 mm diameter shall not penetrate	Tool
4	The access probe of 1.0 mm diameter shall not penetrate	Wire
5	Limited ingress of dust permitted (no harmful deposit)	Wire
6	Totally protected against ingress of dust	Wire
Second numeral		
Protection against ingress of solid objects		*Protection of persons against access to hazardous parts with*
IP	**Requirements**	**Example**
0	No protection	Non-protected
1	Protected against vertically falling drops of water. Limited ingress permitted	Vertically dripping
2	Protected against vertically falling drops of water with enclosure tilted 15° from the vertical. Limited ingress permitted	Dripping up to 15° from the vertical
3	Protected against sprays to 60° from the vertical. Limited ingress permitted	Limited spraying

(Contd.)

Table 11.31 (*Contd.*)

	Second numeral	
	Protection against ingress of solid objects	*Protection of persons against access to hazardous parts with*
IP	*Requirements*	*Example*
4	Protected against water splashed from all directions. Limited ingress permitted	Splashing from all directions
5	Protected against jets of water. Limited ingress permitted	Hosing jets from all directions
6	Protected against strong jets of water. Limited ingress permitted	Strong hosing jets from all directions
7	Protected against the effects of immersion between 15 cm and 1 m	Temporary immersion
8	Protected against long periods of immersion under pressure	Continuous immersion

	Additional letter (*optional*)		
IP	*Requirements*	*Example*	*Protection of persons against access to hazardous parts with*
A For use with first numeral 0	Penetration of 50 mm diameter sphere up to barrier must not contact hazardous parts		Back of hand
B For use with first numeral 0 and 1	Test finger penetration to a maximum of 80 mm must not contact hazardous parts		Finger
C For use with first numeral 1 and 2	Wire of 2.5 mm diameter × 100 mm long must not contract hazardous parts when spherical stop face is partially entered		Tool
D For use with first numeral 2 and 3	Wire of 1.0 mm diameter × 100 mm long must not contract hazardous parts when spherical stop face is partially entered		Wire

12

Electric Generator

The aim of this chapter is to prepare the reader for self selection of AC generators. The principles involved and some of the design limitations are given. This chapter will make the reader well aware of the

1. Various methods of excitation of rotor.
2. Calculation of load.
3. Parallel operation of generators.

12.1 BASIC DESIGN THEORY

12.1.1 General

One of the simplest ways to remember how every AC generator works is to imagine a magnet and a piece of wire. Move one of them with respect to the other, whilst keeping them close together, and a measurable voltage will be induced at the ends of the wire (Fig. 12.1).

Immediately then we see that all AC generators must have the following, before an output voltage can be generated.

1. A magnet—to produce the magnetic field excitation.
2. A piece of wire—usually coils of copper wire.
3. Relative movement between these two—usually a constant rotational speed.

12.1.2 Theory

In its simplest form, an AC generator is diagrammatically shown in Fig. 12.1. All three criteria stated above are met and a voltage output will be produced. In this case the magnetic field produced is at a constant level from a permanent magnet. This type of machine does have practical applications, particularly when supplying a constant load; for example, a pedal bicycle dynamo.

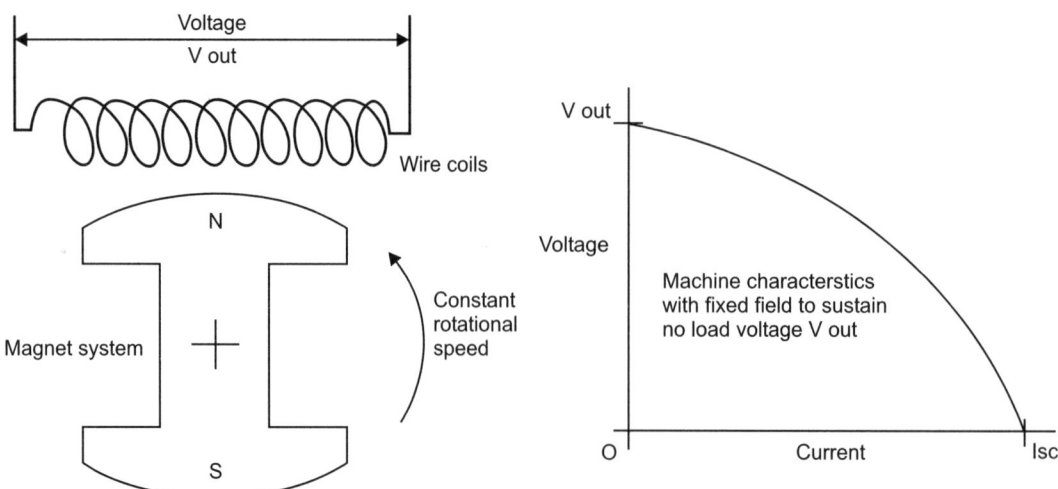

Fig. 12.1 Simple AC generator (Permanent magnet type)

There are two conditions to consider at this point, the no load condition and the on load condition. The no load output voltage level is sustained by the constant magnetic field strength produced and fixed by the permanent magnet. On load, current is drawn from the machine which will cause the output voltage to fall, since the permanent magnet cannot produce a change in the magnetic field strength. The typical relationship between the output voltage and the load current is shown in the graph in Fig. 12.1. Typically the relationship is nearly linear between no load voltage (V) at zero current and the short-circuit current (Isc) at zero voltage. To maintain the output voltage whilst supplying current, the magnetic field strength must be increased as load is applied. It is this requirement that brings us to the next stage of AC generator design incorporating the electrically produced magnetic field system (Fig. 12.2).

Curve 1 is the machine characteristic with minimum fixed field to sustain no load voltage *V*.

Curve 2 is the machine characteristic with maximum available fixed field from variable DC supply.

Curve 3 is the typical desired machine characteristic using variable DC supply to provide variable field.

Again all the basic criteria are met, except that the magnet is now produced electrically from an external variable DC source supply by a coil of wire wound on the magnetic material. To get the supply to the rotating coil, the coil ends have to be brought to slip rings and the static supply connected to brush gear. The magnetic field strength can be increased by increasing the supply current. When this machine is run on load, we can increase the magnetic field strength to maintain the output voltage level. (Curve 3, Fig. 12.2).

From here onwards there are many design improvements that can be made. For example, a second AC generator (an exciter) can be put on the same shaft to make the basic brushless machine; that is a machine without slip rings and brush gear but which still maintains control over magnetic field strength. Also instead of having a manual control over the variable DC supply to the field coil, we can provide an automatic control system. These items are discussed later on in this section.

Now let us consider the factors governing the rotational speed and the output voltage.

12.1.2.1 Speed

There is a simple relationship between speed, output frequency and the number of magnetic poles forming the main field excitation system of the machine.

Output frequency (Hz)

$$= \frac{\text{Driven speed (rev/min)} \times \text{No. of magnetic poles}}{120}$$

Obviously the number of magnetic field poles is decided and fixed by the machine manufacturer. Therefore, we can see that output frequency is directly proportional to driven speed. In other words the only way frequency can change is due to a corresponding change in driven speed.

12.1.2.2 Voltage

The relationships of the machine governing voltage level are more complex. Output voltage (V) depends on.

1. Driven speed (rev/min).
2. Number of turns of copper wire in output winding.
3. Strength of magnetic field produced by the main field excitation magnetic poles.

The number of turns of copper wire in the output winding is fixed by the machine manufacturer. Voltage is also affected by driven speed. In fact with a constant

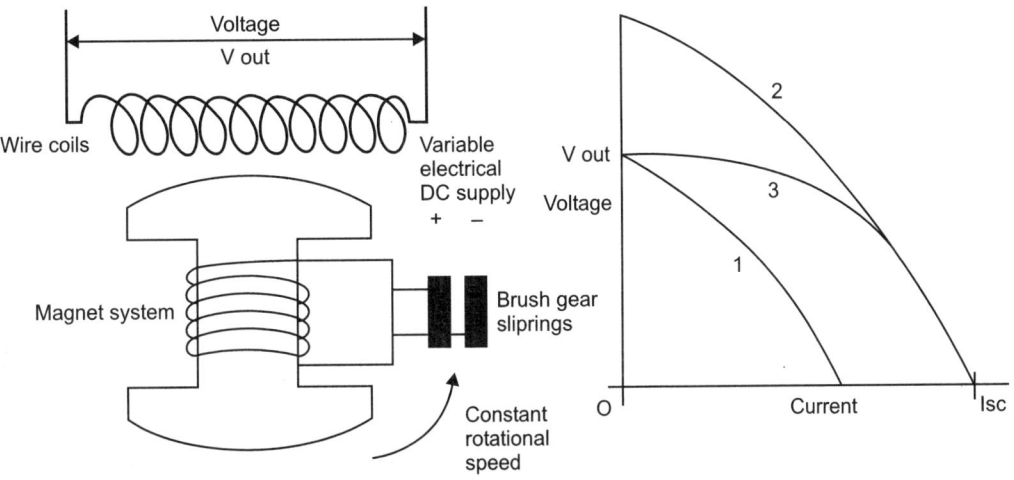

Fig. 12.2 AC generator (Electric magnet type)

magnetic field strength, voltage as well as frequency would be directly proportional to driven speed.

As the speed is fixed to obtain the correct frequency, the only variable left that can be used to change and fix the machine voltage level is the magnetic field strength. This is exactly the parameter that the machine's control system does adjust to set the voltage level, and compensate for both speed and load current changes.

12.1.2.3 Control System

Let us now take a look at the control systems that are readily available.

These are two popular methods of voltage control system for AC generators.

1. Closed loop electronic system.
2. Open loop transformer system.

12.1.2.4 Closed Loop Electronic System

This system continually monitors the output voltage and compares it with a reference voltage level set by the user. Once the reference voltage level is set, the automatic voltage regulator (AVR) will automatically compare the actual output voltage with the reference voltage level and if they are different it will adjust the magnetic field strength to make the output voltage the same as the reference voltage. This system is therefore an accurate closed loop control by using static magnetic amplifiers (combinations of transformers and controlled reactors). This system is very reliable although slow in operation and very bulky compared to the electronic systems which have now virtually superseded it.

12.1.2.5 Transformer Control

This system is really in two parts, firstly adjustment for no load voltage level and, secondly, compensation for load current. This system once set up is not normally adjusted. It is an open loop control system; there is no continuous monitoring to adjust the output voltage. It can only provide the amount of magnetic field strength it has been set up to provide.

12.2 MACHINE TYPES

12.2.1 Rotating Field Type

Previously we have stated that rotational speed must be maintained between the output winding (the 'wire') and the field winding, (the 'magnet'). Also we have assumed that the field winding is rotated with the output winding remaining stationary. This type of machine is logically called a rotating field AC generator. All brushless machines are rotating field types, but as we see in Fig. 12.2 that not all rotating field type machines are brushless.

12.2.2 Rotating Armature Type

The alternative is to rotate the output winding whilst keeping the field winding stationary. This still complies with all the basic requirements for voltage generation and such a machine is called a rotating armature AC generator.

12.3 OPERATING PRINCIPLES

12.3.1 Introduction

In the preceding section, the basic AC generator and voltage control system parameters were discussed. In this part we shall examine several practical machines with their associated control system. In all cases the basic requirements are the same.

1. An output voltage is induced between the ends of a conductor (usually copper wire), when the conductor is adjacent to a magnet and there is constant relative motion between them.
2. The output voltage is controlled by variation of the magnet's field strength to maintain the output voltage under load and speed changes.

12.3.2 Self Excited, Rotating Field, Brushless, AC Generator with Electronic Voltage Control

This machine is self excited as the excitation power is derived from the main output winding of the machine itself (Fig. 12.3). Rotating field indicates that it is the main machine's magnetic field system that the prime mover rotates. Brushless means that there are no slip rings or brush gear needed in this type of machine design.

Let us now consider the necessary motion the mechanical rotational input power. It is this input power that sustains the correct speed under all load conditions. From this input power is derived the electrical power output, the field excitation power and all the machine losses.

Excitation power is supplied to coils of copper wire wound on some magnetic material. A characteristic of these materials is that a magnetic field can be easily set up and controlled in them. In all such materials, there already exists a small amount of magnetism even before any electrical excitation supply is connected. This is called *residual magnetism*.

Referring now to Fig. 12.3, with the machine rotor being driven at the correct speed, we have a small magnetic field set up due to residual magnetism, rotating adjacent to coils of copper wire in the main stator output winding. Therefore, a small voltage is induced at the ends of the winding and is termed 'residual voltage'. The automatic voltage regulator (AVR) will sense this low voltage and compare it with the 'set reference' voltage level. The 'set reference' voltage level is an externally adjustable voltage level, derived by the AVR corresponding to the value of sensing voltage obtained when the machine is running at nominal speed and at rated output voltage. The AVR will find initially that the sensed voltage is considerably lower than the 'set reference' voltage. The AVR will therefore provide such power as is available from the main stator winding to establish the exciter field.

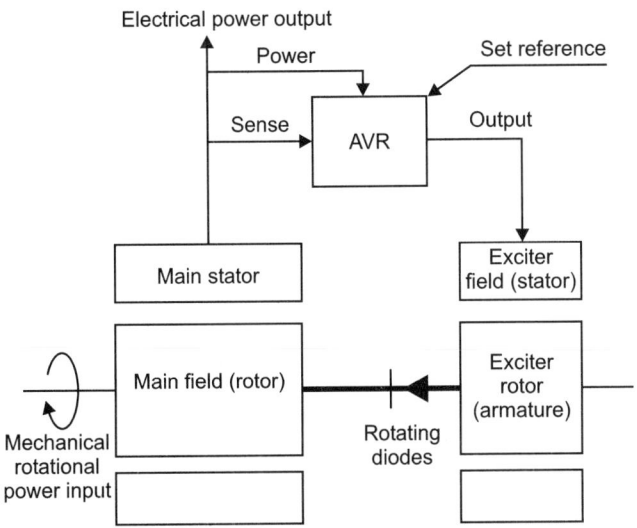

Fig. 12.3 Self excited, rotating field, brushless AC generator with electronic control

The exciter is made from similar magnetic materials as the main machine, so the exciter field will also have a small amount of residual magnetism. The power from the main output winding, which is rectified by going through the AVR now adds to this residual level to produce a greater magnetic field strength in the exciter field. It is worth noting that correct polarity must be observed, since, if incorrect, the additional excitation power will subtract from the residual magnetism until zero magnetic field strength is reached. This means that the output voltage will not build-up but remain at zero until an external DC source is applied to re-establish the exciter field.

With the exciter magnetic field strength increased, the AC output voltage from the exciter rotor will also increase. This voltage is rectified by the rotating diodes to provide additional DC excitation to the main machine field. This extra excitation adds to the residual level of the main field and produces an increase in output voltage from the main stator.

The AVR senses this increase, compares it with the 'set reference' and uses the increased power from the main stator to further increase the exciter field excitation as required. In this way the main stator voltage is progressively built-up until the 'sensed' voltage is the same as the 'set reference' voltage. At this point the exciter field excitation will be stable and

of such a value to just maintain the nominal or rated voltage level. This build-up process in fact starts during the run up of the set. By the time the prime mover speed is stable, the AC generator output voltage will normally be stable and at the correct pre-set level.

Since this is a closed loop electronic voltage control system, a change in output voltage due to load current or speed changes is automatically compensated for by the action of the AVR. This will adjust the excitation under all circumstances to achieve minimum error between the 'sensed' output voltage and the 'set reference' voltage. There are, however, upper and lower limits of stable excitation voltage and power than can be provided by the AVR. These limits must be borne in mind during the design of the field systems.

12.3.3 Separately Excited, Rotating Field, Brushless AC Generator with Electronic Voltage Control

By comparing Fig. 12.3 with Fig. 12.4 above, we see the two types of machine are very similar. The difference is in the source of exciter field excitation power. In this case there is a separate source of exciter field power from a small permanent magnet field AC generator situated on the same shaft as the main machine. Hence, the machine title is 'separately excited'.

When running at the correct speed, excitation power is always available independent of load condition. The permanent magnet produces constant magnetic field excitation and rotates close to the permanent magnet machine's output winding. The constant output voltage is then fed to the exciter field winding through the AVR. By comparing the main output 'sensed' voltage with the 'set reference' voltage, the AVR decides on the proportion of permanent magnet machine output to rectify and feed to the exciter field.

The process of initial voltage build-up is very positive in this system, as residual magnetism is no longer continually depended upon the AVR sensed output voltage, compares it with the 'set reference' voltage and can apply the full permanent magnet output power rectified to the exciter field if necessary. The exciter rotor output would then increase, establishing a strong main field and therefore a marked increased in main output voltage. The AVR senses and compares voltages and adjusts exciter field excitation until, as in the previous machine type, both voltage and exciter field excitation becomes stable.

Another significant advantage of this system is the ability to sustain main field excitation when the main output winding is short-circuited. This means the 'sensing' voltage to the AVR is forcibly held at nearly zero by the applied short-circuit. Since the difference

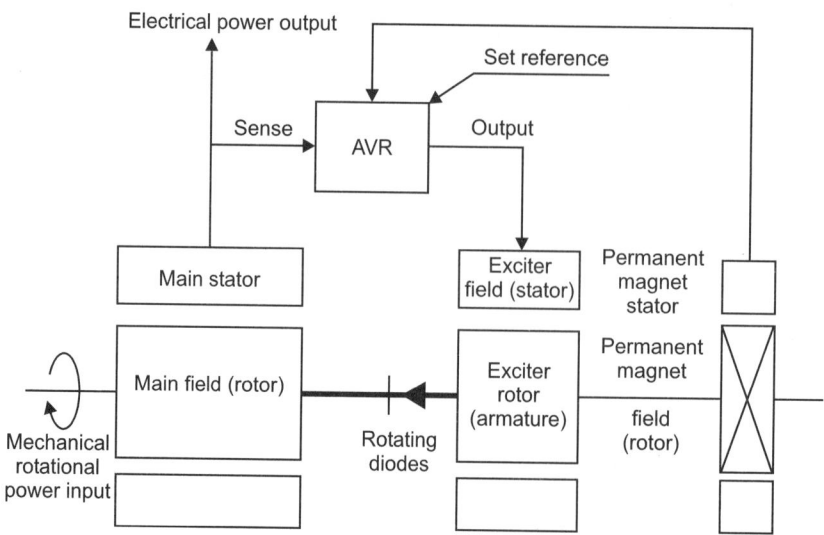

Fig. 12.4 Separately excited, rotating field, brushless AC generator with electronic control

in the 'sensed' and 'set reference' voltages is now large, the full permanent magnet output is rectified and applied to the exciter field. This sustains the main field excitation which in turn maintains the short-circuit current. This facility is advantageous wherever positive voltage build-up, high overload capacity, or short-circuit current fault discrimination is required.

12.3.4 Self Excited, Rotating Field, Brushless, AC Generator with Transformer Control

This combines the brushless style of machine with the open loop transformer type of control system. (See machines described under Figs 12.2 and 12.4). When rotating, the residual magnetism causes an output voltage, a fixed proportion of it being rectified and fed to the exciter field (Fig. 12.5). Output voltage will be built-up in the way described previously until the proportion of output voltage fed to the exciter field is just enough to sustain that voltage. The machine and control gear are designed so that this happens at the nominal output voltage of the machine. Compensation for load current is achieved by using a transformer to obtain a voltage proportional to load current. This voltage is rectified and used to increase the exciter field strength thereby sustaining the output voltage on load.

12.4 POWER RATING

The most fundamental factor governing the correct sizing of an AC generator is the power rating. By consideration of the electrical load likely to be applied to the AC generator, the user can estimate the required power rating. This is usually done by adding together the kW ratings of the individual parts of the load to arrive at a total kW power rating.

Initially every possible load should be included. In addition an allowance for future growth typically between 15% and 20% is common practice. This total kW power rating can now be checked with standard published output lists and an AC generator frame size selected. For standby or emergency service, only the essential loads need to be included.

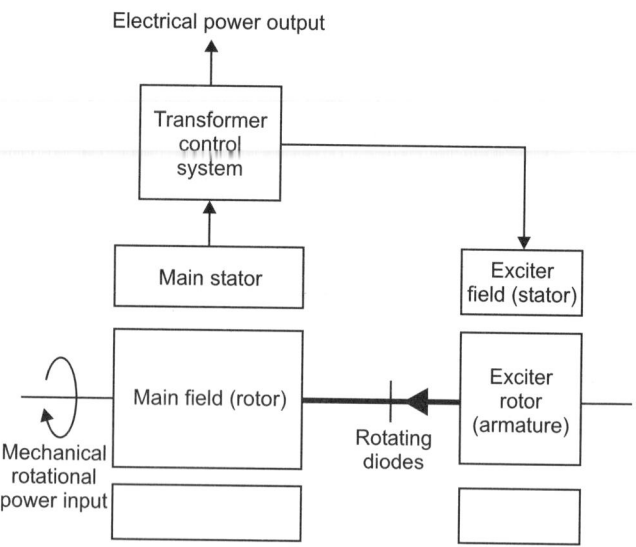

Fig. 12.5 Self excited, rotating field, brushless AC generator with transformer control

It should be noted that standard published output lists usually quote a kVa rating as well as a kW power rating, and to relate them a power factor of 0.8 lagging is assumed, i.e. kW = 0.8 × kVa.

Also general different ratings may be quoted for the same machine. Typically three ratings may be quoted.

a. *Continuous maximum industrial rating.* This is the rating to which all others are referred. All performance criteria and technical data will be based on this rating, unless instructions to the contrary are given. When working at this rating all quoted performance figures will be met.

b. *Standby duty rating.* This is a higher rating than the continuous maximum industrial rating. It permits a greater continuous output from the machine providing that worsened performance criteria, increased temperature rise and reduced lifetime are acceptable.

c. *Continuous marine rating.* This is a lower rating than the continuous maximum industrial rating. All standard performance criteria are met and bettered, but at a 50 °C ambient temperature, rather than the 40 °C ambient temperature considered for industrial applications. The overall actual temperature of the windings in insulation remains constant.

12.4.1 Efficiency and Drive Power

The selection of a suitable size of prime mover for an AC generator is governed by the electrical power output supplied to the load and the efficiency of the AC generator. The relationship is given below.

$$kW_{drive\ input} = \frac{kW_{output}}{\mu_a}$$

Where

"$kW_{drive\ input}$" defines the prime mover power rating in kilowatts.

"kW_{output}" defines the electrical power supplied to the load (usually the continuous maximum industrial power rating of the AC generator).

"μ_a" is the efficiency of the AC generator.

To convert kW input into horsepower (hp) the following conversion constants should be used.

$$hp = \frac{kW_{input}}{0.746}$$

In special circumstances the prime mover size may need to be larger than specified above, to cope with particular overload or transient conditions. Consideration of motor load starting torque can be especially relevant when determining prime mover size.

AC generators are designed and insulated to operate on full load within a maximum permitted temperature. The insulation system must retain its properties over this operating temperature range for the lifetime of the machine.

Table 12.1 (simplified from BS 4999 Part 32) gives the standard insulation classes available and the associated maximum permitted temperature rises, (i.e. actual temperature minus ambient temperature).

Table 12.1 Permitted temperature rise of various class of insulations

Insulation class of material	A	E	B	F	H
Maximum permissible temperature rise (°C) based on an ambient temperature of 40 °C and the standard lifetime period.	60	75	80	105	125
		(measured by change in winding resistance)			

12.4.2 Transient

When a load is suddenly applied to an AC generator the voltage will fall instantaneously to a level dependent upon the amount of load applied. The AVR will monitor this voltage dip and increase excitation to restore voltage level to nearly the original value, within a fraction of a second. Similarly, on load removal, there is a voltage over-shoot and the AVR reacts reducing the excitation (Fig. 12.6).

In certain applications, a voltage dip better than specified above may be required, for example, a 10% voltage dip on application of full load. The most effective way of achieving this is to provide a bigger AC generator. To assist in the correct selection of machines capable of this improved performance, graphs of applied load against voltage dip are given by the manufacturers. Sometimes, when only small improvements to voltage dip are required, a special winding can be designed to give the correct performance without going to a larger machine.

12.4.3 Temperature, Altitude, Humidity

Ambient Temperature

Ambient temperature can be defined as the temperature of the surrounding air at a particular location. The internationally accepted standard value for this is 40 °C. All design work and most ratings of AC generators are based on this figure.

The ambient temperature measured should be that of the cooling medium. In the case of an air cooled machine this may be higher than the surrounding air ambient temperature due to the heat generated by the prime mover within the confined space of an engine house.

It is essential that the total actual temperature does not exceed the limits set by the class of insulation used.

It follows then, that a machine operating in an ambient temperature greater than 40 °C, must be derated to ensure that total actual temperature does not exceed the specified maximum.

The converse of this is also true; that by reducing temperature a greater output can be obtained from an AC

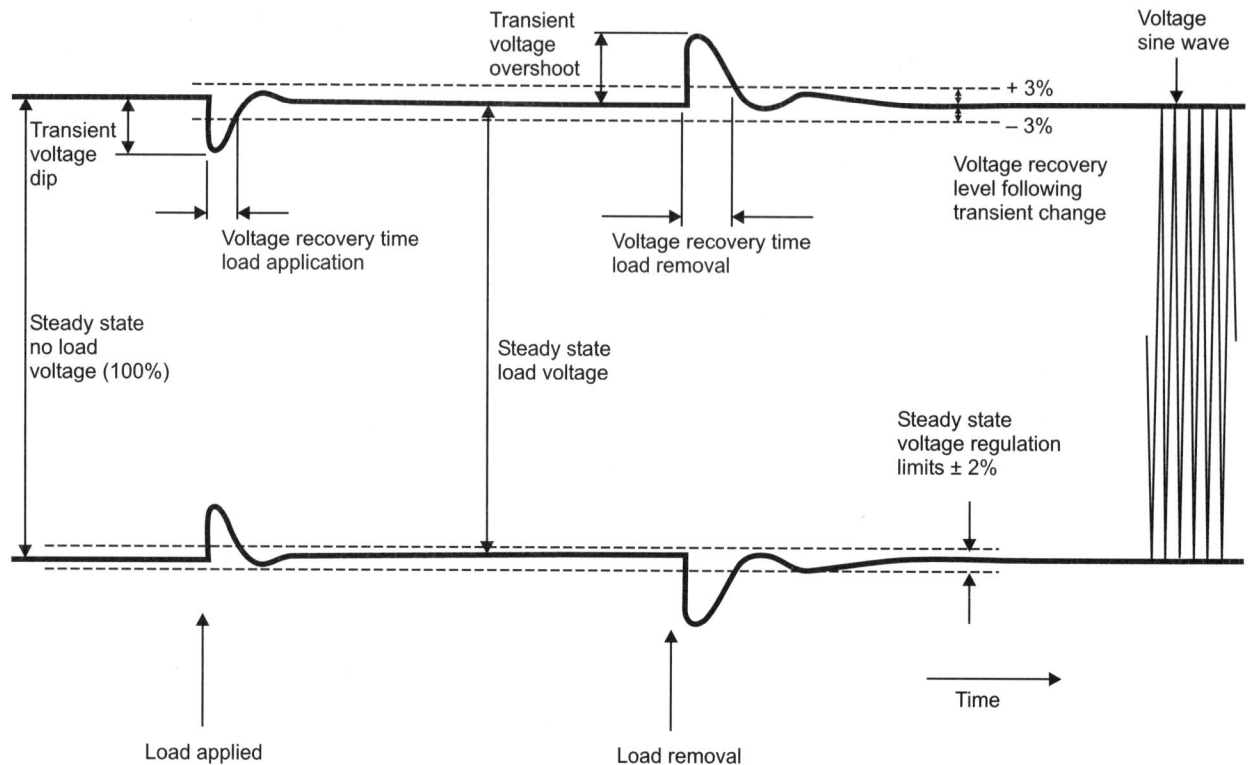

Fig. 12.6 Load *vs* voltage fluctuation

Notes:

1. Voltage waveform envelope is shown as the heavy line.
2. This diagram is not to scale and is intended to bring out the features mentioned in the text.
   a. *Transient voltage dip.* The amount of transient voltage decrease due to the sudden application of a specified load usually expressed as a percentage of the original voltage level.
   b. *Recovery time.* The length of time taken for the voltage level to recover to within 3% of the original value.
   c. *Transient voltage overshoot.* The amount of transient voltage increase due to the sudden removal of a specified load usually expressed as a percentage of the original voltage level.
   d. *Steady state regulation.* A measure of the maximum permitted steady voltage changes over a wide variety of machine conditions (includes machine hot to cold variations no load to full load applied, power factor 1.0 to 0.8 lag).

generator for the same actual temperature. This is permitted in most standards down to an ambient of 30 °C.

Outputs are normally quoted at 40 °C. These outputs must be multiplied by the factors given in Tables 12.2 and 12.3 for higher ambient temperatures.

Table 12.2 Multiplication factor for ambient temperature change

Temperature (°C)	Multiplier
45	0.97
50	0.94
55	0.91
60	0.88

Unlike ambient temperature, the converse is not permitted. No greater output is allowed from a machine operating at sea level to one operating at 1,000 m above sea level.

For altitudes above 1,000 m outputs must be multiplied by the following factors.

Table 12.3 Multiplication factor for altitude change

Altitude (m)	Multiplier
1,500	0.97
2,000	0.94
2,500	0.91
3,000	0.88
3,500	0.85
4,000	0.82

However for humidity anticondensation heaters will be sufficient to control the output power.

12.5 GENERAL COMMENTS ON VARIETY OF LOADS

Some specialised loads such as computers and transmitting stations may demand very specific performance parameters.

There are two basic conditions to check when sizing machines. The steady state condition, which is mainly concerned with normal operation of the machine within temperature rise limits; and the transient condition, which examines voltage variations when high current load is suddenly applied (e.g. during motor starting).

12.5.1 Constant or Steady State Loads

The total constant or steady state load is derived by adding together all the individual constant or steady state loads.

Load ratings can be given in current (A, amperes); horsepower (hp); kVa; or in kW. Conversion formulae are given below.

$$kW_{genr} = kW_{load} = kVa \times pf = 0.746 \times hp$$

$$= \frac{\sqrt{3} \times volts \times current}{1,000} \times pf$$

Where

kW_{genr} = output power from AC generator.
kW_{load} = input power to load in kW.

$$kVa = \frac{\sqrt{3} \times volts \times current}{1,000}.$$

pf = power factor derived from data given or assume 0.8 pf lag.
hp = input power to load in hp.
volts = rated line to line voltage.
current = rated line current.

These formulae give the output power from an AC generator based on the input power required by the load. This is true of all loads except motors. Motors are usually rated for their shaft output power. Therefore, an extra 'motor efficiency' term is required in these cases.

Thus

$$kW_{genr} = 0.746 \times \frac{hp_{(output)}}{\mu_m}$$

$$= \frac{kW_{(output)}}{\mu_m}$$

Where

$hp_{(output)}$ = motor shaft output power rating in hp.
$kW_{(output)}$ = motor shaft output power rating in kW.
μ_m = motor efficiency in per unit.

Having derived the total kW loading on the AC generator required to supply the load, then an estimate of prime mover output can be made, thus:

$$kW_{(prime\ mover\ output)} = \frac{kW_{genr}}{\mu_{genr}}$$

Where, μ_{genr} = AC generator efficiency.

By this calculation the minimum kW rating of an AC generator has been determined.

The effect of power factor must now be checked. If all the individual loads are between 0.8 pf lag and unity pf, then no adjustment to the selected frame is required. If any one load is outside these limits, then the resultant steady state power factor should be calculated by vector additions of all the loads. Again, if this resultant is within the range 0.8 pf lag and unity pf, no adjustment is required. If it is outside this range, then a derate factor must be applied to the calculated minimum kW figure and a new frame size selected, based on the new kW rating figure.

12.5.2 Transient or Motor Starting Loads

Preceding section dealt with constant or steady state loads and were concerned with the temperature rise. In this section, it is the transient performance that is discussed, in particular during motor starting. By far the most common problem in correct sizing of an AC generator concerns the starting of induction motors.

In order for a motor to start to rotate, the magnetic field of the motor must be built-up to create sufficient torque. During the starting period, a very large current is demanded form the power source. This is known as starting or locked rotor current. The level of starting current can vary greatly depending upon the motor design. Six times motor full load current can be considered as usual starting current for the most three phase motors.

In applying this level of load to an AC generator, the output voltage disruption may be quite severe. Momentary transient voltage dips in excess of 40% are possible. Consequent effects of this on other connected loads may be experienced. For example, lighting may dim or even go out altogether; other motors may stop due to insufficient holding voltage on the control contactor coils or release of under voltage protection relays.

Therefore, for most applications a maximum voltage dip ought to be specified. Generally the maximum voltage dip should not exceed 30%; and in the absence of any prescribed limit this is the figure normally assumed.

An AC generator inherently has a characteristic which resists any change in voltage. Time is required for the machine's magnetic field system voltage to rise or decay to compensate for load or output voltage changes. Such long time lags (in order of a second) are not acceptable with today's fast acting voltage control systems.

To reduce this time lag to acceptable limits (in order of 1/10th second), it is necessary to force a lot of current into the field, forcing the voltage to change quickly. This forcing current is about 3 times normal full load field current.

The majority of voltage control systems operate in this way and the effect is called 'field forcing'. For brushless machines this means that the exciter, the main field and the control system must be liberally rated for normal load conditions. Under severe overload conditions, after the initial voltage dip, the high field forcing will rapidly cause the output voltage to return near to the normal value, provided other connected loads can tolerate a high transient voltage dip, this characteristic is ideally suited for loads consisting mainly of direct on line started induction motors.

This starting torque increase should rapidly become sufficient to start the motor rotating. Therefore in this specific situation it is not the voltage dip characteristic that is of most interest as is normal, but the steady state relationship between voltage and load. It is this that must be examined to see if sufficient starting torque will become available. These overload curves are provided by the manufacturers of generating sets.

Figure 12.7 is a graph of typical performance characteristic of an induction motor when operated at a nominal constant voltage level, i.e. a mains supply giving no voltage dip on motor starting.

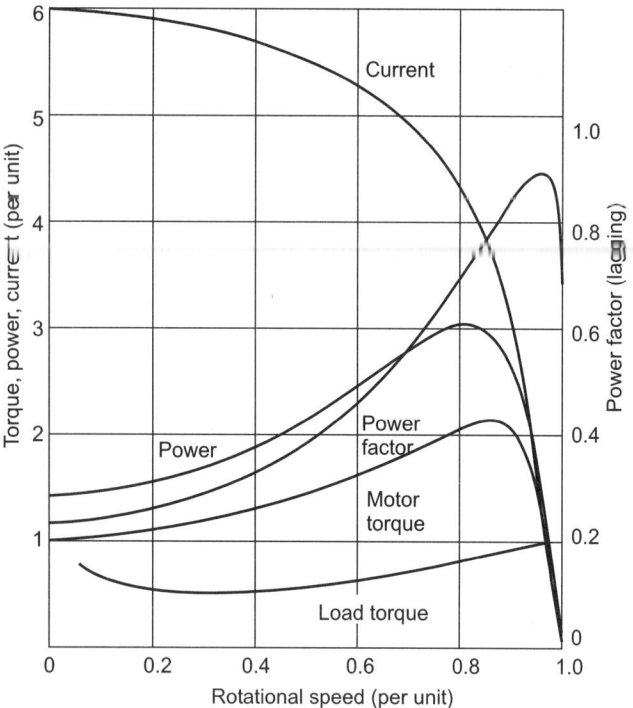

Fig. 12.7 Performance characteristic of an induction motor

From this graph, we see that the motor power factor is very low on start. The AC generator overload capabilities given are valid for any lagging power factor in the range zero to unity. If the machine has a connected base load at a high power factor (say 0.9 lag) and a motor is then started (at around 0.2 lag); then the power factor

of the two loads should be taken into account by vector addition, to correctly estimate any overload during start.

Another point to notice from the graph is the peak power requirement needed from the prime mover. This can be up to 3 times full load power (kW) of the motor being started, and this has to be supplied by the prime mover, instantaneously as the motor passes through about 80% speed. Should the prime mover be unable to develop sufficient power for this purpose, the prime motor being started will crawl low at speed determined by the balance between the power developed by the prime mover and the power required by the motor. Normally the energy stored in the fly-wheel of a diesel engine is sufficient to overcome this problem on starting motors that are small compared to the size of the prime mover. However, on starting comparatively large motors, the difficulties here are complex and it is not always appreciated that the prime mover under these circumstances can be the cause of many motor starting failures. The various methods of starting the motors are discussed in subsequent sections.

12.6 MOTOR STARTING METHODS

There are a number of ways to reduce the dip level on starting motors. Consideration of the best sequence of starting, if there are many motors or groups of motors, is vital. Simply by rearranging the sequence it is sometimes possible to utilize a smaller AC generator. It is possible to design machines specifically for motor starting duty. These are basically low transient reactance machine and may have a lower than normal steady state rating for a given size of AC generator, but this can still be economic where the load is totally direct-on-line started motors. However, simply using a larger machine to reduce voltage dip is not economical when compared with the cost of some form of reduced voltage starting.

12.6.1 Star-Delta Starter

Through two contactors or a change over switch, the windings are initially connected in 'star'; then usually after a preset time delay or when the motor has run up to a steady speed, the windings are reconnected into 'delta'. This is the normal running condition at full line volts (Fig. 12.8).

This mean that:

1. The starting voltage is reduced to $1/\sqrt{3}$ of V_L since $V_L = \sqrt{3}\ V$ line to neutral.

2. Starting current is also reduced to $1/\sqrt{3}$ of the DOL value since $I \propto T_M$.
3. Starting kVa is reduced to 1/3 of the DOL value since kVa $\propto V_L^2$.
4. Starting torque is reduced to 1/3 of the DOL value since $T_M \propto V_L^2$.

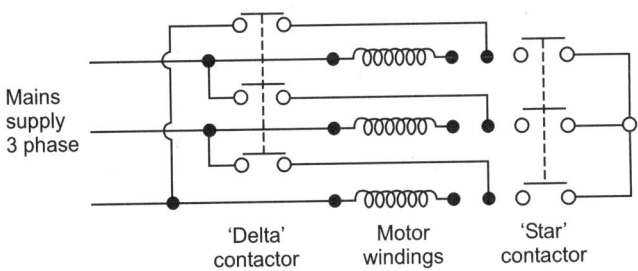

Fig. 12.8 Star-delta starting connections

Whereas, the star-delta system uses contactors and requires a 6 ends motor winding to achieve a fixed, reduced voltage motor start; this method can be used with any type of motor and uses an auto transformer with fixed voltage taps along its winding (Fig. 12.9). The basic idea is that a low line voltage is tapped off the auto transformer and fed to the motor on start. As the motor speeds up, the tap position is changed in any number of steps, increasing the line voltage until the full line voltage is applied directly across the motor terminals.

12.6.2 Direct on Line Starting

In this case the full line voltage is switched directly to the motor terminals. The motor winding normally is connected in delta. The maximum starting torque is available with this method, but a very high starting current is required (Fig. 12.10).

The voltage dip during a three phase DOL motor start should be considered. If starting current is known, this can be converted to starting kVa (multiply by voltage × $\sqrt{3}$ for a star connected motor). The voltage dip curve on the appropriate generator data sheet supplied by the manufacturer may then be used to read off the transient dip. If the dip is beyond a specified limit, a larger generator must be chosen. If starting current is not given, a good working assumption is to take 7.1 × hp rating of the motor (9.5 × kW rating) as the starting kVa value.

12.7 LOAD CALCULATION

This example was devised to consider many different load conditions simultaneously. The supply required is assumed to be 415 V, three phase, 50 Hz.

12.7.1 Load Details

Single Phase 240 V

  a. 72 fluorescent light fittings each 100 W and power factor corrected to 0.95 pf lag.

  b. 7 heaters each requiring 20 A.

  c. 4 × 5 hp motors, started simultaneously direct-on-line (DOL).

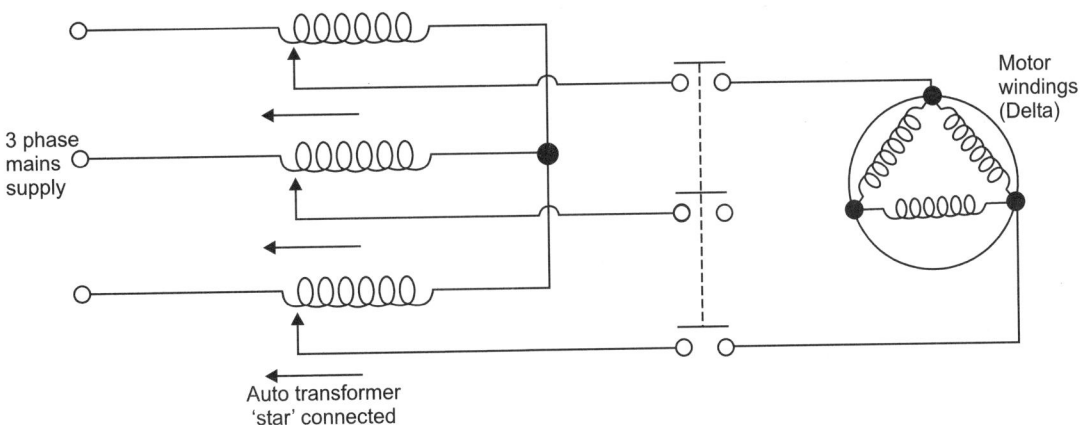

Fig. 12.9 Star-delta starting connections with auto transformer

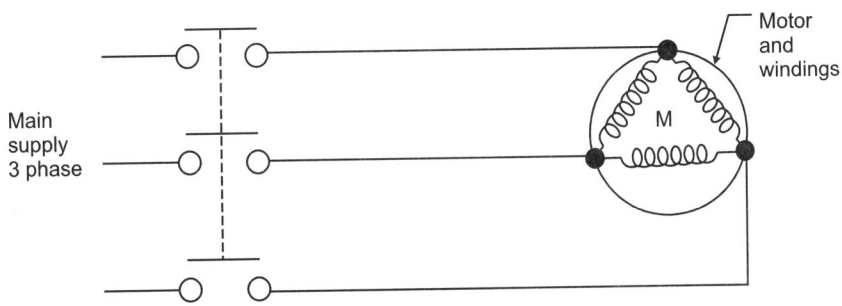

Fig. 12.10 Direct on-line starting connections

Single Phase 415 V

5 welding sets each rated at 11 A primary current 0.4 pf lag.

Three Phase 415 V

a. 3×3 hp machine tool motors started in sequence DOL.

b. 1×80 hp motor started by auto transformer at 80% tap but with a poor full load power factor of 0.6 pf lag.

c. 1×80 kW motor started star-delta with a locked rotor current of 750 A.

12.7.2 Load Analysis

Now let us consider each load in turn. For the single phase loads (a to d), we shall calculate the 'worst case' single phase loading for each item and then multiply by 3 to obtain the equivalent three phase loading. This assumes that one of the three phases always carries the highest single phase load of each item. This is unlikely to occur in practice as the total loads should be connected as balanced over the three phases as possible; however, it is a valid assumption for the purpose of AC generator sizing.

Load (a)

72 fittings over three phases can be balanced at 24 fittings per phase.

Steady State

Hence maximum single phase load is
$$= 24 \times 100 = 2.4 \text{ kW and kVa} = 2.4/0.95 = 2.5 \text{ kVa.}$$

Transient

There is no transient condition given for consideration.

Load (b)

7 heaters over three phases means one phase must supply 3 heaters.

Steady State

Hence maximum single phase load is
$$= 3 \times 240 \times 20/1000 = 14.4 \text{ kVa.}$$

Heaters can be assumed to run at unity power factor; hence load = 14.4 kW.

Transient

There is no transient condition given for consideration.

Load (c)

4 motors over three phases means that one phase must supply 2 motors.

Steady State

Hence maximum steady state phase load is from manufacturer's data; 5 hp single phase motor typical performance figures are.
Efficiency = 78%.
Power factor = 0.8 pf lag.
Steady state load is
$$= 2 \times 5 \times 0.746 = 9.6 \text{ kW} = 9.6/0.8 = 12.0 \text{ kVa.}$$

Transient

For the one phase with 2 motors starting simultaneously, from manufacturer's data; 5 hp single phase DOL starter multiplication factor = 7.1.
Hence start kVa = DOL start factor × hp rating
$$= 7.1 \times 5 \times 2 = 71 \text{ kVa.}$$

Load (d)

5 welding sets across one pair of 415 V lines given a 'worst case' of one pair of lines having to supply 2 welding sets.

Steady State

Hence loading is
$$= 2 \times 415 \times 11/1000 = 9.1 \text{ kVa} = 9.1 \times 0.4 = 3.6 \text{ kW.}$$

Transient

No transient condition details are given for consideration.

Load (e)

3×3 hp three phase motors.

Steady State

From manufacturer's data, the typical full load running performance is

Efficiency = 78%.

Power factor = 0.8 pf lag.

Then the steady state loading is
$$= 3 \times 3 \times 0.746/0.78 = 8.6 \text{ kW} = 8.6/0.8 = 10.8 \text{ kVa.}$$

Transient

For the transient or starting condition, only one of these motors will be started at any one time since they are started in sequence. In the absence of actual starting performance data maximum DOL start kVa is thus
$$= 3 \times 0.746 \times 7.1 = 15.9 \text{ kVa per motor started.}$$

Load (f)

1×80 hp motor.

Steady State

From manufacturer's data.
Efficiency = 91%.
Actual power factor (given) = 0.6 pf lag.

Then steady state loading is $80 \times 0.746/0.91$
$= 65.6 \text{ kW} = 65.6/0.6 = 109.3 \text{ kVa}.$

Transient

The transient or starting kVa for DOL start is thus
$= 80 \times 0.746 \times 7.1 = 424 \text{ kVa}.$
For auto transformer start at 80% tap start kVa
$= 424 \times (0.8)^2 = 271 \text{ start kVa}.$

Load (g)

1×80 kW motor.

Steady state

Again from manufacturer's data we obtain.
Efficiency = 91%.
Power factor = 0.91 pf lag.
Hence, the steady state loading is
$= 80/0.91 = 87.9 \text{ kW} = 87.9/0.91 = 96.6 \text{ kVa}.$

Transient

The DOL starting kVa required is calculated from the locked rotor current given; thus
$= \sqrt{3} \times 415 \times 750/1000 = 539 \text{ kVa for DOL start}.$
For star/delta start, the kVa on start is reduced to 0.33 of the DOL value. Therefore start
kVa = $539/3 = 180 \text{ start kVa}.$

12.7.3 Load Summation

We can now analyse these results and derive a final rating for a suitable AC generator. We shall make some further assumptions, and whenever such a calculation as this is attempted, it is essential that all assumptions made, are clearly detailed. In this case assume.

1. No two loads will be switched on simultaneously.
2. Generally, the loads come on in the sequence given.
3. For sizing purposes, the effort of power factor (vector addition) during transient or motor starting loads is ignored.

Since arithmetic addition of the loads will always give a larger kVa load value than the vector addition.

This total single phase loading Table 12.4 represents the worst possible case, as we have assumed that one particular phase always carried the highest unbalanced load current for each single phase load considered.

Table 12.4 Summation of single phase loads

Load (1)	Steady state			Transient
	kVa (2)	kW (3)		start kVa (5)
a	2.5	2.4	0.95	–
b	14.4	14.4	1.0	–
c	12.0	9.6	0.8	71.0
d	9.1	3.6	0.4	–
Total	38.0	30.0	0.79	

In practice it is likely that the individual loads would be more evenly distributed between the three phases. However, for the purposes of sizing and in the absence of actual load connection data, this is a valid assumption. The steady state total single phase ratings can now be multiplied by 3 to obtain the equivalent three phase rating.

Regarding the start kVa load of (c), this again is the worst case single phase surge kVa. The make up of load (c) is such that on start there is unbalanced surge kVa. Therefore, the total three phase surge kVa on starting load (c) will be less than 3 times the worst case 71 kVa. This product (213 kVa) is in turn less than the highest three phase balanced start kVa which, from the table below, is 271 kVa. Hence, this single phase start kVa may in this case be ignored when considering the three phase summation Table 12.5.

In the previous table we have assumed that all the single phase loads are applied together with the transient condition ignored as discussed above.

The highest load which must be sustained is given in col. 6 on the application of load (f), 437.8 kVa. The machine selected must always have a overload capability greater than the largest kVa value given in col. 6. Indeed the order of load switching can be vital. To achieve the minimum size of AC generator based on 'overload' performance (col. 6), the loads must be applied in descending order of start kVa's (col. 5). That is the largest start kVa load must be switched on first and the smallest

Table 12.5 Summation of three phase loads

Load (1)	Steady state			Transient	
	kVa (2)	kW (3)	pf (4) = (3)/(2)	Start kVa (5)	Total load at start (kVa) (6)
Equivalent three phase for a, b, c, d	114.0	90.0	0.79	–	–
e	10.8	8.6	0.8	15.9	114 + 15.9 = 129.4
f	109.3	65.6	0.6	271	114 + 10.8 + 271 = 395.8
g	96.6	87.9	0.91	180	114 + 10.8 + 109.3 + 180 = 414.1
Total	330.7	252.1	0.76		Worst case is 114 + 10.8 + 96.6 + 271 = 492.4

last. This is the ideal situation for load switching sequence, although rarely achieved in practice.

In this particular example, the 'overload' values (col. 6) are not significantly higher than the steady state total kVa. Therefore in this instance, it is unlikely that the switching sequence is that important. Hence, we shall consider the worst case for 'overload', that is when load (f), the largest start kVa, is switched on last. This would give a total load on start kVa as follows.

Worst case total load
$$= 114 + 10.8 + 96.6 + 271 = 492.4 \text{ kVa}.$$

12.7.4 Derate Factors

It is at this point in the calculation that any steady state derate factors must be applied. In this case the final power factor is below 0.8 pf lag, hence a derate is required. Interpolating from available data for 0.76 pf lag.

Derate factors = 0.98.

Also any derates for high ambient temperature, restricted temperature rise or high altitude must be applied. In this example we shall assume that the ambient temperature is 45 °C; that as the machines are insulated with 'class H' materials, the associated temperature rise (125 °C) is acceptable; and finally that the altitude is below 1000 m. Looking at the available data; the derate factor is 0.97 (Tables 12.2 and 12.3).

Applying these derate factors, the selected AC generator must be capable of supplying a steady state kVa requirement of

$$\frac{330.7}{0.98 \times 0.97} - 347.9 \text{ kVa}$$

It is worth noting that if the derate required for transient performance demands a greater 'kVa' size of machine, then these steady state derate factors do not need to be applied as well. In other words either the steady state derate factors or the transient performance derate factors are applied; whichever gives the larger 'kVa' requirement. It is extremely unlikely that both sets of derate factors should be applied.

12.7.5 Standard Sizing

a. Steady State

From the summation tables we arrive at a total site load requirement of 330.7 kVa, 252.1 kW, 0.76 pf lag.

Applying the steady state derate factor, the kVa requirement was changed to 347.9 kVa. It is this figure that is looked for in the standard output lists given by the manufactures.

b. Transient

Again from the summation tables the maximum start kVa transient requirement is 271 kVa. The 271 kVa transient load must be checked against the voltage dip curve for the generator selected to determine the percentage transient voltage dip.

If this level of voltage dip on starting the largest load is acceptable, then the frame size chosen also remains acceptable from a transient performance point of view.

c. Input Power

The AC generator shaft input power is the same as the prime mover shaft output power and can be estimated. The AC generator efficiency values are given in the data sheet of manufacturer.

Prime mover shaft power output

$$= \frac{\text{AC generator kW output rating}}{\text{Appropriate AC generator efficiency}}$$

12.7.6 Power Factor Correction

In general, any load with total power factor 'leading' needs careful consideration in selecting a suitable frame size of machine to supply it. Power factor correction capacitors operate at almost zero power factor leading and are used to correct the overall low lagging power factor of a complete installation to a value unity power factor but still lagging. This is usually done to reduce tariff charges for industrial consumers of electricity.

Let us look at the effect of adding a 35 kvar capacitor power factor correction bank to the load installation considered in this example.

The final steady state load figures, from the summation tables, are 330.7 kVa, 252.1 kW, 0.76 pf lag.

Hence, the kvar component = 252.1 × tan⁻¹ 0.76 = 215.6 kvar.

With the introduction of 35 kvar of capacitors at zero pf lead, the kvar component reduces to 215.6 − 35 = 180.6 kvar.

Hence, the new steady state loading figures are:
i. The 252.1 kW remains the same, as the real power requirement does not change.

ii. Power factor $= \cos\left[\tan^{-1}\frac{(180.6)}{(252.1)}\right] = 0.813$ pf lag.

iii. The kVa $= \dfrac{252.1}{0.813} = 310.1$ kVa.

Notice now that no derate is required for power factor since, with the power factor correction bank installed, the system power factor lies within the range unity to 0.8 pf lag. Therefore, the required rating must be considered at 0.8 pf lag, since this is the standard condition for rating all AC generators.

That is, the nameplate rating and site rating becomes

$$\frac{252.1}{0.8} = 315.1 \text{ kVa}; 252.1 \text{ kW}; 0.8 \text{ pf lag}.$$

Assuming the same ambient temperature, temperature rise and altitude as before, we arrive at the machine rating of

$$\frac{315.1}{0.97 \times 1.0 \times 1.0} = 324.8 \text{ kVa}.$$

Notice the addition of the power factor correction bank can reduce the size of AC generator required to supply these loads under steady state conditions.

The major problem with power factor correction banks is that, when all other loads are switched off, the power factor correction banks normally remain connected. This represents a purely zero power factor leading load to the AC generator. One effect of having a comparatively large bank connected to an AC generator is for the terminal voltage of the AC generator to rise dramatically. Voltages in excess of 500 V have been recorded from a nominal 415 V machine in such cases.

However, as a general guide, providing the capacitor bank rating not greater than about 10% of the AC generator rating, then no real problems of this nature should be encountered.

In this case, we have

$$\frac{35 \text{ kvar capacitor bank} \times 100}{325 \text{ kVa}} = 10.8\%.$$

Example

To illustrate motor rating vis-à-vis AC generator rating.

Let the supply be 230 V three phase 50 Hz and the load is always a total of 80 kW of motors. However, at different times the actual loads can be one of the following:

a. 4 × 20 kW motors started in sequence DOL.

b. 2 × 40 kW motors started in sequence DOL.

c. 1 × 80 kW motor started DOL.

Voltage dip not to exceed 30%.

Consider Case (a)

In the absence of actual starting data.

Consider maximum DOL start kVa = 9.5 × kW rating = 190 kVa for each motor.

Typical power factor = 0.89 pf lag.

Typical efficiency = 88%.

Now with 3 motors running run

$$\text{kVa} = \frac{3 \times 20 \text{ kW}}{0.89 \times 0.88} = 76.7 \text{ kVa}.$$

Starting the 4th motor gives start kVa = 190 kVa.

Total kVa (neglecting effort of power factor) = 266.7.

For voltage dip limited to 30% or better, look at the voltage dip curves in the individual data sheets of manufacturer.

Check the steady state load to ensure that the selected generator is large enough.

$$\text{Steady state load} = \frac{4 \times 20 \text{ kW}}{0.88} = 91 \text{ kW}$$

Check overload capability.

Maximum demanded by load = 250.7 kVa. Check against the overload capability of the generator selected.

Consider Case (b)

Going through a similar procedure as case (a) we get 40 kW motor DOL start kVa = 380 kVa. Typical power factor = 0.91 pf lag. Typical efficiency = 91%.

Now with one motor running; kVa = 48.3

Starting second motor; kVa = 380

Total (neglecting effect of power factor) = 428.3.

Consider Case (c)

80 kW motor start kVa DOL = 656 kVa. Typical power factor = 0.92 pf lag. Typical efficiency = 92%.

Again the generator can be selected by using the dip curves of the data sheet of manufacturer.

Note: From this it is concluded that a single motor supplied by a single AC generator may demand a much larger AC generator than may be thought, since it must be capable of supplying the large kVa required by the motor starting.

12.8 PARALLEL OPERATION OF AC GENERATORS

12.8.1 Introduction and Theory

No other aspect of the AC generator operation causes more misunderstanding than the parallel operation of two or more AC generators.

This section will explain the reasons for paralleling, the method by which it is carried out, the setting up procedures and possible problems which may arise.

Paralleling operation may be necessary for the following reasons.

i. To increase the capacity of an existing system.

ii. Size and weight may preclude the use of one large unit.

iii. Allow noninterruption of the supply when servicing is required.

To parallel AC generators satisfactorily, certain basic conditions have to be met. These are as follows.

i. All systems must have the same voltage.

ii. All systems must have the same phase rotation.

iii. All systems must have the same frequency.

iv. All systems must have the same angular phase relationship.

v. Systems must share the load with respect to their ratings.

A minimum amount of instrumentation is required to ensure the above information is satisfactorily monitored, comprising an ammeter, a wattmeter and a reverse power relay. A voltmeter is not specified for each system because it is preferred to use one voltmeter on the distribution or synchronising panel with a selector switch for each system. This eliminates any possible meter inaccuracies.

A reverse power relay, although contributing a larger proportion of the instrumentation cost, is essential as any engine shut down, from low oil pressure, over-temperature, etc. will result in other system motoring the failed set with consequent overload on the remaining system.

Only one frequency meter is required with the facility of being switched to the bus bar of the incoming system.

A synchroscope and/or lights, are required to detect the angular phase displacement. If lights are used, three different connections are possible. For paralleling with the lights dim, they must be connected across like phases or like lines (single phases), i.e. U-U, V-V or L1-L1. For paralleling with lights bright they should be connected across unlike phases, i.e. U-V, etc.

If a three lamp system is used with the lamps connected across U-W, V-V and W-U the lamps will 'rotate' and give the indication which machine is running fast. Synchronism is reached with two lamps bright and one dark and in some respects this connection gives a closer visual indication of the point of synchronism. Note the lamps should be rated for at least twice the machine voltage or it will be necessary to connect two or three in series. A preferred method is a resistor/lamp combination.

Figure 12.11 illustrate the connections.

12.8.2 Load Sharing

The most important aspect of parallel operation is load sharing. The total load, comprising a kW or active component and a kvar or reactive component, must be shared by the systems in proportion to their normal ratings.

The kW component is adjusted by purely mechanical means and requires relatively fine speed control of the prime mover. It is advisable to fit a limited range governor to avoid 'misuse' of the speed control.

The kvar component is a function of the AC generator excitation. When machines are in parallel, the magnitude of the field excitation cannot directly influence the output voltage. It does, however, adjust the internal power factor at which a particular machine operates. For instance, an overexcited AC generator will draw lagging current whereas an underexcited AC generator will draw leading current. If a difference in excitation exists, then circulating currents will flow, limited by only the internal machine reactance. This current will appear as a zero pf leading or lagging current, dependant on the machine excitation and will

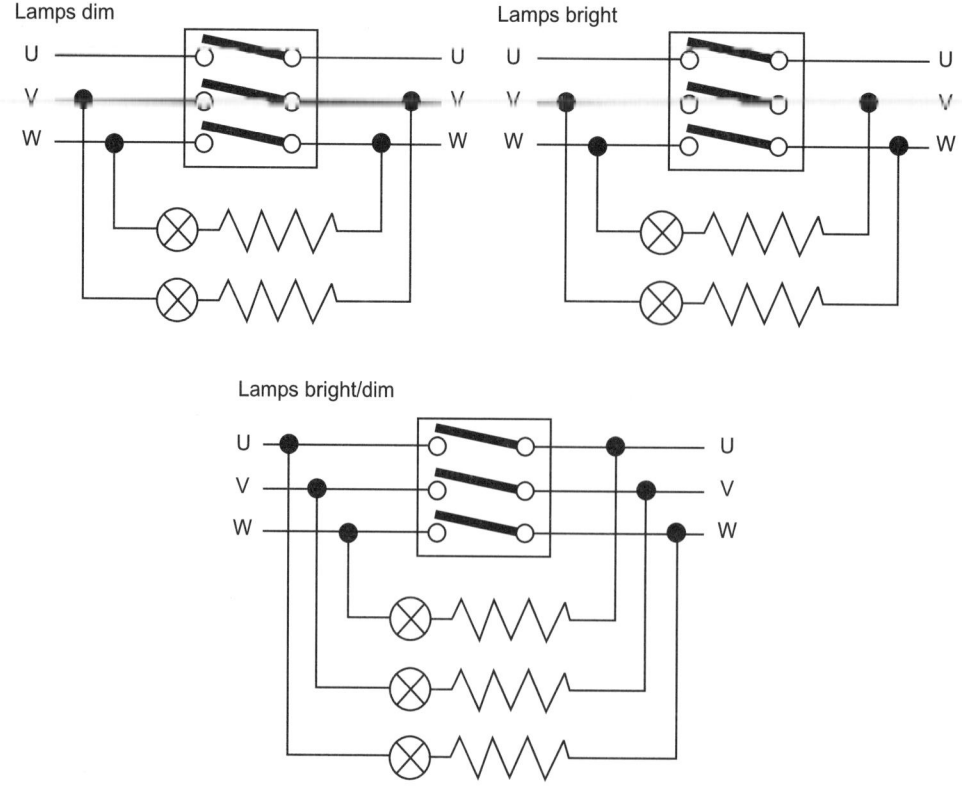

Fig. 12.11   Lamp connections for angular phase displacement for parallel operation of AC generators

either subtract or add to the total current that each machine supplies.

Reactive current, either leading or lagging is, by virtue of the 90° phase displacement, quite commonly described as being in quadrature. Means must, therefore, be provided to sense this reactive current and limit is to an acceptable level. Hence, a quadrature droop kit. This comprises a current transformer (CT) connected to the AVR burden resistor. The following paragraphs and vector diagrams describe and illustrate its action (Figs. 12.12 and 12.13).

On a 2 phase sensed generator, the reference or sensing voltage is obtained from the two phases U and V. The CT is wired in W phase and its output dropped across the burden resistor within the AVR. The burden resistor is connected such that the voltage produced across it adds vectorially with the sensing voltage.

Examination of the vector diagram shows that at 1.0 pf, the small voltage produced across the burden resistor adds at right angles to the sensing voltage. This produces little change in the sensing voltage and therefore little change in the terminal voltage.

The effect is more marked at 0.8 pf but only marginally so at zero pf. However, the additional voltage is in phase with the sensing voltage, producing a much larger change.

The artificially high voltage now seen by the sensing circuit causes the machine excitation to be reduced. Sufficient voltage is produced to ensure that when full load current at zero pf is flowing, the terminal voltage will droop by 5%.

This should be sufficient to limit the circulating current to a satisfactory level which in any case should not exceed 5% of the normal maximum current.

In short, the machine supplying more than its share of the reactive current has its excitation reduced. As an under excited AC generator draws less reactive current, the excitation balance and hence power factor balance is restored.

12.8.3 Setting up Procedure

A simplified step by step procedure is given in pages to follow.

Stable parallel operation and accurate load sharing between no load and full load can only be obtained when the initial voltage settings and the droop kits are correctly set up. It is also most important that the governor characteristics are similar otherwise incorrect

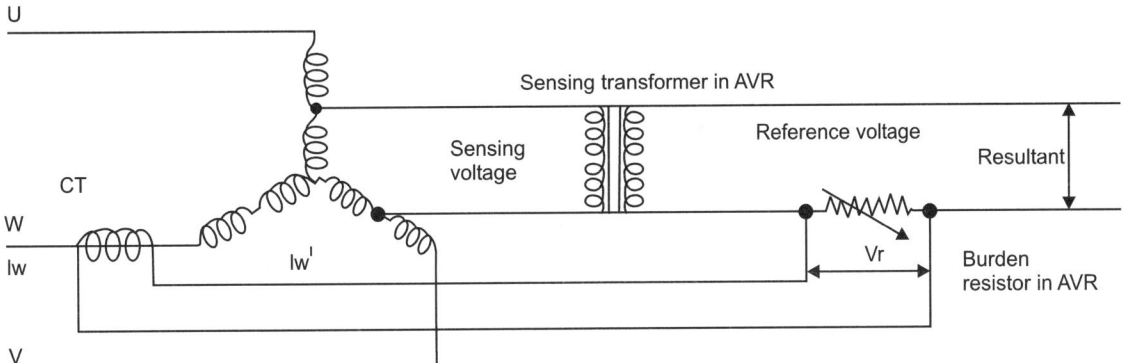

Fig. 12.12 Droop kit wiring diagram

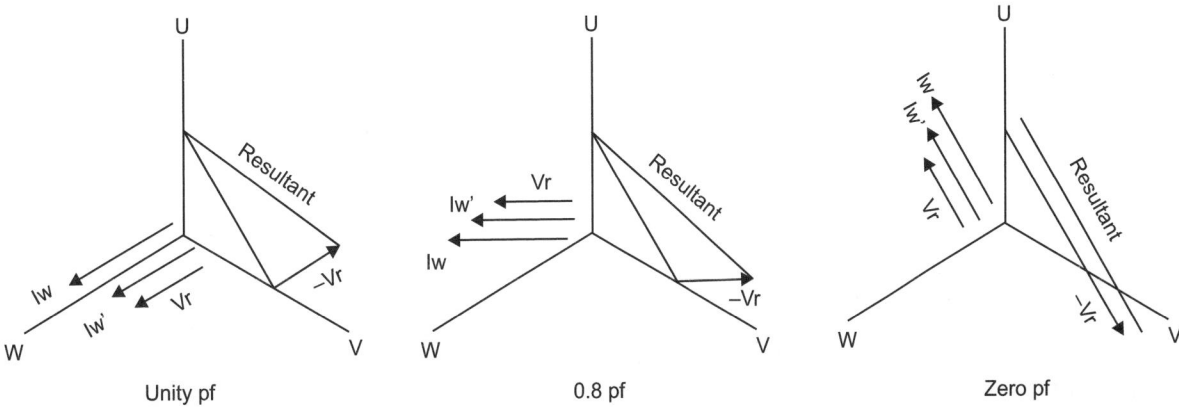

Fig. 12.13 Vector diagrams

kW load sharing can result when either increasing or reducing load.

To check the no load voltage settings, run each machine singly at the normal no load frequency, i.e. 51.5 Hz for 50 Hz operation. The rated voltages should now be set to within 1% of each other.

12.8.3.1 Quadrature Droop Equipment

The most important aspect of initial setting up procedure concerns the droop kit. Most of the troubles allied to poor parallel operation originate from droop kits. They are either incorrectly adjusted for the level of voltage droop or are incorrectly connected such that a rising voltage characteristic is obtained. If a machine is specified for parallel operation at the time of ordering, then the droop kit supplied will have been set up on test. So long as the terminal markings and connections are followed, no problems should result.

Note: Reversal of the transformer or reversal of the secondary connections to the transformer will result in a rising voltage characteristic which is completely unstable during parallel operation.

Where machines have to be modified to incorporate a droop kit at a later date, ensuring a drooping voltage characteristic appears to be of the greatest difficulty. As previously stated, the droop kit is correctly adjusted when the terminal voltage droops 5% with a zero pf or reactive current equal to the full load current being supplied.

12.8.3.2 Setting the Droop Circuit

Remember: The generator droop kit can control only the kvar sharing and circulating current. The sharing of kW load is determined by engine governors.

The droop DT on three phase machines is connected to the droop input terminals S1-S2 and is adjusted by the droop potentiometer which is mounted on the AVR circuit board. The droop can be increased by turning the potentiometer in a clockwise direction.

On single phase machines, the droop CT is connected across a burden choke. The droop is varied by changing tapping on the burden choke.

For both the two phase and three phase sensed machines, the operation is the same, i.e. any circulating current between machines produces a voltage across the burden resistor or choke which directly adds to or subtracts from the sensing voltage fed into the AVR. This makes the excitation system sensitive to circulating currents and ensures correct sharing of the kvar load. The larger droop voltage is set; the more flexible becomes the excitation system to reduce circulating currents and ensure kvar load sharing. In most cases a 5% droop on the output voltage at full load zero pf lag is satisfactory. This droop is only measurable with the machine running alone (i.e. not parallel). This setting

degrades the voltage regulation by about 1% at full load unity pf and about 3% at full load, 0.8 pf. When a unity power factor load (kW) is applied, the voltage produced across the burden resistor/choke adds vectorially at right angles to the sensing voltage and has a minimal effect. When the machines are run individually, the droop circuit can be switched out by short-circuiting auxiliary terminals S1 and S2 to obtain the normal close regulating characteristics of the machine.

Generally a zero power factor load will not be available. A 5% droop at zero pf can be set accurately by measuring the voltage across S1 and S2 on the auxiliary terminal block.

The optimum position for the 'droop' setting potentiometer on the AVR can be found as follows:

1. Apply full load amperes to machine, at any power factor including 1.0.
2. Measure voltage across the AVR terminals S1-S2, which is also the output from the CT.
3. Using the following formula, calculate the position of the droop potentiometer as a percentage rotation from *fully anti-clockwise*.
4. Position of droop potentiometer (0–100%)

$$= \frac{C \times 100}{\text{Voltage across S1-S2}}$$

C is a constant for various AVRs the value of which varies from 0.7 to 2.0.

Note: If the formula produces a figure of more than 100%, check that you have the correct CT for the size of the load current.

12.8.3.3 Single Running Operation of Machines with Parallel Droop

The voltage is required to droop at least 2½% for satisfactory load sharing in parallel. If better regulation is required for single machine operation, a shorting switch must be fitted across the parallel droop current transformer.

This switch should be mounted on the instrument panel, clearly marked *parallel running*, in the open circuit position.

12.8.4 Step by Step Setting Procedure

The points detailed below are meant as a general guide only. If any doubts exist as to the reason for the various tests, further reference should be made to the preceding notes. Obviously, all machines must be correctly wired in accordance with the appropriate connection and wiring diagram.

a. Run no. 1 AC generator on no load at rated speed. Check AC generator voltage and adjust where necessary.
b. Check phase rotation of no. 1 AC generator.

c. Run no. 2 AC generator and proceed as items (a) and (b).

d. With nos. 1 and 2 AC generators running on no load, switch in synchroscope or lights.

e. Adjust speed until synchroscope rotates very slowly or lights slowly brighten and dim.

f. Check finally that voltages are equal or within 1% of each other. Adjust as required.

g. Close breaker at synchronism; observe ammeters for circulating current. If in excess of 5% recheck no load voltage settings and droop kits for polarity (reversal).

h. Increase load until full load appears on each AC generator when in parallel. Some adjustment to one engine governor may be required to ensure balanced kW meter readings.

i. Check the ammeter readings with the kW meters equal. They should be within 5% of each other.

j. If the ammeter readings are outside 5%, the machine with highest current is over-excited and therefore requires more droop to compensate. Increase the droop resistance.

k. With full load current on each AC generator, reduce the external load in 20% increments. At each loading, observe kW meter and ammeter readings down to 20% full load. Any variation of either instrument beyond 5% of each other requires correction.

l. Unequal kW sharing implies a faulty prime motor, most likely the governor.

m. Unequal ammeter readings at the full load end of the range imply incorrect levels of droop.

n. Unequal ammeter readings approaching the no load condition imply incorrect voltage settings.

12.8.5 Working Procedure

The most likely procedure that occurs in practice concerns the paralleling of additional machines to already loaded sets. For instance, if a set is supplying a load equal to 75% of its output and further load is anticipated, the resident engineer may decide to spread this load over two sets. A procedure somewhat on the lines of the following is required.

The incoming set is started and run at the no load frequency. The synchroscope/light switch is closed connecting the incoming machine and the bus bar via the synchroscope or lights. As the incoming machine is fast, the synchroscope will rotate in the 'fast' direction or the lights will brighten and dim at a rate dependent on the frequency difference. The speed of the incoming machine should be reduced by actuating the motorised governor in the slow direction.

When the frequencies are nearly equal, the speed of rotation of the synchroscope or the changes in brilliance of the lights will slow enough to enable the set contactor to be closed when the voltages are in synchronism. This will be at the twelve O'clock position on the snychroscope or with lights bright or dim dependent on which connection is used.

The incoming machine may now take its share of the load, the governor control should be held in the speed raise position until the desired load is indicated by the kW meter and ammeter. Conversely, if too much load is applied it can be reduced by holding the governor control in the speed lower position. It is most important that the total load be shared in respect of their normal ratings and the meter readings should be compared with the nameplate data.

In any event, unequal load sharing requires correction to avoid mechanical problems which occur when diesel engines are run light for any considerable time.

It is important to differentiate between unbalanced loadings caused simply by the operator failing to spread the load equally over the two sets, and by circulating currents unbalancing the ammeter reading.

For example: Consider two—100 kVa AC generators in parallel with no circulating currents supplying 150 kVa 0.8 pf.

With load distributed equally, meter readings would appear as follows.

	Volts	AMP	kW	kVa	pf
Machine no.1	400	108	60	75	0.8
Machine no. 2	400	108	60	75	0.8

If the load were distributed unequally, again with no circulating currents, the following figures could appear.

	Volts	AMP	kW	kVa	pf
Machine no. 1	400	144	80	100	0.8
Machine no. 2	400	72	40	50	0.8

If now the same unequally distributed load is being supplied, but circulating currents are present, meter readings something on the lines of the following could be observed.

	Volts	AMP	kW	kVa	pf
Machine no. 1	400	192	80	133	0.6 lag.
Machine no. 2	400	62	40	43	0.93 leading

Machine no. 1 is now supplying 133 kVa at 0.6 pf considerably in excess of its normal rating. Continued operation under this loading would cause the overload protection circuit to trip or the main stator and the rotor to fail. Machine no. 2 is operating so under-excited that it is operating at leading power factor and at a much reduced kVa.

This will in no way damage no. 2 AC generator, but in itself implies that no. 1 AC generator is very heavily overloaded.

A leading power factor condition is particularly difficult to detect except when individual power factor meters are fitted. The normal instrumentation of ammeter, voltmeter and kW meter cannot indicate such a load condition.

12.8.6 Difficulties

Some of the paralleling problems that can occur are detailed below. Probable causes are also shown.

a. Oscillating kW meter, ammeter and voltmeter.

Cause: Engine governing; replace by known serviceable unit.

b. Unbalanced ammeter readings. kW meters balanced and stable.

Cause: Circulating current through incorrect voltage settings, droop kit connections reversed or insufficient droop.

c. Unbalanced ammeter readings on no load or rapidly rising currents as soon as the breaker is closed.

Cause: Incorrect voltage settings or droop kit connections reverse.

d. Unbalanced kW and ammeter readings as load increased or decreased.

Cause: Dissimilar governor speed regulation.

e. Unbalanced ammeter readings as load increased. kW meters balanced.

Cause: Droop circuit settings not identical or one droop kit reversed.

Apart from the above problems, certain peculiarities may exist which are in no way detrimental to the operation of the sets. They may, however, confuse the operator into thinking a fault exists. The most common query results from voltage oscillation during the initial paralleling procedure. When an additional set is being connected to the bus bars with the synchroscope/lights switch in the on position, a point may be reached where the incoming machine voltage starts to fluctuate. This only occurs when the frequency difference is at its greatest. As the frequencies approach each other, no further instability is noticed. This is not, however, a function of the stability circuit within the AVR, but relates to 'pickup' problems associated with the switchboard wiring.

12.8.7 Neutral Interconnections

It should be noted that paralleling of all the system neutrals can under certain circumstances lead to overheating or possible stator burn outs.

This is particularly evident when machines of a dissimilar type/manufacture are paralleled. Differences in generated wave shape may cause large harmonic circulating currents through the inter-connected neutrals.

The neutrals of dissimilar machines must, therefore, never be connected. On the other hand, neutrals of like machines may be connected.

13

Pumping System

13.1 INTRODUCTION

This chapter is about the detailed method to select the appropriate pump for the specific application of the user so as to give required discharge with least consumption of power. The selection of pump mainly depends on the pumping system being followed. The designer is required to determine quantity of flow, head to be achieved and the friction losses. The determination of type of pump, i.e. reciprocating, centrifugal, vertical or horizontal along with specifications, drives and accessories are integral to the process of pumping system design.

13.2 FLUID PROPERTIES

Before any pump is selected, it is very necessary to determine the type of fluid which is to be pumped and then determine its density, specific weight, specific gravity and viscosity. These properties of the fluid are important as they affect the selection of material of impeller, power required to rotate the impeller and determining whether flow is linear or turbulent.

Specific weight (γ) of a substance is the weight per unit of volume. It is expressed in kg per cubic meter or pound per cubic meter or pound per cubic feet.

Density (ρ) is the mass of the substance per unit of volume. The density and specific weight are related to each other through a factor of gravity 'g', i.e.

$$\rho = \frac{\gamma}{g}$$

viscosity (μ) is the property of a fluid to resist the motion imparted to it. The viscosity acts opposite to the direction of flow and is the ratio of shearing stress between adjacent layers of fluid to the rate of change of velocity with distance perpendicular to the direction of motion.

$$\mu = \frac{\tau}{\left(\dfrac{dv}{ds}\right)}$$

τ = shearing stress

v = velocity of fluid

s = distance perpendicular to flow

Absolute viscosity (or dynamic viscosity) is measured in grams per centimeter second and its unit is poise. It is commonly expressed in the smaller unit called centripoise (1/100 of a poise). Water viscosity is given mostly in kinematic viscosity (v). The relation between absolute viscosity and kinematic viscosity is

$$v = \frac{\mu}{\rho} = \frac{\mu g}{\gamma} \text{ and the unit is stokes.}$$

13.3 TYPES OF FLOW THROUGH PIPES

Osborne Reynold identified that the flow through a pipe can be either laminar or turbulent. When the flow of the fluid is in a straight line with little or no motion taking place perpendicular to the direction of flow, then the flow is called laminar flow. In this case, the maximum velocity is at the center of the pipe and almost zero near the inner walls. Reynold determined that a relation exists between density, velocity, diameter and viscosity which is a non dimensional number and determines if the flow is laminar or turbulent. The relation is:

$$R_e = \frac{dv\rho}{\mu}$$

According to Reynold, if the Reynold number is less than 2000 then the flow is laminar and it is turbulent if Reynold number is more than 4000. The flow is not predictable in between these two values and this is called critical zone.

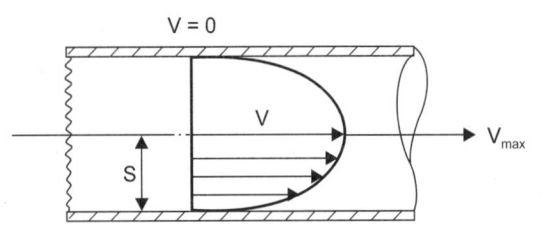
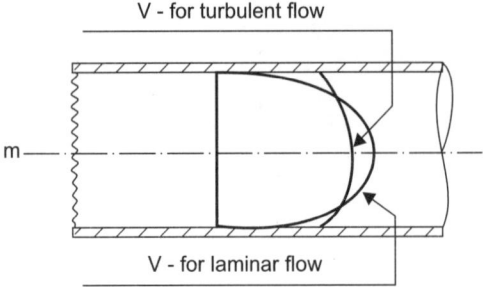

Fig. 13.1 Velocity profile in pipe flows

In turbulent flow, the motion is not steady and changes in both magnitude and direction and results in irregular motion. The turbulent flow is not desirable as it causes wear, erosion, noise and excessive pressure on the system.

As seen in Fig. 13.1, the velocity distribution is parabolic in case of laminar flow. But for most calculations, the mean velocity is used and determined as below:

$$\text{Mean velocity} = \frac{\text{Rate of flow}}{\text{area}}$$

13.4 HYDRAULIC TERMS AND EQUATIONS

13.4.1 Hydraulic Mean Depth (*m*)

It is the ratio of area (A) of flowing fluid to wetted perimeter (S).

Thus $\qquad m = \dfrac{A}{S}$

$$= \frac{\pi}{4} d^2 \times \frac{1}{\pi d} \text{ (for circular pipes)}$$

$$= \frac{d}{4}$$

or $\qquad d = 4m$

When the fluid is flowing in a pipe, it may not be running full. In such situations, the friction loss will be proportional to wet perimeter and will thus effect the determination of Reynold number.

13.4.2 Bernoulli's Theorem

The energy can neither be created nor destroyed and is called the principle of conservation of energy devised by Bernoulli and used in fluid flow through a pipe. According to Bernoulli, the total energy at a point in a pipeline comprises of pressure energy, kinetic energy and potential energy and remains constant (though friction loss takes place but is ignored being very small).

Bernoulli's equation for a pipe at point 1 and 2 is:

$$\frac{p_1}{\rho} + \frac{v_1^2}{2g} + z_1 = \frac{p_2}{\rho} + \frac{v_1^2}{2g} + z_2 + h_f$$

where $p_1 p_2$ = pressure at Point 1 and 2

ρ = density of fluid

v = mean velocity

g = acceleration due to gravity

z_1, z_2 = elevation of point 1 and 2 above a datum level

h_f = friction loss between point 1 and 2

13.4.3 Continuity Equation

In a steady flow, the rate of flow of fluid at any point is constant. The weight of fluid flowing per sec is

$$\text{Weight} = \gamma AV \text{ (constant)}$$

This is known as continuity equation. Since the specific weight (γ) remains constant, the equation becomes:

$$Q = AV = \text{constant}$$

Thus, as the cross sectional area of the pipe increases, velocity of fluid decreases and vice-versa.

13.4.4 Loss of Head in Pipe Flow

There are various formulae available for use in calculating velocity, loss of head in pipe flow. Hazen-William formula is used for pressure conduits, Manning's formula is used for free flow conduits.

a. Hazen-William formula:

$$v = 0.849 \, c \, r^{0.63} \, s^{0.54}$$

$$v = 4.567 \times 10^{-3} \, c \, d^{0.63} \, s^{0.54} \text{ (for circular pipes)}$$

$$q = 1.292 \times 10^{-5} \, c \, d^{2.63} \, s^{0.54} \text{ (for circular pipes)}$$

b. Manning's formula:

$$v = \frac{1}{n} r^{0.67} s^{0.5}$$

$$v = \frac{3.968 \times 10^{-3} d^{0.67} s^{0.5}}{n} \text{ for circular pipes}$$

$$q = \frac{8.661 \times 10^{-7} \times d^{2.67} \times s^{0.5}}{n} \text{ for circular pipes}$$

where:

q = discharge in m^3/hr

d = dia of pipe in mm

v = velocity in mps

r = hydraulic radius in m

s = slope of hydraulic gradient line

c = Hazen-William coefficient

n = Manning's coefficient of roughness

These empirical formulas were developed as the simple $q = av$ method did not give accurate velocity and flow on account of various factors like wetted parameter, type of flow, type of material used thus incorporating different friction values. Darcy Weisbach suggested first dimensionless equation for pipe flows to determine head loss as below:

$$h = \frac{flv^2}{2gd}$$

h = Head loss due to friction

f = dimensionless friction factor

l = length of flow in meters

v = velocity in mps

d = diameter of pipe in meters

g = acceleration due to gravity

when flow is laminar and $Re \le 2000$, then

$$f = \frac{64}{Re} \text{ for pipes}$$

$$f = 0.89 \times \frac{64}{Re} \text{ for square channel}$$

$$f = 1.5 \times \frac{64}{Re} \text{ for annulus}$$

But when flow is turbulent, f depends on Re and relative roughness (k/d) of surface of pipe wall, where k is absolute roughness of pipe surface.

The valves of c, n and f are given in Tables 13.1 to 13.3 respectively. The coefficient of roughness depends on Reynolds number (hence on velocity and diameter) and relative roughness (d/k). For Reynold numbers greater than 10^7, the friction factor 'f' (and hence 'c' value is relatively independent of diameter and velocity. However, for normal ranges of Reynolds number of 4000 to 10^6, the friction factor 'f' (and hence the c value) do depends on diameter, velocity and relative roughness.

Since water supply pipes are always running full, Darcy Weisbach formula is used to calculate frictional head loss. However in drainage, sewerage, storm water system, the Hazen-William, Mannings and Colebrook formulas are used.

Table 13.1 Hazen-Williams coefficient 'c'

Pipe Material	Recommended 'c' values	
	New pipe	Design purpose
Unlined Metallic Pipes		
Cost Iron, Ductile Iron	130	100
Mild Steel	140	100
Galvanized Iron above 50 mm dia.*	120	100
Galvanized Iron 50 mm dia and below used for house service connections.*	120	55
Centrifugally lined Metallic pipes		
Cast Iron, Ductile Iron and Mild Steel Pipes Lined with cement mortar or Epoxy		
Up to 1200 mm dia	140	140
Above 1200 mm dia	145	145
Projection Method Cement Mortar Lined Metallic Pipes		
Cast Iron, Ductile Iron and Mild Steel Pipes	130*	110**
Non Metallic Pipes		
RCC Spun Concrete, Prestressed Concrete		
Up to 1200 mm dia	140	140
Above 1200 mm dia	145	145
Asbestos Cement	150	140
PVC, GRP and other Plastic pipes	150	145

* For pipes of diameter 500 mm and above; the range of c values may be from 90 to 125.

** In the absence of specific data, this value is recommended, However, in case authentic data is available, higher value upto 130 may be adopted.

Table 13.2 Manning's coefficient of roughness 'n'

Type of lining	Condition		n
Glazed coating of enamel timber	In Perfect Order		
	a.	Plane boards carefully laid	0.010
	b.	Plane boards inferior workmanship or aged	0.014
	c.	Non-plane boards carefully laid	0.016
	d.	Non-plane boards inferior workmanship or aged	0.018
Masonry	a.	Neat cement plaster	0.013
	b.	Sand and cement plaster	0.015
	c.	Concrete, Steel troweled	0.014
	d.	Concrete, wood troweled	0.015

(Contd.)

Table 13.2 (contd...)

Type of lining	Condition	n
	e. Brick in good condition	0.015
	f. Brick in rough condition	0.017
	g. Masonry in bad condition	0.020
Stone work	a. Smooth, dressed ashlar	0.015
	b. Rubble set in cement	0.017
	c. Fine, well packed gravel	0.020
Earth	a. regular surface in good condition	0.020
	b. In ordinary condition	0.025
	c. With stones and weeds	0.030
	d. In poor condition	0.035
	e. Partially obstructed with debris or weeds	0.050
Steel	a. Welded	0.013
	b. Riveted	0.017
	c. Slightly tuberculated	0.020
	d. Cement mortar lined	0.011
Cast iron and ductile iron	a. Unlined	0.013
	b. Cement mortar lined	0.011
Asbestos cement		0.012
Plastic (smooth)		0.011

Table 13.3 Values of absolute roughness 'K' to determine horizontal factor 'F'

Sl. No.	Pipe Material	Value of 'k'	
		New@	Design
1.	Metallic Pipes—Cast iron and ductile Iron	0.15	
2.	Metallic pipes—Mild steel	0.06	
3.	Asbestos cement, cement concrete, cement mortar or epoxy lined steel, C I and D I pipes	0.025	0.025
4.	PVC, GRP and other plastic Pipes	0.003	0.003

When the fluid flowing in the pipe is viscous, then the head loss can be calculated using the following formula:

$$h_v = \frac{k_v L v^2}{2gd}$$

where k_v is coefficient for the effect of viscosity.

13.4.5 Loss of Head in Valves and Fittings

Having calculated loss of head in straight run of pipe without any valves or bends, now is the time to determine friction loss in valves and bends.

Valves offer resistance to flow due to swinging plates, globe or stem of the valve and is considered a friction loss. Various types of valves offer different resistance to flow and the valves analyzed over a period of time is listed in Table 13.4.

Table 13.4 Values of coefficient of friction (K_f) for valves

S.No.	Type of valve	Co-efficient of friction for Valves (K_f)
1.	Gate Valve without contraction	0.2
2.	Globe Valve	
	CI with vertical stem	3.8–4.0
	Forged Steel –Vertical stem (upto 50 mm NB)	6.5
	Cast Steel- Vertical stem (above 50 mm NB)	3.8–4.0
3.	Y- Valves	1.3–1.6
4.	Diaphragm valve (smooth Bore)	1.5–2.0
5.	Diaphragm Valve (Web Bore)	3.0–3.5
6.	Angle Valve	2.8–3.0
7.	Non-Return Valve (Vertical Stem)	5.0–6.0
8.	Non-Return Valve (Inclined Valve)	2.5–3.0
9.	Suction Strainer with foot valve	2.2–2.5
10.	Swing check Valves (Horizontal)*	
	NB 40 mm and velocity 4.4 m/s	0.5
	NB 50 mm and velocity 3.1 m/s	0.4
	NB 65 and velocity 3.9 m/s	0.3
	NB 80 mm and velocity 2.3	0.7
	NB 100 mm and velocity 1.6 m/s	0.6
	NB 125 mm and velocity 2.8 m/s	1.0
	NB 150 mm and velocity 1.6 m/s	0.9
	NB 175 mm and velocity 2.2 m/s	0.5

* The value of coefficient varies with velocity of flow and pipe position (horizontal or vertical)

The friction loss due to bend in pipes is a major contributor of the total loss of head. The loss due to bends is calculated by converting it into equivalent length of pipe fitting. Thus total loss of head will be calculated as if frictional loss is caused due to flow in straight run of pipe. The equivalent lengths will be added to the length of straight run to obtain total equivalent length of pipe system. This length is then used in Darcy Weisbach or any other formula to determine the total loss of head.

Table 13.5 lists equivalent length of pipe fittings for various bend angles while Table 13.6 lists equivalent length of various type of valves.

13.4.6 Limitation and Comparison of various Formulas

Here we describe the limitation of various formulas devised by various researchers:

a. **Hazen William formula:** The constant 'c' for Hazen William formula has been calculated considering hydraulic radius of 1 foot and friction slope of 1/1000. However, for other diameters and friction slopes, this formula results in an error of upto ± 30% in the calculation of velocity and ± 50% in calculation of frictional loss. This formula is

Table 13.5 Equivalent length of pipe fittings

NB mm	90°– big bend	90°– small bend	45°– bend	180°– bend	Equal Tee →	Equal Tee ↑	Equal Tee →	Short piece
25	0.20	0.25	0.10	0.25	0.20	0.50	0.70	0.50
32	0.25	0.35	0.15	0.35	0.25	0.70	0.90	0.75
40	0.30	0.45	0.20	0.45	0.30	0.85	1.2	0.90
50	0.45	0.60	0.25	0.60	0.45	1.2	1.6	1.2
65	0.55	0.75	0.30	0.75	0.55	1.4	2.0	1.5
80	0.70	0.95	0.40	0.95	0.70	1.9	2.6	1.9
100	0.95	1.3	0.55	1.3	0.95	2.6	3.7	2.7
125	1.3	1.8	0.70	1.7	1.3	3.5	4.9	3.6
150	1.6	2.2	0.90	2.2	1.6	4.3	6.1	4.4
200	2.2	3.1	1.2	3.0	2.2	6.0	8.3	6.2
250	2.8	3.8	1.5	3.8	2.8	7.6	10.6	7.7
300	3.6	5.0	2.0	5.0	3.6	9.8	13.6	10.1
350	3.8	5.5	2.2	5.4	3.9	10.6	14.8	11.1
400	4.7	6.6	2.7	6.5	4.7	12.8	17.8	13.3
450	5.5	7.6	3.0	7.7	5.5	15.0	21.0	15.0
500	6.1	8.4	3.4	8.3	6.1	16.6	23.0	17.0

Table 13.6 Equivalent length of valves

NB mm	Gate valves	Globe valves	Swing type check valves	Lift type check valves	Angle valve	Foot valves
25	0.10	6.8	1.4	9.9	2.6	15
32	0.15	9.6	2.0	14	36	21
40	0.20	12	2.5	17	4 4	26
50	0.30	16	3.4	23	6.0	35
65	0.35	20	4.2	29	7.5	43
80	0.45	25	5.5	38	9.8	56
100	0.60	36	7.7	51	14	79
125	0.80	47	10	69	18	104
150	1.0	60	12	86	22	129
200	1.4	83	17	121	31	181
250	1.6	103	22	150	39	225
300	2.2	135	28	197	51	296
350	2.6	147	31	215	56	323
400	3.1	176	37	257	67	386
450	3.5	200	44	290	77	450
500	3.8	225	50	330	87	510

dimensionally inconsistent. It depends on the units used in the formula. However it is independent of pipe diameter, velocity of flow and viscosity.

b. **Darcy Weisbach formula:** This formula is dimensionally consistent. To represent true condition of flow in a pipe, this formula takes into account the relative roughness of pipe and Reynold number. A comparison between estimates of Darcy-Weisbach friction factor *'f'* and its equivalent value computed from Hazen-Williams *'c'* for different pipe materials bring out the error in estimation of *'f'* upto ± 45% in using Hazen-William formula. For higher *'c'* values (new and smooth pipes) and larger diameters, the error is less while it is appreciable for lower *'c'* values (old and rough pipes) and lower diameters at high velocities.

c. **Colebrook-White formula:** To avoid limitations of Hazen-William formula, the present day trend is to use Colebrook-White equation for estimation of friction factors. This calculated friction factor is then used in Darcy-Weisbach formula for estimation of head loss due to friction in pipelines. This method leads to much accurate results.

d. **Modified Hazen-William formula:** The limitations of Hazen-William formula in calculating frictional head loss, the modified Hazen-William formula was developed from Darcy-Weisbach and Colebrook-White equation. Since friction coefficient depends on relative roughness of pipe and Reynold Number, C_R values have to be varied for various diameter and velocity

combination to give correct estimation of frictional resistance. The average c_r values are given in Table 13.7. The modified Hazen-William formula as derived is:

$$v = 143.534 \, c_r \, r^{0.6575} \, s^{0.5525}$$

$$h = [l(q/c_r)^{1.81}]/914.62 \, d^{4.81}$$

v = vel. of flow in mps

c_r = pipe roughness coefficient

r = hydraulic radius in m

s = frictional slope

d = internal diameter of pipe in m

h = frictional head loss in m

l = length of pipe in m

q = flow in pipe m^3/sec.

13.5 TYPES OF PUMPS

Pump is a mechanical device generally used for raising liquid from a lower head. A low pressure is created which causes the liquid to be sucked from the lower level and energy is imparted to this liquid to push it to the higher level. This process requires work to be done on the liquid by employing an outside force which is provided by a prime mover or simply a motor. This work done gets converted into mechanical energy and imparts pressure energy to the flowing liquid.

The pumps are divided into two categories based on the basic principle of operation. Figure 13.2 shows various categories and classification of pumps.

Table 13.7 Average C_r values in modified Hazen-Williams formula (at 20°C)

Sl. No.	Pipe material	Diameter (mm)		Velocity (mm)		CR value when new	C_R value fo design period of 30 years
		From	To	From	To		
1.	RCC	100	2000	0.3	1.8	1.00	1.00
2.	AC	100	600	0.3	2.0	1.00	1.00
3.	HOPE and PVC	20	100	0.3	1.8	1.00	1.00
4.	CI/DI (for water with positive Langlier's index)	100	1000	0.3	1.8	1,00	0.85*
5.	CI/DI (For waters with negative Langelier's index)	100	1000	0.3	1.8	1.00	0.53*
6.	Metallic pipes lined with cement mortar or epoxy (for water with negative Langelier's index)	100	2000	0.3	2.1	1.00	1.00
7.	SGSW	100	600	0.3	2.1	1.00	1.00
8.	GI (for waters with positive Langelier's index)	15	100	0.3	1.5	0.87	0.74

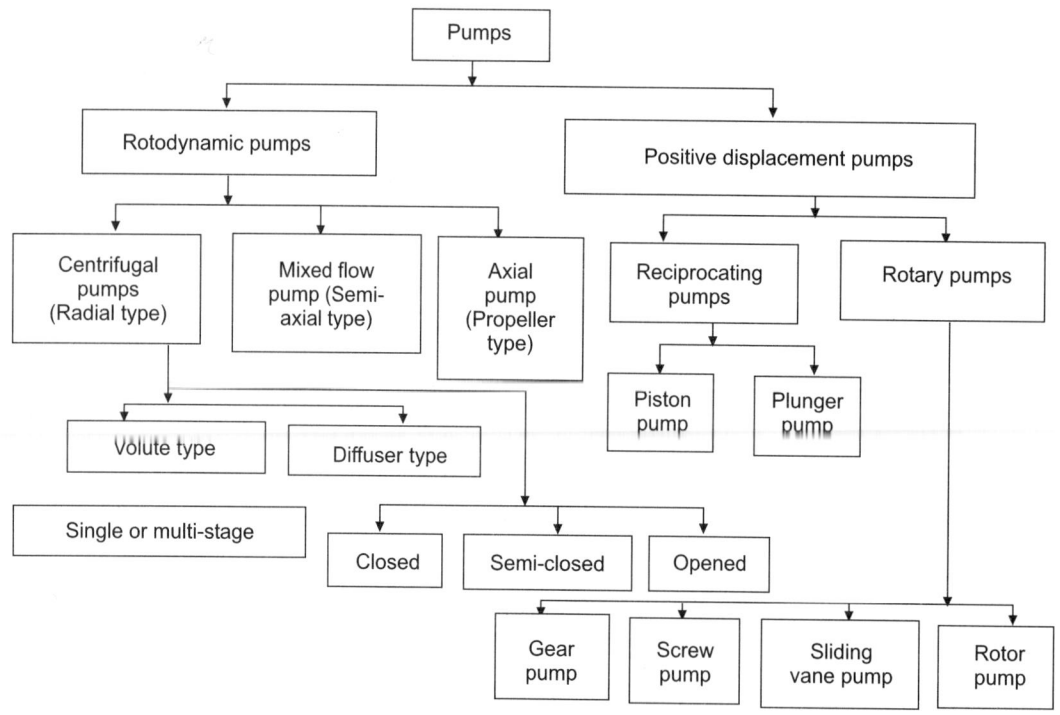

Fig. 13.2 Classification of pumps

13.5.1 Centrifugal Pump

As the name indicates, the centrifugal pumps are the machines which employ centrifugal force to lift liquid from a lower level to a higher level by developing pressure. It consists of a shaft, impeller and a volute casing. The liquid sucked from the sump (Fig. 13.3) enters the center of the impeller, called 'eye' of the impeller and is imparted acceleration to a high velocity by centrifugal force the high velocity of the rotating impeller. The liquid is then discharged into the volute casing causing vacuum in the suction pipe. This vacuum causes more water to be lifted. The discharging liquid is high in kinetic energy which is not desirable as it is discharged. Thus volute casing is designed with a continuously expanding diameter so as to convert kinetic energy into pressure energy. Whenever the suction head is negative, i.e. lower than the center of the impeller, the priming of the pump is required so as to remove the air from the pipes. The liquid is filled in the suction pipe, casing a portion of delivery pipe, i.e. upto delivery valve. This is necessary as pump cannot produce pressure in the presence of air which has very low density. The pressure generated by pump is directly proportional to density of liquid to be pumped.

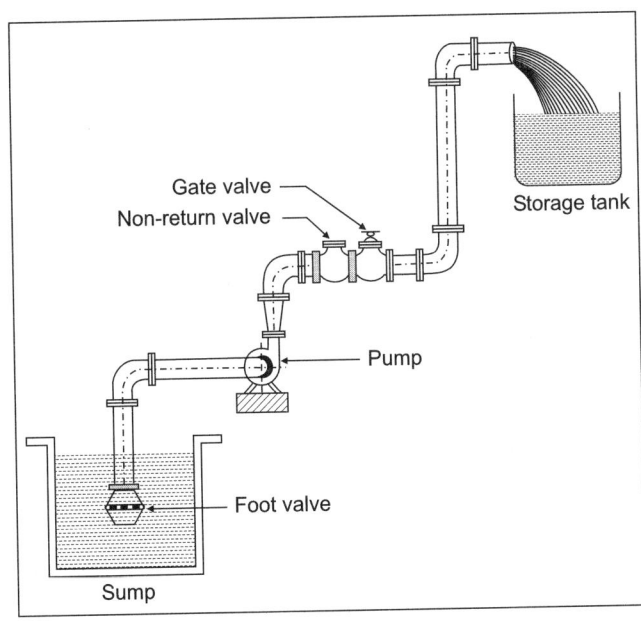

Fig. 13.3 Centrifugal pump arrangement

13.5.1.1: The impellers are classified on the basis of their construction, i.e.:

    a. Closed or shrouded impeller

    b. Semi open impeller

    c. Open type impeller

Closed or Shrouded Impeller: This type of impeller is used for pumping pure liquids (without any contamination) like water, oil, acid, alkali, etc. The material selected for the impeller shall be such that it can resist the reaction of the liquid to be handled. The various designs of shrouded impeller are given in Fig. 13.4.

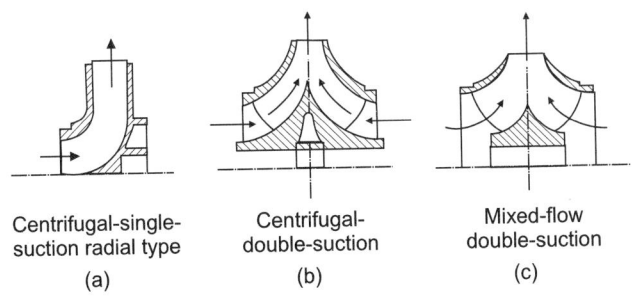

Centrifugal-single-suction radial type (a)   Centrifugal-double-suction (b)   Mixed-flow double-suction (c)

Fig. 13.4 Closed or shrouded impeller

Semi-open impeller: In this type, the vanes are fixed on one shroud only. These impellers have less number of vanes but have more length to avoid clogging of impeller. This type of impeller is used for pumping liquids containing small debris such as sewage water, paper pulps, etc. (Fig. 13.5).

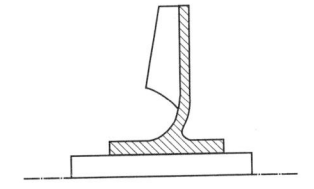

Fig. 13.5 Semi-open impeller

Open-type impeller: When the vanes are directly fixed on the web without any shroud, they are called open-type impeller. This type is used to pump slurry or liquids having large solid particles like sand, pebble, etc., i.e. for very rough duty and should stand wear and impact resistance (Fig. 13.6).

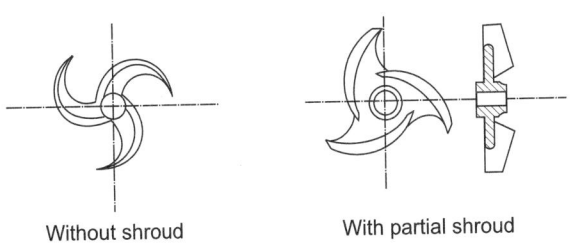

Without shroud   With partial shroud

Fig. 13.6 Open-type impeller

13.5.1.2 Pump casing

A casing is provided for housing the impeller and is connected with the suction and delivery pipelines. The pump casing design influences the efficiency with which a pump operates. An inappropriate design will cause eddy formation and result in loss in energy imparted by impeller to liquid. The common type of casings are:

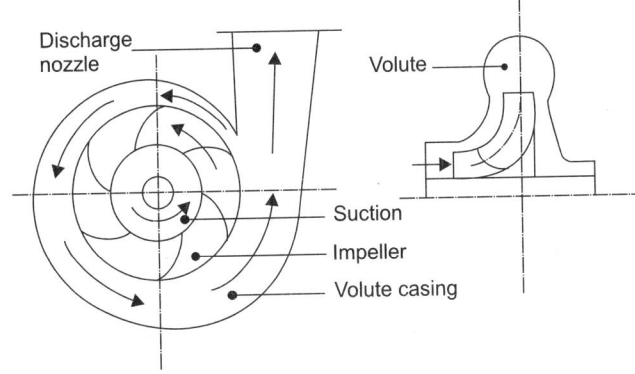

Fig. 13.7 Volute type pump casing

  a. **Volute type casing:** This Type is very simple in construction. The water discharged by the impeller enters into a progressively expanding spiral casing. The cross section increases gradually (Fig. 13.7) to the point of discharge pipe. This casing gradually reduces the velocity of liquid as it flows from impeller to discharge pipe, thus converting velocity head into pressure head. The disadvantage of this casing is that a large amount of velocity head is lost in eddies and thus these pumps can produce comparatively low heads.

  b. **Vortex or whirlpool casing:** In this type of casing, an annular space is provided, known as vortex or whirlpool chamber, between impeller and volute (Fig. 13.8) which forms part of casing with parallel side walls and serves as a diffuser without guide vanes.

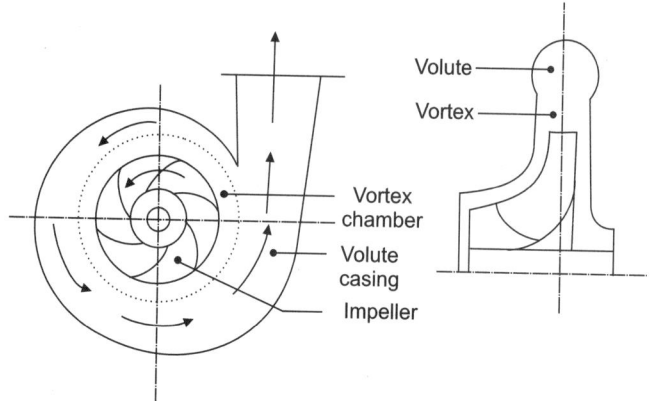

Fig. 13.8 Volute and vortex chamber

Since no work is done on the water while in this chamber, its energy remains constant except slight loss by friction. The reduction in velocity being accompanied by a rise in pressure, the chamber adds to the efficiency of pump.

c. **Diffuser (Ring) type or turbine casing:** In this system, the impeller is surrounded by a series of stationery guide vanes or by a diffuser ring with guide vanes, which by their divergence furnish gradually expanding passages for the liquid to follow after leaving the impeller. In this process, the direction of flow is changed, and velocity head is converted into pressure head before the liquid enters the volute. This design is very efficient to convert velocity head to pressure head than in the volute type (Fig. 13.9).

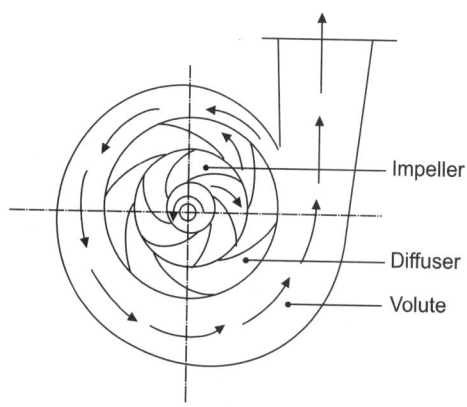

Fig. 13.9 Diffuser (ring) type pump casing

The turbine casing has higher efficiency than volute type but costly to manufacture due to complicated design. This type of design is preferable for designed rate of discharge and velocity as the efficiency goes down with variable rates of flow and velocity. These (turbine pumps) are commonly used in power plants, deep wells and mines, etc.

13.5.2 Positive Displacement Pumps

The positive displacement pumps are those in which liquid is sucked and then pushed forward due to the thrust exerted on it by a moving member. These pumps are divided into two categories:

a. **Reciprocating type:** These pumps are simply cylinder and piston arrangement powered by a prime mover. The motion of piston being reciprocating in nature, the pump is named as reciprocating pump. These pumps have almost become obsolete due to their high capital and maintenance costs.

b. **Rotary pumps:** These pumps have two closely fitting elements that revolve continuously in a fixed casing and deliver the fluid continuously. These pumps are not suitable for high pressure operation as internal losses are more than in case of reciprocating pumps. Gear pumps, screw pumps, sliding vane pumps, rotor pumps, etc. are the varieties of rotary pumps which positively displace fluids.

13.6 SPECIFIC SPEED, PUMP TYPE, PUMPING POWER

Specific speed is the speed of rotation of a geometrically similar pump of such a size that it ensures a delivery of 75 LPS at a head of one meter. If D, H, n and Q are known for a pump, the specific speed for this pump can be derived as below:

$$\frac{H_1}{H_2} = \left(\frac{n_1}{n_2}\right)^2 \left(\frac{D_1}{D_2}\right)^2 \text{ equation for similar pumps}$$

$$\frac{H}{1} = \left(\frac{n}{n_s}\right)^2 \left(\frac{D}{D_s}\right)^2 \qquad \dots (1)$$

But $$\frac{Q_1}{Q_2} = \left(\frac{n_1}{n_2}\right)\left(\frac{D_1}{D_2}\right)^3$$

$$\frac{Q}{0.075} = \frac{n}{n_s}\left(\frac{D}{D_s}\right)^3 \qquad \dots (2)$$

Solving the above two equations

$$H^{3/2} = \left(\frac{n}{n_s}\right)^3 \left(\frac{D}{D_s}\right)^3$$

or $$\eta_s = 3.65\frac{n\sqrt{Q}}{H^{3/4}}$$

Specific speed is not affected by the specific weight of the liquid being handled and refers to a single impeller of a multistage pump or one side of a double suction impeller. If Z is the no. of stages in a multistage pump, the relation between specific speed of single impeller, specific speed n_{sp} of whole pump is given as:

$$n_s = (Z)^{3/4} \cdot A_{sp}$$

Specific speed is a dimensionless parameter

Pump type: The pump type required for an application can be decided from its specific speed. Pumps with closed valves of equal specific speed will have many similar features.

Similarly low specific speed impellers (60–80) have far higher exit diameter than inlet diameter. These pumps are used for high heads.

Affinity laws: These laws establish a mathematical relationship between variables involved in pump performance such that the performance characteristics of a pump at different input values can be derived , if the performance at other parameters is known, These laws are as below:

1. With diameter D of impeller being constant:

   i. $\dfrac{Q_1}{Q_2} = \dfrac{n_1}{n_2}$

   ii. $\dfrac{H_1}{H_2} = \left(\dfrac{n_1}{n_2}\right)^2$

   iii. $\dfrac{P_1}{P_2} = \left(\dfrac{n_1}{n_2}\right)^3$

2. With speed n of impeller being constant:

   i. $\dfrac{Q_1}{Q_2} = \dfrac{D_1}{D_2}$

   ii. $\dfrac{H_1}{H_2} = \left(\dfrac{D_1}{D_2}\right)^2$

   iii. $\dfrac{P_1}{P_2} = \left(\dfrac{D_1}{D_2}\right)^3$

Where Q_1, Q_2 = Flow rates in m³/hr

            n_1, n_2 = Pump speed in RPM

            H_1, H_2 = Total head in m

            P_1, P_2 = Power required in metric $H.P.$

Thus if speed of the pump is reduced from n_1 to n_2, its flow is reduced in proportion to speed, head in proportion to square of speed and power as cube of speed.

But if reduction in speed is not possible, diameter of impeller can be machined so that reduction in diameter alters flow proportional to diameter, head in proportion to square of diameter and power in proportion to cube of diameter.

Pumping Power

The pumping is needed because a liquid is to be transported from one location to another, either at the same level or at different elevations. The power of the motor shall be sufficient to transport the liquid and overcome various losses encountered on the way. The pump power shall take care of the following (Refer Fig. 13.10 and 13.11):

i. **Static suction head (h_{ss}):** This value is negative, if surface of source of supply is below the center line of the pump. This is also called static suction lift, otherwise static suction head is positive.

ii. **Static discharge head (h_{sd}):** when pump discharges liquid in an open vessel, the vertical distance between center line of pump and point of free discharge or surface of liquid in the vessel is known as static discharge head.

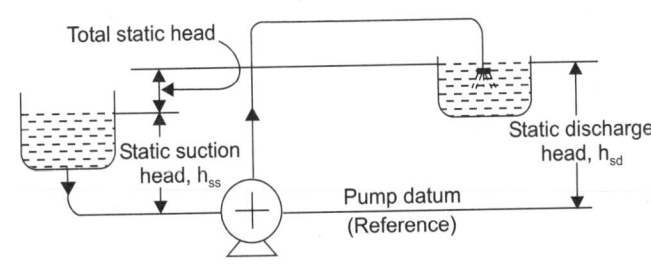

Fig. 13.10 Installation with positive suction head

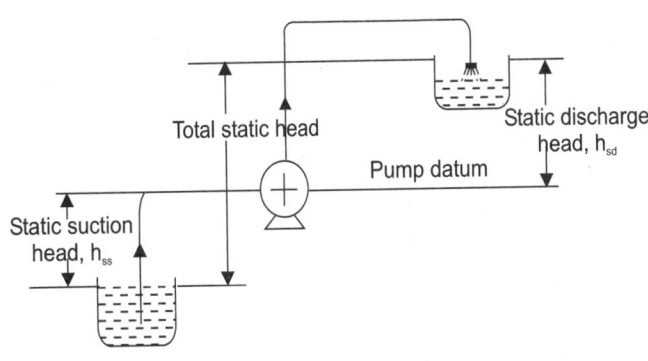

Fig. 13.11 Installation with suction lift (-ve head)

iii. **Total static head:** The vertical distance between free level of source of supply and point of free discharge or free surface of discharged liquid.

iv. **Velocity head (h_v):** The kinetic energy imparted to the liquid equivalent to velocity with which liquid flows out at discharge point. This head is a major factor in low head installations.

v. **Friction head (h_{fs}):** The resistance offered by the pipe, valves and bends to the flow of liquid shall be over come by the pumping power. This is called friction head.

vi. **Total dynamic suction head (h_s):** Sum of static suction head and velocity head at pump suction flange reduced by friction head in suction line.

$$h_s = h_{ss} + \frac{V^2}{2g} - h_{fs}$$

vii. **Total dynamic discharge head (h_d):** Sum of static discharge head and velocity head at pump discharge line.

$$h_d = h_{sd} + \frac{V^2}{2g} + h_{fd}$$

viii. **Total pumping head or total dynamic head (TDH):** This is the head against which pump is to be designed so as to deliver the liquid in the discharge tank.

$$H = h_d - h_s \qquad \text{(with a suction head)}$$
$$= h_d + h_s \qquad \text{(with a suction lift)}$$

ix. **Manometric head:** If two pressure gauges are installed, one at the suction flange and another at discharge flange, then the difference of readings is the total head produced by the pump and is called manometric head:

$$H_m = H_{md} - H_{ms}$$
$$= h_d + h_s \qquad \text{(with a suction lift)}$$
$$= h_d - h_s \qquad \text{(with a suction head)}$$

The power required to pump the liquid is thus given by the following equation:

$$P = \frac{QH\gamma}{75\eta_p}$$

where P = brake horse power of pump in HP.

Q = quantity of liquid to be pumped in m^3/sec.

γ = density of liquid in kg/m^3

H = head in meters

η_p = efficiency of pump

The calculate the motor input power, the horse power so calculated shall be divided by the efficiency of motor.

$$P_m = \frac{P}{\eta_m}$$

where

P_m = power input to the motor in HP

P = power output of motor (or brake horse power)

n_m = efficiency of motor

The overall efficiency of installation η_o consisting of pump, prime mover, i.e. motor and an intermediate drive can be expressed as:

$$\eta_o = \eta_p \times \eta_m \times \eta_{gear/belt}$$

13.7 GENERAL CHARACTERISTIC CURVES

A centrifugal pump operating at a constant speed can deliver wide range of flows at wide range of head. This flow can become zero at the maximum head

deliverable by the pump. Similarly the power consumed by the pump varies widely with variation in head or quantity. This means pump has variable efficiency at different combinations of head and quantity. This also means that the pump shall be selected for operation at best efficiency to pump a determined quantity or liquid at determined head which shall be calculated carefully. In order to select a pump for a particular applications, the manufacturer provides general characteristic curves which are common curves plotted for head, power, efficiency and *NPSH* against flow capacity at constant speed and labeled as $Q–H$, $Q–P$, $Q–\eta$ and $Q–NPSH$ curves. The designer can then select the pump with best efficiency. The understanding of these curves thus becomes very important in pump selection. These curves are explained below:

i. **$Q–H$ Curves:** The centrifugal pump is a self regulating pump with regard to flow, quantity and total head. This property is depicted by $Q–H$ curve. This curve has four basic forms:

a. **Stable curve:** In this curve, the head continuously falls as flow increases.

b. **Unstable curve:** The curve first rises as the delivery increases and then falls.

c. **Steep curve:** The curve falls steeply as discharge increases marginally or vice versa.

d. **Drooping curve:** The head rises from shut off point and then falls steeply.

e. **Flat curve:** The head remains constant (except for minor variations) for any change in discharge

The $Q–H$ curve is shown in Fig. 13.12.

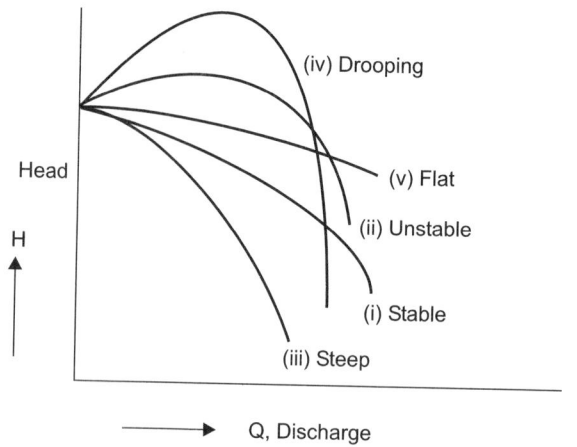

Fig. 13.12 *Q–H curves for centrifugal pumps*

ii. **$Q–P$ Curve:** This is the curve which projects power consumption, efficiency of pump and head at varying quantities of flow. This curve rises with increasing delivery and when the pump has a low specific speed but it is nearly horizontal when the pump has a higher specific speed (Fig. 13.13).

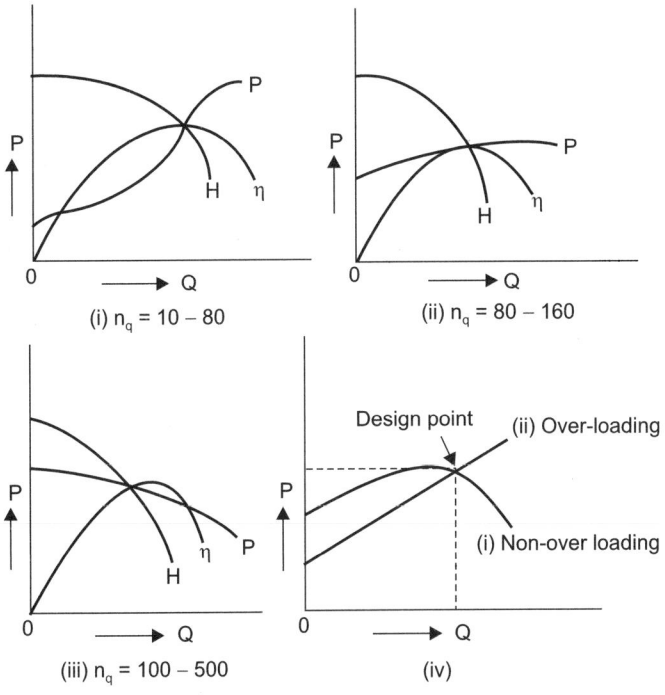

Fig. 13.13 *Q–P* curves

These curves are also classified as non-overloading curves which rise to a limiting height upto the designed point and then fall as the discharge increases. The overloading curve is one which increases continuously with the increase in capacity.

iii. **Q–η Curve:** The efficiency curve starts from zero at zero capacity and rises to a maximum and then falls to zero again. This is the curve which gives the point of operation where maximum efficiency occurs (Fig. 13.14).

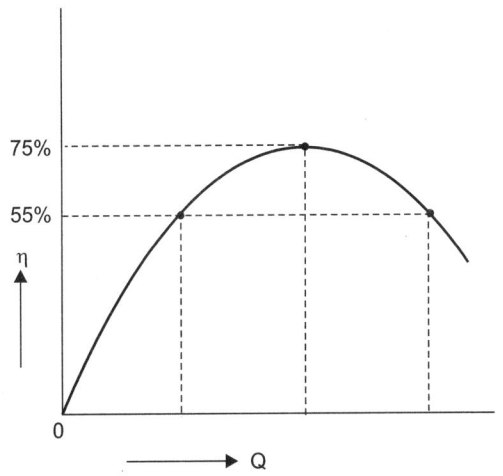

Fig. 13.14 *Q–η* curve

13.8 CONSTANT EFFICIENCY CURVE OR ISO EFFICIENCY CURVE (MUSCHEL CURVES)

We have seen above the variation in performance of pump with change in flow quantity but at constant speed. Now we will discuss the performance of pump in terms of efficiency when the speed changes. These curves are plotted on a common graph for various speeds and are called constant efficiency curves. These curves cross the constant speed lines. Another curve which rises upwards is the line of maximum efficiency (Fig. 13.15).

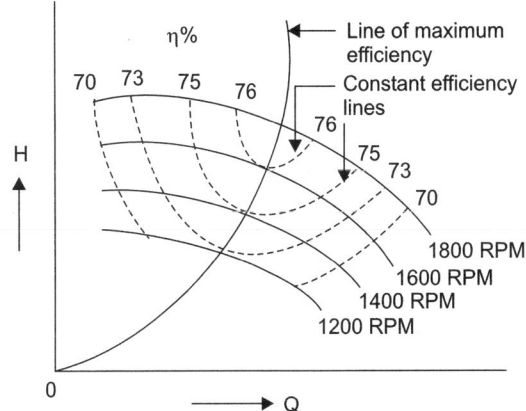

Fig. 13.15 Iso-efficiency curves at different speeds

The iso-efficiency curves are also plotted for different diameters of the impeller (Fig. 13.16). The line of maximum efficiency helps in determining the best efficiency situation for a particular impeller and flow-head combination.

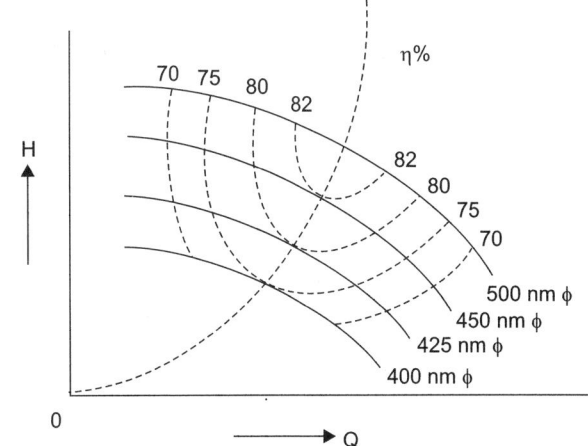

Fig. 13.16 Iso-efficiency curves for different diameters of impellers

All these curves described above help in determining the exact point of operation, i.e. flow and head combination to get maximum efficiency and minimum power consumption.

The next topic explains the effect on the performance of the pump due to wrong selection of head.

13.9 PUMP PERFORMANCE AT VARIOUS HEADS

The total head against which a pump must operate is a combination of frictional losses, potential head and kinetic head in the system. The frictional head varies as the square of discharge, thus the curve Q–H_f is a parabolic curve (Fig. 13.17) and is called System head curve.

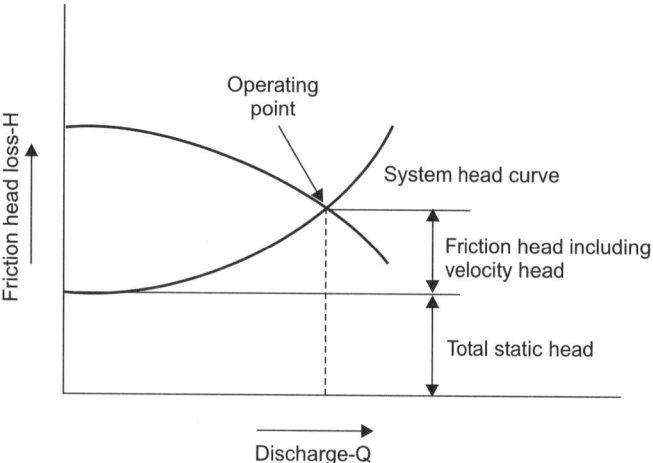

Fig. 13.17 System head curve

If Q–H curve is also imposed on it, then the point of intersection of the two curves is the point at which pump will operate. The operation slightly towards right of this point will make the pump discharge more liquid causing motor to get overheated and burn. This is because system head was estimated to be higher while in actual it was lower thus causing pump to operate against lower head and in turn pump more liquid. The solution in such cases is to throttle the delivery valve and reduce discharge (Fig. 13.18a).

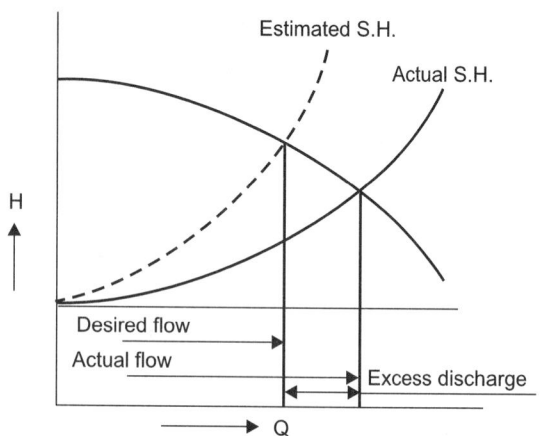

Fig. 13.18a *Q-H* Curve: Estimated S.H. more than actual

If on the other hand, estimated System Head is less than the actual head against which pump has to operate, then the pump will discharge less (Fig. 13.18b).

In both the cases, the efficiency of the pump selected suffers and results in more power consumption. Though functional performance can still be achieved either by changing the pipe diameter or changing with a new pump, it is the duty of designer to make accurate calculation of system head and then select the appropriate pump capacity.

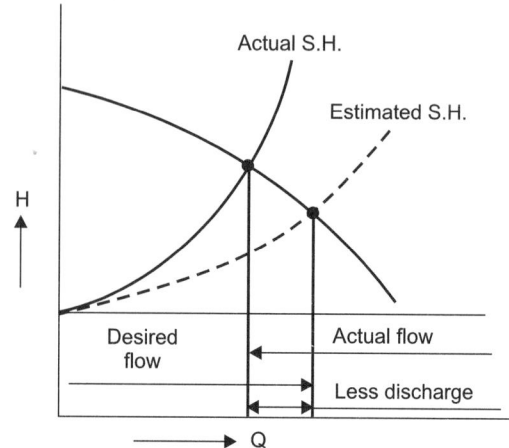

Fig. 13.18b *Q-H* Curve: Estimated S.H. less than actual

13.10 PUMPS OPERATING IN SERIES

There is a limitation in the head that can be developed by a single pump vis-a-vis the quantity of liquid to be pumped. The manufacturers of pumps give a tabled data of the flow and the head upto which the liquid can be pumped. Selecting a pump with a higher head for low quantities will make the pump operate to in the region of lower efficiency (ISO-efficiency Q–H curves).

To overcome this problem, pumps are installed in series. Liquid from one pump is brought to the inlet of another pump to further boost the head. Another variation of series pumps is called multistage pumps in which two or more impellers are housed inside the same casing (Fig. 13.19).

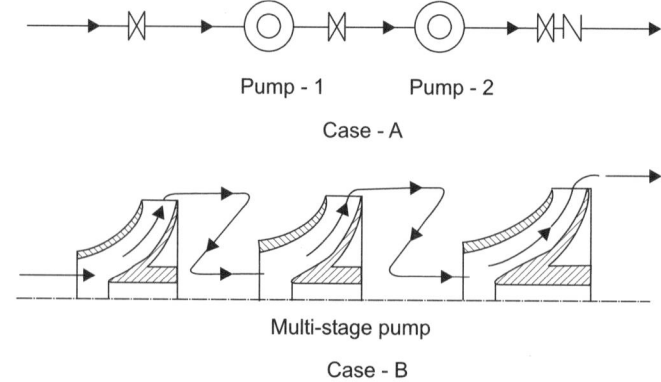

Fig. 13.19 Series operation of pumps

The Q–H curves of series operation is combination of Q–H curve of individual pumps (Fig. 13.20).

$$H = H_1 + H_2$$

13.11 PUMPS OPERATING IN PARALLEL

If the quantity of liquid is more and cannot be pumped by a single pump for a defined head, then two or more pumps are operated in parallel (Fig. 13.21).

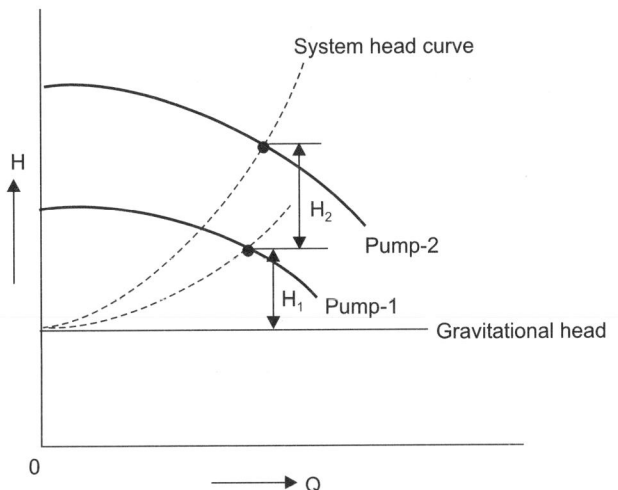

Fig. 13.20 Q–H curves for pumps operating in series

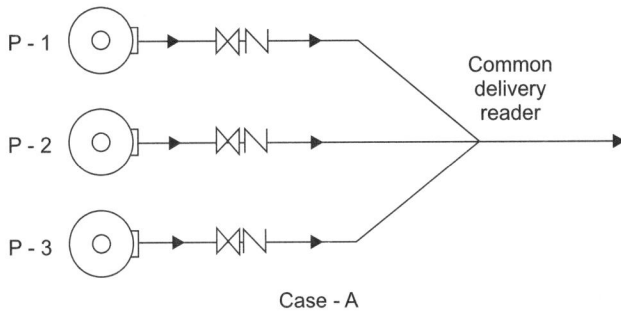

Case - A

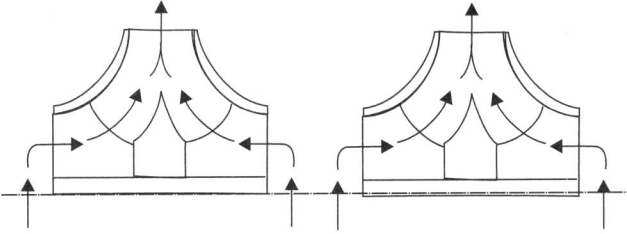

Case - B Typical impeller arrangement

Fig. 13.21 Parallel operation of pumps

There are five variations in parallel operation which cause different Q–H curves in such applications (Fig. 13.22a to e).

a. **Pumps of same design and rating:** The Q–H curve of pumps operating in parallel is obtained by adding individual flow quantities at the same head.

b. **Pumps with different capacities:** As can be seen from the Q–H curve, the operating point of pumps with different capacities shift towards left when operating point of group is projected horizontally. Thus the efficiency of each pump and the quantity actually discharged can be found out.

c. **Parallel operation of pumps with flat and steep characteristic curves:** The flow rate of each pump drops below the normal flow rate obtained when run alone. This is more prominent when Q–H curve is flatter.

d. **Parallel operation with unstable characteristic curves:** For a centrifugal pump with unstable Q–H curve, the head developed by it at $Q = 0$ must be greater than the static head for its smooth starting against completely filled discharge lines.

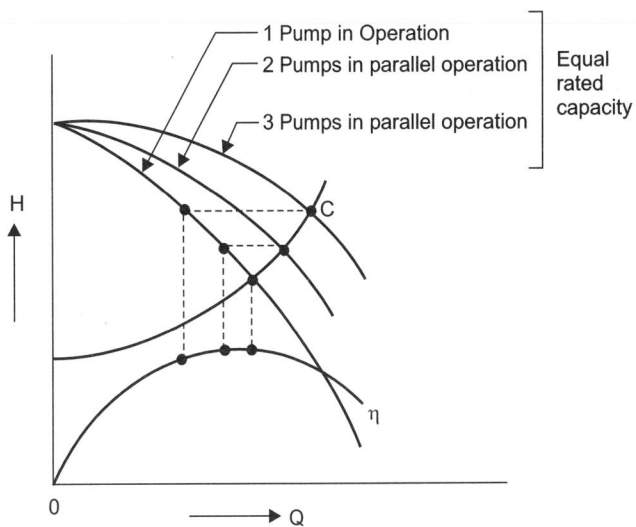

Fig. 13.22a Q-H curves for pumps of same capacity operating in parallel

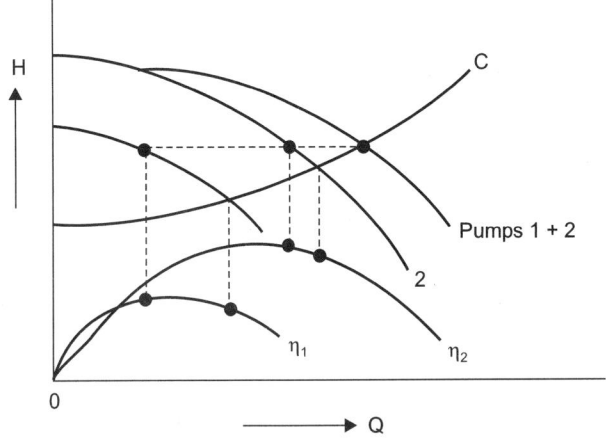

Fig. 13.22b Q–H curves for pumps of different capacities operating in parallel

If the situation is such that the head developed at $Q = 0$ is less than the static head, the characteristic curve will intersect the Q–H curve at two points and the pump will be unable to start against completely filled discharge line. In such a case, pipeline must be partially emptied till static head is brought down below head developed by pump at $Q = 0$.

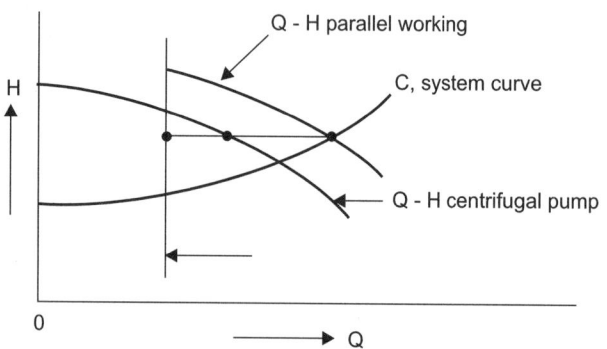

Fig. 13.22c *Q–H* curve for a reciprocating pump operating in parallel with a centrifugal pump

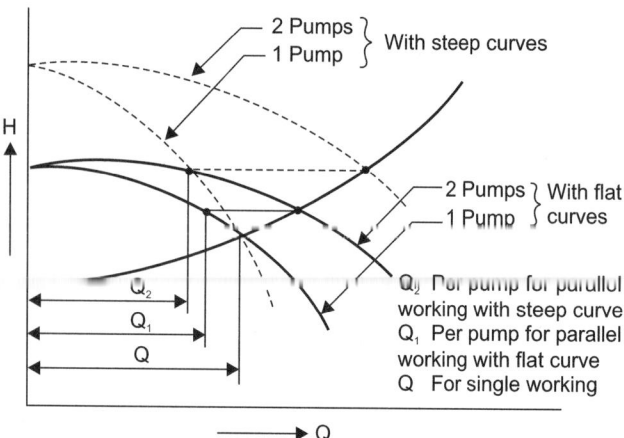

Fig. 13.22d *Q–H* curve for parallel operation of centrifugal pump having various types of characteristic curves

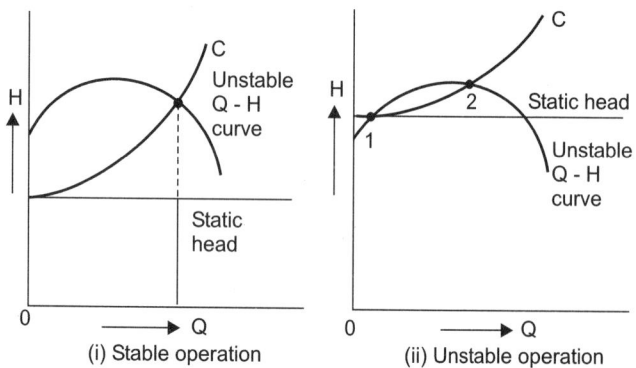

Fig. 13.22e *Q–H* curves for parallel operation of centrifugal pumps with unstable characteristic curves

13.12 THROTTLING OF PUMP DISCHARGE

When the quantity of liquid to be pumped is less than the rated quantity of pump, throttling method is adopted. This increases frictional loss causing increase in discharge head and in turn reducing the quantity to be discharged.

The speed of pump can also be regulated to achieve required flow rate below normal discharge point and thus shifting the operating point along characteristic curve. This will avoid throttling and efficiency of pump may not be compromised to that extent as in case of throttling.

However, throttling is achieved only with the help of orifice plate in the discharge line. The diameter 'd' of orifice plate can be obtained from the equation:

$$d = k\sqrt{Q/H}$$

where d = dia of orifice
Q = flow rate
H = head
K = a constant called throttling coefficient (Fig. 13.23a and b)

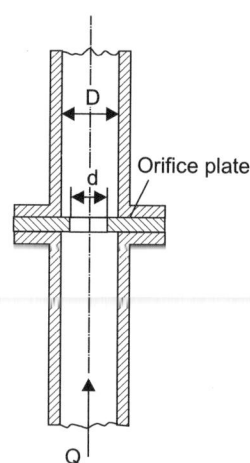

Fig. 13.23a Operation of pump with orifice plate in discharge line

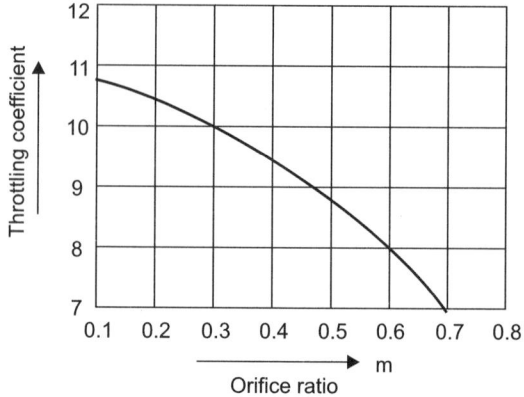

Fig. 13.23b Operation of pump with throttling

13.13 NET POSITIVE SUCTION HEAD (NPSH)

The total suction head in meters at the inlet flange of the pump corrected by potential head of pump inlet and the vapour pressure head of liquid at pumping temperature is called Net Positive Suction Head (NPSH). In other terms, the total energy available at pump suction is NPSH. This is the positive head required at pump suction to overcome pump internal head losses like turbulence, losses in suction passage, vanes and also to maintain liquid above its vapour pressure.

Pump manufacturers generally publish information of required NPSH along with pump characteristic curves. The available NPSH as per the design shall be more than NPSH required, otherwise liquid is vaporised by reduction of pressure below vapour pressure of liquid. The bubbles of vapour so formed at the eye of the impeller move along the impeller vanes to a high pressure area and collapse. This collapse of bubbles is so rapid that the resultant in rush of liquid into the cavities causes a rumbling noise. The forces are developed during the collapse and may cause fatigue failure of impeller vanes. This also causes reduction in capacity, reduced and unstable head and erratic power consumption. This phenomenon of vaporisation of liquid and collapsing of bubbles causing failure of equipments is called *cavitation*.

To ensure that cavitation does not occur and to ensure satisfactory performance of the pump, it must be ensured while designing the system that available NPSH is more than that required by the pump (as mentioned by manufactures).

The following illustration describes the typical cases of available NPSH (Fig. 13.24).

$$\text{Case I} \quad : NPSH_a = h_a - h_s - h_{vp}$$
$$= h_a - h_{ss} - h_{fs} - h_{vp}$$
$$\text{Case II} : NPSH_a = h_a + h_{ss} - h_{fs} - h_{vp}$$
$$\text{Case III}: NPSH_a = p_s - h_{ss} - h_{fs} - h_{vp}$$
$$\text{Case IV}: NPSH_a = p_s + h_{ss} - h_{fs} - h_{vp}$$

where

$NPSH_a$ = available NPSH

h_{ss} = head between liquid level and shaft center (in horizontal pumps) or mouth of suction bell (in case of vertical pumps) in meters

h_{fs} = friction loss in suction pipe in meters

h_{vp} = vapour pressure head of liquid at pumping temperature in meters

$NPSH_a$ can also be calculated as below:

$$NPSH_a = \frac{10}{\gamma}(P_a - P_v) + h_{ss} - h_{vf}$$

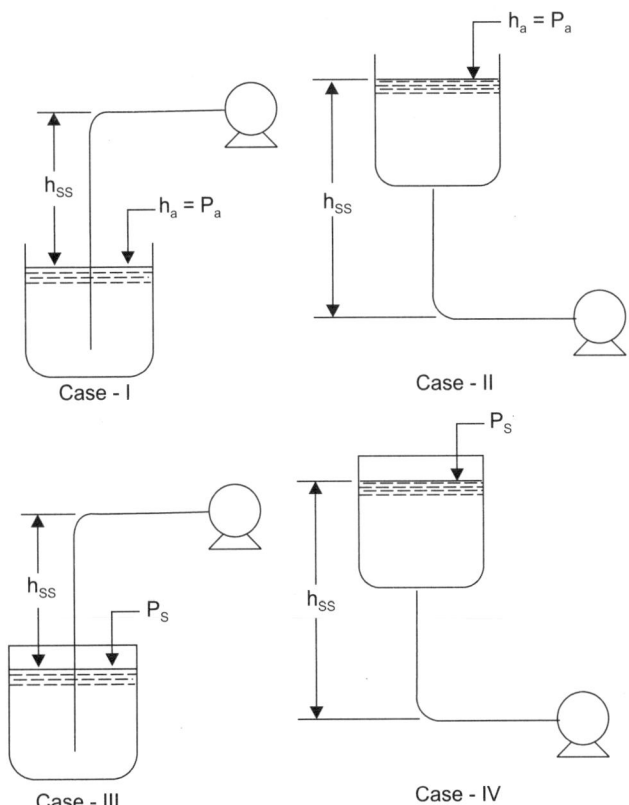

Fig. 13.24 NPSH$_a$ of Centrifugal pumps

where

P_v = vapour pressure at temperature of pumping liquid kg/cm^2

γ = density of liquid in gm/cc

$$h_{vf} = \frac{v_s^2}{2g} + h_{fs} \text{ in meter where } v_s \text{ is the velocity of}$$

liquid at the impeller inlet in m/s

h_{ss} is positive when pump has a suction head and negative when it has a suction lift.

13.14 WATER HAMMER

In a running pumping system, if a valve is suddenly closed at the end of the discharge line, the moving water column is brought to a sudden stop and the kinetic energy imparted to water column by the pump gets converted to internal energy by compression of liquid. If this pressure rise is sufficiently high, it may burst the pipe. The pressure wave so created travels back upstream continuing its dissipation of energy within pump and pipe wall. This phenomenon is called *water hammer* as it causes a series of shocks sounding like hammer blows. The pressure rise due to sudden valve closing is

$$h_{max} = \frac{av}{g}$$

where

a = velocity of presssure wave in m/s

v = velocity of water just before valve closure in m/s

g = acceleration due to gravity, 9.81 m/s^2

The magnitude of velocity of pressure wave is determined by following formula:

$$a = \cfrac{1}{\sqrt{\cfrac{\gamma}{g}\left(\cfrac{1}{k} + \cfrac{d}{t.E.}\right)}}$$

γ = specific weight of liquid

K = bulk modulus of liquid

d = pipe diameter

t = pipe wall thickness

E = modulus of elasticity of pipe material

13.15 SOURCES OF WATER AND METHODS OF PUMPING

When an area is developed on a large scale, say a water supply scheme is being designed for a town or city, a source of water has to be found first. The source may be a river, stream, lake, well, etc. But the most important aspect is that this source should provide enough supply of water to the city throughout the year and for years to come.

The sources of water may be classified as:

i. Surface water

ii. Ground water or subsurface water

The difference between the above two categories is that surface water is the quantity of water remaining on the surface after all losses and is also called run-off while water that percolates into the soil and collects underground is called the ground water.

13.15.1 Categories of Surface Water

i. Lakes/ponds

ii. Impounding reservoirs like dams

iii. Rivers/streams, irrigation canals

The surface water is not dependable form of supply of water except impounding reservoirs like dams. The lakes and ponds have limited capacities while rivers capacity go on reducing as it flows down the course. Rivers and ponds are also subject to impurities, industrial waste, climatic conditions and seasons of the year. Dams are characterized by the capacity of water they store so as to be made available during lean season. They serve as a primary source of supply for big cities, irrigation and power generation. But dams can be constructed only across valleys to tap the capacity of river having perennial flow of water. But other cities or towns cannot be supplied their annual demand of water from any of the sources as above.

13.15.2 Categories of ground water

Cities have to depend on under ground water for continuous supply of water for their need. The underground sources of water and their characteristics must be understood before pumping system can be designed. These sources are:

i. **Aquifier:** The surface of earth consists of alternate layers of pervious and impervious soils. A portion of rainfall percolates into the soil until it reaches an impervious stratum. It then moves into lateral direction towards some outlet. The pervious layer in which water moves laterally is called aquifer. If the aquifier is composed of sand and gravel, water can be easily drawn (Fig. 13.25).

Water table: The free surface of water in the aquifer is called water table. When the aquifer is overlaid by an impervious layer such as clayey soil, then the water table is the free surface of water in the topmost layer of soil. This table falls in summer and rises during rains.

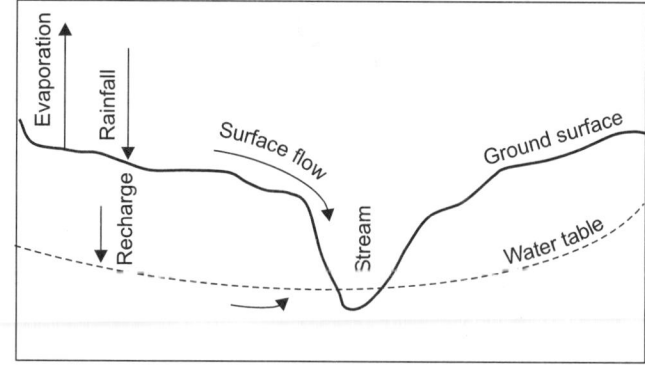

Fig. 13.25 Profile of ground water table

ii. **Springs:** Whenever an aquifer or an underground channel reaches the ground surface such as a valley or a side of a cliff, water starts flowing naturally and this natural flow is called a spring. These springs may form a lake, a creek or even a river. The quantity and velocity of a spring flow depends on the aquifer size and the position of the spring relative to the highest level of the water table. The larger the aquifer, more is the flow; the lower the spring site (than the water table), the higher is the velocity and vice versa.

iii. **Wells:** The water requirement for city or towns are usually met by drawing water from wells as springs are rare. A well is a bore drilled in ground to draw water from the aquifer. Deeper wells have less turbidity, more dissolved minerals and less bacterial count than shallow wells. Shallow wells have less natural filtration of water due to less depth of soil. Well water is very good for potable use and free residual chlorine is sufficient to cause disinfection. The well shall be adequately protected and located at least 500 feet from sewage treatment plant, animal feed lots and sanitary land fill.

13.15.3 Developing a Well

A well is developed by pumping out the mud continuously until clear discharge of water is obtained (Fig. 13.26). The distance between the water level and ground surface is called static level or the height to which water rises when not pumping is called static level.

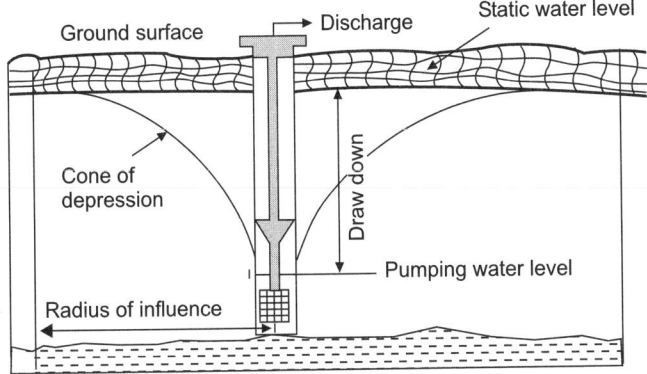

Fig. 13.26 A well while pumping

When water is drawn from the well with the help of a pump, the water level drops to a level called pumping level. The difference between static level and pumping level is called *drawdown*. The well is developed and pump is so designed that water reaches back static level once the pumping is stopped. If the water does not reach original static level, the short fall in the distance is called *residual drawdown*. The capacity of well per foot draw down is called specific capacity. As an example, if the capacity of well is 20000 LPM and its drawdown is 5 meters, then the specific capacity of the well is 20000 LPM/5 meter or 4000 LPM/meter.

Well yield is another term to define the capacity of a well and expressed as rate of water production by a well in litre per minute. Well yield is proportional to draw down and is more if drawdown is more. The yield is also proportional to depth of well in aquifer. If a well penetrates 5 meters into the aquifer, the yield will be doubled if the depth is increased to 10 meters with the same drawdown.

The wells are classified and developed according to quality and quantity of water to be drawn. They are classified as under:

13.15.3.1 Shallow wells

These wells are constructed in the uppermost layer of the earth's surface. They draw water from the ground water table (Fig. 13.27). The diameter of shallow well varies from 2–6 meters and may be lined or unlined. The unlined wells are generally constructed upto a depth of about 7 meters, but for greater depths, the lining becomes essential. These wells are also known as draw wells, open wells or dug wells. The disadvantage with these wells is that quantity of water is limited as they draw water from uppermost layer of earth only. They dry up sometimes in summer, and to ensure continuous supply of water, they are taken much below the surface of ground water tables so that even if the water table falls, some water will always be available. But, certainly these wells cannot be used for large water requirements. Moreover, these wells are easily susceptible to contamination from nearby septic tanks, etc. These wells are used for water supply for small villages, undeveloped municipal towns, isolated buildings and camps.

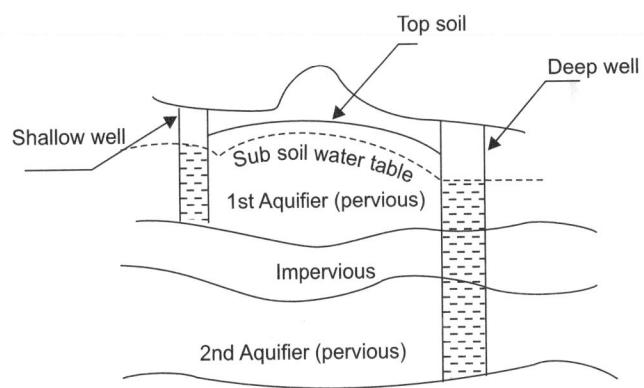

Fig. 13.27 Shallow and deep wells

13.15.3.2 Deep wells

These wells draw water from an aquifer below an impervious layer (Fig. 13.27). The water enters the aquifer from a place called "out crop" which is exposed to atmosphere. The water then reaches the site of deep well. Since the water is contained in lower aquifer, the water is at a pressure greater than atmosphere and deep wells are therefore also referred as pressure wells.

13.15.3.3 Tube wells

A tube well is a deep well and draws water from a number of aquifers. This well is of the diameter varying from 50 mm to 200 mm. The blind pipes are placed against the impervious layers (Fig. 13.28). The construction of tube well involves following steps:

a. Bore is drilled in the ground larger than diameter of tube well and obtain information about various layers of soil. If diameter of tube well is 150 mm, the diameter of bore is kept as 300 mm.

b. The depth of tube well is decided. It depends on quantity of water to be drawn, material of aquifier, etc. Thus depth is usually 30 to 50 meter but in some dry areas, it may go up to 300 meters as well.

c. The pipe for tube well, comprising of strainers and blind sections, is then inserted into the bore hole. A strainer is a metallic filter, provided to allow only water to pass through it. The pumping is then started. The entrapped mud in the strainer is removed either by allowing water from a high tank under pressure or by blowing air under pressure or by reversing the direction of flow. This process of removal of fine particles from the strainers is called *well development*.

The quantity of water available from a tube well is generally sufficient and more or less reliable. The discharge from a tube well does not exceed 40 to 50 litres per second.

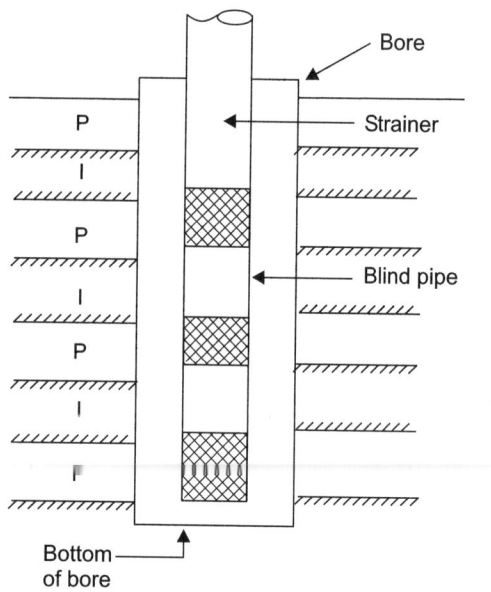

Fig. 13.28 Tube well

The water from tube well or wells is generally sufficient to meet the needs of residential colonies, towns, cities, etc.

13.15.4 Tube Well Pumping Station

A typical tube well pumping station is shown in Fig. 13.29. The pump is lowered in the tube well and the pump room is constructed above ground level. The suction pipe is connected to tube well pipe through a non return valve which allows water to go up only and prevents reverse flow when pump is removed for repairs. The pumped water is taken out through delivery pipe. A bottom plug is constructed at the foundation level of the station to prevent the entry of water from ground water table. A ladder is provided to connect the floor of pump room with the bottom of the station.

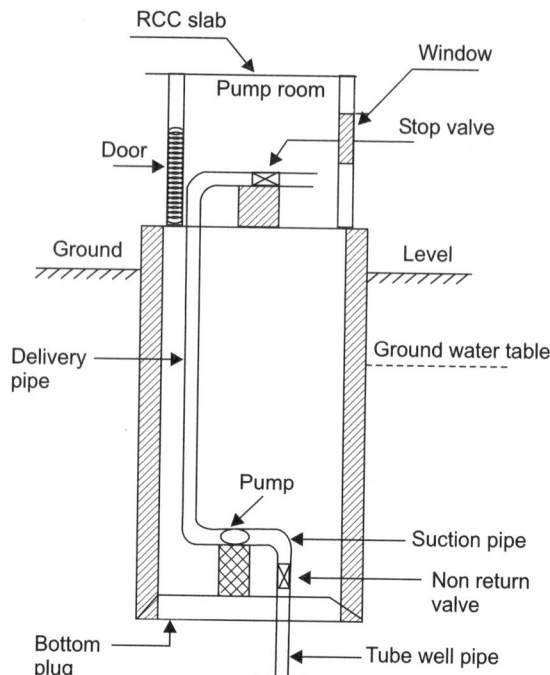

Fig. 13.29 Tube well station

The following information is required to select a suitable pump for a tube well:

  i. yield in litre per minute
 ii. static and pumping levels in well
iii. well internal diameter
 iv. Discharge head..

Example

A pump is to be selected along with motor of three phase for operation on 3 Phase supply with the following information obtained on development of tube well:

Given data	
Yield:	3500 LPM
Delivery head:	45 m
Pumping level: (Suction head)	65 m
Inside diameter of well:	300 mm
Pump RPM:	3000
Power supply:	3 Phase/50 HZ/415 V
Liquid to be pumped:	fresh water

Solution

1. Total dynamic head (H) = Suction head + Delivery head + friction loss in pipe + loss in valve

  a. Suction head = 65 m
  b. Delivery head = 45 m

Since diameter of bore is 300 mm, pipe of 200 mm is selected. From chart 7.2 for open pipes, the friction is calculated as 2.2 m/100 m and velocity as 1.67 m/s which is very reasonable (maximum 5 m/100 m of friction should be allowed). Thus friction loss in 65 + 45 = 110 m of pipe is (2.2 ÷ 100) × 110 = 3.42 m

c. friction loss in pipes = 3.42 m

d. loss in 8″ check valve = 0.72 m

$$H = 65 + 45 + 3.42 + 0.72$$

$$= 114.14 \text{ m}$$

2. Impeller selection

a. No. of stages $= \dfrac{H}{\text{Head/stage}}$

Assuming 40 m head per stage or the same can be selected from any manufacturer's catalogue

$$= \frac{114.14}{40}$$

$$= 2.85 \text{ stages (say 3 stage)}$$

Also, note down the efficiency from catalogue for a three stage pump, say 75.5%

b. Pump BHP $= \dfrac{QH\gamma}{75\eta_p}$

where $Q = $ Quantity to be pumped in m^3/s

$= 3500 \text{ LPM} = 0.058 \text{ m}^3/\text{s}$

$\gamma = $ density of liquid in kg/m^3 (1000 in case of water)

$H = $ head in meters

$\eta_p = $ efficiency of pump = 75.5%

$$\text{BHP} = \frac{0.058 \times 114.14 \times 1000}{75 \times 75.5\%}$$

$$= 116.91$$

c. Motor selection

$$\text{Motor HP} = \frac{\text{BHP}}{\text{Efficiency of motor}}$$

$$= \frac{116.91}{0.96} \text{ (assuming motor efficiency of 96\%)}$$

$$= 121.78 \text{ HP}$$

13.16 WATER DEMAND CALCULATION

The environmental hygiene committee has suggested certain optimum service levels for communities based on population groups. In the IS code 1172–1983, 'Basic Requirements of Water Supply, Drainage and Sanitation' and NBC–2005; a minimum of 135 LPCD has been recommended for all residences provided with full flushing system for the disposal of excreta.

The calculation of water required by different consumers, say hospitals, air posts, industries, etc. have to be met by public water supply. Based on the objective of meeting the above requirements for domestic and other essential non-domestic needs, the following values shall be adopted to determine the total quantity of water needed and accordingly to design a suitable and effective pumping system:

a. Domestic and non-domestic needs

The recommended values for domestic and non-domestic needs are given in Table 13.8.

Table 13.8 Recommended per capita water supply levels for designing schemes

S. No.	Classification of town/cities	Recommended maximum water supply levels (LPCD)
1.	Towns with piped water supply but without sewerage systems	70
2.	Cities provided with piped water supply where sewerage system exist/contemplated	135
3.	Metropolitan and mega cities with piped water supply where sewerage system exists/contemplated	150

Above figures include requirement for commercial, institutional and minor industries. Bulk supply shall be assessed separately.

b. Institutional needs

The water requirement for institutions shall be as below in addition to indicated in a. above, if they are of considerable magnitude and not covered in provisions already made. The individual requirements would be as given in Table 13.9.

c. Industrial needs

The per capita water requirements mentioned above will normally include the requirement of water by small scale industries existing in a town. But the water requirement to meet the need of processes to be carried out in a large industry will have to be worked out specific to that type of industry. The forecast of this demand can be made per unit of production. The Table 13.10 below lists very broadly the water consumption pattern of various type of industries. The consumption may be less, if waste water can be recycled. Thus, it is essentially required that every type of industry is studied in isolation and demand of water is worked out.

Table 13.9 Recommended per capita water demand for institutions

S. No.	Institutions	Litres Per capita per day (LPCD)
1.	Hospitals (including laundry)	
	a. No. of beds exceeding 100	450 (per bed)
	b. No. of bed less than 100	340 (per bed)
2.	Hotels	180 (per bed)
3.	Hostels	135
4.	Nurses homes and medical quarters	135
5.	Boarding schools/colleges	135
6.	Restaurants	70 (per seat)
7.	Air ports and sea ports	70
8.	Junction stations/bus stations	70
9.	Terminal stations	45
10.	Intermediate stations	45 (can be reduced to 25, if bathing facility not provided)
11.	Day school/colleges	45
12.	Offices	45
13.	Factories	45 (can be reduced to 30 where no bath rooms are provided)
14.	Cinemas, concert halls and theatres	15

Table 13.10 Industry-wise water requirement guidelines

S. No.	Industry	Unit of production	Water requirement in KL per unit
1.	Automobile	Numbers	40
2.	Distillery	KL of Alcohol	122–160
3.	Fertilizer	Tonne	80–210
4.	Leather	100 kg (tanned)	4
5.	Paper	Tonne	200–400
6.	Special quality paper	Tonne	400–1000
7.	Straw board	Tonne	75–100
8.	Petroleum refinery	Tonne (crude)	1–2
9.	Steel	Tonne	200–250
10.	Sugar	Tonne (cane crushed)	1–2
11.	Textile	100 kg (goods)	8–14

d. Fire fighting needs

The fire fighting water demand is calculated based on the formula $100\sqrt{P}$ where P is the population in thousands for communities larger than 50,000. As per NBC-2005, it is mandatory to keep one-third of fire fighting water requirement as service storage. The balance may be kept in various tanks at strategic points. Thus the concept of underground storage and overhead storage in buildings have been developed. Normal convention is to provide overhead tanks of one-third capacity and underground reservoir of two-third capacity.

The normal fire-fighting tanker can cope with fires upto 25 m height of buildings. Beyond such heights, the fire fighting tanker cannot fight the fire and thus water storage with pumping system is to be provided. The pumping system shall be capable of delivering 2280/2850 LPM. The storage tank capacity may be 1,00,000 litres so that stored water quantity is sufficient to fight fire for 45 minutes. During this period, water supply can be recouped by receiving supply from municipal mains or from fire tanker filling the storage tank.

The pumping system usually in the basement, has to be designed to deliver 2280/2850 LPM water at 3 kg/cm^2 at the topmost floor (terrace).

In the dry riser systems the overhead tank of 5000/10000 litre is provided, as per NBC guidelines, for low rise buildings. A booster pump is provided on the terrace to boost the pressure of water received from overhead tank and deliver 900 LPM at 3 kg/cm^2.

The pump set has to be provided with pressure vessel and pressure switches, to keep the system pressurised. The pumps start operating, as soon as the fall in pressure is detected by the pressure switches. In order to prevent non-operation of pump in case of power failure or intentional shutting down the supply in case of fire, a dedicated pump of full capacity operated by generating set shall also be provided in the pump room and this pump shall automatically come into operation on detecting failure of power.

e. Peak values of water requirement

The values of per capita water requirement given in the preceding paragraphs indicate the average water consumption per day per person averaged out over a period of one year. In real life scenario, the water demand varies with season, month, day and even an hour. It is the hourly variation in consumption which is a cause of concern for pump system design. This variation is accounted for by considering peak rate of consumption, which is equal to average rate multiplied by a peak factor, as rate of flow for which distribution/pumping system shall be designed. The variation in demand will be more pronounced in small towns with less population and will even out in large towns with more population.

The following peak factors may be adopted for various population communities:

For population less than 50,000	3.0
For population range of 50,000–2,00,000	2.5
For population above 2,00,000	2.0
For small water supply schemes	3.0

f. Residual pressure

Distribution system should be carefully designed to deliver water at correct pressure to the consumer. The following pressure shall be maintained at the ferrule points:

Single storey buildings	7 m
Two storey buildings	12 m
Three storey buildings	17 m

Multi storey buildings shall be provided with booster pumps to achieve required pressure. The water in the tap shall be supplied at 0.7–1.2 kg/cm^2. In case of high rise buildings, the water pressure from over head tanks becomes too excessive for lower floors and this should be reduced by using pressure reducers after 3-4 floors from top and repeating the same every three floors.

13.17 TYPES OF VALVES

Valves are used in pumping system to isolate and drain pipe sections for testing, installation and repairs and to build pumping pressure at delivery. These are explained in brief as below as per the function performed by them.

13.17.1 Line Valves

These valves are provided to stop and regulate the flow of water in course of ordinary operations and during emergency. The principle consideration in deciding the location of valves is accessibility and proximity to special points such as branches and equipments like pumps, reservoir, etc.

13.17.1.1 Sluice valves

These are also called gate valves and are used for isolating a portion of pipeline. These valves can seal very well under pressure. These are rising spindle type (which rotates in a screwed attachment in the gate). The gate may be parallel sided or wedge shaped. For low pressure, gun metal seat may be used, but for higher pressures, stainless steel should be used. These valves shall not be used for continuous throttling for fear of erosion of seat and cavitation. Though these valves are simple, they need high force to operate against high unbalanced pressure. Power operated or manual operated actuators are also used to overcome this problem.

Some variants of sluice valves are needle valve and butterfly valve.

13.17.1.2 Butterfly valves

These valves are sometimes cheaper than sluice valves, occupy less space and are easy to operate. These valves produce slightly higher head loss than sluice values and the seal is not as effective as sluice valves specially at higher pressures. They offer higher resistance to flow as the disc obstructs the flow even when rotated to fully open position. Sluice valves as well as butterfly valves are not suited for operation in partly open positions as the gates and seats would erode rapidly. The butterfly valves with fixed liner design can withstand higher pressures and are available upto 16 kg/cm^2 rating.

13.17.1.3 Globe valves

These valves have a circular seal connected axially to a vertical spindle and hand wheel. The seat is a ring perpendicular to the pipe axis. The flow changes direction through 90° twice thus resulting in high head losses. These valves are normally used in small bore pipe works and as taps or control valves.

13.17.1.4 Needle or cone valves

Needle valves are more expensive than sluice or butterfly valves and are well suited for throttling flow. They have gradual throttling action as they close. They are generally used with counter balance weights, springs or actuators to maintain constant pressure conditions either upstream or downstream of valve to maintain a constant flow. The cone valve is a variation of needle valve. These valves are not used commonly in water supply but are occasionally used as water hammer release valves when coupled to an electric or hydraulic actuator.

13.17.2 Pressure Relief Valves

These valves are also known as overflow towers and keep the pressure in the line below given value by causing the water to flow to waste when the pressure builds up beyond the design valve. Usually they are spring or weight loaded and are not sufficiently responsive to rapid fluctuations of pressure.

13.17.3 Check Valves

These are also called non-return valves and automatically prevent reversal of flow in a pipeline. They prevent backflow when pump shuts down. The closure of the valve shall be such that it will not set up excessive shock conditions within the system. Dual plate check valves employ two spring loaded plates hinged on a central hinge pin. When the flow decreases, the plates close by torsion spring action. The dual plate check valves are more accurately designed to suit varying load conditions.

13.17.4 Ball Valve or Ball Float Valve

These valves are used to maintain a constant level in the reservoir. In both cases, the float follows the water level and permits the valve to admit more water till it reaches a predetermined level.

13.18 SELECTION OF A PUMP

The right selection of a pump is the key for the successful operation of the system for which it is designed. If it is simply drinking water or flushing water application, the correct design will ensure that all occupants get sufficient water throughout the usage period whereas if the application is related to some industrial process or HVAC application, pumping of correct quantity of water at desired pressure means fail proof operation of the system.

The following steps are involved while selecting a pump:

i. Determine quantity of water to be pumped and the pumping head against which the water is to be pumped.

ii. Select the pump either manually using pump curves, or with softwares. For the purpose of this chapter and also to understand the concept clearly, it is first necessary to learn to use manual selection procedure. Generally all manufacturers provide softwares to accurately select the pump.

iii. Get family curves of different pump manufacturer's (Fig. 13.30). Every pump manufacturer publishes a family curve of pumps manufactured by them. These curves make it possible to do a preliminary pump selection by looking at a wide range of pump casing sizes for a specific impeller speed. These curves helps in narrowing down the choice of pumps that will satisfy the system requirements. These curves are not detailed ones—they are simply a road map to tell you which specific pump fit your flow and head requirement.

As can be seen from the curve, there can be various pumps suitable for pumping 2.5 LPS at varying heads of 9, 12, 14 meters. A pump pumping 2.5 LPS at 12 meters head is selected.

iv. Get typical pump performance curve Fig. 13.31 for the pump selected from family curve. A lot of information is crammed into one chart and this can

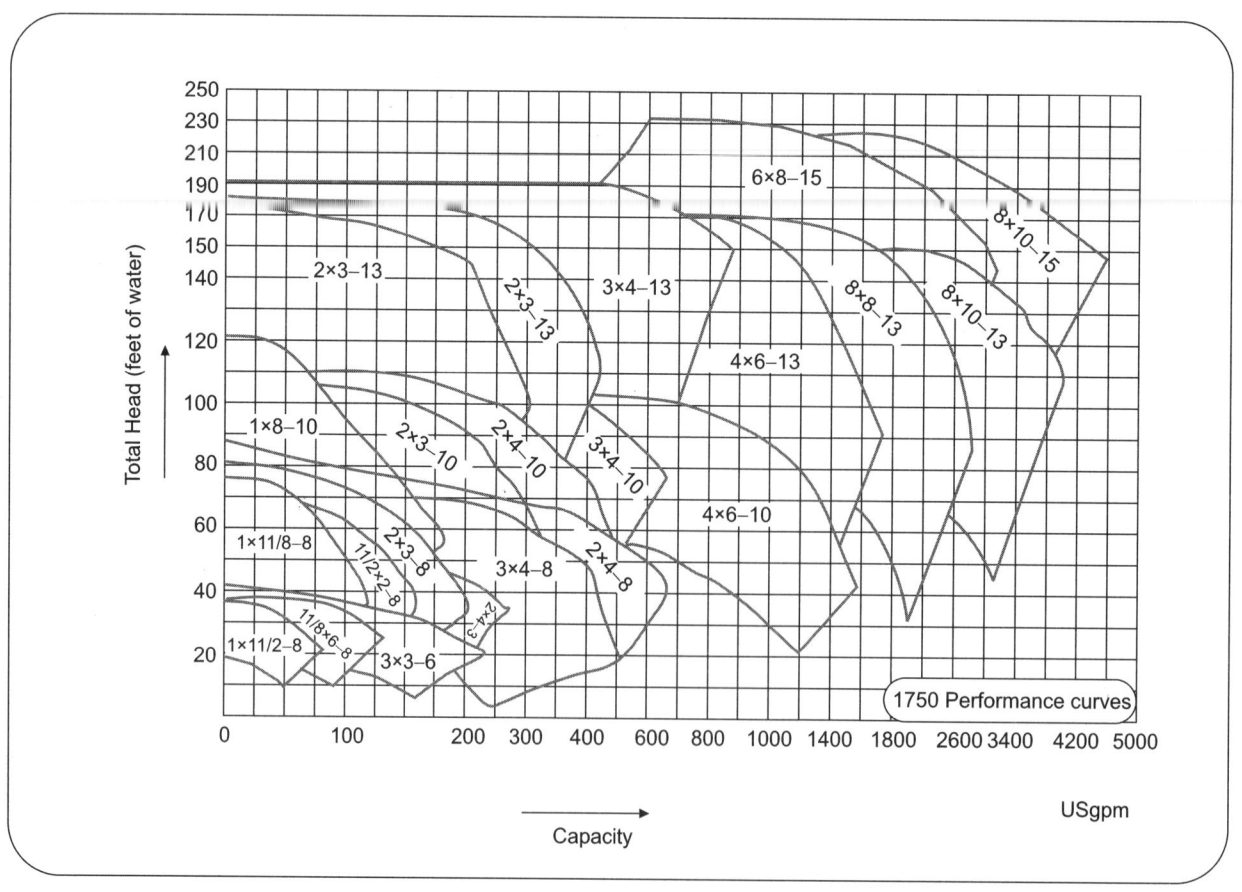

Fig. 13.30 Family curve of pumps

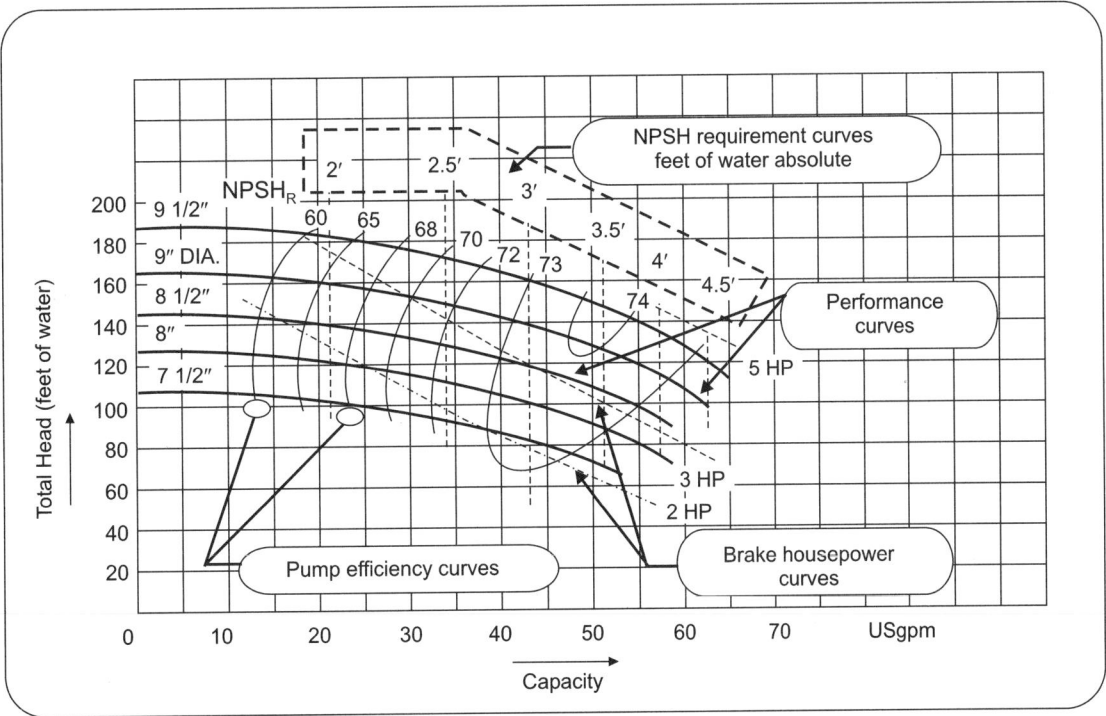

Fig. 13.31 A typical pump performance curve

be confusing at first. The performance curve covers a range of impeller sizes which are shown in increments of 1/2″ from 7.5″ to 9.5″. It may be noted that manufacturers usually manufacture impellers to the largest size and sell them by trimming their diameters to the required one.

Performance Curve: The curve sloping down wards from left to right represents the flow versus head performance with different impeller diameters. At zero flow, the head is called shut off head.

Efficiency Curve: The U-shaped curves are the pump efficiency curves and represent the points of constant efficiency for various combinations of flow, head and impeller diameters.

BHP Curve: The performance curves also show BHP curves (dashed and sloping downward from left to right)

System Curve: The calculation of total head at different flow rates produces a plot of total head versus flow and is called system curve. The operating point is the point where system curve intersects performance curve.

v. Plot the parameters and evaluate selection.

Plot the flow and head on pump performance curve and determine the required impeller diameter, BHP and efficiency of pump. The impeller diameter can be determined by inter-polation on the curve.

The selection of pumps impeller diameter, efficiency and BHP from other series or manufacturer's curves shall also be performed to select the best efficiency pump.

The rule of thumb shall be as follow:

a. When several pumps can do the job, the most efficient one should be selected. Select the pump operating at best efficiency point (BEP).

b. Do not extrapolate beyond the largest impeller diameter shown on the curve. This is the largest diameter that can fit in a casing.

c. Do not extrapolate beyond the right end of the curve. Pump operation will be erratic with unstable flows and heads.

d. Do not extrapolate below the minimum impeller sizes. A minor deviation can be allowed but add false head by partially closing a balancing valve so as to be able to use this diameter. But significant reduction from minimum diameter shall not be used.

e. Select the BHP of the motor required to operate the pump, e.g. all points on the performance curve to the left of 2HP will require a motor of 3HP.

f. Meet NPSH requirement. The pump manufacturer's specify a minimum NPSH requirement (refer Fig. 13.32) in order for the pump to operate at its design capacity. These are the vertical dashed lines in Fig. 13.32. The

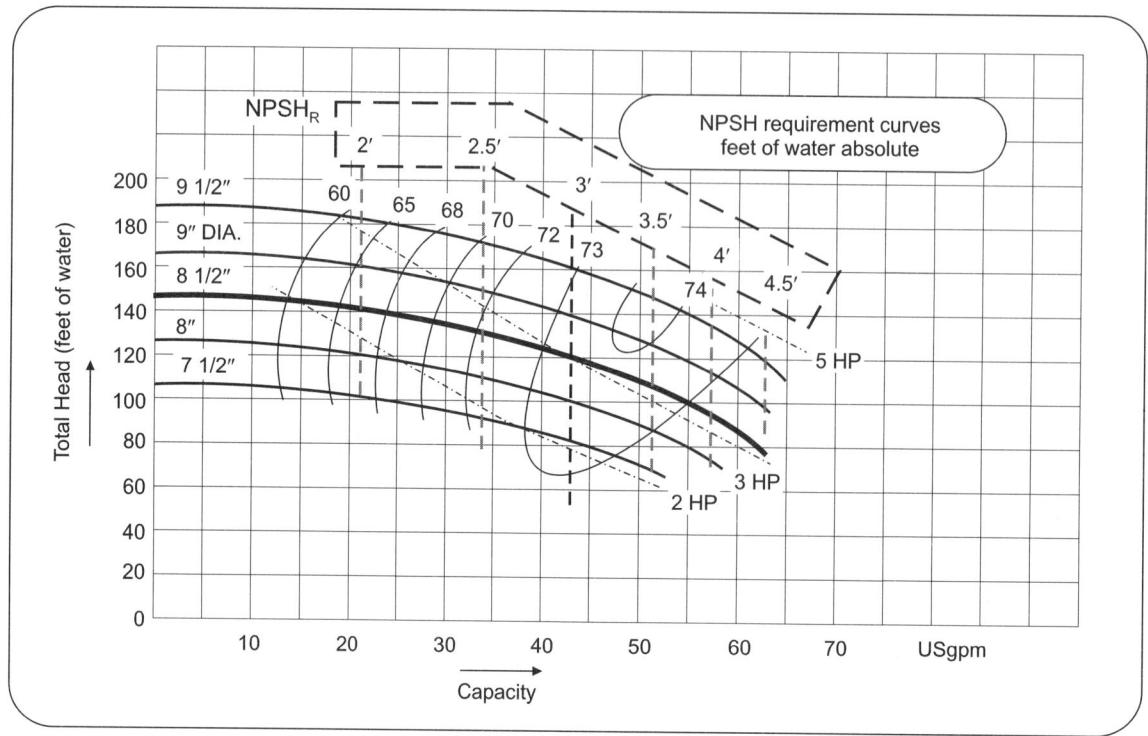

Fig. 13.32 NPSH requirement curves

requirement of NPSH becomes higher as flow increases and vice versa. This typically means that more pressure head is required at pump suction for higher flows than lower flows. NPSH is a head term and independent of fluid density. Meeting NPSH requirement is more stringent when pumping hot water than while pumping cold water as water's boiling point is reached earlier when pumping hot water.

g. Select an impeller between 1/3 and 2/3 of the impeller range allowed for that casing. This will help in assuring future expansion simply by replacing the impeller to a higher diameter without having to replace pump casing and motor.

Also, ensure not to go too left or right of the BEP (Fig. 13.33). A guideline is to locate operating point between 110% and 80% of the BEP flow rate with an operating point in the desirable impeller selection area.

h. Obtain the system curve for the pump (Fig. 13.34) and superimpose on the perform-ance curve (Fig. 13.35). The dashed line represents performance curve of pump and pump can operate only on its performance curve. The system (to take care of friction head, equipment head and velocity head, all of which are flow dependent) can run some where on

system curve. The intersection of these two define operating point which is the only point where pump and system can operate.

vi. Safety factor on total head or capacity:
If the flow and head as calculated are plotted on performance curve, (Fig 13.36), the operating point will be point 1 on impeller curve A (on system curve a). However, it is a good idea to apply a safety factor of 10% to take into account error in frictional losses or some unforeseen losses. In this case, impeller 'C' is selected. The operating point is now at point '2'. The system curve has also changed to 'c'. However, if it turns out that there was no error in calculating losses, and we are operating with curve 'a', then the operating point will shift from point '2' to '3' on impeller curve 'C'. If we need to get back to flow corresponding to point '1' for process reasons, then throttling a pump valve at pump discharge will change the operating curve to match curve 'C' and bring the operating point back to point '2'.

Similarly, if safety margin is applied on capacity, say 10%, we will have to select impeller 'B'. The system curve in reality is curve 'd'. But if our original flow estimate was correct, then the operating point will shift from point 4 to 5. To get back to the original flow, we have to throttle back so that we shift to point 6, and we will operate on a new system curve 'b'.

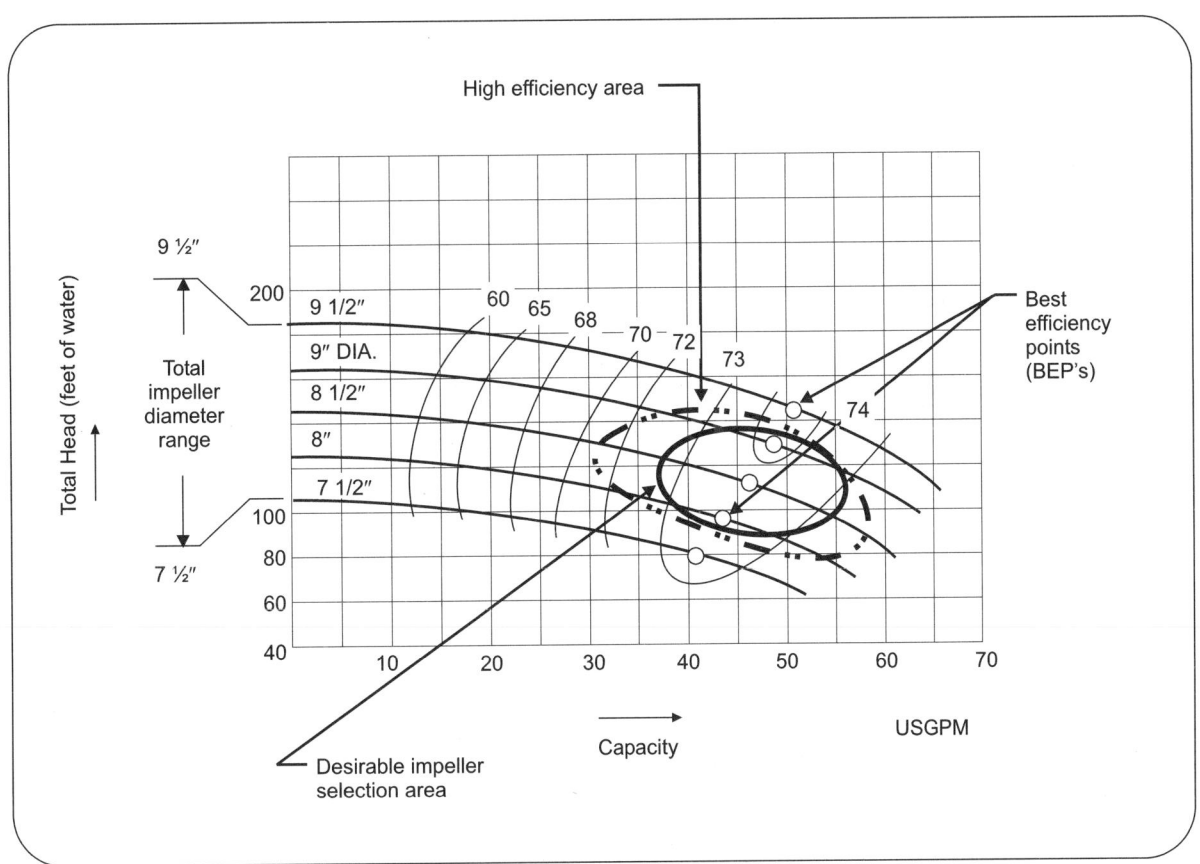

Fig. 13.33 Desirable pump selection area

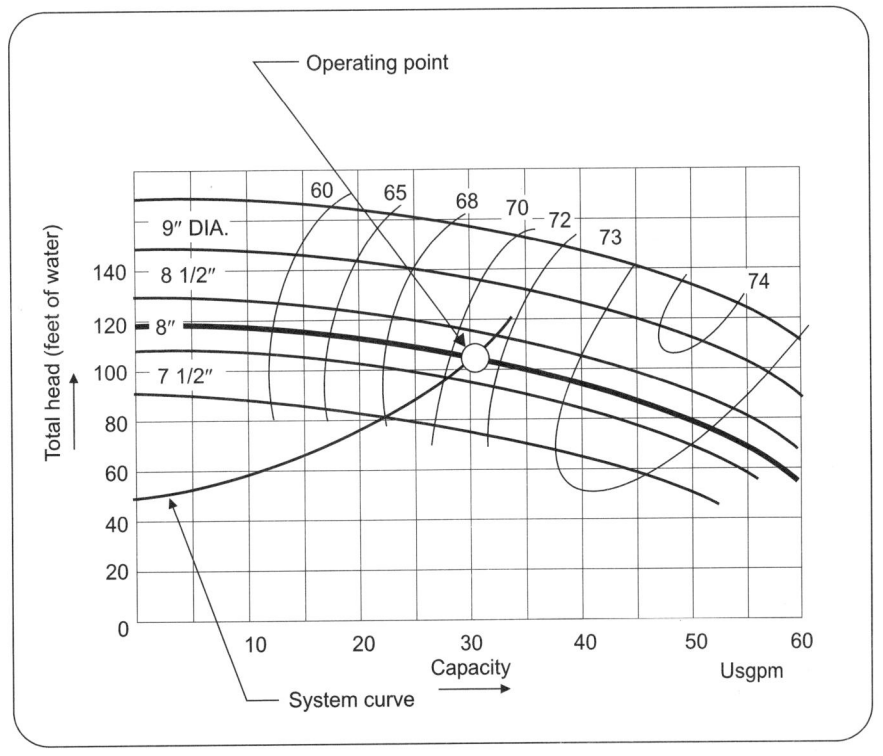

Fig. 13.34 System curve

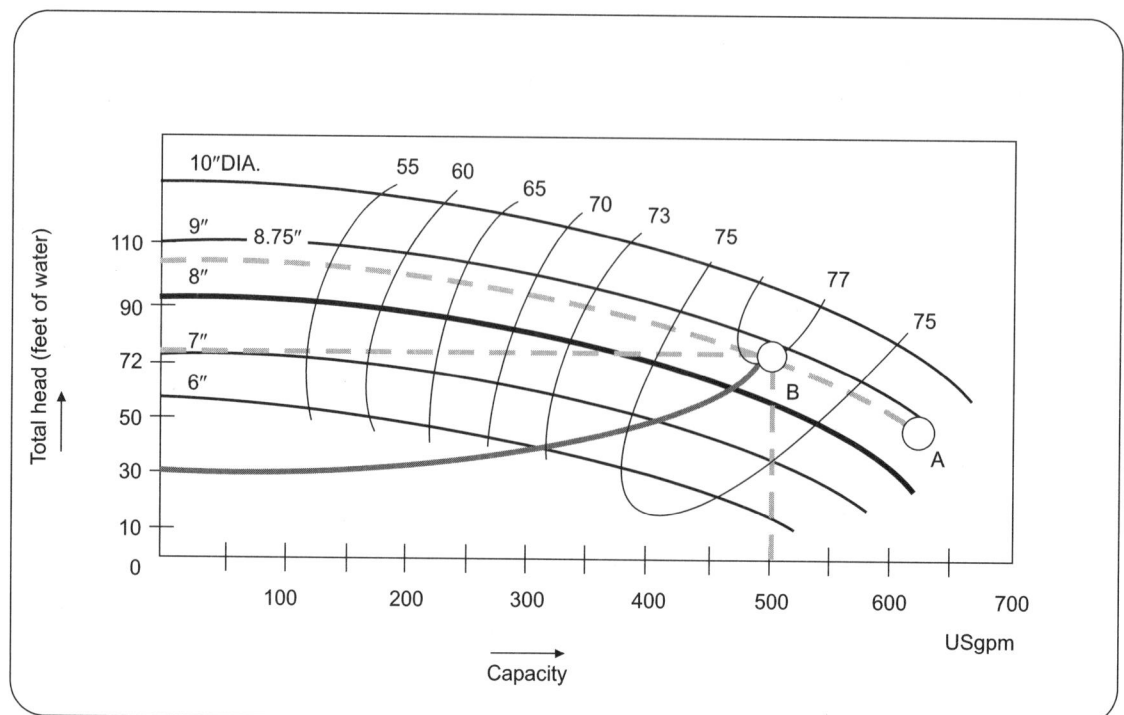

Fig. 13.35 Location of operation point

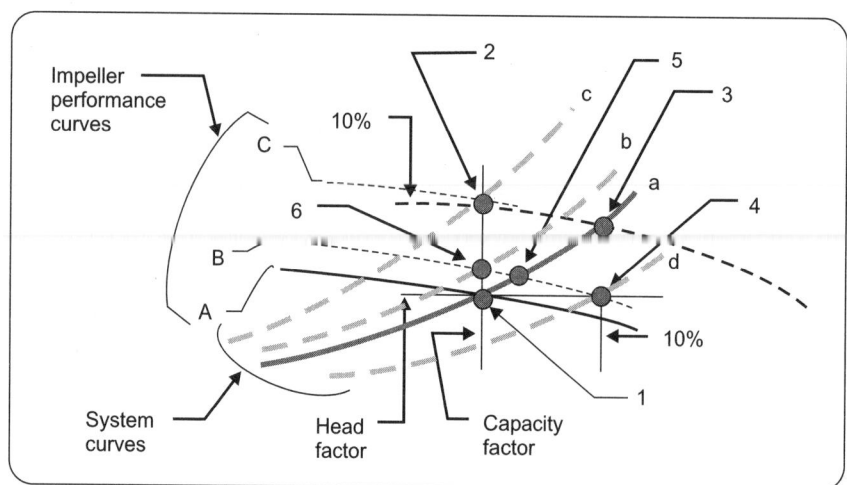

Fig. 13.36 Curve with 10% safety factor

13.19 METHODS OF STARTING A PUMP V_s OPERATING POINT

There are two methods with which a pump can be started but reaching the operating point is altogether a different scenario.

a. **Discharge valve closed:** This method is normally used for pumps with discharge more than 30 LPH. The discharge valve is kept closed while the pump is started. Immediately after start, the discharge head rises to 'D' (Fig. 13.37). The system curve is vertical (on y-axis). As the valve is gradually opened to full open position, the point 'D' travels

along performance curve and settle at point 'C'. The system curve's shape gets progressively changed during this process.

b. **Discharge valve open:** When the pump is started keeping the discharge valve opened (Fig. 13.38) the pump produces little head and flow. As the pump accelerates, it will intersection the system curve (fixed in this case) at point A at 200 RPM, point B at 500 RPM and finally at point 'C' at the full speed of pump and motor. This happens very fast and in a matter of few seconds. The disadvantage with this arrangement is that it leads to water hammer,

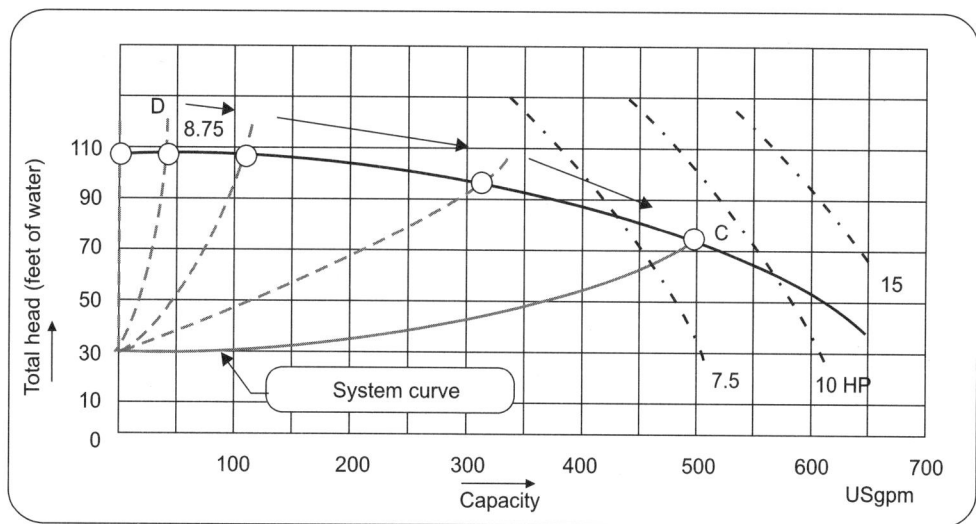

Fig. 13.37 Starting the pump with discharge valve closed

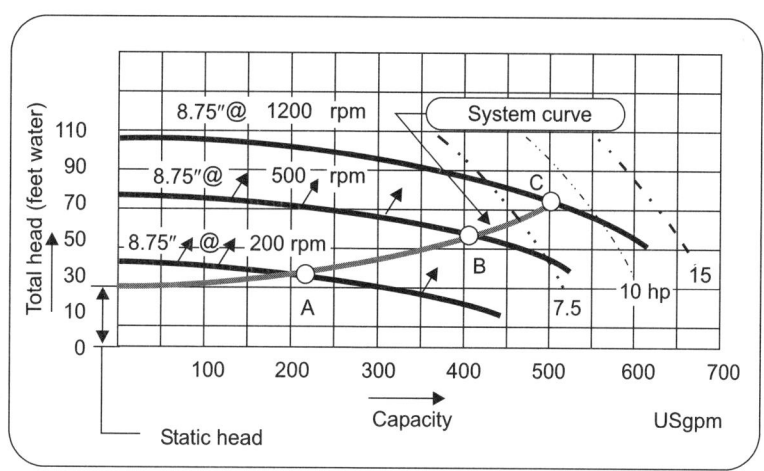

Fig. 13.38 Starting the pump with discharge valve open

shaking and bursting of pipes and equipments. This arrangement may be preferable for smaller sized pumps.

13.20 ACHIEVING EFFICIENCY AND ECONOMY IN PUMP SELECTION

Use of pumps in any building, industry or town for water/sewerage pumping is the source of maximum power consumption. The pump motors capacity in a typical commercial building or a high rise apartments for HVAC and water pumping needs range from 75 Kw to 200 Kw. Unfortunately, the scale of economy that can be achieved by paying little attention during designing or while system has run for some time, is often ignored. The result is that the end user unknowingly bears the electricity cost associated with the uneconomical or inefficient design.

This section describes in detail, the various parameters to be looked into while designing and operating the pumping system.

13.20.1 Effect of Speed on Pump Selection

The system designer shall first understand the effect of speed of pump motor on selection of a pump. The common relationship between speed, flow, head and power is

a. flow increases in proportion to speed increase
b. head increases as the square of speed increase
c. power consumption, the most critical element, varies by cube of speed increase.

When speed increase is so annoying in terms of power consumption, then why not to give a careful thought, if application under consideration can be served with low RPM motor. Let's examine case by case:

i. **Primary pumps in HVAC system:** The primary pumps in HVAC system circulate high quantity of cold water within AC plant room and thus is an application of low head, high flow as there are no equipments like heat exchangers on this circuit which may cause friction loss.

Thus low speed pumps (1450 RPM) will serve the most economical selection.

ii. **Secondary pumps in HVAC system:** A secondary pump has to pump the water through chiller, Air Handling Units and on the way encounter friction loss in long length of pipes, valves and above all achieve a high static head. Thus it is an application of high flow and high head. Thus a high speed pump (2900 RPM) will be more suitable. However, the option will have to be weighed between the scale and height of building. It may, at times, be possible to use low speed pump for smaller and low rise buildings.

iii. **Economics of 1450 *Vs* 2900 RPM motor and pumps:** The motors with 1450 or 2900 RPM don't vary much in cost, but 2900 RPM pump requires a smaller casing and smaller impeller diameter. The overall cost difference will favour 2900 RPM. The factor against 2900 RPM motor selection is noise criteria and NPSH requirements. In chiller room, boiler rooms, and mechanical ventilation like applications, noise is never a criteria, but it may be major source of noise if installed close to sitting or work area. If installed in AHU, then acoustic treatment and elastomer piping connection are required to reduce transmitted noise. Lower RPM require lesser NPSH and hence preferred to prevent cavitation problems when handling hot water or any such application where vapour pressure can be reached easily.

v. **Methods and need to control speed:** In systems with highly variable loads, the pumps that are sized to handle the largest loads may be oversized for normal operating loads. In such situations, the variable speed motors or variable speed drives may provide a good solution to conserve power, though multiple pumps operating in parallel can also be used in some situations. The same situation of oversizing is encountered where safety margins are intentionally introduced either for future expansion or fear of wrong calculations. Since power is related to third power of speed, decreasing the speed from 1800 to 1200 rpm results in a 33% decrease in flow and in turn effecting 70% decrease in power.

For future expansions, the pump casing and impeller diameters are so selected that there is room for future expansion by simply replacing with a larger diameter impeller and also selecting motor which shall be suitable for the largest diameter impeller to be used in future.

Though flow can be controlled by using control valves but energy consumption is not reduced. The speed control has other benefits too like:

a. bearing life is increased as the forces are reduced by square of speed and bearing life is improved by seventh power of speed.

b. vibration and noise are reduced with an increase in real life, provided that duty point remains within an allowable range of best efficiency point (BEP).

The speed of motor can be controlled by two methods. One is to use multi speed motor and the other but better method is to use variable speed drives (VSD). These drives provide speed adjustment over a continuous range, avoiding the need to jump from one fixed speed to another as in the case of first option.

To understand how the duty point changes with speed variation. The performance and system curve are considered. There can be two situations, one having low static head compared to friction head and the other having high static head.

Case I: High friction head

Consider Fig. 13.39 where reduction in speed causes the operating point to move along the line of constant efficiency. Thus operating point, relative to its best efficiency point, remains constant and the pump continues to operate in the ideal region. The affinity laws are valid here which means there is substantial reduction in power. Thus the variable speed control method is the ideal choice in this case.

Case II: High static head

In the pumping systems, where the static head is considerably high compared to friction head, the operating point of the pump moves across the lines of constant pump efficiency on changing the speed (Fig. 13.40). The reduction in flow is no longer proportional to speed variation. A small reduction in speed reduces the flow and efficiency by a large margin. This may create a situation of pump damage if it ran for an extended period of time even at a lower speed. At the lowest speed (1184 RPM), the pump does not generate sufficient head to pump any liquid into the system (40 m as against 45 m required). At this point, pump efficiency and flow are zero while energy is still being consumed by the pump. The pump behaves like a water heater, churning the same water again and again, causing temperature to rise quickly to damaging limits.

As is seen here, there are no economical benefits achieved in case II scenario by controlling the speed when static head is high. Usually it is advisable to select pump to be operating to the right of BEP, so that on reducing the speed, the efficiency will first increase and

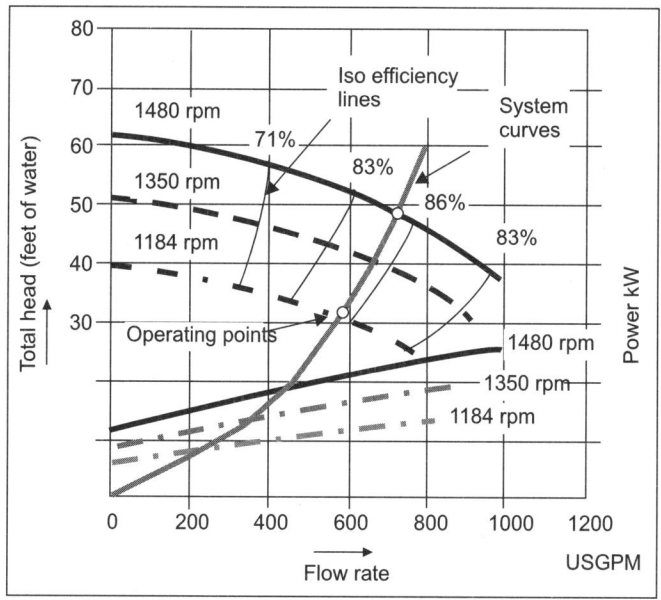

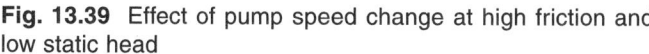

Fig. 13.39 Effect of pump speed change at high friction and low static head

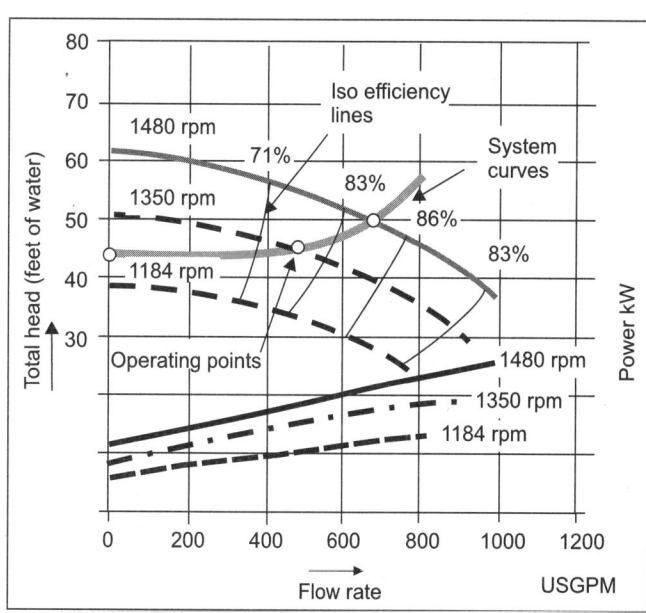

Fig. 13.40 Effect of pump speed change at high friction and relatively high static head

then reduce. This can extend the useful range of variable speed operation in cases of high static head systems.

13.20.2 Operating Pumps in Parallel

Significant saving in energy can be achieved by operating the pumps in parallel by turning off one pump when demand is lower. Parallel pumps are useful when static head is more than 50% of total head. Figure 13.41 shows pump curves for one, two and three pumps operating in parallel. The combined pump curve is obtained by adding the flow rates at a specific head. The system curve is usually not affected by the number of pumps running. The operating point on their

performance curves moves to a higher head when more pumps are started and hence lower flow rates per pump. The flow rates with two pumps running is not double that of a single pump. But if the system head were only static or a large proportion of the total head, then flow rate would be proportional to the number of pumps running.

13.20.3 Stop/Start Control

In this method, the flow is controlled by switching pumps ON or OFF. The system comprises of a storage tank to give cushion for water supply during off period. The pump can thus be selected to operate at the best

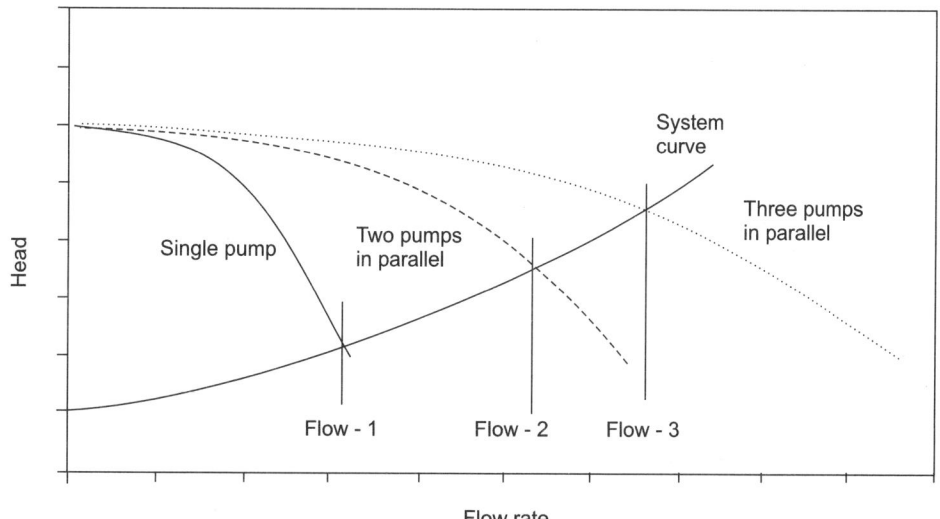

Fig. 13.41 Performance curves for pumps in parallel

duty point to give least power consumption and there will be no consumption when pump is OFF. If this approach is acceptable, then this is the best energy saving option available.

The pump can be selected at as low flow rates as system permits. Low flow rates minimises frictional losses in the pipe and a small pump can be installed. Reducing the pump flow capacity to half but operating twice as long can reduce energy consumption to a quarter.

13.20.4 Eliminating Discharge Valve (Throttling)

In this method, the pump runs continuously while flow is adjusted by opening/closing the discharge valve at the outlet of pump. Though this method reduces the flow, but energy with motor running at constant speed, is not reduced. Figure 13.42 shows how the operating point moves upward and to the left when a discharge valve is half closed.

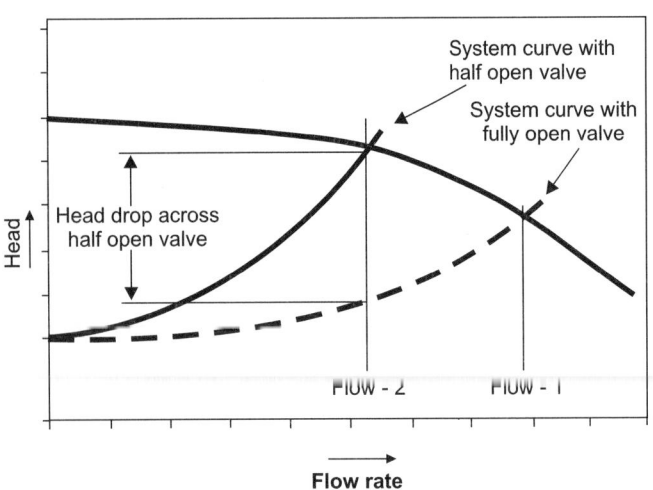

Fig. 13.42 Adjustment of flow with discharge valve

13.20.5 Eliminating by-pass Control

This method of flow control also suffers with the same disadvantage as throttling method since reducing the flow does not reduce the energy consumed by the motor because reduction in demand of water is achieved by bypassing the pumped water.

13.20.6 Impeller Trimming

The affinity laws are applicable to impeller diameter also. A change in impeller diameter changes impeller's peripheral velocity. The variation of performance with impeller diameter 'D' takes place as follow:

$$Q \propto D$$
$$H \propto D^2$$
$$P \propto D^3$$

Changing impeller diameter is an efficient way to control pump flow rate and also energy consumed. But this option applies to changing of diameter within a fixed casing which is a standard practice for making small permanent adjustments to the performance of a centrifugal pump.

The diameter can be reduced to a maximum of 75% of original beyond which vibration due to cavitation takes place and also reduces efficiency. The impeller diameter reduction by 25% effects reduction in flow by 50%.

This situation is different from the one where speed is controlled to control the flow. Reduction of speed by 50% hardly reduces the efficiency by 1 to 2% as mechanical losses in seals and bearings (< 5% of total power) are proportional to speed, rather than speed cubed.

Impeller trimming is the process of machining the diameter to reduce the energy added to the system fluid. This method is particularly useful when a conservative design may add to system load and power consumption which can be corrected later by trimming to reduce the power consumption. This method is economical also as compared to buying a new impeller of smaller diameter. At times, the smaller impeller diameter may be too small for the pump load and thus trimming remains the only option (Fig. 13.43).

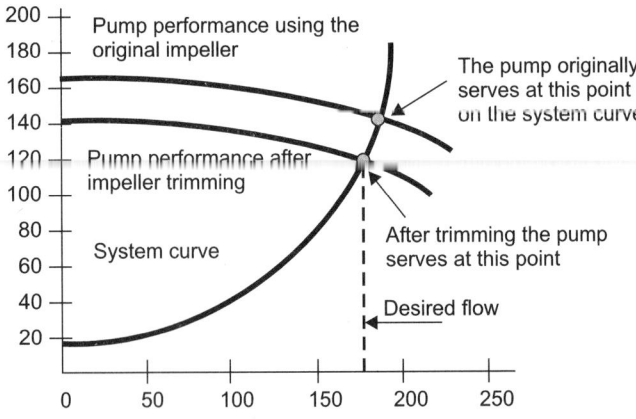

Fig. 13.43 Pump performance with change of impeller diameter

13.20.7 Smaller Steps in Energy Conservation

There are number of steps one can take to conserve energy by improving the efficiency of pumps and pumping systems.

i. Pump should be operated near their best efficiency point.

ii. Comply with NPSH requirements.

iii. Use booster pumps for small loads requiring high pressures.

iv. Balance the system to minimize flow and ultimately reducing power consumption.

v. In multiple pump operations, carefully combine operation of pumps to avoid throttling and by passing.

vi. Replace old pumps with energy efficient pumps.

vii. Reduce number of stages in multistage pumps if margins in pressure exists.

viii. Check vibration, bearing damage, unbalance and see if pump is operating away from its BEP which may by the cause of this unbalance.

ix. Repair seal and packing to reduce water loss by dropping.

x. Reduce system resistance by carefully selecting pipe material (having low friction), bends, no. of valves and other accessories thus reducing head requirement of a pump.

13.20.8 Comparison of different Energy Conservation Techniques

A comparison of energy consumption with various methods of control is given in Table 13.11.

Table 13.11 Comparison of various methods of control

Parameter	Discharge control valve	Trimming of impeller	Variable speed drive
Impeller diameter	430 mm	375 mm	430 mm
Pump head	71.7 m	42 m	34.5 m
Pump efficiency	75.1%	72.1%	77%
Rate of flow	80 m³/hr	80 m³/hr	80 m³/hr
Power consumed	23.1 kW	14 kW	11.6 kW

14

Illumination Engineering

The most beautifully designed building will not come to life unless it is illuminated effectively from inside as well as outside Facade lighting of monuments, buildings of national importance and other high rise buildings has become a trend of the day to invite tourism.

Comfort illumination for providing an efficient working atmosphere, gallery lighting for painting exhibitions, departmental store lighting and special effect lighting are gaining more and more importance to get more output from an employee and to attract more customers.

With increased focus on development of infra-structure, highways have given way to express ways allowing for high speed. This has given rise to the requirement of effective illumination for vehicles driving at 100–140 km at ease along highways. Inter-section lighting has to be carefully designed to avoid accidents.

The concern all over the world to reduce night lighting pollution now requires designing lighting luminaires that restrict the light spread on road surface and do not let the light spread in upward direction. Energy conservation norms have given rise to development of technological inventions like highly efficient reflectors, LED lights, SCADA system for dimming the lighting levels in the late night hours.

This chapter outlines the designing principles for calculation of illumination in a room, on roads and spacing of luminaires. Though calculation of illumi-nation, uniformity of illumination is a complex procedure and is well taken care by the software easily, but it is important for design engineers to understand how the lighting engineering has evolved that led to the development of softwares.

14.1 TERMINOLOGY

1. **Light:** A form of energy produced in the form of radiation by the heating or burning of a heat source say coal, wood, etc. The light travels at a velocity of 3×10^8 meters per second. Figure 14.1 shows the spread of wavelength of light energy. Radiation emitted in the short wavelengths ranging from 380 nm to 780 nm are of great importance for lighting engineers because it is this wavelength which produces impression of an object in the eye. This range lies in between ultraviolet and infrared radiations, both of which are invisible. The most important aspect of visible radiation is the creation of light as well as colour impression. If a ray of light is passed through a prism, a colour band of spectrum is produced with red colour at one end having longest wavelength and violet on the other having shortest wavelength.

2. **Photometry:** The measurement of light is known as photometry. Various parameters to measure light and their symbol, formulae are given under-neath:

   a. **Candle power:** This is the amount of light pro-duced by a whale oil candle.

   b. **Luminous flux, F:** This is the rate at which light energy flows from the source represented as energy per second and expressed in lumens (symbol lm).

   c. **Steradian:** Solid angle is measured in stera-dian. On the surface of a sphere of radius 1m, imagine an area of 1 sq meter. The shape of this surface is immaterial. If the radius of this sphere is allowed to follow the perimeter of this area, then radius describes a cone (Fig. 14.2) which encloses a solid angle defined as

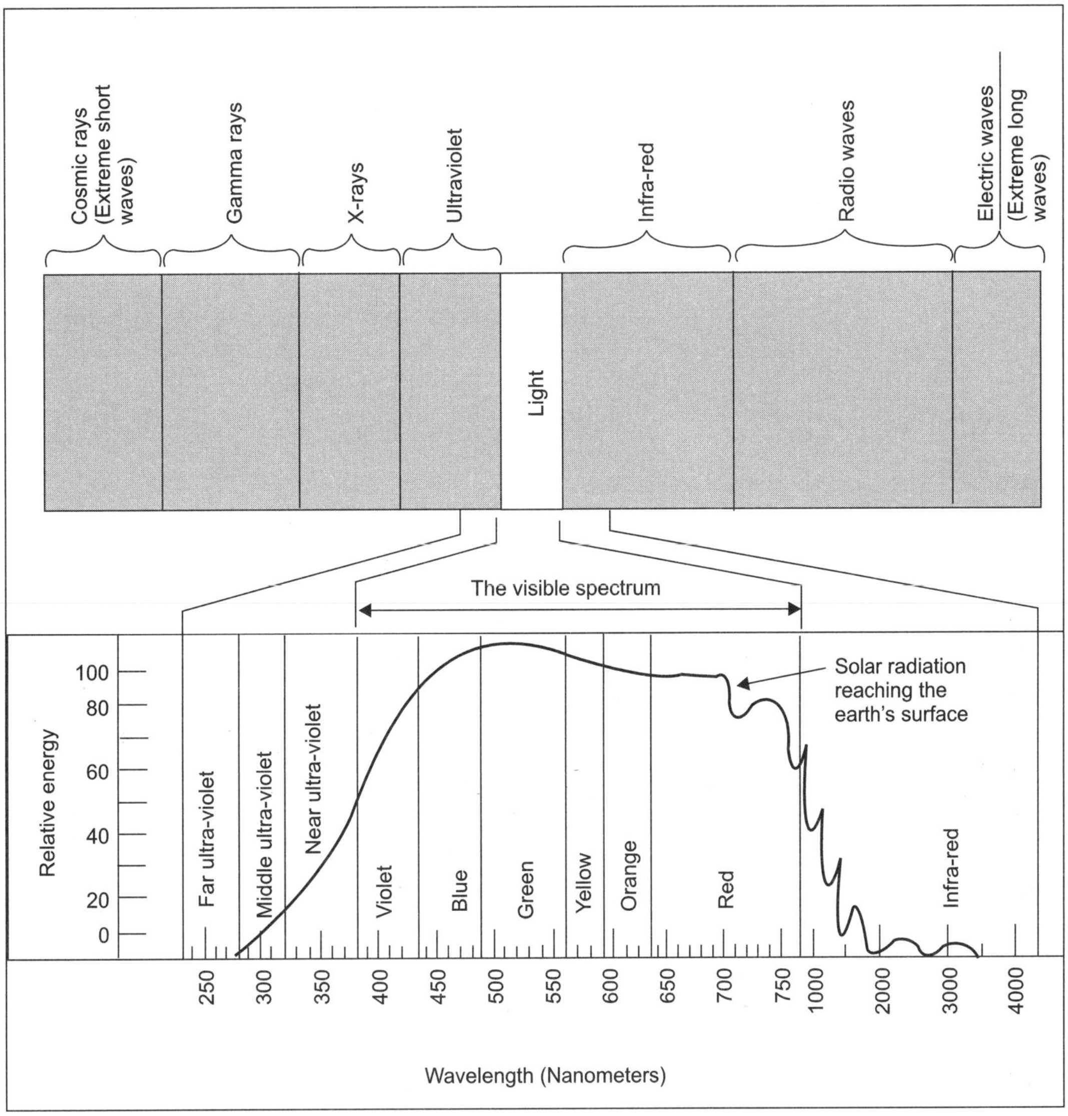

Fig. 14.1 Light spectrum

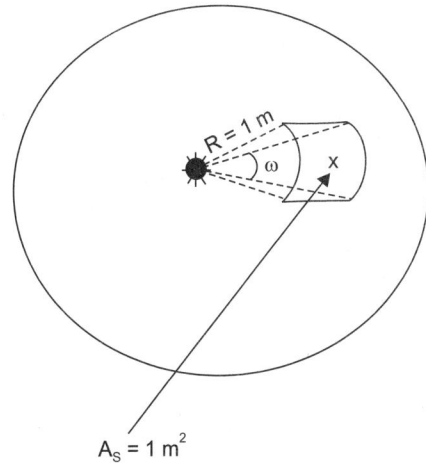

Fig. 14.2 Solid angle in steradian

$$\omega = \frac{A}{r^2}$$

d. **Luminous intensity, *I*:** It is the measurement of intensity of light in a given direction measured in candela or candle power (symbol cd). It is the luminous flux radiating per unit of solid angle in a given direction (Fig. 14.3) defined as:

$$I = \frac{F}{\omega}$$

where, I = luminous intensity in candela
F = luminous flux in lumens
ω = solid angle in steradian (st)

In the Fig. 14.4
$\theta_1 > \theta_2$

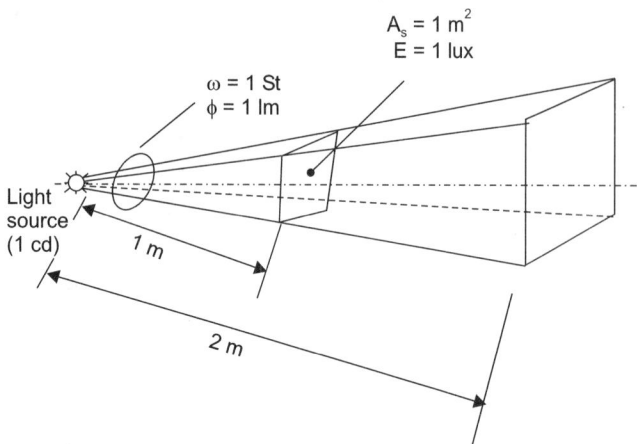

Fig. 14.3 Luminous intensity on a segment of unit sphere

But $\quad I = \dfrac{F}{\theta}$

then, $\quad I_2 > I_1$

But, $\quad \omega = \dfrac{A}{r^2}$

and $\quad I = \dfrac{F}{\omega}$

then $\quad I_1 > I_2$

$\angle$ AOB (radians) $= \dfrac{AB}{R}$

ω (solid angle) $= \dfrac{\pi}{4}\dfrac{(AB)^2}{R^2}$

If $\theta = 90°$ or $\dfrac{\pi}{2}$ radians, ω is 2π steradian

If $\theta = 180°$ or π radians, ω is 4π steradian

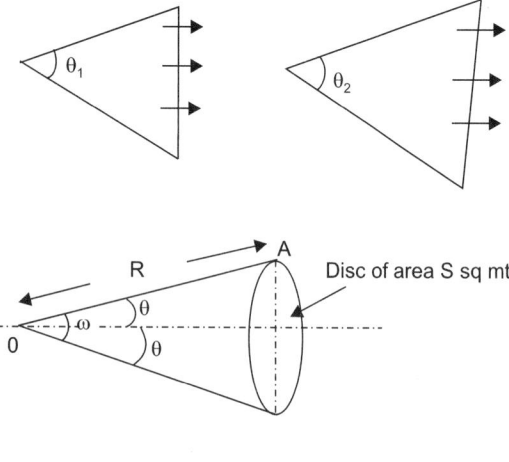

B

Fig. 14.4 Illustration of solid angle

e. **Illuminance, E:** The lighting level at a point on a surface, defined as luminous flux falling on a unit area of the sphere under consideration is called illuminance. Illuminance is measured in lumens per square meters (lm/m²) or simply lux. In FPS system, the unit of illuminance is foot candles or lumens per square feet as illustrated in Fig. 14.5.

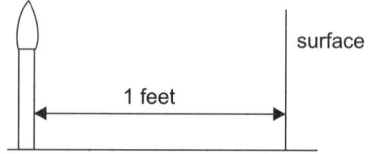

Fig. 14.5 Illustration of foot candle

$$E = \dfrac{F}{A}$$

where, E = illuminance in lm/m² or lux

$\quad F$ = luminous flux

$\quad A$ = Area of surface

But $I = \dfrac{F}{\omega}$

and $\quad \omega = \dfrac{A}{r^2}$

Thus, $\quad I = \dfrac{F}{A/r^2}$

or $\quad F = \dfrac{IA}{r^2}$

$$E = \dfrac{IA}{r^2 \times A}$$

$\Rightarrow \quad E = \dfrac{I}{r^2}$ (Inverse square law of illumination)

This equation is known as inverse square law of illumination. If the light falls on a surface at an angle θ as in Fig. 14.6, then the illumination on the surface is given by:

$$E = \dfrac{I}{r^2}\cos\theta \dots \text{(Lambert's cosine law)}$$

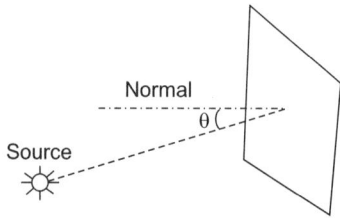

Fig. 14.6 Illustration of Lambert's cosine law

This is known as Lambert's cosine law. Let's consider the application of inverse square law of illumination.

In Fig. 14.7, the light falls on three surfaces which are at a distance of d, $2d$ and $3d$ from the light source.

Thus,
$$E_B = \frac{I}{d^2}$$
$$E_C = \frac{I}{4d^2}$$
$$E_D = \frac{I}{9d^2}$$

Thus light illuminance E on surface C and D is one-fourth and one-ninth respectively of illuminance on surface B. But the area C and D illuminated by the same ray of light is four times and nine times respectively of the area of surface B.

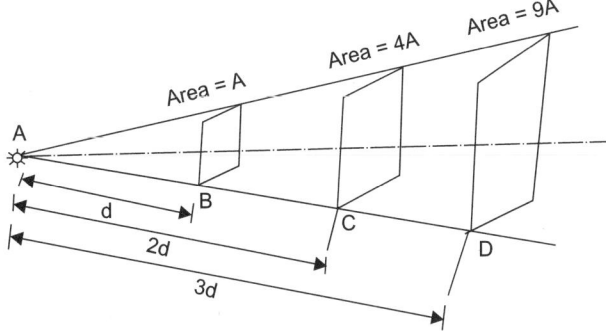

Fig. 14.7 Principle of inverse square law of illumination

Surfaces at right angle to ray of light are termed as normal to ray and illumination is termed as normal illumination. The illuminance E on a book held normal to ray of light is equal to intensity I of lamp source divided by square of distance in this angular direction. Note that I value is the value in angular direction considered (Fig. 14.8a).

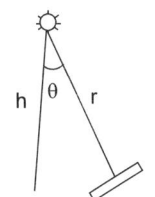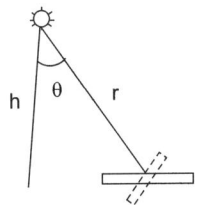

(a) Ray normal to object (b) Ray at an angle to object

Fig. 14.8 Ray normal and angular on an object

Thus, $E = \dfrac{I}{r^2}$

But, if the book is not placed normal to ray, but lying at an angle to ray, say horizontally on a table, then the illuminance is calculated as before considering the book at normal to the ray but multiplying the result with cosine of angle between ray of light and perpendicular drawn from source of light to illuminated surface (Fig. 14.8b). Illuminance on horizontal surface is:

$$E = \frac{I \cos \theta}{r^2}$$

Note: I value is taken from polar curve in the angular direction considered

But, $h = r \cos \theta$

$$r = \frac{h}{\cos \theta}$$

$$E = \frac{I}{r^2} \cos \theta$$

$$= \frac{I \cos^3 \theta}{h^2}$$

f. **Luminous efficiency:** Relationship existing between light emission and power consumed from an output of a light source expressed as lumens per watt is called luminous efficiency. The radiation emitted in yellow and green areas of spectrum is said to have high luminous efficiency because it produces many lumens per watt of energy consumed.

g. **Luminance of a light source, L:** This is defined as the luminous intensity per unit projected surface area of the light source in that direction, i.e.:

$$L = \frac{I}{A}$$
$$= \frac{F}{\omega \times A}$$

The unit of luminance of light source is cd/m².

h. **Luminance of an illuminated surface, P:** It is defined as luminous intensity per unit area of the surface in that particular direction.

$$P = \frac{r \cdot E}{\pi}$$

where, P = luminance of a surface, cd/m²
E = illuminance of that surface in lux
r = reflectance factor of the surface.

14.2 TYPES OF LAMPS

Thomas Alva Edison, the great scientist who discovered filament, incandescent lamp, gave the most beautiful gift to humanity in 1879 as it lights millions of millions of houses today. Over a period of time different technologies have developed which light up homes, streets, signboards, architectural lights for monuments and in numerous other applications. Let us explore the variety of lamps available these days very quickly:

1. **Incandescent lamp:** Modern day lamp is made of tungsten filament having melting point of 3400° C. The luminous efficiency (lumen per watt), which is proportional to 5–6th power of absolute temperature of surface, is also very high (15 lumens per watt). The life is approximately 1000 hours. The

lamp is still considered to be the cheapest option available as far as first cost is considered but consume more power as compared to other options described further ahead.

2. **Halogen lamps:** These lamps are made with tungsten coil inserted in a cylindrical quartz envelope which reduces the size considerably. The envelop contains argon gas and a small quantity of iodine or bromine. The evaporated tungsten is returned back to filament in a chemical process. Thus there is no blackening of tube and the lamp gives a constant light output throughout its life. Lamp life is determined by filament thickness and its evaporation rate. This also means a relatively low resistance of filament, hence low wattage lamps (50 or 100 watts) are used with low voltages only (12/24 volts). But in general, halogen lamp of 500 W or more are normally used at 230 V. These lamps are excellent in colour rendering and are extensively used in floodlighting, car lights, photography, projectors and sports lights.

3. **Fluorescent lamps:** These are low pressure gas discharge lamps. The electrical energy is converted to ultraviolet radiation which falls on fluorescent powder, coated on the inner walls of discharge tube, which glows to produce white light. The free electron moves from cathode to anode and thus produce a flow of current. The mercury (gas filled inside the tube) ions move slowly in opposite direction. This movement of free ions create a discharge creating illuminance of fluorescent powder. The fluorescent tube lights produce upto 80 lumen/watt these days. A 40 watt FTL is the most popular and cheapest form available producing 3200 lumens. Other options like 36 W, 26 W (T-5) options are also available, though costly but more energy efficient.

4. **T-5 fluorescent lamps:** These lamps are 16mm in diameter, more compact and produce high brightness. These are available in 14, 21, 28, 35 watts options and have high lumen output of 104 lumen per watt. These lamps work only on electronic ballasts. These days circular T-5 lamps with 24 W and 225 mm dia going upto 54 W and 300 mm diameter are also available. These lamps are also available in lower wattage ratings and high life of 20000 hours. These lamps are almost double the cost of convention FTL but save on energy and perform longer (four times the life of FTL).

5. **Compact fluorescent lamp (CFL):** These lamps have gained popularity, more as a substitute of incandescent lamp in energy consumption. The lamp ratings of 9 W, 13 W, 18 W and 25 W were developed to replace GLS lamps of 40 W, 60 W, 75 W and 100 W respectively. The luminous efficacy varies from 45–50 lumens/watt which is much higher than 15 lumens per watt of GLS lamp.

6. **High pressure mercury vapour lamps:** This is a variety of high intensity discharge lamp which includes metal halide and high pressure sodium lamps as well. These lamps contain mercury and argon gas in vapour form at low pressure. The mercury lamps usually take 3–4 minutes to reach full light output. This is basically due to time taken by voltage and current through discharge tube to stabilize from the moment it is switched 'ON'. The mercury lamp takes about 10 minutes to re-ignite once they are switched 'OFF' or due to voltage flicker. This is due to the reason that resistance to passage of current is proportional to vapour or gas pressure which further depends on temperature. Thus a hot lamp must cool down before the re-ignition can take place. The time taken to cool down depends on ambient conditions and usually re-ignites after 7–10 minutes.

The light emitted by mercury vapour lamps gives our eyes a blush-green impression. This is due to the notable absence of red wavelength. Thus, the mercury arc has both colour rendering and colour appearance properties. Mercury arc also emits invisible ultra-violet radiations which can be utilised to generate a red component to improve both colour appearance and colour rendering by coating the inside of outer bulb with suitable phosphor.

These lamps are available in 80 W, 150 W, 250 W, 400 W and 1000 W ratings. The lower wattage upto 250 W are normally used for street lighting while higher ones are used for industrial applications where luminaires are mounted at higher levels. The luminous efficiency of high pressure mercury lamp is usually 50 lumens/watt.

7. **Metal halide lamps:** As mentioned earlier, the colour rendering of high pressure mercury vapour lamps is not so good as red wavelength is missing in addiction to low luminous efficacy. These deficiencies were improved by the addition of Thallium, Sodium and Indium in the form of their iodides in correct quantities. When ignition take place, the metal iodides vaporize and dissociate at the arc discharge temperature (approx 5500° C). The metals take over the function of light emission owing to their low ionization voltage. In the spectrum, the blue wavelength (400 nm) comes from indium, the green (550 nm) and red (700 nm) from thallium and the yellow (600 nm) from sodium.

Metal halide lamps are a light source with high luminous output of 95 lumens per watt and generate powerful beam of light. They are widely used on sports ground, training fields and for TV broadcast due to their simple installation, quick ignition (3–5 minutes), good colour rendering and high luminous efficacy

8. **Low pressure sodium vapour lamps:** These lamps are widely used to light up large open areas such as railway yards, harbours, motorways, traffic intersections and dangerous spots like inter-sections where abundant light with strong contrast is important. These lamps have the highest luminous efficacy (200 lumens/watt) out of lamps introduce so far.

In the discharge tube of these lamps, there is sodium and rare gas mixture of 99% Ne + 1% Hg under low pressure. The optimum saturation vapour pressure is obtained at discharge tube temperature of approx. 260° C at which there are just sufficient sodium atoms in the discharge to be excited. The tin oxide, Indium oxide are applied which act as infrared reflecting layer. These lamps are available in the ratings of 18 W, 35 W, 55 W, 90 W and 180 W.

The principle of light generation is by the evaporation of solid sodium when discharge takes place generating heat. Sodium emits light much rapidly than Neon and thus colour of light becomes yellow, i.e. the characteristic of sodium. The inner side of outer envelope is coated with indium oxide which allows passage of visible radiation but blocks infrared radiation. The LPS lamps radiate light in yellow spectrum, these are less likely to be surrounded by swarms of insects which are attracted by blue spectrum. The mono-chromatic light of LPSU lamps is impervious to scattering glare and discomfort glare to a minimum. These lamps re-ignite within minutes if switch "OFF" due to power outages or flickers.

9. **High pressure sodium vapour lamps (HPSV):** In contrast to LPSU lamps, as the sodium vapour pressure is increased, the luminous efficacy drops as the self absorption of resonance radiations in the outer shell of sodium atoms increases. With further increase in pressure, sodium light spectrum broadens with strength of other lines like green and blue. Typical sodium lines even disappear due to self absorption. The HPSV lamps thus generate "warmer light" with improved colour rendering. The luminous efficacy of HPSV lamps is 100–120 lumens per watt.

At the high vapour pressure in HPSV lamps, the envelope temperature rises to 100° C, hence quartz hard glass used for the making of arc tube in LPSV lamps could not be used here. Instead poly crystalline aluminium oxide sintered alumina with translucence of 90% and resistance to sodium upto 160° C was used.

The re-strike time in case of HPSV lamp is very less as compared to other high pressure discharge lamps. The initial discharge takes place in the rare gas. Heat is thus dissipated, causing some of the sodium to evaporate. The sodium and mercury with their lower excitation potentials then take over the discharge and after a short time, the lamp burns stably, emitting its characteristic golden-white light. Around 75% of the rated lumen output is reached within 3 minutes. In case of interruption, the lamp re-ignites within 60 seconds.

The HPSV lamps are available from 70 W, 150 W, 250 W and 400 W outputs and are extensively used for road light flooding and in interior applications where colour discrimination is not of primary importance.

10. **Light emitting diodes (LED's):** LED's are solid state semi conductor devices converting electrical energy directly to light. LED is basically a PN function semi conductor diode emitting light when operated in a forward direction. The LED is housed in a frame primarily of lead. It contains an encapsulated epoxy resin which is either clear or coloured depending on the effect desired.

The conductor is made of Aluminium Indium Gallium Phosphate (Alln Ga P) or Indium Gallium Nitride (In Ga N). When the current flows across the junction, light is generated and the composition of material determines the wavelength and thus colour of the light.

LEDs are specified in term of their peak intensity, in milli candelas (mcd) and are highly directorial light sources. They consume very less power and are available as low as 1 W capacity. The luminary efficacy of LEDs is 20 lumens/watt. The directionality of LED is one weakness preventing the full utilization of flux generated to do a useful work. The colour of an Alln Ga P LED ranges from red, red-orange, orange, amber, white that of In GAN LED ranges from green, blue-green, blue and white. The LEDs have a life of about 1,50,000 hours and are being increasingly used for traffic signal, display devices, monuments/flyover/building illumina-tion strip lights, street lights, auditoriums, etc. With more amid more research going on, LEDs are going to be the lighting source of future.

14.3 UNDERSTANDING A POLAR CURVE

A source of light dissipates light in all directions, i.e. transverse and longitudinal. The intensity of light varies with direction. The luminous output falling on a work plane is affected by the distance from the light source already described. Thus there is a need to measure the output of light in a standard format.

The luminaire's light output is measured in several planes, called 'c' planes around the luminaire (Fig. 14.9a). A longitudinal axis is drawn from the lamp vertically and cutting through it a measurement plane is drawn termed as $c = 0°$, $c = 180°$ across the longitudinal axis. On one such plane, several γ angles are drawn at every 5° internal (Fig. 14.9b). Another measurement plane is drawn in the direction of longitudinal axis of the lamp termed as $c = 90°$, $c = 270°$ plane. The luminaire's luminous intensity in candela's is plotted on both these planes, at every 5° interval.

The light distribution curve so obtained is called **polar diagram**. This diagram shows the luminous intensity in different directions as a function of viewing angle in one or more planes. There can be many other planes in between 0°–180° and 90°–270° planes and luminous intensity can be plotted on each of them.

In Figs. 14.10a and 14.10b, the luminous intensity curves on plane 0°–180° and 90°–270° are plotted on the same diagram called polar curve. The solid lines on the polar curve indicates distribution of light perpendicular to light source's longitudinal axis, i.e. 0°–180° plane while dashed line refers to distribution of

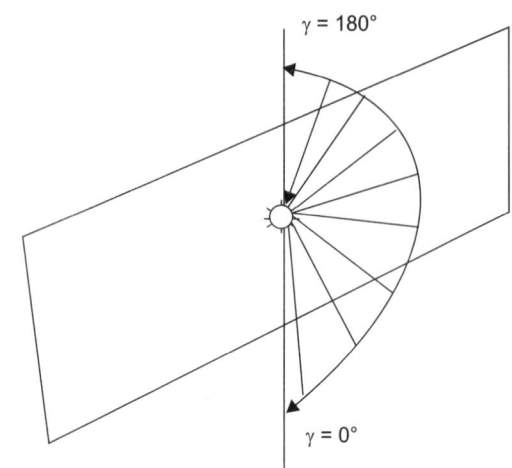

Fig. 14.9b Measurement of light output in 0–180° planes

light in longitudinal direction, i.e. 90°–270°. The concentric circles indicate the value of luminous intensity in candela per 1000 lumens. (cd/1000 lm or cd/klm). The radial lines originating from the light source refer to the γ angle ranging from 0° to 180° on both sides. In the case of symmetrical luminaire, the curve will be symmetrical on both sides of zero degree radial line (Fig. 14.10a) while it will be asymmetric in case of asymmetrical luminaire (Fig. 14.10b). Thus luminaire's with different outputs can be compared on a common polar curve.

Sometimes, manufacturers show only one side of a polar graph as the other side is identical in case of symmetrical luminaire. But with asymmetrical luminaire, curves in number of planes are needed to fully illustrate the light distribution. Figure 14.10c is example of one such curve showing the candle power distribution in the form of a polar curve. In the adjoining table of luminous intensity (Table 14.1) of a 1 × 28 W T-5 mirror optic luminaire, the values of luminous intensities for vertical angles 0-180° and horizontal angles 0–90°, i.e. c–0° to c–90° planes and planes in between are listed. Manufacturers also give spacing criteria along and across the direction of luminaire that helps in determining the spacing of luminaire in both directions.

Distance between luminaires = SC × MH

where SC = spacing criteria

MH = Mounting height

Let's say SC = 1.2 (along)

= 1.4 (across)

MH = 8 feet above working platform

Thus luminaire spacing (along) = 1.2 × 8 = 9.6 feet

Luminaire spacing (across) = 1.4 × 8 = 11.2 feet.

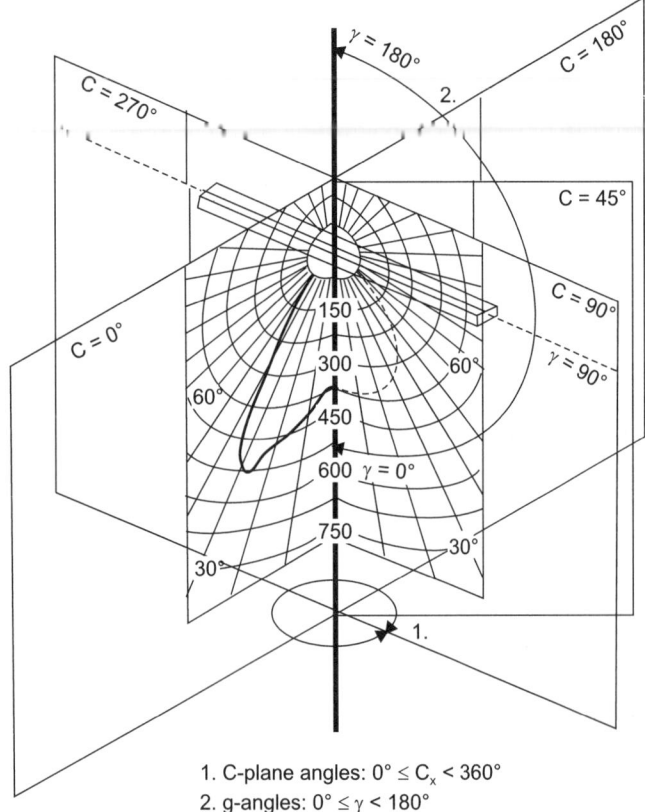

1. C-plane angles: $0° \leq C_x < 360°$
2. g-angles: $0° \leq \gamma < 180°$

Fig. 14.9a Measurement of light output in 'C' planes

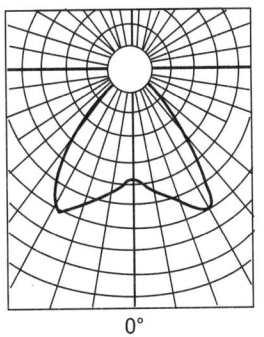

(a) Symmetrical light distribution

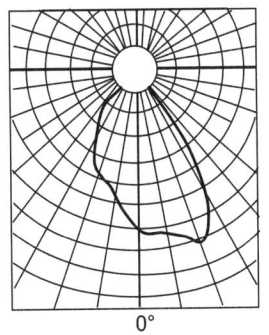

(b) Asymmetrical light distribution

Fig. 14.10a and b Polar curve

Table 14.1 Candela distribution

Vertical angle	Horizontal angle					Zonal lumens
	0	22.5	45	67.5	90	
0	2589	2589	2589	2589	2589	
5	2581	2599	2637	2669	2673	257
15	2376	2509	2667	2788	2820	743
25	1932	2172	2442	2538	2559	1070
35	1564	1721	1713	1686	1694	1039
45	832	891	816	659	693	603
55	126	131	170	148	113	149
65	38	35	31	40	33	38
75	13	15	18	17	15	18
85	0	0	0	0	0	5
90	0	0	0	0	0	
95	38	149	139	154	170	155
105	153	332	441	567	565	431
115	260	399	568	602	664	505
125	363	455	613	786	816	545
135	421	556	598	790	922	506
145	475	573	696	793	831	426
155	532	603	747	807	776	323
165	567	612	663	707	706	186
175	548	557	571	584	587	58
180	542	542	542	542	542	

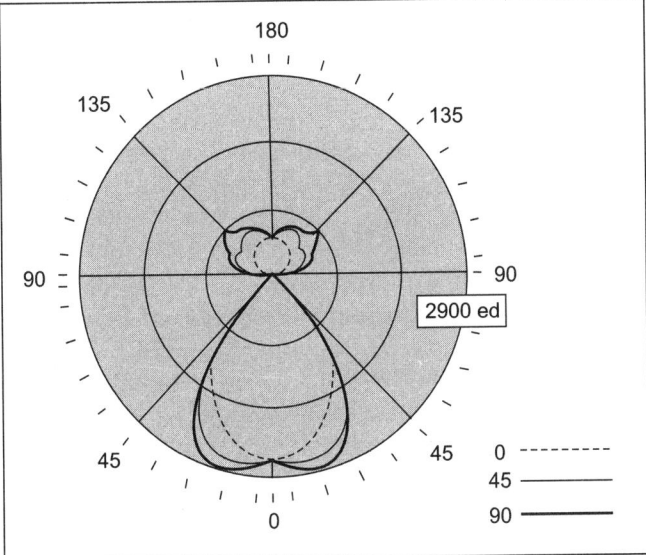

Fig. 14.10c Polar curve and candela distribution

14.4 INTERPRETATION OF DATA GIVEN BY MANUFACTURER'S OF LUMINAIRES

Manufacturer's give various parameters alongside polar curve. What do they mean and how they can be utilized for further calculations is clear from Figs. 14.11 and 14.12. The various parameters given in the diagram are defined here as under:

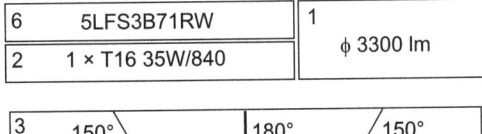

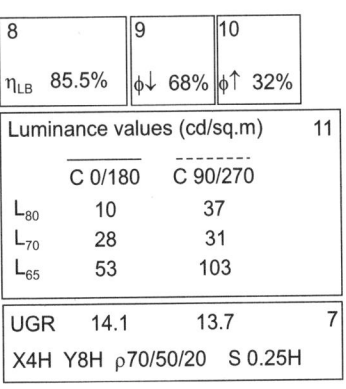

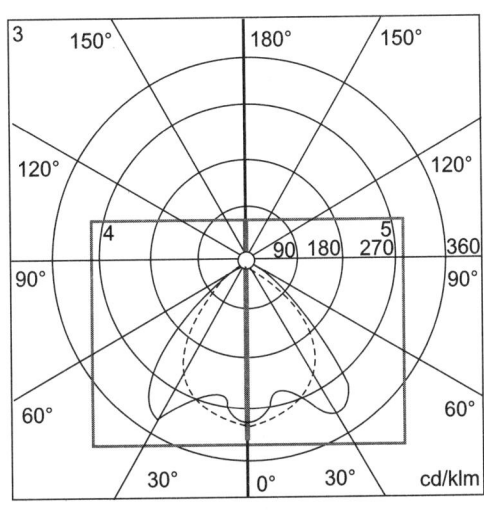

Fig. 14.11 Manufacturer's data for a polar curve

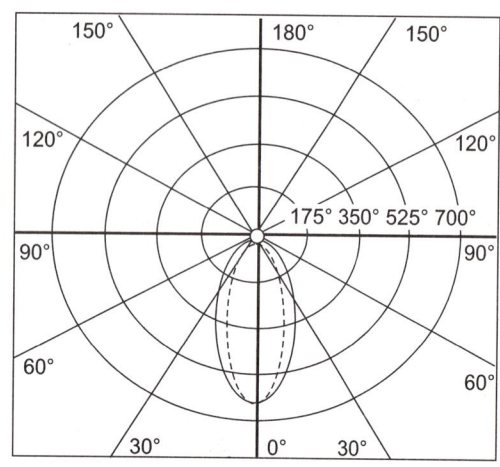

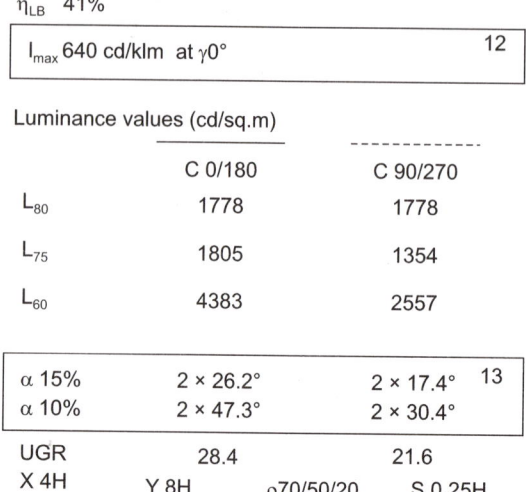

η_LB 41%

I_max 640 cd/klm at γ0°			12

Luminance values (cd/sq.m)

	C 0/180	C 90/270
L_80	1778	1778
L_75	1805	1354
L_60	4383	2557

α 15%	2 × 26.2°	2 × 17.4°	13
α 10%	2 × 47.3°	2 × 30.4°	

UGR	28.4	21.6	
X 4H	Y 8H	ρ70/50/20	S 0.25H

Fig. 14.12 Photometric data for a downfighter

1. φ: luminous flux of lamp: 3300 lm

2. Quantity, type and light colour of lamp

3. Light distribution polar curve, values being calculated as luminous flux in lm divided by 1000.

4. This rectangle determines luminous intensities on c–180° and c–270° plane by continuous and dotted lines respectively.

5. This rectangle determines the luminous intensities on c–0° and c–90° plane by continuous and dotted lines respectively.

6. Luminaire code

7. Unified Glass Rating (UGR). This parameter is the specification for direct glare control with indoor luminaire. The notation means room width = $4H$, room length = $8H$, reflection factor $\rho_{ceiling}$ = 70% ρ_{wall} = 50% and ρ_{floor} = 20% while H is the distance between eye level of a seated viewer (1.2 m) and luminaire level. UGR value on left signifies that luminaire/lamp axis is parallel to X, across direction of view. Similarly, value on right signifies UGR value with luminaire/lamp axis parallel to Y and along direction of view (Figs. 14.13a and b). $S = 0.25H$ is the inter axis distance between two or more luminaire. The UGR process can be used with luminaires with less than 65% indirect component.

8. Luminous light output ratio (η_{lb}%): This specifies how much of luminaire flux φ of lamp actually leaves the luminaire.

9. φ↓: Component of luminous flux emitted to lower hemisphere (direct component γ ≤ 90°) out of total luminous flux leaving the luminaire.

10. φ↑: Component of luminous flux emitted to upper hemisphere (direct component γ > 90°) out of total luminous flux leaving the luminaire.

11. Luminous levels at luminaire in (cd/sq mt) on planes $c = 0°$ and $c = 180°$ or $c = 90°$ and $c = 270°$ with various γ angles.

    L_{80} = luminance with γ = 80°

    L_{70} = luminance with γ = 70°

    L_{65} = luminance with γ = 65°

12. Maximum value of luminous intensity and the angle γ at which it is emitted.

13. $\alpha_{50\%}$: Peak half divergence in degrees. This specifies the angle γ at which the luminous intensity still consists of 50% of maximum value.

    $\alpha_{50\%}$: One tenth peak divergence in degrees. This specifies at which angle γ, the luminous intensity still consists of 10% of maximum value. Angle specification multiplied by two defines that $\alpha_{50\%}$ or $\alpha_{10\%}$ exists in both the planes.

Cartesian diagram

Figures 14.14a and b depicts the luminous intensity (cd/klm) on a X–Y graph in simple terms of value V_s angle γ.

Light cone diagram

Figure 14.15 is helpful in determining the illuminance values occurring on the measuring plane with respect

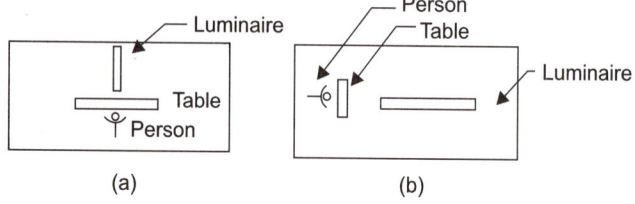

Fig. 14.13a and b Significance of UGR value

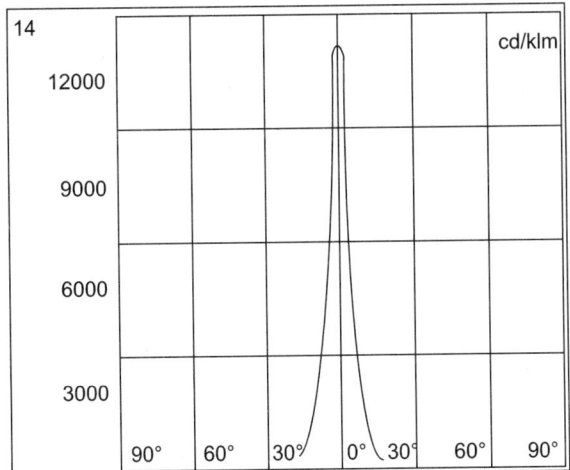

Fig. 14.14a Photometric data for narrow distribution on spotlight

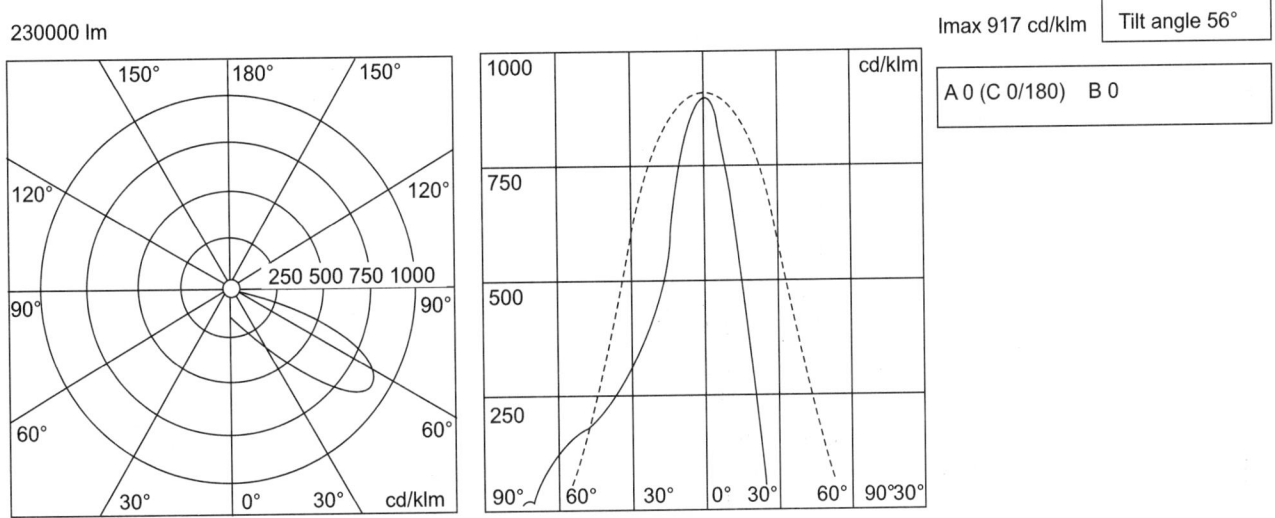

Fig. 14.14b Photometric data for asymmetric floodlight

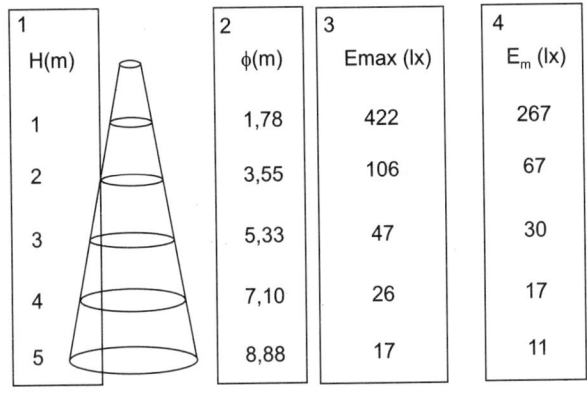

M.F. = 0.8

1. Distance of plane to luminaire in meters
2. Diameter of light cone in meters
3. Maximum horizontal illuminance E_{max} within light cone (lx)
4. Medial horizontal illuminance E_m within light cone (lx)

Fig. 14.15 Light cone diagram

to distance of luminaires. This is specific to roto symmetrical luminaires. The intersection between cone and plane are described by half peak divergence, i.e. angle at which light output is still 50% of maximum light output. Within these circles, the median illuminance and the maximum illuminance are specified with maintenance factor of 0.8.

Isolux curves

This curve shows a defined area, inside which the horizontal illuminance is marked in varying intensities (Figs. 14.16 and 14.17). The points with same illuminance are connected with one curve and the value of horizontal illuminance marked over it. This value is measured at a virtual plane 0.75 m above the floor (i.e. working level) while in case of outdoor luminaires, these values are marked at ground level. The median horizontal illuminance can also be specified.

Monting height 10 m   Tilt 0°

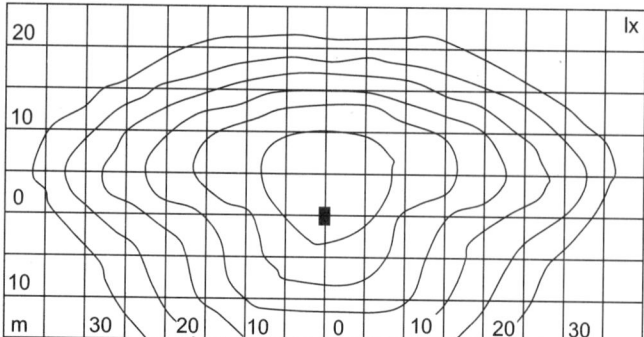

Fig. 14.16 Isolux curve

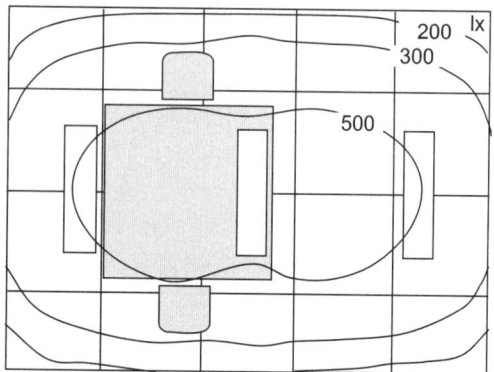

Reflection grade 70/50/20, maintenance factor 0.8
Em Room 481 lx Em table 535 lx

Fig. 14.17 A typical office layout

According to DIN EN 12464-I, the ratio between E_{min}/E_m shall be greater than 0.7 on the table and greater than 0.5 in the room.

Thus $\dfrac{E_{min}}{E_m}$ for room $= \dfrac{200}{475} = 0.42$

$\dfrac{E_{min}}{E_m}$ for table $= \dfrac{500}{535} = 0.93$

It can be thus seen that $\dfrac{E_{min}}{E_m}$ for room is not > 0.5 and luminaires need adjustment.

14.5 CALCULATION OF ILLUMINATION AT A POINT

Before we start to calculate illumination at a point from various sources of light present, it is important to understand the candle power radiated from a luminaire in all directions, its measurement and graphical depiction.

Let us take a lamp of 16 candle power (incandescent lamp) and measure its candle power in all directions in vertical plane with the help of a photometer and mark this on a graph paper. The curve so obtained is called a photometer curve or polar curve. The distance of the

curve in various directions from the lamp represent the candle power in that direction. It can be seen that the distribution varies greatly with different angles, while it is 16 at horizontal, it gets as low as 6.6 underneath the lamp (Fig. 14.18).

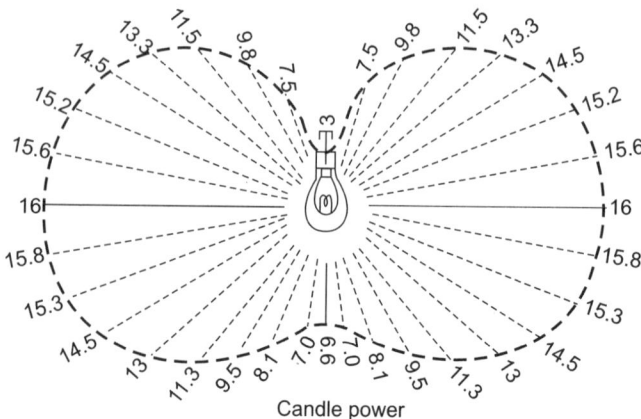

Fig. 14.18 Candle power of 16 candle incandescent lamp

Some lamps do not give same illumination when viewed from different directions in horizontal plane. It is thus customary to make similar photometric curve for horizontal plane showing such variations in illumination.

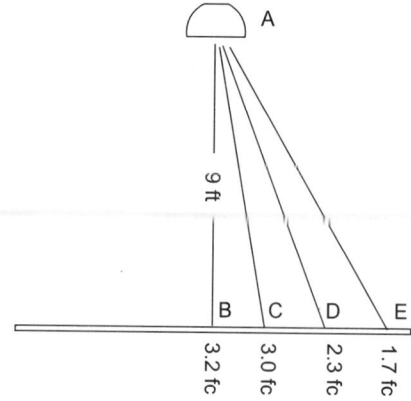

Fig. 14.19 Calculation of horizontal illumination

To draw a horizontal photometry curve, fix the lamp position and draw lines AB, AC and AD every 10 degrees (Fig. 14.19). The illuminance at various angles on a horizontal surface 3 feet below the lamp vertically calculated as shown in Table 14.2. The process can be repeated for other positions on horizontal surface.

Table 14.2 Horizontal illuminance calculation method

Point under consideration	Angle from vertical	Luminous intensity (cd) ray	value	Illuminance (Foot candle) $E = (I/h^2)\cos^3\theta$
B	0°	AB	29.3	3.2
C	10°	AC	29.0	3.0
D	20°	AD	25.5	2.3

14.6 INDOOR LIGHTING DESIGN

The lighting design is not simply the result of one lamp illuminating the entire room. The illumination is achieved by a series of luminaires arranged aesthetically on the ceiling, wall, various combination of direct/indirect light, etc. and the combined effect of all luminaires at a point should be calculated. As an example, consider a hall of 80 feet by 50 feet (Fig. 14.20) with lamp hanging 15 ft from the floor and 12 ft from work surface.

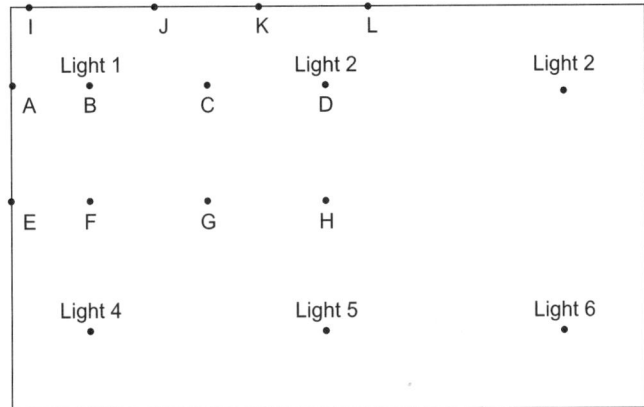

Fig. 14.20 Plan of 80 × 50 feet hall

Consider the polar curve on a horizontal plane (Fig. 14.21). For simplicity, only one fourth curve is shown. Mark lamp at a height of 12 feet from a horizontal surface. Divide surface with points marked equidistant. Measure the values of candle power from the polar curve (Refer Table 14.3) and calculate the illuminance (*E*) at various points, the distance from lamps being already known.

Table 14.3 Horizontal illuminance values at various points

D	α	$\cos^3\alpha$	c.P.	Horizontal I
0	0	1.000	160	1.11
1	4.5	0.990	165	1.13
2	9	0.963	174	1.16
3	14	0.913	185	1.17
4	18	0.860	195	1.16
5	22	0.797	196	1.10
6	26	0.726	188	0.950
7	30	0.649	183	0.830
8	33.5	0.580	182	0.734
9	36.5	0.519	182	0.657
10	39.5	0.459	183	0.570
12	45	0.353	189	0.463
14	49	0.282	202	0.396
16	53	0.218	216	0.327
18	56	0.175	225	0.274
20	59	0.137	235	0.224
24	63	0.0936	244	0.159
28	66.5	0.0633	250	0.110
32	69.5	0.0430	256	0.0765
36	71.5	0.0320	260	0.0596
40	73	0.0250	267	0.0464

$$\text{Horizontal } I = \frac{\text{c.P.}}{h^2} \times \cos^3 \alpha$$

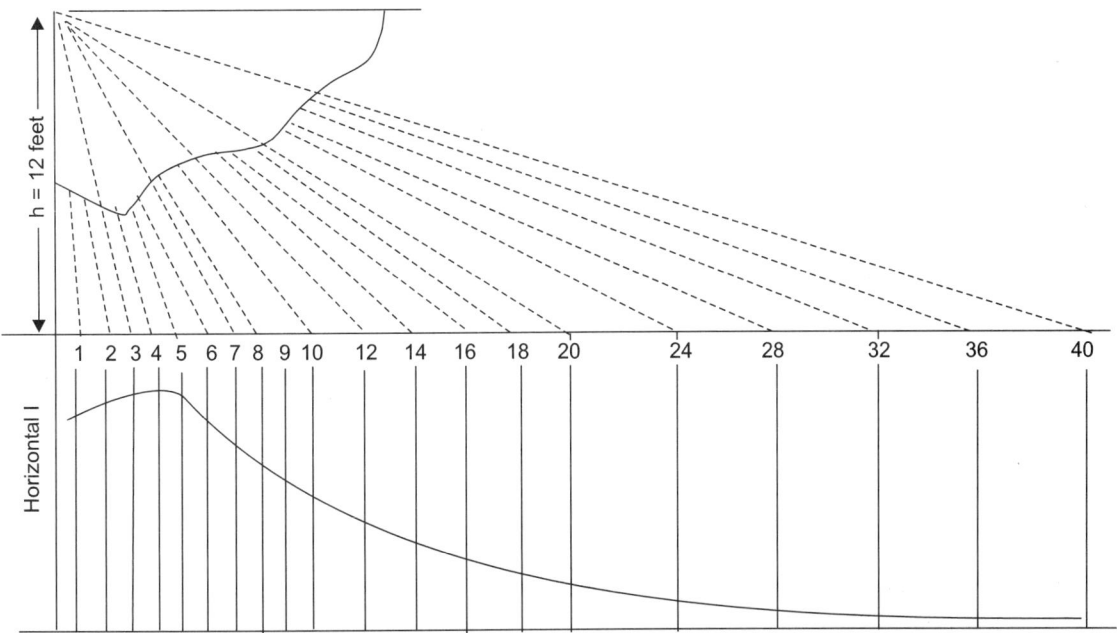

Fig. 14.21 Photometric and illumination curve

Table 14.3 Values of horizontal illumination at the 12 points indicated in Fig. 14.20

	Ft.	A	Ft	B	Ft	C	Ft	D	Ft	E	Ft	F
Light 1	10	.570	0	1.11	15	.362	30	.093	18	.274	15	.362
Light 2	40	.046	30	.093	15	.362	0	1.110	42	.040	34	.067
Light 3	–	–	–	–	45	.030	30	.093	–	–	–	–
Light 4	31	.084	30	.093	33	.072	42	.040	18	.274	15	.362
Light 5	–	–	42	.040	33	.072	30	.093	42	.040	34	.067
Light 6	–	–	–	–	–	–	42	.040	–	–	–	–
Total		.70		1.34		.90		1.47		.63		.86
	Ft	G	Ft	H	Ft	I	Ft	J	Ft	K	Ft	L
Light 1	21	.208	34	.067	14	.396	10	.570	18	.274	32	.076
Light 2	21	.208	15	.362	41	.043	32	.076	18	.274	10	.570
Light 3	–	–	34	.067	–	–	–	–	46	.028	32	.076
Light 4	21	.208	34	.067	41	.043	40	.046	43	.037	–	–
Light 5	21	.208	15	.362	–	–	–	–	43	.037	40	.046
Light 6	–	– ·	34	.067	–	–	–	–	–	–	–	–
Total		.83		.99		.48		.69		.64		.77

In Table 14.3, the combined effect of all luminaires, mounted inside the room at all points under consideration is shown. Whenever the value of illuminance falling on a point is less than 5% of the maximum illumination on the same from any lamp, the same have been neglected and shown by dash.

The graph below polar curve is a representation of the illuminance at different points on the horizontal work plane. As can be seen, the illuminance goes on decreasing with distance away from the lamp, but luminous intensity in candela does not follow this pattern. This is because of the division by square of distance from lamp and cosine of angle formed.

Similar curves and tables can be drawn to show illuminance in the crosswise direction. The illumination along the sides of the hall is considerably lower than that at center. This value will in effect be considerably raised due to reflection of wall, depending on the colour. For light coloured walls, this increase may be upto 50%.

If the illuminance is to be calculated on walls instead of horizontal surface, known as vertical illumination, the polar curve should be simply rotated by 90° and instead of height above the work plane, horizontal distance from light source to wall should be considered. But where luminaires are used around the lamp, as is normally the case, then polar curve in the vertical plane should be used.

The illumination level, number of lighting fixture for an indoor area are designed using "Lumen method". This method is applied for a uniform lighting scheme where flexibility of working locations are other activities is required. This method is applied to square or rectangular rooms with a regular matrix of luminaires.

This method calculates average illuminance over a horizontal working plane. The formula used to calculate average illuminance is:

$$E = \frac{n \times N \times F \times UF \times LLF}{A}$$

where,

E = average illuminance over the horizontal working plane

n = number of lamps in each luminaire

N = number of luminaires

F = Lighting design lumens per lamp (flux)

UF = Utilization factor for the horizontal working plane

LLF = Light loss factor

A = Area of horizontal working plane

If the average illuminance to be provided is known, the no. of luminaires can be determined with the same formula.

Light loss factor is to account for the decrease in output of a luminaire due to fall in lamp output with usage, deposition of dirt on luminaire and decrease in reflectance of surface over a period of time. This factor is expressed in percent and is a matter of experience. Author personally prefers to take 75% as the LLF.

Utilization factor is a measure of effectiveness of lighting scheme and takes into account the proportion of luminous flux emitted by a lamp that actually reaches the working plane. This factor is influenced by effectiveness of luminaire reflector and its design, room proportions, room reflectances, spacing and mounting height of luminaires.

Room proportion or room index (RI) is the ratio of room plan area to half the wall area between the working and luminance plane.

$$RI = \frac{L \times W}{H_m \times (L + W)}$$

where, L = length of room
  W = width of room
  H_m = Mounting height of luminaire above working plane.

Spacing to height ratio is the ratio of distance between adjacent luminaires to their heights above working plane.

$$SHR = \frac{1}{H_m}\sqrt{\frac{A}{N}}$$

where A = Floor Area
  N = Number of luminaires
  H_m = mounting height

The uniformity of illumination on a working plane can be achieved either by reducing the spacing between luminaires or increasing the height of luminaires for a given spacing.

SHR should not exceed maximum spacing to height ratio of the given luminaire as specified by manufacturer.

Another check to be performed is the geometric mean spacing to height ratio of luminaire which should be within the nominal SHR of given luminaire prescribed by manufacturer.

i.e. $\sqrt{SHR_{axial} \times SHR_{transverse}} = SHR_{nom.} \pm 0.5$

Example: Let us now design a lighting installation for a room having dimensions 12 m long × 8 m × 3.2 m height having working plane 0.7 m above floor. The other given factors are as below:

Reflection factors:

Ceiling   : 70%
Walls   : 50%
Working plane   : 20%
Light loss factor : 0.779
Luminaire type : 1800 mm twin tube with opal diffuser ceiling mounted
Downward light output ratio: 36%
SHR$_{max}$: 1.60 : 1
SHR$_{nom}$: 1.50 : 1
Dimensions : 1800 mm long × 200 mm wide
Lamps   : 1800 mm, 75 W plus white
Initial lumen per lamp: 5800
Lamps per luminaires : Two
Desired average illuminance on working plane   : 500 lux

Solution: Let us proceed step by step to design the lighting installation

a. Calculate room index : L

$$RI = \frac{L \times W}{H_m \times (L+W)}$$

$$= \frac{12 \times 8}{2.5(12+8)} = 1.92$$

b. Determine utilization factor from manufacturer's data sheet using room index and effective surface reflectances.

For this type of luminaire, $UF = 0.5336$ (reader may refer data sheet of some manufacturer)

c. Calculate numbers of luminaires:

$$N = \frac{E \times A}{n \times F \times UF \times LLF}$$

$$= \frac{500 \times 12 \times 8}{2 \times 5800 \times 0.5336 \times 0.779}$$

$$= 9.95$$

Therefore, the number of luminaires are 10

d. Perform spacing to mounting height ratio check:

$$S/H_m = \frac{1}{H}\sqrt{\frac{A}{N}}$$

$$= \frac{1}{2.5}\sqrt{\frac{12 \times 8}{10}}$$

$$= 1.24 : 1$$

From manufacturer's data supplied in question, SHR$_{max}$ is 1.6 : 1.

Thus, number of luminaires as calculated suffice this test

e. Design a layout

A 5 × 2 array is proposed (alternatively 10 × 1)

f. Check proposed layout by calculating total length of luminaire along long axis and width along short axis.

For long axis $S = \dfrac{12}{5} = 2.4$ m

$$\frac{S}{H_m} = \frac{2.4}{2.5} = 0.96 : 1$$

(which is less than 1.6 : 1, hence acceptable)

For short axis $S = \dfrac{8}{2} = 4$ m

$$\frac{S}{H_m} = \frac{4}{2.5} = 1.6 : 1$$

(equal to given 1.6 : 1, hence acceptable)

g. Calculate SHR$_{nom}$

SHR$_{nom}$ ± 0.5 = 1.5 ± 0.5 = 1.0 to 2.0

Check geometric mean spacing to height ratio:

$$= \sqrt{SHR_{axial} \times SHR_{transverse}}$$

$$= \sqrt{0.96 \times 1.6}$$

= 1.24 (which is within the range of SHR$_{nom}$ hence correct)

Note that if the checks calculated above were unsatisfactory, then number of luminaires and array should be remodelled and checks should be conducted again.

h. Arrange luminaires in 5 × 2 array so that luminaire distance from walls is half of inter-luminaire distance for aesthetics and to take advantage of reflectance of wall surface.

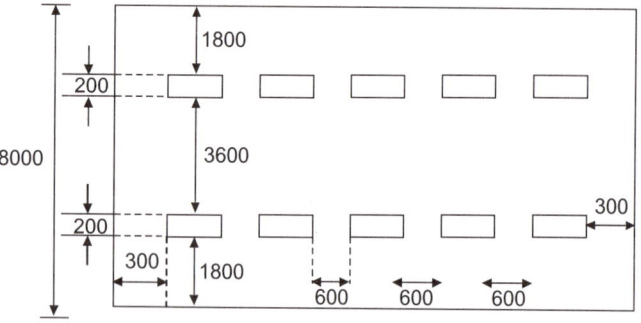

14.7 RECOMMENDED ILLUMINANCE RATIO AND SURFACE REFLECTIONS

The following table shows the illuminance ratios and surface reflections which can be adopted in various design strategies:

Illuminance ratios

a.	minimum/average illuminance on task plane	0.8 min.
b.	In an interior with general lighting	
	Avg illuminance on ceiling/task plane	0.3–0.9
	Avg illuminance on wall/task plane	0.5–0.8
c.	In an interior with localized lighting, the ratio of illuminance on task area to that around task area	3:1max.

Surface reflectances

a.	Ceiling cavity	0.6 min
b.	Principal walls	0.3–0.7
c.	Window wall surfaces	0.6 min
d.	Floor cavity	0.2–0.3
e.	Equipments and furnishings in work interiors (e.g. desk top)	0.2 min
f.	Immediate background to a task	Matt

14.8 POLAR CURVE AND LUMEN CALCULATION

The method described above is a good method to determine average illuminance without any consideration for uniformity of illuminance and is commonly used as a handy tool to determine number of luminaires, like in a room, seminar hall where identical luminaires are repeated.

However, the mathematical calculation involved in this calculation must be understood so as to be able to read polar curve and derive illuminance at any point on horizontal working plane.

The most common method is to plot a rectangular area of illumination with the help of polar curve of a single lamp. When there are more than one lamp, the illumination obtained from various sources are added, so that total illumination at a point could be found. This is called point-by-point method (calculations explained later).

One lumen is the amount of light falling on one square foot of area when it is evenly illuminated to an intensity of 1 foot candles (unit of light). A sphere of 1 feet radius has 12.57 sq ft area. If a lamp emitting one candle power is placed in the centre of sphere, each square feet of sphere would receive one lumen. Thus, total lumens falling on sphere would be 12.57 lumens.

To determine the average illumination at a point from a given polar curve, a Rousseau diagram is plotted from a polar curve. In Fig. 14.22, such polar curve of a typical lamp is shown. As can be seen, the candle power at 10° from vertical is 65. On a graph paper, below polar curve, vertical lines are drawn corresponding to various degrees shown on polar curve. Since, the length of radial line, on a polar curve, represent the candela power these values are marked on the X–Y graph on the corresponding vertical lines. The curve so obtained is called Rousseau diagram.

The area of this diagram between any two vertical lines corresponds to the total flux of light in lumens given out in the zone bounded by those degrees. As an example, the area between zero and ninety degrees (being horizontal) represent the lumens in the lower hemisphere. Similarly, area enclosed by whole curve from zero to 180° represent total number of lumens given out. The mean spherical candle power is determined by measuring the average height of curve.

In the polar curve referred above, let us calculate the total number of lumens in the lower hemisphere.

First of all calculate the candela values on the photometric curve at the radial lines of 10°, 20° and upto 90°.

Radial line	Candela value
10	70
20	80
30	92
40	85
50	60
60	45
70	28
80	24
90	24
	508

$$\text{Average out the calculated candela value} = \frac{508}{9}$$

$$= 56.44 \text{ cd}$$

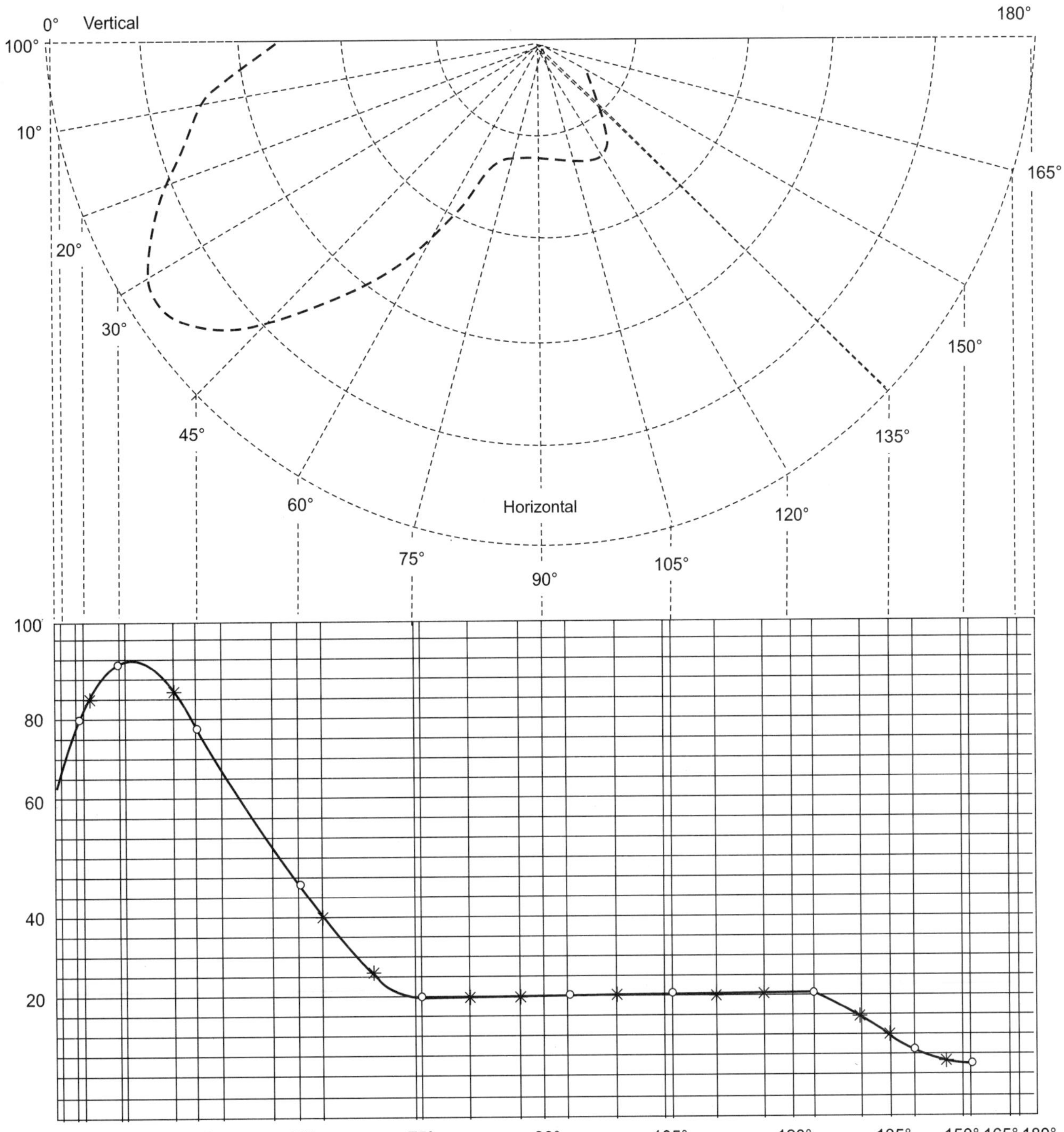

Fig. 14.22 A typical Rousseau diagram

The area of the zone of a sphere is calculated as below:

$$L = 2\pi (1 - \cos\alpha)$$

where L is the length of arc/zone of sphere and α is the angle between vertical and the edge of zone (Fig. 14.23).

Thus, for lower hemisphere (90° zone), the total lumens are

$$l_m = 56.44 \times 2\pi (1 - \cos 90°)$$
$$= 56.44 \times 6.28$$
$$= 354.47 \text{ lumens}$$

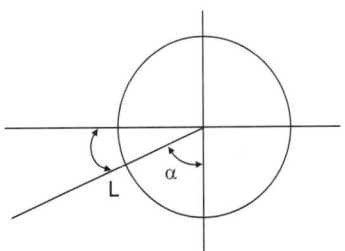

Fig. 14.23 Zone length of a sphere

Similarly, the number of lumens for 10°, 20° and other zones can be calculated. The only precaution to be taken is to sum up candela power values relevant for that zone only, e.g. for 10° zone calculation.

$$l_m = 70 \times 2\pi (1 - \cos 80°)$$
$$= 363 \text{ lumens}$$

14.9 OUTDOOR LIGHTING DESIGN

The outdoor lightening involves illumination of colony roads, highways, sports training field, stadiums, parks, facade lighting, architectural lighting and numerous other applications.

The luminaires are usually located around the perimeter of the area to be illuminated, e.g. on the foot path, illumination of parks, high mast towers located around the stadium and usually projecting above the seating arena, below the parapet of roof of seating arena parallel to the track in stadiums, etc.

The lamp types generally used are halogen, high intensity discharge lamps, sodium vapour and mercury vapour lamps varying with the application.

The design process for illumination of the application area involves following steps:

a. Calculation and survey to mark location of flood lights, type of lamp and luminaire and the associated type of distribution (Polar diagram). Light source characteristics suitable for a particular application.

b. Lumen calculation to find out number of lamps, average illuminance.

c. Point by point calculation to determine aiming pattern of flood lights for the required uniformity satisfying average to minimum and maximum to minimum illuminance ratio.

14.9 .1 Lumen Calculation Method

The lumen method is the most simple method and gives the average lux level achieved over the area under consideration. The basic formula is:

$$E = \frac{N \times l_m \times MF \times L \times U}{A}$$

where,

E = average illuminance on the area under consideration

N = no. of lamps

l_m = lamp lumens

MF = maintenance factor usually 0.6–0.7 for long cleaning intervals and 0.8–0.9 for frequently cleaned or effectively sealed.

L = atmospheric absorption loss factor

UF = utilization factor of flood lights

A = area to be lit by flood lights

Atmospheric absorption loss factor is caused by air borne moisture and solid particles. This will vary with time of day, mounting height, length of throw, location, season, etc. In a football installation having 30–45 m high masts, this factor will be 0.7 to 0.8 on a clear night. Calculations shall be performed on a clear night.

Utilization factor is the ratio of lumens available on the area to be illuminated to the total lumens from the lamp. Zonal flux diagram published by the manufacturer is used to determine utilization factor of flood light as explained in example below.

Example: An area measuring 20 m × 20 m has to be illuminated using a simple flood light mounted on a 8 m column as shown in Fig. 14.24. The luminaire is a floodlight of 1 × 150 W SON–T lamp of 1600 lumens. Zonal flux diagram is shown in Fig. 14.25. Peak intensity for flood lighting is aimed at a distance 11.9 m along line EF, i.e. at point P. Calculate the total flux falling in the area, utilization factor and average horizontal illuminance.

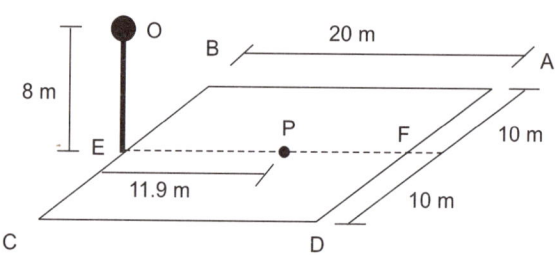

Fig. 14.24 Zone length of a sphere

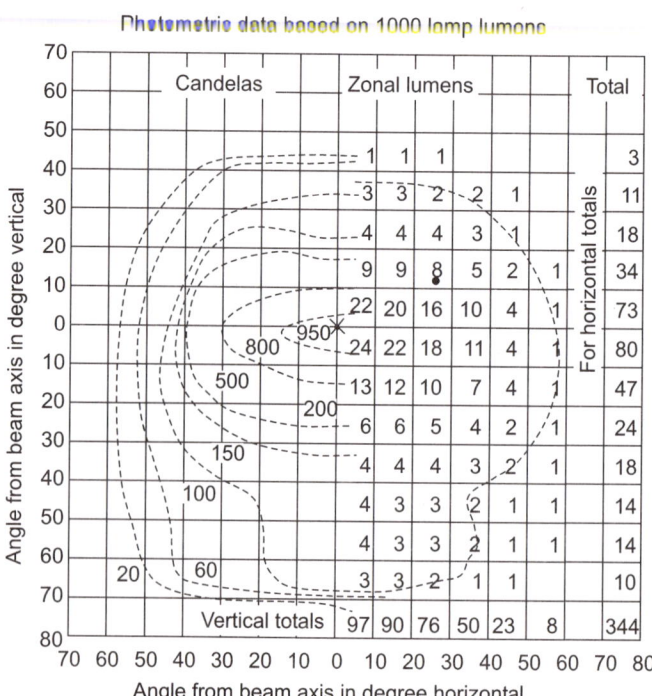

Fig. 14.25 Zonal flux diagram

Solution: Normally flood light aiming is about 2/3rd the distance along centre line (EF). Figure 14.26 depicts the horizontal and vertical angles formed by four corners of plot with top end of column. Figure 14.27 is a combined zonal flux and associated isolux diagram.

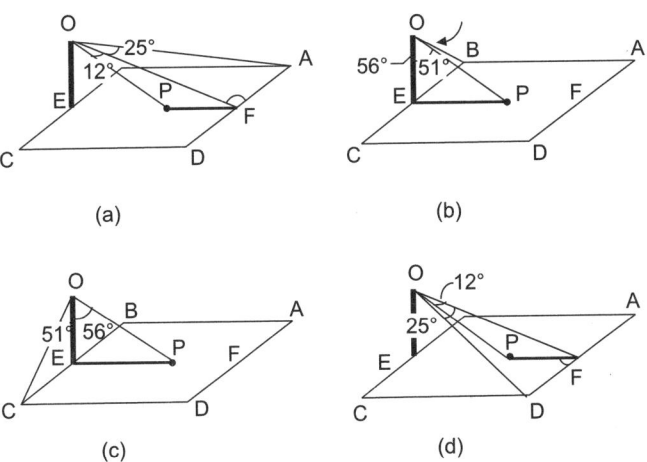

Fig. 14.26 Horizontal and vertical angles of corner points A, B, C, D

Both the zonal flux and isolux are shown only for half position of the luminaire through the plane of symmetry. As can be seen, the figures on the right side are the zonal flux which are totalled under the total column indicating total flux for the half portion of luminaire. Since similar half exists on the other side,

this total figure of 344 lm is to be doubled, i.e. 2 × 344 = 688 lux created by luminaire. Now we have to find out how much of flux out of this 688 lumens is captured by the field of 20 m × 20 m dimension. For this calculation, first determine the horizontal and vertical angles (Table 14.4), with respect to beam axis, at which the corner points A, B, C, and D of the field are located as below in Fig. 14.26 (a to d).

The angular coordinates of point A, B, C and D of the area are plotted on zonal flux diagram as shown in Fig. 14.27 with the help of above table. As can be seen in this diagram, some parts of them to be lit falls outside the lowest iso candela contour. In such situation it is recommended to increase the number of flood lights or use a wide beam or increase mounting height.

In Fig. 14.27, the lumens falling inside the plotted area are summed up. If the line cuts only a part of the box, the zonal lumen value within the line is estimated. The total value of lumen is 2 × 222 = 444 lumens

The utilization factor $= \dfrac{444}{1000}$ (isolux diagram is based on 1000 lamp lumens)

Average horizontal illuminance $= \dfrac{0.444 \times 16000}{20 \times 20}$

$= 18$ lux

It is now important to conduct a uniformity check to determine illuminance at a particular point and also to check ratio of average to minimum and maximum to

Table 14.4 Horizontal and vertical angles of a point

Point	Angle from beam axis (horizontal) degrees	Angle from beam axis (vertical) degrees
A	$\angle POF = \tan^{-1}\dfrac{EF}{OE} - \angle EOP$ $= \tan^{-1}\left(\dfrac{20}{5}\right) - 56° = 12°$	$\angle FOA = \tan^{-1}\left(\dfrac{AF}{OF}\right)$ $= \tan^{-1}\left(\dfrac{10}{20}\right) = 25°$
B	$\angle EOP = \tan^{-1}\left(\dfrac{EP}{OE}\right)$ $= \tan^{-1}\left(\dfrac{11.9}{8}\right) = 56°$	$\angle EOB = \tan^{-1}\dfrac{EB}{OE}$ $= \tan^{-1}\left(\dfrac{10}{8}\right) = 51°$
C	$\angle EOP = \tan^{-1}\left(\dfrac{EP}{OE}\right)$ $= \tan^{-1}\left(\dfrac{11.9}{8}\right) = 56°$	$\angle EOC = \tan^{-1}\left(\dfrac{EC}{OE}\right)$ $= \tan^{-1}\left(\dfrac{10}{8}\right) = 51°$
D	$\angle POF = \tan^{-1}\left(\dfrac{EF}{OE}\right) - \angle EOP$ $= \tan^{-1}\left(\dfrac{20}{8}\right) - 56° = 12°$	$\angle FOD = \tan^{-1}\left(\dfrac{FD}{OF}\right)$ $= \tan^{-1}\left(\dfrac{10}{20}\right) = 25°$

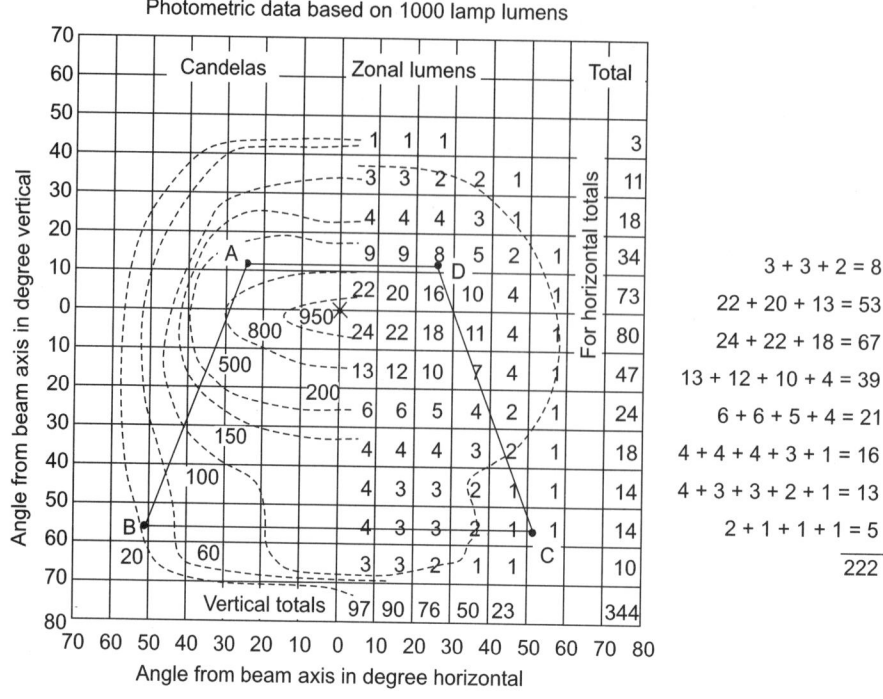

Photometric data based on 1000 lamp lumens

$3 + 3 + 2 = 8$
$22 + 20 + 13 = 53$
$24 + 22 + 18 = 67$
$13 + 12 + 10 + 4 = 39$
$6 + 6 + 5 + 4 = 21$
$4 + 4 + 4 + 3 + 1 = 16$
$4 + 3 + 3 + 2 + 1 = 13$
$2 + 1 + 1 + 1 = 5$
$\overline{222}$

Fig. 14.27 Combined zonal flux and isolux diagram

minimum illuminance satisfying the criterias mentioned in Table 14.5 and 14.6. This check is performed using point by point method.

Table 14.5 Illuminance range

Illuminance range (lux)	Critical plane	Application
1–10	Horizontal	Amenity, general storage area
	Vertical	Security, casual sport training
10–50	Horizontal	Stock and cargo handling
		Non-critical work areas
		Car parks
50–100	Horizontal	Critical work areas
		Sports practice
		Play grounds
	Vertical	Aircraft service areas
		advertising-unlit roads.
100–500	Horizontal	Club and tournament sports,
		sales area
	Vertical	Advertising-lit roads, spectator
		sports

Let us apply inverse square law and cosine law to calculate illuminance at a point from intensity data (Fig. 14.28).

$$\text{Horizontal illuminance} = \frac{I}{d^2}\cos\theta \text{ (already described)}$$

$$h = d\cos\theta$$

Thus,

$$E_H = \frac{I}{h^2}\cos^3\theta$$

taking maintenance factor and atmosphere loss into account, the equation can be written as:

$$E_H = \frac{I \times L \times MF}{h^2}\cos^3\theta$$

Let

$$L = MF = 0.9$$

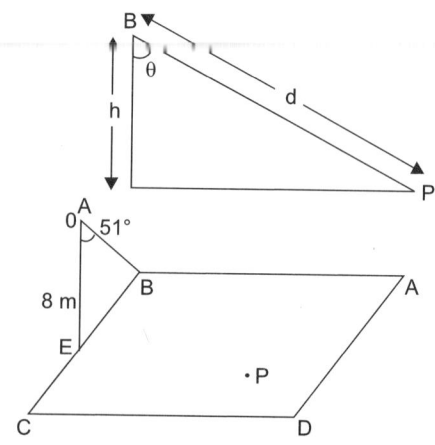

Fig. 14.28 Horizontal and vertical angles of a point

From isolux curve, the luminous intensity of floodlight in direction B is 20 candelas. The horizontal illuminance of point B is calculated as:

$$E_H = \frac{16000 \times \frac{20}{1000} \times 0.9 \times 0.9}{8^2} = \cos^3 51 = 1 \text{ lux}$$

$$= 1 \text{ lux}$$

Table 14.6 Uniformity values

Application	Uniformity in critical plane		Minimum distance over which 20% change in illuminance occurs
	Max : Min	Av.: Min	
Non critical areas	50:1		
Parks gardens amenity lighting			
Working areas			
– Most building facades	20:1	10:1	2
– Sports training area			
Even lighting of plain light coloured surfaces, spectator, sports area	10:1	5:1	3
Filming and television	3:1	1.5:1	4

On isolux curve, the point of maximum candela falls on 950 cd curve, at 0° vertical and 12° horizontal.

$$E_H = \frac{16000 \times \frac{950}{1000} \times 8.9 \times 0.9}{8^2}$$

$$= 192 \text{ lux}$$

$$\text{Ratio } \frac{\text{Max}}{\text{Min}} = \frac{192}{1} = 192$$

$$\text{Ratio } \frac{\text{Average}}{\text{Min}} = \frac{18}{1} = 18$$

Thus, it is seen that the maximum to minimum ratio is too high and will be supplemented by providing more flood lights at perimeter of the field.

The same method can be applied for calculation of illumination for external lighting, say street lighting. In case of street lights vertical illumination is as important as horizontal illumination. This is due to the reason that vehicles, pillars, obstruction will not be properly lit up if illuminance on vertical plane is poor and may be a cause of accident. The illuminance on the horizontal plane is important for the driver to see the road, spot any object lying on it, pot holes, dividers and turnings, etc.

Vertical illumination is defined as the measure of light delivered at a sufficient height from the ground so that the people can see the faces of other pedestrians. Areas suffering from high levels of street crime and robbery benefits from high value of vertical illuminance.

It is important to note that horizontal illuminance requires only a single measurement at each point, whereas vertical illumination necessitates deciding in which direction to make the measurement and recording both, the direction and the measurement (Fig. 14.29).

Horizontal illuminance is of importance in office and other areas where work is to be carried out on a horizontal plane. But it will not be a good measurement from the view point of motorist or a pedestrian where good vertical illumination is required.

Some of the recommended values of illumination levels by IESNA are given in Table 14.7a, b, c and 14.8.

An example below explains the method of calculation of illuminance levels, both horizontal and vertical for a road.

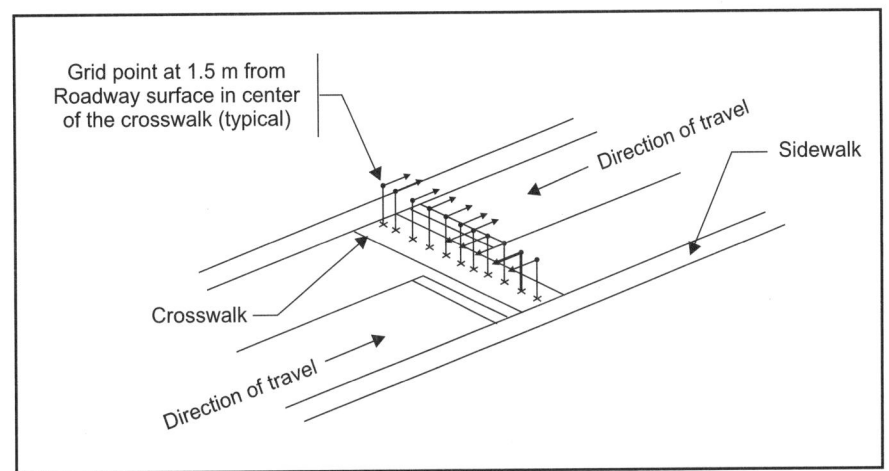

Fig. 14.29 Vertical illumination in the direction of travel

Table 14.7 IESNA recommended values of illumination

a. For low pedestrian conflict areas for walkways/bikeways

	E_H (lux)	$E_{V\,min}$ (lux) (1.5 m above road)	E_{avg}/E_{min}
Rural/ semi rural areas	2.0	0.6	10.0
Low density residential	3.0	0.8	6.0
Medium density residential	4.0	1.0	4.0

b. For medium pedestrian conflict areas for walkways/bikeways

	E_H (lux)	$E_{V\,min}$ (lux) (1.5 m above road)	E_{avg}/E_{min}
Pedestrian areas	5.0	2.0	4.0

c. For high pedestrian conflict areas for walkways/bikeways

	E_H (lux)	$E_{V\,min}$ (lux) (1.5 m above road)	E_{avg}/E_{min}
Mixed vehicle and pedestrian	20	10	4
Pedestrian only	10	5	4

Table 14.8 Lighting design criteria for different categories of roads

Road type	Pedestrian activity	Av lum. cd/m²	Av to min. uniformity ratio	Max to min. uniformity ratio	Max to Av. veiling luminance ratio
Freeway	–	≥ 0.6	≤ 3.5	≤ 6.0	≤ 0.3
Partial lighting of interchange on ramp/off ramp	–	≥ 0.6	≤ 3.5	≤ 6.0	≤ 0.3
Expressway	High	≥ 1.0	≤ 3.0	≤ 5.0	≤ 0.3
Highway	Med	≥ 0.8	≤ 3.0	≤ 5.0	≤ 0.3
	Low	≥ 0.6	≤ 3.5	≤ 6.0	≤ 0.3
Arterial	High	≥ 1.2	≤ 3.0	≤ 5.0	≤ 0.3
	Med	≥ 0.9	≤ 3.0	≤ 5.0	≤ 0.3
	Low	≥ 0.6	≤ 3.5	≤ 6.0	≤ 0.3
Collector	High	≥ 0.8	≤ 3.0	≤ 5.0	≤ 0.4
	Med	≥ 0.6	≤ 3.5	≤ 6.0	≤ 0.4
	Low	≥ 0.4	≤ 4.0	≤ 8.0	≤ 0.4
Local/ Alleyway	High	≥ 0.6	≤ 6	≤ 10	≤ 0.4
	Med	≥ 0.5	≤ 6	≤ 10	≤ 0.4
	Low	≥ 0.3	≤ 6	≤ 10	≤ 0.4

Consider a road (one way) 10 meter wide which is to be illuminated with walkway mounted pole light, mounted at a height of 10 meter and 1.5 m overhang

with a luminaire having 1×150 W HPSV lamp and polar curve shown (Figs. 14.30 and 14.31). Maintain $\dfrac{E_{av}}{E_{min}} \leq 3.0$, and $E_{av} \geq 0.8\,\text{cd/m}^2$, $\dfrac{E_{max}}{E_{min}} \leq 5.0$, $E_{V\,min} \geq 15$ at 1.5 m above road surface.

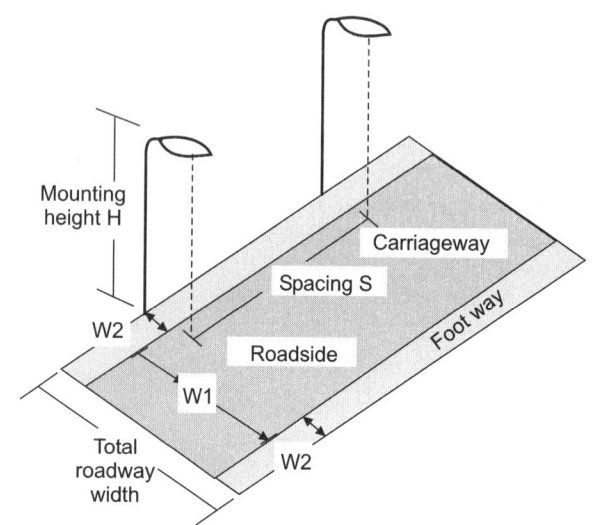

Fig. 14.30 Road and luminaire location

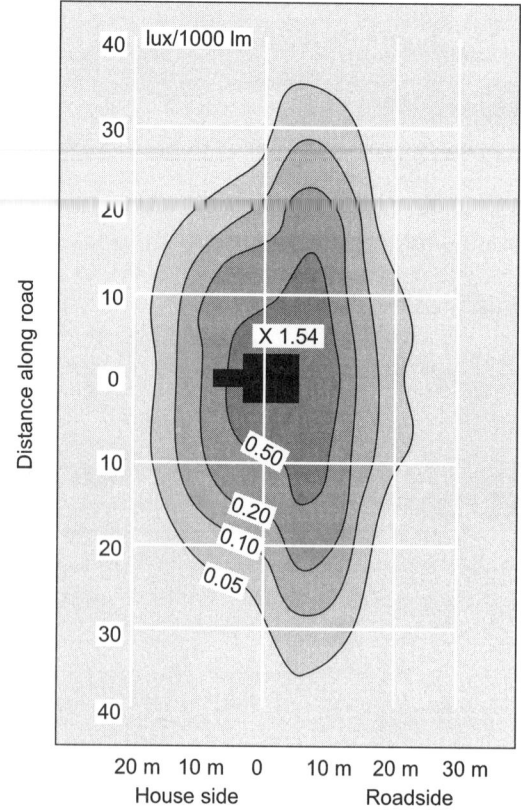

Fig. 14.31 Isolux curve of 150 W HPS luminaire

Draw a grid of points on a plan of road 10 m wide. The grid should be such that maximum spacing of grid points is one-tenth of pole spacing not exceeding 5 meters. The sidewalk grid point shall be matching the road grid points. The sidewalk vertical illuminance shall be designed at 1.5 m above sidewalk (in the plane indicated by arrows). The direction of vertical plane is the direction of traffic. Thus, there will be two isolux curves for two poles mounted luminaires. The distance between poles should be so adjusted that E_{av}, E_{max}/E_{min} and $E_{v\,min}$, E_{av}/E_{min} are within desired limits.

14.9.2 Level of illumination

The lighting installations have been classified as group A_1, A_2, B_1 and B_2 with prescribed illumination levels as below in Table 14.9 (IS: 1944 : 1970).

Table 14.9 Classification of installations on the basis of level of illumination

Classi-fication of lighting installation	Type of road	E_{av} lux	E_{min}/E_{av}	Transverse variation (uni-formity) of illumination (%)
Group A_1	Important routes with fast traffic	30	0.4	33
Group A_2	other main roads with mixed traffic	15	0.4	33
Group B_1	Secondary routes with considerable traffic	8	0.3	20
Group B_2	Secondary roads with light traffic	4	0.3	20

As a general rule, the following number and spacing of luminaire can be adopted.

1. Single side installation: Mounting height equals width of carriage way.
2. Both side installation (Zig-Zag): When width of carriage way is more than (1) above but not more than 1.5 times mounting height.
3. Both sides installation (Facing each other): When width of carriage way is more than 1.5 times the mounting height.
4. Central verge installation: For narrow roads, the width of which does not exceeds mounting height. It may not be used in high traffic situations as this system gives more illumination at the verge than

sides which are the points of obstacles. However, with modern computer softwares, this deficiency may be avoided by accurately selecting lamp, watts, lumens, spacing and overhang of poles.

14.10 MOUNTING HEIGHT AND SPACING

The mounting height and spacing will depend on power of lamp, polar curve of luminaire and area geometry. The mounting height should be selected to avoid glare to driver and also to achieve transverse uniformity. Generally, 9–10 meters mounting height has been recommended by IS 1944 for group A roads and 7.5–9 meters for group B roads.

The spacing to height ratio has been specified based on category of luminaire (Table 14.10):

Table 14.10 S]pacing to height ratio

	Type of luminaire	Maximum spacing to height ratio
1.	Cut off	3
2.	Semi cut off	3.5
3.	Non cut off	4.0

14.11 COMPUTER AIDED LIGHTING DESIGNS

Computers are being extensively used in lighting design, particulars roads and highways so that designers design with utmost accuracy and the user has confidence in the illumination level so achieved while using such high speed expressways. These softwares have been developed due to practical impossibility of carrying out complex manual calculations involved for multi luminaires and their combined effect on various points and planes in longitudinal and transverse directions, horizontal and vertical illumination. The softwares help in examining various combinations of lamps, luminaires and mounting heights more accurately and rapidly.

Before a computer design is undertaken, thumb rule calculation (based on lumen method) are to be carried out which forms the basis of computerized calculations.

The intent of this chapter was to make a designer aware of the basics of illumination engineering and the steps, thumb rules that go in the preparation of a final implementable design. There are various software tools available on the internet, some are free and some chargeable. Almost all lamp, luminaire manufacturers have softwares which they share with the potential clients.

Section D

Other Services

The chapters forming part of this section are:

1. Green Buildings in Simple Steps   Chapter 15
2. Fire Alarm System   Chapter 16
3. Automatic Sprinkler System   Chapter 17
4. Lift Design: Important Factors and
   Traffic Constraints   Chapter 18

Chapter 15

In the current scenario of rapid urbanization and infrastructure development where natural resources are getting depleted fast, it has become necessary to think of methods of conservation of resources by efficiency in design and utilizing renewable sources of power generation. The energy consumed in transporting can be conserved by using locally available material and in turn supporting local economy. Such buildings which affects environ-ment to the minimum started to be called as Green Buildings.

This chapter has been designed to unlock the concept for the benefit of designers who can think about their designs in the larger perspective of saving environment and incorporate small-small steps which are environment friendly. So far green buildings have been thought of as a very complex mechanism which can be designed and constructed only after a special training. This myth has been broken in this chapter for the usage by even a common man.

Chapter 16

This chapter describes in detail the various kinds of fire, various detectors suitable in each case, the location and the distance at which these detectors shall be placed. This chapter explains in a step-by-step approach to divide the complete building into an array of smoke, heat, and combination detectors.

Chapter 17

This chapter illustrates the various kinds of sprinklers, the designing of arrays of piping network. The minimum quantities of water, pressure requirements have been illustrated in the form of easy to understand tables. The whole concept has been explained with the help of case diagrams so that the designer can easily understand the concepts and methodology of the design.

Chapter 18

This chapter is the backbone of designing an efficient passenger transportation system. Any building howsoever large it may be needs a well-managed lift/elevator network to send the occupants of the building to their destination. The most important part of the building design but usually the most neglected one has been explained in detail to let the designer understand what goes into the complexity of elevator design and why designers do not pay much attention to design the elevator system with proper attention. This chapter simplifies the procedure involved in the designing and illustrates how to calculate the traffic pattern, handling capacity, car size, number of cars. In addition, the various methods of group control have been described so as to reduce the waiting interval to the minimum.

15

Green Buildings in Simple Steps

The construction sector is considered to be the biggest polluter of environment. The pollution starts from the stage when land meant for agriculture or forest is acquired, trees are cut and greening is destroyed. The land which was fertile gets converted to high rise towers. The soil retaining capacity, natural slope of earth is disturbed causing water run off to rivers and ponds to be reduced. The construction activities cause sand and dust levels to rise in air and thus increasing the suspended particulate matter beyond permissible limits. The water is drawn using tube wells on large scale, thus depleting the ground water table and further affecting the greening or agricultural activity. Management and disposal of construction and solid waste is another challenge posed by construction industry.

Once the construction is over, there is a heavy demand on public supply company to supply electricity, water and dispose off waste and sewage generated therein. The generation of electricity causes more and more pressure on fast depleting fossil fuels like coal and gas. The treatment of sewage and final disposal becomes another problem. The air conditioning has become an essential part of life but has also created associated problems like release of CFC gases in the atmosphere causing depletion of ozone which is the ultra violet radiation protecting layer of atmosphere.

In India, the construction sector is growing at the rate of 10% every year. Energy consumption is also following this growth proportionately. The residential and commercial buildings alone consume 50% of power produced for all sectors combined. Per capita water consumption cannot be more than 1725 m^3 per capita per year by the time we reach 2025 as against 2500 m^3 in 1990. Similarly, waste water generated in 2003-04 from class-I cities and class-II towns (CPCB report) was 26254 million litres per day (MLD) as against waste treatment facilities available for 7044 MLD only.

All the above activities have caused or have started to cause imbalance in the water cycle resulting in missed rains or excessive rains causing floods witnessed in Uttrakhand in 2013, melting of glaciers causing rising sea levels and posing threat to human habitats, heat island effect due to replacement of green cover with hard scape like concrete pavements, roads and buildings, causing rise of local temperature upto 10°F.

We as human being in a rush to consume and earn more today have to be careful about the future generations and have to leave sufficient for them to consume. Natural resources like petrol, diesel, natural gas and coal that we have consumed in last one century has been created by nature which we call "Mother Earth" over millions of years. We have to keep an eye on our deeds and misdeeds towards environment and if we do not wake up now, our future generations are never going to forgive us. As engineers, designers and builders; the foremost responsibility lies on us to design most efficient building systems which consume less energy, less water, generates less waste and thus pave way for preserving more fertile land, forests and natural resources.

15.1 WHAT IS A GREEN BUILDING

The fast depleting green cover giving way to construction of high rise buildings, residential and commercial townships causes major environmental impact. This impact is not a one time effect but continues during the life time of building.

The construction process requires energy intensive manufacturing of bricks, cement, tiles, steel in various forms. This energy, though not consumed directly at site of construction, is consumed in kilns, quarries and factories which may be hundreds of kilometers away from construction site. The farther these manufacturing utilities from site, more the energy consumed in transporting the construction material to site by consumption of non-renewable energy sources, i.e. petrol and diesel. The construction activity consumes large amount of electricity in running concrete mixers, hot asphalt plant and bitumen plants, cranes and a variety of construction equipments and tools. The consumption of water during construction cannot be overlooked. It is drawn from underground natural reservoirs or rivers thus depleting the water table. The debris generated from construction is disposed off, sometimes in rivers, which is another matter of concern as rivers change course and enter the habitable areas causing floods.

Once the building has been constructed, it consumes energy to fulfill the need of occupants in running air conditioning, providing illumination and power elevators and other devices. The power back up, relying heavily on generator power, consumes energy in the form of diesel and natural gas. Landscape is another concern which consumes large amount of water to provide greening and washing.

Green building is one which uses less water, optimises energy efficiency, conserves natural resources, generates less waste and provides healthier spaces for occupants as compared to a conventional building.

Now-a-days, the concept has been extended further to develop 'Net-Zero' buildings and to certain extent "Power-Surplus" buildings. Net-zero buildings are those which do not consume any outside power but generates within the complex for its use. The waste water generated is treated within the building using STP and biogas generators. The treated water is utilized in cooling towers, flushing and land scaping purposes.

Power-Surplus buildings on the other hand are not only self sufficient for their survival but export power and treated water for utilization by others.

The green buildings utilize technology to reduce reliance on outside resources, e.g. hi-tech control equipments like light and motion sensors reduce power consumption on lights, intelligent controls in air conditioning system help to run the equipments according to load demand, building management system (BMS) controls the power consumption of various devices and switch them off by monitoring the space occupancy, usage pattern and running them at varying load by sending signals to various controls.

The green buildings attempt to reduce off site power demand on manufacturing of construction materials and their transportation. Materials which consume less energy in manufacturing are used. They are obtained from nearby resources also.

Thus green building is a concept which has to be comprehensively developed jointly by architects, civil, electrical and mechanical engineers as explained in detail in next few pages. They critically examine the negative impact of their design on environment and modify/moderate it into a positive impact.

In short, the following aspects of a building design are considered that result in the construction of a green building:

a. Site Planning
b. Wholistic building design (also called envelope design)
c. Use of renewable energy resources
d. Electrical, air conditioning and lighting design efficiency
e. Selection of environment friendly materials
f. Water and waste management and treatment
g. Indoor thermal, visual and air quality

15.2 BENEFITS OF GREEN BUILDINGS

Though green buildings are comparatively higher in cost of construction, they have lesser operational cost which recovers the higher cost of construction in a short span of time of 2–3 years. The green buildings offer following advantages to the owner and the occupants.

1. They consume 40–60% less electricity compared to conventional buildings by utilizing architectural design improvements, use of intelligent building technologies to reduce power consumption and high efficiency materials like insulations on terrace and walls to prevent heat penetration.

2. These buildings use solar and other non conventional power production techniques to become self sufficient and independent of grid power and associated outage problems.

3. The water consumption is reduced by a large margin due to design of rain water harvesting system, use of plumbing fixtures which use less water, water less urinals, reuse of treated waste water in cooling towers and land scaping, and rain water harvesting system.

4. Green buildings have their own waste treatment system which generate biogas or use them as manure or compost.

5. Green building develops erosion and sedimentation control plan for all the activities associated with the project, during construction as well as post construction.

6. Green buildings increase the brand value and hence marketability.

7. Green buildings use more of natural light, building orientation which allow lesser heat inflow which gives a feeling of openness and comfort to occupant. The lesser operational cost results in more profit.

8. Green buildings also provide superior indoor air quality and thus enhances health, well being and productivity of the occupants.

15.3 EVALUATION OF A GREEN BUILDING

Since green building is a comprehensive concept to minimize the impact on environment, there has to be a mechanism to evaluate the closeness with which building has been constructed and its impact on environment. Evaluatory parameters have been developed on the basis of which buildings can be evaluated to have achieved desired closeness to being environment friendly.

On the basis of the number of parameters achieved, different agencies evaluate and audit the project right from the concept stage and certify the building in different rating patterns which vary from country to country and with certifying agency due to country specific bye-laws, codes and standards and their cultural behaviour. The most coveted rating is considered to be from united states Green Building council (USGBC) and the buildings certified by them are judged as Platinum, Gold, Silver and buildings certified by Leadership in Energy and Environmental Design (LEED).

Different countries have formed their own rating systems as well and have pre-defined parameters based on their culture and constraints. In India, Indian Green Building council (IGBC) has adopted LEED system and has launched LEED India version for rating of new constructions, homes, factories, among others. Another agency, The Energy Research Institute (TERI) has developed its own tool for evaluation of buildings on green parameters. This tool, popularly known as "Green Rating for Integrated Habitat Assessment" (GRIHA) is a five star rating system.

Other popular evaluation tools developed by various countries are listed below:

1. **United Kingdom:** Building Research Establishment's Environmental Assessment Method (BREEAM)
2. **Japan:** Comprehensive Assessment System for Building Environmental Efficiency (CASBEE)
3. **Hong Kong:** The Hong Kong Building Environmental Assessment method (HK-BEAM)
4. GB Tool developed by International Framework committee for the Green Building challenge, an international project that has involved more than 25 countries.

Every agency has defined performance criterias for their own rating system. Every performance criteria has points associated with it. Some of the criteria may earn more points as compared to other depending upon the effectiveness associated with environmental impact. Based on the number of points earned, a building or project gets rating in terms of standard fixed by the rating agency. The rating certificate based on the number of points earned for two rating agencies of India are given in Table 15.1.

Table 15.1 Classification of green building certification by Indian agencies

S. No.	Name of agency	Points earned	Certification awarded
1.	GRIHA	50–60	One Star
		61–70	Two Star
		71–80	Three Star
		81–90	Four Star
		91–100	Five Star
2.	IGBC	50–59	Certified
		60–69	Silver
		70–79	Gold
		80 or more	Platinum

15.3.1 Procedure of Evaluation

Any building which is intended to be constructed as a green building is assessed based on its performance, which is predicted in advance through various calculations and undertakings, during the entire life cycle of building, i.e. from the stage of inception till operation. The various stages where a project is evaluated by the certifying agency are

1. Evaluation at pre-construction stage
2. Building design and planning stage
3. Construction stage
4. Operation and maintenance stage.

Every owner or developer of the building has to make sure of his intentions to construct a building with green certification and chose the agency from whom he intends to obtain certification. The project has to pass through following stages:

1. The project has to be registered with the certifying agency. Once the project is registered, the certifying agency imparts complete training on the criterias, procedure to be followed and the list of documents to be submitted by the client from time to time.
2. A certified professional is engaged by the client to undertake coordination work with the certifying agency after collecting various inputs from consultants who are responsible to produce documents in the desired format.
3. The certifying agency refers the project to "Third party evaluators" for evaluation of documents who certify the number of points likely to be achieved for the project and also comments on specific criteria, if required.
4. The client shall then incorporate the changes, suggestions given by evaluator and agency team and make final submission.
5. Once the project is evaluated by 'Third party evaluators', the certifying agency does the site audit of project and awards the final rating.

6. The agency representatives visit the site and audit the compliance with various criterions.

7. Once the project is completed, it is evaluated by the committee of certifying agency for compliance of all points related to pre-construction and construction phase according to provisional rating allotted. The final rating is awarded which is valid for a period of few years from the commissioning of building. During this period, the compliance with energy and water consumption requirements and other parameters is recorded.

It is thus very essential for the owner of the building to strictly adhere to the guidelines and regulations of the certifying agency as little slackness can result in downgrading or no rating being imparted at all.

15.4 SIMPLE STEPS TOWARDS GREEN BUILDING

Making of a green building is not that complicated or a tedious work as has been considered by a major segment of industry and engineering professionals. It is only that we have drifted so far during the course of technological development that present generation do not know the association our ancestors enjoyed with nature and were still resource surplus. They extracted everything from the nature for their present needs without depleting the resources of nature reserved for future generations. The concept of green building is basically to utilize those ideas, designs and techniques with which the present generation can harness the resources of nature without disturbing the balance maintained by it for future generations.

In the following sections, simple techniques and basic fundamentals have been explained which will help to construct an environment friendly building or the so called green building.

15.4.1 Site Selection and Development

15.4.1.1 Site Selection

The intent of this criteria, in awarding requisite number of points for green building rating, is to check the uncontrolled growth of urbanization by sacrificing natural habitats of local species of plants and animals and green lands used for farming. The following steps and precautions will help while selecting the land:

a. It should not be fertile land used for farming.

b. It shall not fall under land specified by Wildlife Institute of India.

c. Development plan and various regulations shall be consulted and use pattern allowed there shall be strictly followed.

d. The land shall also comply with coastal zone regulations, heritage area regulations and water body zone regulations.

e. In order to check the infrastructure development requirement, it will be a simple technique to select the land which is nearer to, say within half to one kilometer radius of an existing market place, bus stand, railway station, school, college and metro station.

15.4.1.2 Protection of Landscape

a. Comply with the provisions of NBC, Part 10, Section 1, Chapter 4 – Protection of Landscape during Construction.

b. Top soil of an area is very important for growing local plant species. Hence this should be protected all throughout the construction. It may not be washed off by storm water run off and/or wind erosion. This wash off can cause choking of drains and sewers due to sedimentation. Various techniques are in force like contour trenching, mulching and soil stabilization methods.

c. Construction activity shall be so timed that external digging and development work gets finished either before rain or is started after rains. During rain period, internal work can be carried out. In addition, effective sedimentation and erosion control measures shall be taken to prevent soil erosion.

d. Segregation bins to prevent mixing of waste and spilling on the site shall be provided. This also helps in treatment of waste and their final disposal as per guidelines. This also helps in efficient circulation of man and material on site.

e. A very important point which can help protect environment is by preserving existing mature trees or transplanting them at a different location. Regulation specifies to compensate the loss of trees by planting new trees in the proportion of 1:3 out of which 25% can be planted within the site premises.

15.4.1.3 Development of Site

a. Once the site has been selected, the architectural designs shall ensure to harness the wind and solar energy so as to minimise dependence on air conditioning and artificial lighting. This is what we see in old monumental buildings where enough of air movement and lighting was allowed to give comfortable environment to the occupants.

b. The direction and orientation of a building has a great bearing in achieving thermal and visual comfort. VAASTU designs were simple architecture of olden times to trap the resources of nature for one's advantage. Table 15.2 shows the solar radiation in summer and winter for various latitudes. It is evident that radiation is maximum in east and west while it is minimum in North and south. It is thus obvious that building shall be so oriented that

Table 15.2 Total solar radiation (direct plus diffused) incident on various surfaces of buildings in W/m²/day for summer and winter seasons

Orientation		Latitude					
		9°N	13°N	17°N	21°N	25°N	29°N
(1)	*(2)*	*(3)*	*(4)*	*(5)*	*(6)*	*(7)*	*(8)*
North	Summer	1494	1251	2102	1775	2173	1927
	Winter	873	859	840	825	802	765
North-East	Summer	2836	2717	3144	3092	3294	3189
	Winter	1240	1158	1068	1001	912	835
East	Summer	3344	3361	3475	3598	3703	3794
	Winter	2800	2673	2525	2409	2211	2055
South-East	Summer	2492	2660	2393	2629	2586	2735
	Winter	3936	3980	3980	3995	3892	3818
South	Summer	1009	1185	1035	1117	1112	1350
	Winter	4674	4847	4958	5059	4942	4981
South-West	Summer	2492	2660	2393	2629	2586	2735
	Winter	3936	3980	3980	3995	3892	3818
West	Summer	3341	3361	3475	3598	3703	3794
	Winter	2800	2673	2525	2409	2211	2055
North-West	Summer	2836	2717	3144	3092	3294	3189
	Winter	1240	1158	1068	1001	912	835
Horizontal	Summer	8107	8139	8379	8553	8817	8863
	Winter	6409	6040	5615	5231	4748	4281

Source: National Building Code-2005

its usable areas are in the north and with sufficient shades in south directions to get maximum sunlight while less used areas like store, staircase, bathroom, etc. are placed in west directions.

c. The usage of glass has increased manifolds in recent times for giving a neat external appearance to the building and basically to use it as a marketing tool. Customer identifies the neatness of exterior of the building with sparkling glass without realizing the environmental concerns associated with its usage. Glass allows radiation heat through it requiring more air conditioning which is the biggest power consumer in the building segment. The glare inside the building becomes too high and is controlled by providing venetian blinds or other light dampening devices which could have been avoided in addition to high air conditioning load. Buildings of earlier period never used glass on facade.

d. Steps can be taken to reduce hard paving at the site. Hard paving increases the heat island effect by absorption of heat by hard paved surfaces. This increases the surrounding temperature. At best, shaded hard paved surfaces can be provided. Net imperviousness of the site shall not exceed the factor prescribed by NBC 2005, Part 9 Plumbing services.

e. Total surface parking shall be within the local bye-laws prescribed limits. Paving shall be pervious, open grid or grass paver, can be shaded, or made of material finish having high solar reflectance, say 0.5 or higher. Carpool can be promoted by providing preferred parking, providing alternate fuel vehicles such as those run on CNG and alternative fuel-stations.

f. Safety, sanitation, clean drinking water, urinals and toilets shall be provided for construction workers as prescribed in NBC 2005 and other codes.

g. The construction activity creates lots of dust in the atmosphere which is an area of concern for people working in surrounding areas. The air pollution shall be avoided by taking adequate measures.

h. Development foot print can be reduced even beyond the limits imposed by local bye-laws.

i. Natural water flow slopes shall be disturbed to a minimum, due to construction activity and imperviousness shall be increased.

j. Reduce night sky pollution due to light. Use light fixtures which reduce sky glow, improve night sky visibility through glare reduction. Reduce lighting power densities for exterior lighting below that defined in ASHRAE and IESNA standards.

15.4.2 Water Conservation

Water, the gift of nature to mankind, is the lifeline on which living beings survive. Over a period of time, the requirement of water has increased tremendously from simple agricultural and domestic use to industrial processing, air conditioning and recreational activities. The water table is getting depleted with increased drawl by use of pumps. The more use of water has associated problems like generation of waste water and its disposal causing pollution of rivers, etc.

In the green building concept, emphasis has been laid on conservation techniques, waste water treatment and reuse of treated water. Let us examine the methodologies with which this can be achieved, again with the help of common knowledge which all of us possess:

15.4.2.1 Landscaping

a. Use re-cycled water for landscaping instead of potable water. Alternatively, equipments which use less water and provide more coverage, captured rain water can be used. Some examples are micro-irrigation, moisture sensors and clock timers, etc.

b. Retain or plant native species which consume less water. These plants are less water consuming and grow rapidly.

c. Install rainwater collecting system on the roof or underground in areas which have sufficient rainfall to make their installation viable.

15.4.2.2 Other Water Conserving Methods

a. Water efficient air conditioning where potable water can be replaced with re-cycled water and captured rainwater. More reliance can be placed on air cooled chillers to eliminate cooling towers.

b. Using recycled water or grey water for flushing in domestic use by installing two pipe system and thus reducing dependance on municipal supply.

c. Install on-site waste water treatment system for treating waste water to tertiary standards. On-site treatment reduces costs associated with transportation, energy and chemical use.

e. Design a proper and effective rain water harvesting to charge the underground acquifers.

d. Consider developing wetlands and aquaculture system for waste water treatment where plants, fish and bacterias naturally remove contaminants from waste water. Aerobic and anaerobic waste treatment plants are available in abundance in market.

f. Use water efficient fixtures in toilets, dry fixtures like composting toilet systems and waterless urinals. Shower heads consuming less than 10 LPM, bathroom faucets which are effective with as little as 3.8 LPM, self closing or electronic sensor faucets are widely available these days and are very effective in reducing the consumption of water. Water closets are the most water consuming devices but use of pressure assisted and dual flush closets can easily reduce the water demand to a great extent. Composting toilets mix human waste with organic material to produce compost which can be used as manure.

15.4.3 Energy Consumption, Generation and Atmosphere

In order to earn green building rating, reduction in energy consumption below the baseline, use of renewable energy resources, reducing dependence on fossil fuel based energy sources and protecting atmosphere from greenhouse gases and ozone depleting gases are important factors which earn maximum points. In the following articles, various techniques are listed which can save the energy consumption. Some of these techniques are utilized by various owners and occupants of the building, but the most important aspect of green building is the implementation of all the strategies in an integrated manner.

a. One of the foremost requirement before earning points under this head is to have a building system commissioning plan in place. This is to make sure that the systems are designed and installed as required. There shall be qualified professionals with high level of experience in energy calculations, installation, operation and commissioning procedures. They are well versed with automation controls which go a long way in achieving the desired savings in energy consumption. The owner can thus rest assured that all activities will achieve the desired goals and are being well documented for sub-mission to rating agency.

b. Achieving minimum energy performance levels as set in the beginning of a project. These levels shall be set as prescribed by ECBC code. Any further reduction in energy consumption fetches additional points. Some rating agencies, particularly those following USGBC norms, prefer to adopt the standards set by ASHRAE/ISHRAE.

Let's examine the methodology set by ECBC for various energy devices and their applicability.

i. The ECBC code is applicable to buildings with a connected load of 100 kW or more or a contract demand of 120 KVA or greater. Generally, buildings or complexes having conditioned area of 1000 m² or more will fall under this category. This code is presently voluntary for adoption in India.

ii. The intent behind developing this code was to develop and prescribe various energy efficient practices which could reduce energy consumption requirements of a building. The provisions of this code applies to
   - Building envelopes, except unconditioned storage spaces or ware houses.
   - Mechanical systems and HVAC
   - Service hot water heating
   - Interior and exterior lighting
   - Electrical power and motors

iii. The code prescribes mandatory and prescriptive approach to reduce energy consumption for new as well as existing buildings. The foremost requirement of the code is compliance with mandatory provisions. But to allow flexibility to designers while designing building components, code specifies specific levels for each individual component of the building systems called prescriptive requirements and if these individual components are achieved, compliance with the code is done.

Further flexibility has been allowed in the code by a Trade-off option that allows trading-off the efficiency in the thermal performance of one envelope element with another to achieve the overall efficiency level required by code. But this trade-off option can be exercised only within building envelope components like roof, wall, fenestration, overhangs, etc. This trade-off is not against improvement in lighting and HVAC systems.

Next level of design flexibility allowed in the code is with "Whole Building Performance" (WBP) method which is more complex than prescriptive method in approach. In this method, whole building is analyzed as an integrated unit with the help of energy simulation softwares and the energy usage in various building components and systems (like envelope, HVAC, lighting and other building systems) is optimized to achieve the most cost effective solution. In this method, a computer software prepares a "Proposed design", calculates its annual energy requirement and compares it with "standard design" which is based on the upper limits of energy use allowed by the code for the particular set of conditions. If the energy use in proposed design is lower or equal to the standard design, the compliance to the code is achieved.

From the above, it will be seen that building a model on prescriptive approach is the first step in any building design. The next is to demonstrate that the design complies with mandatory provisions and the final is to develop a model using simulation software and compare with the model prepared in the first step.

c. Halocarbon gases, popularly CFC gases used in air conditioning and fire suppression appliances are a great threat to ultraviolet protective layer of atmosphere called ozone layer. The depletion of ozone is causing global temperature rise resulting in melting of glaciers and increasing of sea levels. The green building norms prescribe zero use of CFC gases in the construction of new buildings or a comprehensive policy of a time based phase out of CFC gases in the existing HVAC equipments.

d. Methods which help to reduce dependency on fossil fuels and encourage renewable sources of energy like solar power, solar water heating, wind and water turbines, geothermal and biomass power generators earn extra points for green building rating.

e. Once the building has been commissioned, it is important that it operates at the same efficiency levels for which it was designed and at which it was commissioned. Thus a comprehensive operation and maintenance plan must be put in place, ensure that contract guidelines and conditions are very clear and a protocol, to measure and verify the energy performance parameters, is put in place by installing necessary metering equipments, providing training to engineers and operators and compare the performance to baseline performance, at least for a period of one year.

f. To encourage investment in off-site renewable energy projects, additional points are allowed where such green power is exported to grid.

15.4.4 Regional, Renewable and Recycled Material

Green buildings consciously imbibe materials which are environmentally friendly. The steady increase in demand for green buildings has brought down the cost of green products, services and materials, thus making green buildings affordable.

The extraction of material, their processing and transportation to distant sites of construction has a great environmental impact, for it causes air and water pollution, destroys natural habitat, consumes energy and depletes natural resources.

To minimize the impact of construction activities, the best strategy would be to reuse the existing buildings with minor modifications, utilize waste material from land fills, thus reducing burden of extraction of more and more natural resources, utilize local produce as this will reduce transportation cost and support economy for the local population. In addition, utilization of renewable materials minimize the impact on natural resources. Various measures that can be taken to achieve the objective are:

a. Reduction in waste generation to reduce load on land fill sites.

b. Segregation waste materials at site by providing separate storage bins so that recycling can be done at site itself but if required to dispose off, then it is convenient to recycle off-site.

   Reuse of waste material can be done either by using recyclable equipments, or diverting them to manufacturing or best to other construction sites where they can be reused.

c. Install recycling equipments to recycle aluminium cans, steel, plastic, wood and paper.

d. When renovating a building or constructing a new building, retain, identify and utilize existing building with modifications wherever required to reduce burden on natural resources and energy intensive manufacturing. This can be done in a variety of ways, say by utilizing outer shell, partly utilizing outer shell and partly non-shell, or furniture and furnishing items. Various items such as wood flooring and panelling, doors and frames, bricks, beams, decorative items, antique light fixtures, chandeliers can be reused.

e. Train workers and staff on the benefits and procedure of segregation for recycling.

f. In the bill of material, specify those items which are made from recyclable material.

g. Specify those materials in contract which are produced within the region of construction. Thus supporting local economy and reducing energy on transportation.

h. Plan to use those materials which are fast renewable compared to those which are not e.g. bamboo, cotton, wool, linoleum, wheat board, etc.

Green building movement in India is encouraging lot of innovations in the field of eco-friendly materials and technologies. Some of them are:

1. High performance glass
2. Fly ash brick
3. Aerated concrete blocks
4. Insulated roofs
5. Pre-fabricated and pre-engineered buildings
6. Light pipe
7. Radiant cooling
8. Building integrated photo voltaic (BIPV)
9. Wind turbines
10. Living walls
11. Geothermal technology
12. Solar air conditioning system

15.4.5 Indoor Environment

The importance of Indoor environment need no over emphasis because it is for the comfort and health of human beings who occupy the buildings after they are constructed. As maximum amount of time is spent by a person indoors, it is important that indoor environment is healthy without any pollutants, algae, bacterias and it provides comfortable visual and thermal comfort as well. These concerns can be addressed with effective ventilation techniques utilising high efficiency filtration, moisture management strategies and control of contaminants. The various measures which can be adopted are:

a. The indoor environment parameters have been specified by ASHRAE 62.1 and NBC-2005. The mechanical ventilation system can be designed to meet these minimum ventilation parameters. A well defined IAQ plan need to be put in place.

b. The outdoor air quality might get affected due to nearby polluting industries, landfills, traffic pollution, construction sites and thus intake points needs to be identified.

c. Air conditioning cooling towers, ducts, standing water, sanitary exhausts are other areas of concern which breed algae and bacteria. It is important to take care of these areas through a well defined plan at the design stage itself.

d. Specify materials which are free from volatile organic compounds (VOC) as in paints and solvents and earmark designated areas for smoking. Smoking areas shall be at a negative pressure differential of at least 5 pascals compared to surrounding areas. There should not be any recirculation of smoke contaminated air.

The smoking areas can be ventilated at higher ventilation rate compared to non smoking areas.

e. Install equipments to monitor outdoor air quality, like CO_2 sensors, being delivered in the building.

f. Indoor environment is not only concerned with air quality, but visual and thermal comfort as well. High level of lighting levels as specified in NBC-2005 are needed for more productivity and well being. Individual lighting controls, task lighting and efficient light fixtures can be provided. Sensors like day light sensors or occupancy sensors, etc. can be provided to save on energy.

g. Thermal comfort levels can be enhanced by providing individual thermal controls to suit individuals requirements. Operable windows must meet the requirements of ASHRAE and NBC. Various types of sensors to monitor temperature, air flow and humidity levels shall be installed to achieve a desired comfort level and energy consumption.

h. An important aspect of green building is to assess the declared parameters over a period of time after occupancy to establish the achievement. A plan to assess the thermal and visual comfort levels shall be put in place extending over a period of one to two years by providing measuring equipments and recording the measurements. A survey can be conducted to assess the satisfaction level of occupants and if more than a defined percentage (generally 20%) is found to be dissatisfied, the plan needs corrective action.

i. An important part of indoor environment is to connect the individual with outdoor environment. This can be done by allowing daylight into the building and allowing viewability of outdoor environment by providing direct line of sight through glazing in 90% of regularly occupied spaces. A computer simulation will be needed to simulate daylight conditions to prevent excessive outdoor radiation heat load through glazing. The software produces daylight patterns of interior spaces over the entire period of sunlight.

15.4.6 Innovative Designs

Manufacturers, designers and NGOs are constantly working to improve design and methods to achieve results which can reduce the environmental damage. Extra points are allowed by all rating agencies where the new design greatly enhances any of the criterions as specified.

The intent of this chapter was to illustrate that by adopting small measures, environment damage can be lessened. Those designers who had no exposure to green building designs can now at least take into account these simple techniques in every building even if it is not green. Advance methods, computer simulations and cost intensive methods can definitely be weighed against the pay back period and acceptability within the overall budget.

16

Fire Alarm System

A fire detection and alarm system is a key element among the fire protection features of any building. Properly specified, designed, manufactured, installed, maintained, tested and used, a fire alarm system can help limit property fire losses in buildings, regardless of occupancy. The major cause of worry in high-rise buildings is the deaths resulting from fire and hence the need of fire detection and alarm system in buildings has been made mandatory under law.

The function of fire alarm system should be not only to detect the fire but also to perform major functions of fire cut-offs, area isolation, raising of alarm at city fire service center and host of other valuable prevention and protection features. With this fundamental approach in mind, the designer shall be able to answer himself the question "what are the protection goals?". The goal of the owner as well as designer is to protect the lives of hundreds of innocent people who come to work in the building based on the firm belief that adequate arrangements have been made by them in this regard.

The goal must be defined by manufactures of fire system components. The various international laboratories have defined various stringent norms for the testing of the system components and these laboratories have gained accreditation to ISO by certifying agencies.

The laboratories like Underwriters Laboratories Inc. (UL) and Factory Mutual Research Corporation (FMRC) are examples. The products being designed and installed must have been tested by these laboratories so as to have enough reliability.

Just as the components of fire alarm system must be listed by internationally recognised laboratories for the specific use, similarly the way fire alarm system is used must conform to the requirements of national building code, CPWD code of practice, various BIS requirements. By carefully working through the steps, i.e. specify design, manufacture, install, maintain, test and use, the user of the fire alarm system will have a facility that will serve him for many years to come.

16.1 FIRE ALARM SYSTEM BASICS

Fire alarm system are supposed to have the following basic features.

a. A system control unit.
b. A primary or main power supply.
c. A secondary or standby power supply.
d. One or more initiating device circuits or signaling line circuits to which manual fire alarm boxes, sprinkler water flow alarm initiating device, automatic fire detectors, and other fire alarm devices are connected.
e. One or more fire alarm notification appliances circuits to which audible and visible fire alarm notification appliances such as horns, hooters, stroboscopic lamps, speakers are connected.
f. Many system may have an off-premises connection to a central station or a public fire service communication center by means of an auxiliary fire alarm system.

The basic philosophy of connection of above features is depicted in Fig. 16.1.

Based on these features, there are many devices that form the complete fire alarm system. These components are described as below.

16.1.1 Components of a Fire Alarm System

a. **Fire detector** is the most essential component of any fire alarm system. The fire detector senses the occurrence of fire by sensing the heat, smoke, flame, light intensity. Before using any kind of detector for a particular application, it is important to understand the stages in which fire take place which is described in the next article.

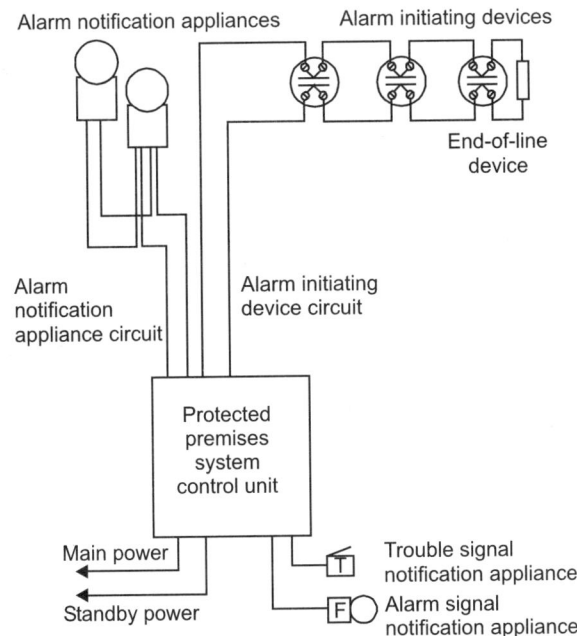

Fig. 16.1 Interconnection of various fire alarm devices

b. **Manual call box (MCB)** is a round or square box with glass front (approximate face area 50 sq cm) that keep a push button pressed such that in the event of breakage of glass cover, the push button is released to actuate an alarm in the control panel. This is mounted at a height of approximately 1.2 m from floor level. The MCB shall be installed at a conspicuous place so that in the event of fire, these can be noticed from either direction.

c. **Response indicator (RI) or spot indicator** is an electronic circuit mounted on the back of a plate 4″ × 3″ size with LED on the front. Any detector that detects the fire causes shorting of the circuit and sends signal to RI for LED to glow. The RI shall be mounted on the outside of room or below the false ceiling (for detectors mounted and hidden inside the false ceiling like heat detectors). This enables the fireman to locate the room where fire has taken place from the corridor itself.

d. **Talk back unit (TBU)** comprises a mini speaker and microphone housed in a box approximately 8″ × 6″. This enables the fireman or any person who detects the fire to contact control room. This unit helps the person near fireplace to communicate information to and receive instructions from control room.

e. **Control and indication panel** is a panel, which gives the alarm in the form of a hooter and beeping LED and also indicates the zone where fire has taken place. These are of following types based on master and slave relationship.

i. **Zonal panel** is mounted on the outside of each zone for which it is designed to give signal and monitor fire. This panel gives the location, i.e. room where fire has taken place. This panel transmits fire signal to sector panel.

ii. **Sector panel** is a panel that is master to several zones, which are very near to the location where sector panel has been located. This panel gives the zone location where fire has taken place. This panel transmits fire signal to C and I panel.

iii. **Main control and indication panel (C and I panel)** is located at the main entrance of the building to guide the fireman to the sector where fire has taken place. This panel is the head of the family of all panels described above.

iv. **Repeater panel** is similar to C and I panel and is usually mounted in a central control room and gives all the indications given by C and I panel.

f. **Sounders** shall be installed in the corridors, staircase and outside the building. These comprise of low intensity sounders and high intensity sounders. High intensity sounders are installed outside of building to warn people of fire inside the buildings.

g. **Mimic diagram** is an illuminated piece or simply a sketch showing the locations of various zones and sectors and is installed at the entrance of the buildings next to main C and I panel or repeater panel.

h. **Public address system** is provided to enable transmission of announcement and instructions to occupants in case of fire. It consists of microphone, amplification equipments and shall be part of C and I panel.

Some of these components are shown in Fig. 16.2.

16.1.2 Type of Circuits

Fire alarm systems have three basic types of circuits.

a. Initiating device circuit (IDC).

b. Notification appliance circuit (NAC).

c. Signaling line circuit (SLC).

Initiating device circuit connects conventional (non-addresseable) fire alarm and supervisory initiating devices to the system control unit.

Notification appliance circuit connects notification appliances (audible and visual) to the system control unit.

Signaling line circuit is a term used to define circuits over which two-way communication takes place.

The communication can take place between an addresseable device and system control unit or the system control unit and the off premises connection, such as a central station, a public fire service communication center, a remote station or a proprietary central supervising station.

A typical circuit used in many fire alarm system consists of a two wire circuit with an end of line resistor.

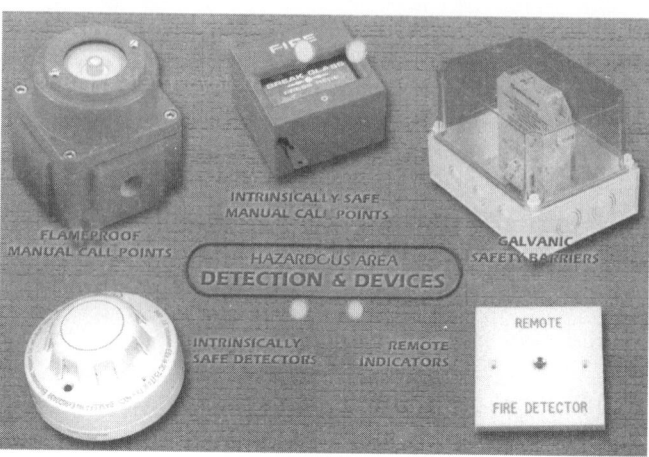

Fig. 16.2 Various components of a fire alarm system

Initiating devices, with normally open contacts are connected in parallel. A small amount of electrical current flows through the wire to monitor the wiring integrity. A break in the wire causes a trouble condition at the system control unit. Everything electrically beyond the break in the wiring is out of service until repairs are made to the circuit. Operation of an initiating device shunts the resistor, thus increasing the current which causes the system control panel to respond in the form of an alarm signal.

Many fire alarm systems have more than one initiating device circuit, so that fire location can be indicated on an annunciator panel by floor, wing or room. The annunciator can be built into the control or located in a lobby, maintenance area, telephone switchboard room or some other location where it is accessible to building and fire service personnel.

A typical circuit diagram of a FAS is shown in Fig. 16.3.

16.1.3 Functions to be Performed by FAS

The main purpose of a fire alarm system is to activate local audible and visible alarm so as to notify the occupants of the building that they must evacuate the protected building.

The system could also interface with other protective system that would help make the building safer in case of fire. Such features are listed below and depicted in Fig. 16.4.

    a. Operation of a fire extinguisher or suppression system.

    b. Recall of elevators.
    c. Unlocking of doors (in case of smart building).
    d. Closing of smoke barrier in air-conditioning ducts return supply.
    e. Control of heating, ventilations and air-conditioning (HVAC) equipments.

In a protected premises system, the alarm is not relayed automatically to a fire department, rather someone has to intimate them personally through some other means say telephone, etc.

Whereas in a central station FAS, the signals are received from a protected premises by an operator to be provided by the company who provided FAS in the premises.

This kind of service is provided only for industrial and commercial facilities of relatively high value. Signals may be transmitted from a protected premises using a variety of transmission technologies such as telephone lines, optical fiber cable, etc. The operator on receiving the signal must inform the designated people at the protected premises or at their homes.

In a remote station system, the outputs from a building fire alarm signal are transmitted to a remote location, i.e. to public fire service center using the same communication means as described above.

16.2 STAGES OF FIRE AND DETECTION

The manifestation of fire usually appears in a sequence of four stages, viz.

    i. Incipient stage. This is the first stage of fire in which invisible particles of combustion are

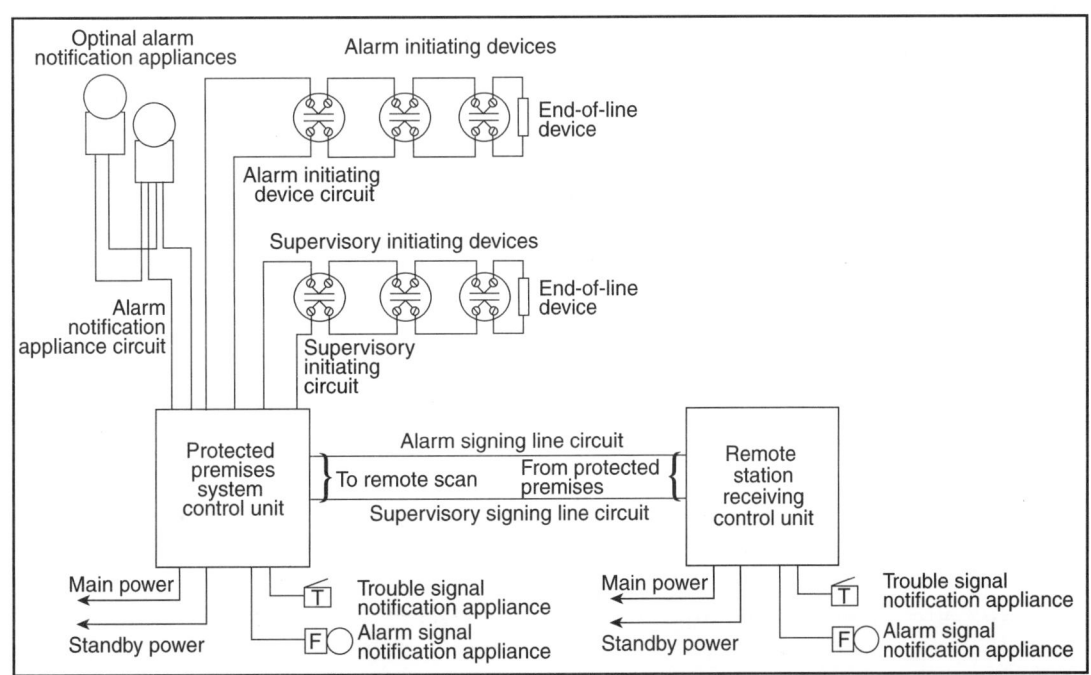

Fig. 16.3 Typical circuit diagram of FAS

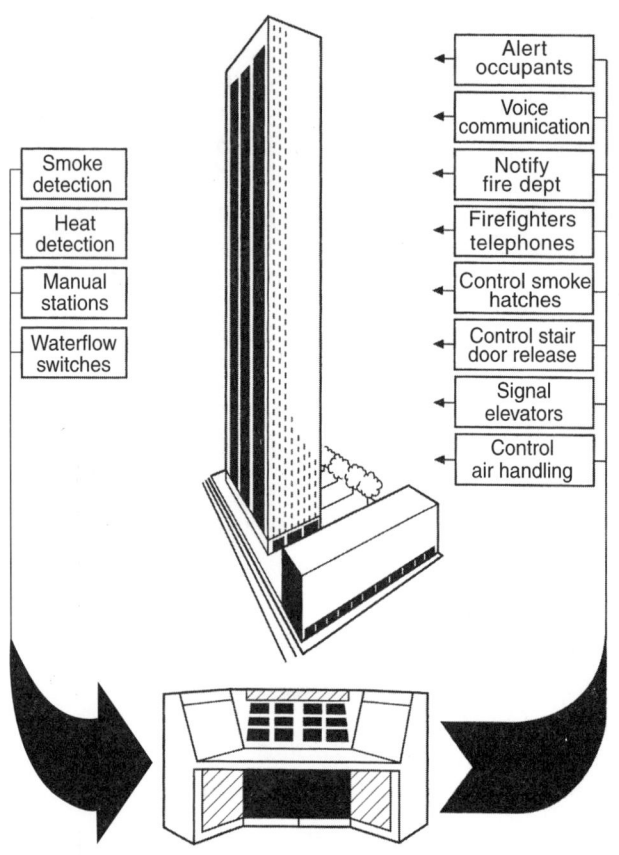

Fig. 16.4 A modern computer controlled FAS

emitted. Approximately 90% of particles of combustion are invisible and remaining 10% are large enough to be seen and recognised as smoke and largely consist of unburnt carbon and sulphur with water vapours.

ii. **Smoldering stage.** In this stage of fire, the products of combustion become very large and are visible as smoke.

iii. **Flame stage.** This is the advanced stage of fire in which actual flame has started though appreciable heat is still not present.

iv. **Heat stage.** This is the last stage of fire where uncontrollable heat and rapidly burning air forms the dangerous combination in addition to smoke and flame.

The detectors, which are used, shall be chosen intelligently depending on the type of material likely to be stored in the place being safeguarded. Accordingly fire detectors have been classified as follow.

16.2.1 Detectors for Incipient Stage

These detectors are called smoke detectors. These detectors utilize principle of ionization.

As a class, smoke detectors using the ionization principle provide somewhat faster response to high energy fires, since these fires produce a large numbers of the smaller smoke particles. While smoke detectors operating on photoelectric principle respond faster to the smoke generated by low energy fires (smoldering fire) as these fires produce more of the large smoke particles.

The conventional smoke detectors sensitivity is dependent more on the alertness of the operator regarding its cleanliness, sensibility of designer to ensure that the sensitivity level of the detector matches the worst environment in the facility to be provided so that false alarms are not generated unnecessarily causing panic.

With the use of microprocessor in FAS, the new detectors use analog technology to measure the conditions in the area or space protected and transmit that information to the computer based fire alarm control unit. The new sensor in these detectors can report when it is too dirty to function properly or is getting too sensitive due to any number of conditions in the protected space. Analog sensors provide a system free from false alarms.

16.2.1.1 Ionization Smoke Detectors

The detectors used are called **ionization detectors** that operate on the principle of sensing a change in electrical properties of a radioactive material housed in the detector chamber. The radioactive material emits alpha emissions and keep the air in ionization chamber ionized and cause a small current to flow.

As soon as the combustion products enter the chamber, they change the chamber impendence and cause a voltage shift, which is sensed by unit's ionization circuitry and results in an alarm signal.

The ionization detectors are of utility where small aerosols are present, yet no flame is present. As a general practice in any air-conditioned area like office complex, computer center, control room or any other similar area ionization smoke detector is used. Even in non air-conditioned area where costly or sensitive equipments are placed and where much of heat may not be generated for long period, ionization type smoke detectors are used. They are also sensitive to burning of certain plastics. They are less sensitive to optically dense smoke like the one generated by burning of PVC. Here optical smoke detectors respond fast.

Smoke/ionization detectors are thus suitable for both slow as well as rapid fires as long as the fire generates invisible and/or visible aerosols.

Smoke detectors are not suitable for dusty areas or humid areas, which may cause false alarm. Optical smoke detectors are sensitive to extraneous light and may give false alarm. Smoke detectors are not suitable where air velocity is high exceeding 5 m/s. Here optical detectors can be used. Both the detectors can be used where cross zoning is required for initiating fire alarm.

The principle of operation and cross-sectional view are shown in Fig. 16.5.

16.2.2 Detectors for Smouldering Stage

As the products of combustion become larger and visible, the detectors used for this purpose utilizes photoelectric principle and are therefore called **photoelectric detectors**.

These detectors utilize the principle of change in intensity of light received by a photoreceptor, resulting in decrease in resistance of photo-cells and resultant increase in current is electronically amplified to produce voltage for alarm signaling.

This effect can be utilized to detect fire in two-ways.

1. Obscuration of light intensity over the beam path.
2. Scattering of the light beam.

16.2.2.1 Light Obscuration Principle

This system consists of a light source, a light beam collimating system and a photosensitive device. When dense smoke obscures part of the light beam, the light reaching the photosensitive device is reduced and this initiates alarm (Fig. 16.6.)

Most of these detectors are of beam type and are used to protect large open areas. They are installed with the light source at one end of the area to be protected and photosensitive device at the other end.

Photoelectric detectors are used mainly in combination with ionization smoke detectors in areas like cinema halls where due to height of building the ionization detector response will be very late. Hence, photoelectric detectors installed on walls at a comparatively lower height sense the larger smoke particles by variation in the intensity of light and sends signals. These detectors are also used in cable galleries. Limits of heights and distances for optical beam smoke detectors is given in Table 16.1.

16.2.2.2 Light Scattering Principle

When smoke particles enter a light path, scattering of light results. This principle is utilized in photoelectric detectors of second variety. They contain a light source and a photosensitive device so arranged that the light rays do not fall on the device. When smoke particles enter the light path, the light is scattered and falls on the photosensitive device causing an alarm (Fig. 16.7).

16.2.3 Air Sampling Smoke Detectors

There is another category of smoke detectors called *air sampling smoke detectors*. The detectors of this type use an air pump to draw a sample of air from the protected

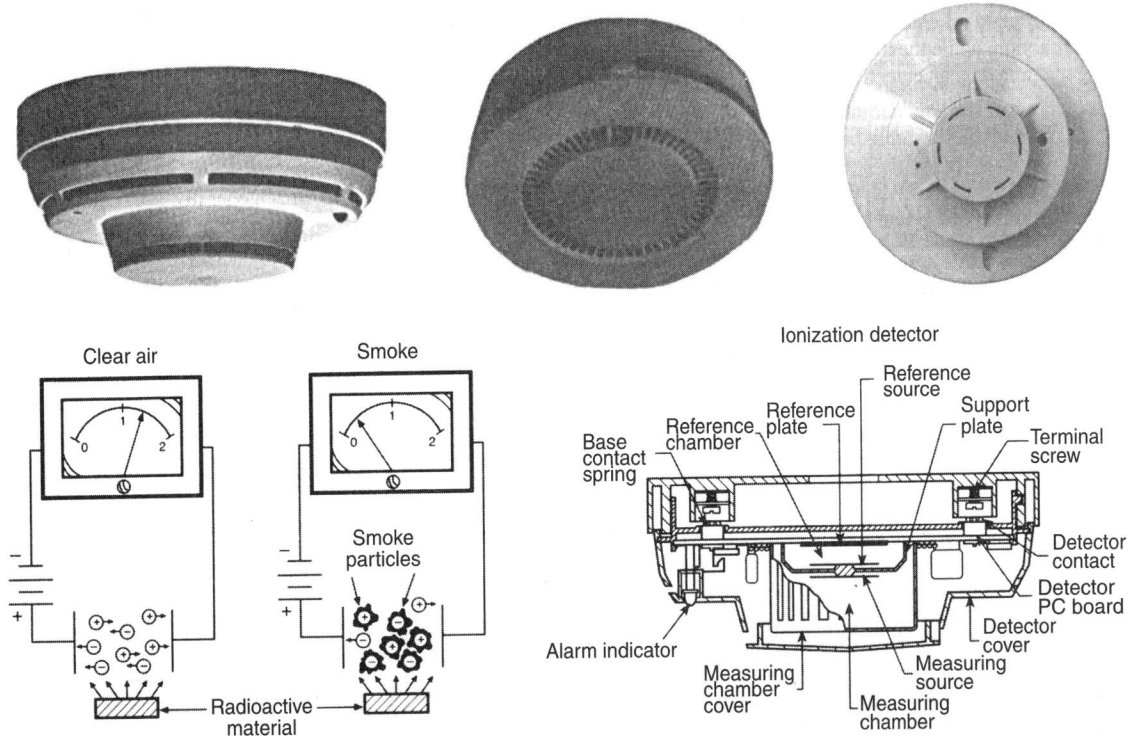

Fig. 16.5 Principle of operation of Ionization smoke detector. Cross-sectional view of an ionization smoke detector and ionization smoke detector

areas into a high humidity chamber within the detector. After the air sample has been raised to a high humidity, the pressure is lowered slightly. If smoke particles are present, the moisture in the air condenses on them

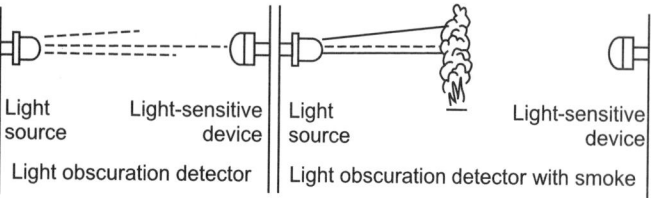

Light source | Light-sensitive device | Light source | Light-sensitive device

Light obscuration detector | Light obscuration detector with smoke

Fig. 16.6 Principle of operation of photoelectric detector

forming a cloud in the chamber. The density of this cloud is then measured by the photoelectric principle. The detector responds when the density is greater than a pre-determined level.

In another variation of this detector called continuous air sampling smoke detector, there are other smoke detection devices in addition to cloud chamber smoke detection devices mentioned above. The air sampling system consists of sampling pipes spaced uniformly over the ceiling together with two supplemental pipes arranged to sample the return air exiting from the monitored space. Each one of the ceiling pipe drops is capped and has a small sampling hole drilled in the cap to draw in a sample of air from that location. The network of piping is connected to detector/control unit where there is fan or aspirator that creates a flow of air in the piping network and this flow causes the pressure

inside the pipe to be less than the local atmospheric pressure.

Inside the detector is a very intense light source that irradiates the sampled air. If there are smoke particles in the sampled air, the device, which can sense smoke particles in extremely low concentrations, will activate the alarm system. These systems are used in electronic data processing areas and museums or where equipment survival is paramount to continuity of operations.

16.2.4 Detectors for Flame Stage

The detectors used for this stage shall be suitable for rapid fire conditions in which flame is also present like fuel loading platforms, explosive gases area, etc. and are called **radiation detectors**.

In this kind of fire, both infrared and ultraviolet rays are present. In detectors operated on ultraviolet principle, rays fall on the gas filled tube across which suitable potential is applied. These rays ionize the gas and causes flow of current that causes a fire alarm. These detectors are used where distance between detector and area of hazard is relatively short.

Detectors based on infrared principle utilize focusing of infrared rays on photovoltaic cell, change of current and hence causing a fire alarm (Fig. 16.8).

These detectors are used where highly inflammable material is stored and any delay in detection may cause the material to explode. These detectors can be used in

Table 16.1 Limits of height and distances for optical beam smoke detectors

	Minimum	Maximum
Height of optical beam above floor	2.7 m	25 m
Horizontal distance between optical beams measured at right angles to a beam	–	14 m
Optical beam length	10 m	100 m
Distance of optical beam from a flat ceiling or apex of roof	0.3 m	0.6 m

Note: The limits as per this table shall apply to detectors conforming to BSS. For detectors to other foreign standards, the recommendations of the respective standards shall apply.

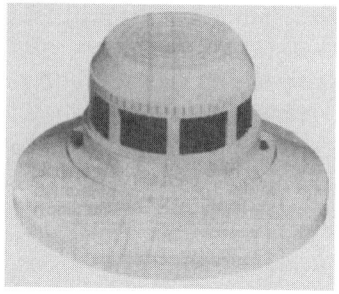

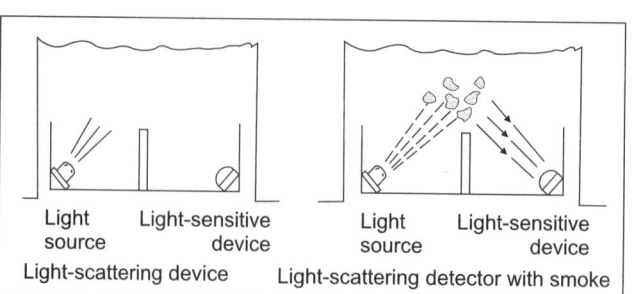

Light source | Light-sensitive device | Light source | Light-sensitive device

Light-scattering device | Light-scattering detector with smoke

Fig. 16.7 Principle of operation and view of photoelectric scattering smoke detector

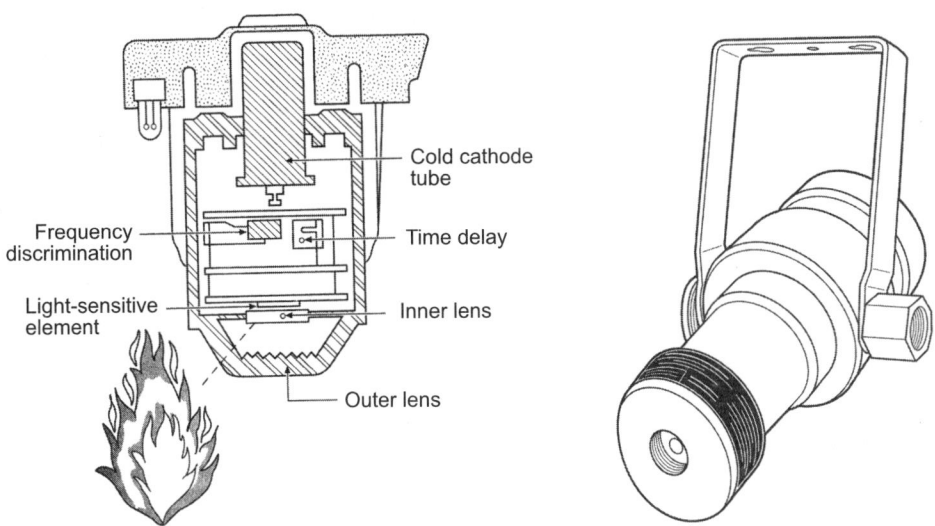

Fig. 16.8 Cross-section and view of infrared flame detector

combination with smoke detectors, so as to oversee and monitor fire over large areas having a high ceiling like hangers, warehouse, etc.

16.2.5 Detectors for Heat Stage

The detectors used for this stage of fire are classified into two categories, viz. those which respond when the detection element reaches a predetermined temperature, i.e. **fixed temperature detector** and those which respond to a rate of increase in heat which is greater than some predetermined value (rate of rise) called rate of rise heat detector. Normally rate of rise detectors are combined with fixed temperature detectors and are called **combination detectors**.

16.2.5.1 Fixed Temperature Detectors

These detectors raise an alarm when the temperature of operating element reaches a specified point. The air temperature at the time of alarm is usually higher than the rated temperature because it takes time for the operating element temperature to rise to its set point. This condition is called thermal lag. Fixed temperature detectors are available to cover a wide range of operating temperatures from about 57 °C and higher. These detectors are further classified according to their principle of operation, i.e.

a. **Fusible element type.** An eutectic metal is used to actuate the heat detector. This metal secures a spring under tension. When the element fuses, the spring action closes contacts and initiates an alarm (Fig. 16.9).

b. **Bimetallic type.** In these types of detectors, principle of bimetallic strips having different coefficients of expansion is applied. The low expansion metal commonly used is Invar, an alloy of 36 percent nickel and 64 present iron.

As the bimetallic strip is heated, it deforms in the direction of contact point. This closes the alarm contacts and raises an alarm.

16.2.5.2 Rate of Rise Heat Detectors

The flaming fire has the effect of increasing the temperature of air above fire rapidly. Fixed temperature heat detectors will not initiate an alarm until the air temperature near the ceiling exceeds the design point. The rate of rise detector, however, will function when the rate of temperature increase exceeds a predetermined value, typically around 12 to 15 °F (7 to 8 °C) per minute. In a pneumatic fire detector, air heated in a tube or chamber expands, increasing the pressure in the tube. This exerts a mechanical force on a diaphragm that closes the alarm contacts. These detectors have a small orifice to vent the higher pressure that builds during slow increase in temperature or during a drop in barometric pressure. If these vents are

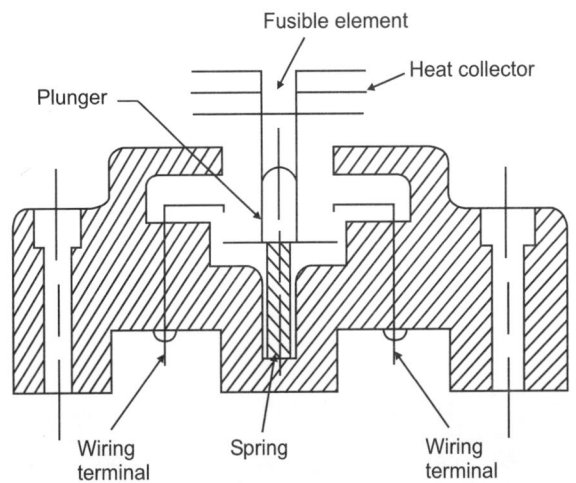

Fig. 16.9 Fixed temperature, fusible element type heat detector

not provided, then slow increase in temperature in an air tight compartment would cause the detector to initiate an alarm regardless of the rate of temperature change. When the temperature rise exceeds 12 to 15 °F (7 to 8 °C) per minute, the pressure is converted to mechanical action by a flexible diaphragm.

16.2.5.3 Combination Heat Detectors

These detectors contain more than one element that responds to a fire. Thus, these detectors operate on both, the rate of rise and fixed temperature principle. The advantage of these detectors is that the rate of rise element will respond quickly to a rapidly developing fire, while the fixed temperature element will respond to a slowly developing fire (Fig. 16.10).

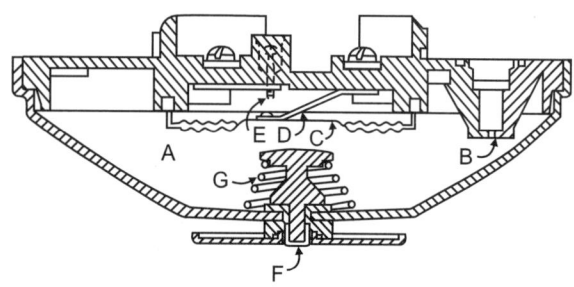

Fig. 16.10 A spot type combination rate of rise, fixed temperature device

Heat detectors are used where one can afford to loose time by allowing fire to attain the conditions of uncontrollable heat. These are suitable for fire that release heat rapidly and are also used in dusty locations where due to dust and dirt, the smoke detectors may give false alarm.

Limit of ceiling heights up to which heat detectors can used is given in Table 16.2.

Table 16.2 Limits of ceiling heights for (point type) fire detectors

Detectors type	Ceiling height (general limits)
Heat detectors to BS 5445 part 5	
Grade 1	9.0
Grade 2	7.5
Grade 3	6.0
Smoke detectors	10.5
High temperature heat detectors BS 5445 part 8	6.0

Notes:

1. When examing use of detectors to the other foreign standard, the respective recommendations will apply.

2. Where there is any section of the ceiling which exceeds the maximum ceiling height specified, such a higher section shall be protected by an independent detector, provided (i) the area of such a section does not exceed 10% of the ceiling area, and (ii) the height of ceiling in such a section does not exceed 10.5 m where heat detector are used.

Because hot air is diluted by colder air, the heat detectors should be spaced closer on high ceiling to achieve the same response time that they would provide on a 3 m ceiling. To determine how much closer, various tests were conducted using fires of various sizes and taking into consideration ceiling height, ambient temperature and fire development.

A typical chart for determining spacing of detectors proportionate to various ceiling heights is shown in Fig. 16.11.

As can be seen from the chart, that with the change in fire intensity or the heat release rate, the spacing will change for the same ceiling height. A comparison of heat and ionization detector is given below.

Heat detector	Ionization type
1. Suitable for fire that releases heat rapidly.	1. Suitable for fire of initial stage where aerosols are present.
2. Suitable for dusty locations and where normal work process releases smoke like laboratories, garages.	2. Not suitable for dusty and high humidity locations as it can give false alarm.
3. Suitable for non air-conditioned areas.	3. Suitable for air-conditioned area.
4. When used in AC area, it should be in combination with smoke detector as cool air reduces the heat development and this detector may not respond well in time. However, above false ceiling this detector can be used.	4. This detector operates on smoke generated. Hence cool air has no effect.
5. Suitable where air movement is high.	5. Not suitable for high air movement or large number of air changes (exceeding 5 m/s) because it causes dilution of smoke which cannot be sensed by detector.
6. Used in kitchen, boiler rooms, furnace kiln and areas above false ceiling or for non air-conditioned space in combination with smoke detectors for air-conditioned space where temperature may be high or rise suddenly.	6. Used where highly valuable items are stored such as computer rooms, archives and where early detection can be done by sensing smoke.

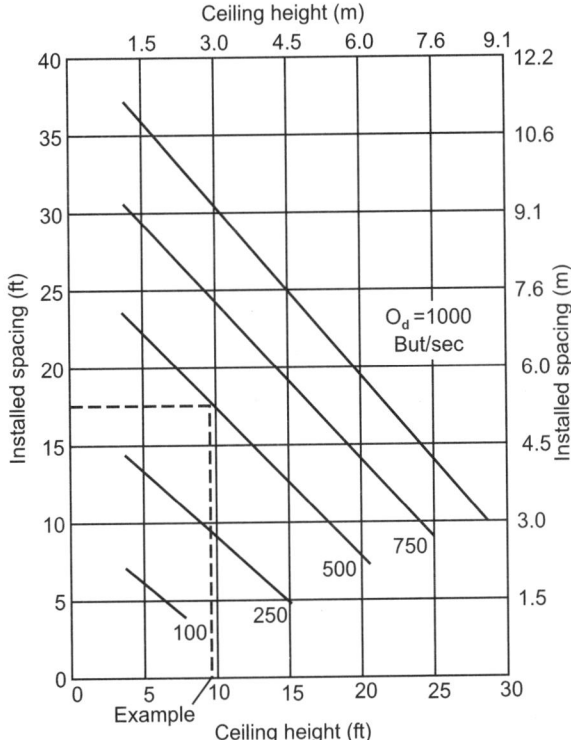

Fig. 16.11 Design curve for a fixed temperature heat detector with a listed spacing of 30 ft (9.1 m) for slow fire

16.3 AREAS TO BE MONITORED BY HEAT DETECTORS

The fire detectors shall be installed to provide total coverage of all areas in a building required to be provided with an AFAS. All rooms, halls, storage areas, corridors, basements, attics, space above false ceilings, below false floors, lift shafts and other shaft/chutes, stair cases, and lantern lights should be covered by detectors. Detectors shall also be provided in lift machine rooms, weather maker rooms, air-conditioning plant rooms, boiler rooms, electric substation, cable tunnels and near electric distribution boards.

16.4 CALCULATION OF NUMBER OF DETECTORS

Since the type of detector has already been decided. The calculation of number of detectors for various types of roofs, rooms is done as below. Since heat and smoke travel upward, heat and smoke detectors are mounted on the ceiling of area.

16.4.1 Effect of Ceiling Height

The increase of ceiling height is counterbalanced by reducing the spacing as given in Table 16.3. Limit of height and distances indicated in Tables 16.1 and 16.2 shall not be exceeded.

16.4.2 Supply Air Duct

Supply air duct of AC system shall be monitored by a detector within 1.5 m of the outlets.

16.4.3 Return Air Duct

Detector shall be mounted within 25 cm of inlet of return air duct.

16.4.4 Computer Installations

Both ionization type and optical type smoke detector in the ratio of 2 : 1 shall be used. The number of detectors shall be suitable enhanced so as to provide an effective coverage of 30 sq m per detector.

16.4.5 Calculation of Number of Heat Detectors and their Spacing

The calculation of number of detectors depends on the type of ceiling, size of room, height, etc. Since heat and

Table 16.3 Reduction factors on spacing of spot type detectors due to ceiling heights for heat detectors conforming to BSS and other foreign standards and smoke detectors

Ceiling above	Heights up to	Percentage of listed spacing
0 m (0′)	3 m (10′)	100
3 m (10′)	3.6 m (12′)	91
3.6 m (12′)	4.2 m (14′)	84
4.2 m (14′)	4.8 m (16′)	77
4.8 m (16′)	5.4 m (18′)	71
5.4 m (18′)	6.0 m (20′)	64
6.0 m (20′)	6.6 m (22′)	58
6.6 m (22′)	7.2 m (24′)	52
7.2 m (24′)	7.8 m (26′)	56
7.8 m (26′)	8.4 m (28′)	46
8.4 m (28′)	9.0 m (30′)	34

Note: In the case of sloping ceilings, the ceiling height indicated in this table refers to the peak height where the slope angle is less than 30 degrees and to average height where it exceeds 30 degrees.

smoke travels upward, heat and smoke detectors are mounted on ceiling of room unless height of room is very large to warrant installation of optical detectors.

16.4.6 Basic Concepts

The following information's have been extracted from different sources, viz. ISS, NBC, BSS and NFPA. There may be areas of apparent contradictions from one another. However generally acceptable principle are as follow.

The maximum coverage of a detector heat or smoke is the area of a circle whose radius is 0.7 D (Fig. 16.12), where D is the maximum recommended spacing for the detector. No part of an area to be protected should be at a distance more than 0.7 D from the detector. The maximum distance of detector from walls and deep beams is 0.5 D.

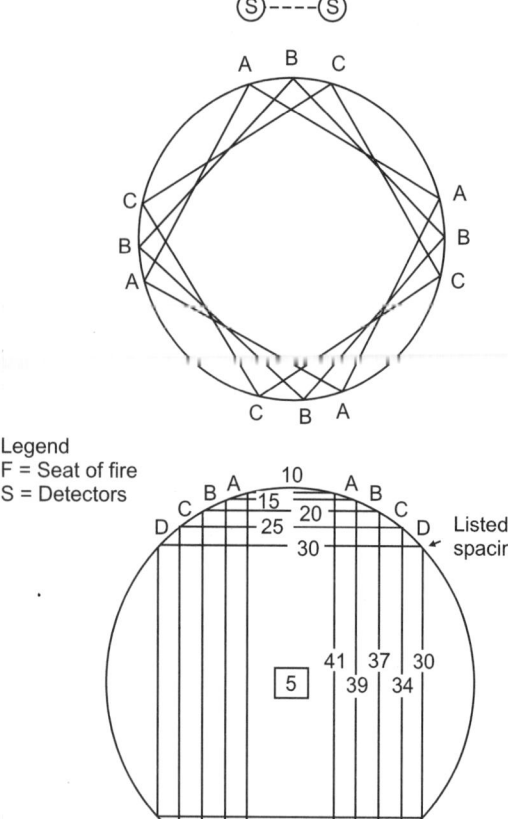

Legend
F = Seat of fire
S = Detectors

Fig. 16.12 Coverage of detector

16.4.7 Heat Detectors

The fixed temperature and combination heat detectors conforming to BSS shall not be spaced at more than 7.5 m

for ceiling height up to 3 m and the minimum number of heat detectors shall be calculated as below.

16.4.8 Minimum Number of Heat Detectors

The number of heat detectors to be installed shall be calculated by dividing the total area to be protected by 50.

Detector spacing should be reduced according to increase in ceiling height and as indicated in Tables 16.3 and 16.4.

16.4.9 Smoke Detectors

The maximum spacing of detectors shall be 9 m for ceiling height up to 3 m. The minimum number of detectors shall be calculated as below.

Minimum number of smoke detectors

$$= \frac{\text{Area in sq m}}{100}$$

In corridors, this spacing can be increased up to 13 m.

16.4.10 Effect of Air Dilution

Smoke detector's response is reduced by dilution of smoke as the number of air changes increase and closer spacing of detectors as indicated in Table 16.5 shall be followed.

16.5 EFFECT OF CEILING CONSTRUCTION ON HEAT AND SMOKE DETECTOR LOCATION

Considerations applicable to both heat detectors and smoke detectors.

  i. **Effect of ceiling projections.** Ceiling projections like beams and joists up to a depth of 15 cm may be ignored. Where this depth exceeds 15 cm, but is less than 10% of the floor to ceiling height, detector spacing (D) should be decreased by double the depth of projection. Where the depth exceeds 10% of the floor to ceiling height, such projections should be treated as walls for the design of detectors layout.

  ii. **Effect of waffles.** Waffles up to 15 cm depth may be ignored. When the depth exceeds 15 cm, each waffle should be considered as a separate compartment, if detectors are to be installed inside the waffles. For installation of detectors on the ridge of the waffles, the maximum spacing permitted is 0.7 D for 15 to 50 cm depth of waffles, and 0.5 D for depth more than 50 cm.

  iii. **Distance from walls and beams.** Detectors should not be placed closer than 50 cm from beams, walls or any such obstructions to the flow of hot gases/smoke towards the detectors.

  iv. **Effect of slope in ceilings.**
    a. Where the ceiling is sloping, a row of detectors shall be installed within a maximum

Table 16.4 Heat detector spacing

Grade as per IS 2175	Ceiling height (m)							
	Up to 3.5	4.0	5.0	6.0	7.0	8.0	9.0	10.0
Grade 1	7	No chance	6	5	5	4	3	Nil
Grade 2	6	5.5	5	4	3.5	3	Nil	Nil
Grade 3	5	4.5	4	3	3.5	Nil	Nil	Nil

Remarks: Spacing from the boundary wall should be kept S/2.
Refer IS 2189: 1999 for further details.

horizontal distance of 0.6 m from the apex. Projections on horizontal plane shall then be considered for designing the remaining detector locations Fig. 16.13.

b. Where applying correction factors due to ceiling heights in sloped ceilings the height up to the apex shall be considered for slopes up to 30° and the average height for above 30°.

v. **Effect of height of partitions/racks.** Rooms divided into sections by partitions, racks, stacks, etc. reaching to within 45 cm of the ceiling shall have detectors for each such section and passage way.

vi. **Provision on racks.** For high rack storage areas, smoke or heat detectors should be installed above each aisle and additionally at intermediate levels on the racks, staggered in adjacent rows. Detectors spacing on racks shall be 2 m vertical and 3 m longitudinal for smoke detectors and 1.5 m vertical as well as longitudinal for heat detectors.

vii. **Wall openings.** All openings such as doors, windows, ventilators, etc. shall be provided with detectors. The detector(s) shall be placed above the opening(s) within 1.5 m distance from the same. The spacing shall be 2 m (for heat detectors) covering the width of the opening.

viii. **Shafts.** At least one detector at each floor shall be placed at a distance not more than 1.5 m away from hoists, lifts, stair wells, and similar flue-like openings. Where only heat detectors are provided

Table 16.5 Heat detector spacing for high air movement areas as per IS 2189 – 1999

Air changes per hour	Multiplying factor (area of coverage)
7.5 or less	1 (1)
8.6	0.95 (0.91)
10	0.91 (0.83)
12	0.83 (0.70)
15	0.74 (0.55)
20	0.64 (0.40)
30	0.50 (0.25)
60	0.38 (0.15)

Refer IS 2189: 1999 for further details.

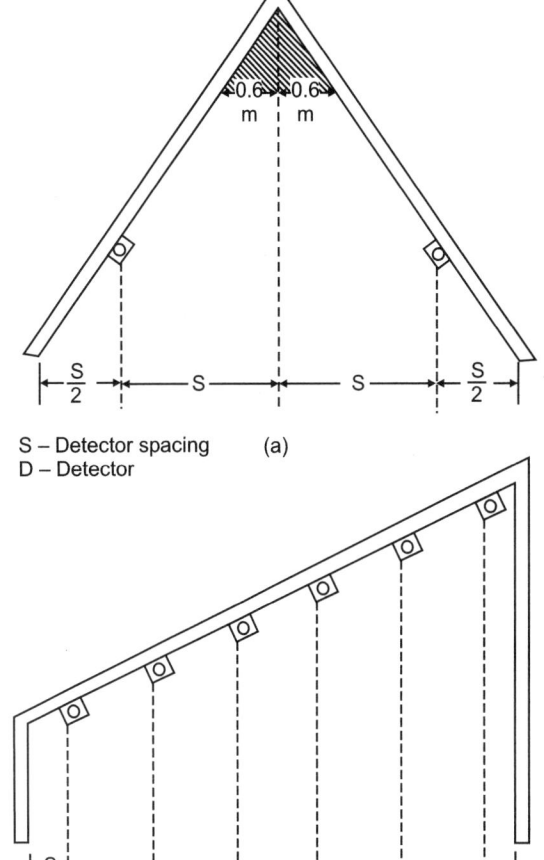

S – Detector spacing (a)
D – Detector

Fig. 16.13 (a) Fire detector spacing for two side sloping roof, (b) Fire detector spacing for one side sloping roof

these shall be one detector for every 2 m width of such openings. Detectors shall also be provided on the top, inside all flue-like openings.

ix. **Detectors on walls.** Detectors when installed on walls shall be located at a distance of more than 10 cm but less than 30 cm from the junction of the wall and the roof.

16.6 DESIGNING OF CONTROL AND INDICATING PANELS

The basic function of control and indicating panel is to constantly monitor the system and respond instantly to actuation of any trigger device connected to it or any fault in the system and to provide audio and visual indications of the same. Such indications should provide unambiguous information about the area (zone) from where alarm has originated so that any person responding to the same can readily identify and reach that area to take appropriate action.

The area to be monitored shall be divided in zones in such a way that the person responding to signal can reach easily to the zone without having to pass through other zones. The typical arrangement is shown in Fig. 16.14.

The floor area of a zone should not exceed 1000 sq m also.

1. A zone should not cover more than one floor including mezzanine floor.
2. A zone should not extend beyond a single fire compartment.
3. Stair walls, lift shafts and other flue like structures within one fire compartment may be considered a separate zone.
4. The search distance, i.e. the distance to be travelled by a searcher inside the zone to fire location should not exceed 30 m. Thus, if a zone can be visually examined from its entrance, the area limitation will be applied, but when there are racks/partitions, etc. the search distance limitation will also apply.
5. While deciding on the extent of a zone. The total number of detectors in that area should not exceed the following:
    a. Detectors connected in series circuit 100 number.
    b. Detectors connected in parallel circuit 20 number.

6. If the total floor area of a building is such that the total number of detectors is less than 20, the entire building irrespective of number of floors may be considered as a single zone.
7. When there are more than one zone in a floor, controlled from a main control and indicating panel, such zones in a floor collectively constitute a section. Sectionalizing is not necessary where there are not many zones in a floor.
8. The main control and indicating panel shall be located conspicuously on the ground floor and in immediate vicinity of building likely to be used by fire brigade and never inside a closed space.

16.7 SOUNDERS

The audible signals must be louder than the ambient noise in a space. Secondly, it should be audible in every part of the building to be protected.

Two different ambient noise levels are required to complete a system design.
    a. A 24 hour average decibel.
    b. Maximum sound level decibel that lasts 60 seconds or more.

The fire alarm system must produce a sound pressure level that is 5 dBA more than ambient noise that lasts more than 60 seconds or 15 dBA above the 24 hour average, whichever is greater. These values can be used for public places. While for private places, the respective values shall be taken to be 5 dBA and 10 dBA respectively.

However, in places like hospitals where sleeping mode exists, a minimum sound level of 70 dBA shall be used.

The maximum allowable sound pressure shall not exceed 120 dBA at the place nearest to sounder. This is

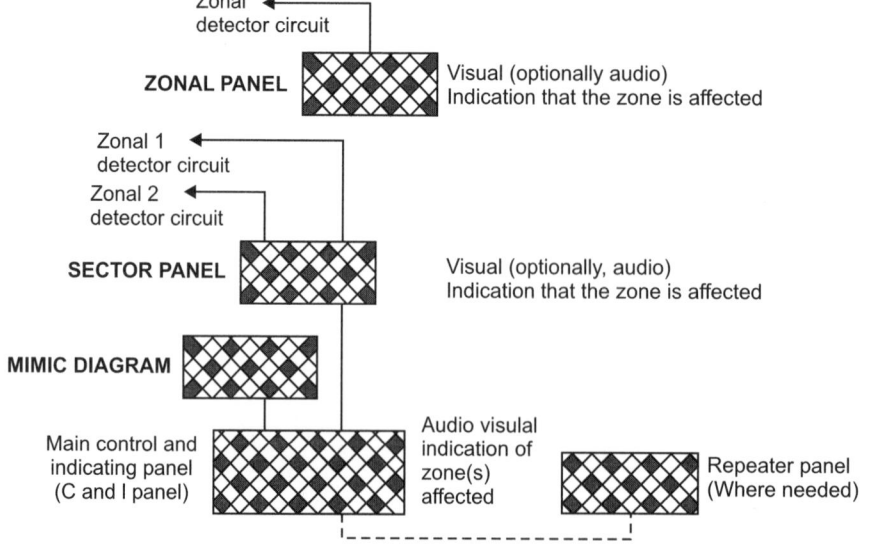

Fig. 16.14 Schematic representation of a FAS distribution

because the threshold of pain occurs at about 130 dBA and can damage hearing power.

It is often possible to achieve fire alarm audibility requirements by using a few very loud appliances or a large number of units at a lower decibel level. In general, system reliability is better with a large number of low decibel units.

16.7.1 Number of Fire Alarm Sounders

16.7.1.1 Low Intensity Sounders

The number of low intensity fire alarm sounders, in a building should be designed considering the following:

a. The sound from each sounder shall be audible at every part of the section of the building covered by it, above the prevailing ambient noise. It is preferable to distribute a number of quieter sounders than a few comparatively louder sounders, so as to prevent the noise in some areas from becoming excessive.

b. At least one sounder shall be provided for each fire compartment.

c. A minimum of two sounders shall be provided even in a small building.

d. Where fire alarm is likely to cause panic as in hospitals or in exhibitions, it may be desirable to provide an alarm signal supplemented with suitable visual signal placed suitably for recognition only by the staff. This aspect shall be examined in consultation with the fire brigade and the client department, as the staff in such premises shall have to be sufficient in number and properly trained to conduct other occupants (patients in particulars) to safety in the event of an alarm.

16.7.1.2 High Intensity Sounders

One high intensity sounder shall be provided for every FAS.

Public address system (PA system).

i. PA system shall be provided as part of FAS for use in making announcements during a fire emergency.

ii. The built-in microphone, amplifier, change over switch and loud speaker switching equipment shall form part of the C and I panel. The fire alarm sounders in the building shall be used as loud-speakers of the PA system.

iii. PA system for FAS shall be independent of any other PA system in the building.

16.8 POWER SUPPLY EQUIPMENTS

16.8.1 Requirements of Power Supply to FAS

i. Power supply to FAS refers to the input power to the main control and indicating panel (C and I panel) of the FAS from the standby generator where provided, or from mains where standby generator set is not provided, and from standby batteries.

ii. Power is required for operating all the control and indicating panels in the system, fire detectors and sounders. Repeater panels do not require independent power supply as they are operated from the C and I panel.

iii. The power input to the C and I panel shall be sufficient to supply the heaviest load under fault and fire conditions, handled by that panel.

iv. The total power demand for FAS even for a large installation is meager compared to the other building power needs, yet the reliability of supply has to be very high.

16.8.2 Standby Battery Supply

i. Standby battery shall be provided with C and I panel to automatically feed the FAS, in the event of variation of input AC voltage beyond preset values on high and low sides.

ii. a. The capacity of battery supply shall be such as to sustain the normal load of the FAS for a period of atleast 48 hours followed by the load of sounders in all zones, and (all) control and indicting panel(s) for a period of atleast 30 minutes thereafter.

b. Provisions in (a) above shall apply to installations where there is no generating plant. Where an automatically started standby generator is provided for the building to feed the FAS among other loads, the battery capacity shall be such as to sustain the normal load of the FAS for a period of atleast 12 hours followed by the load of all sounders for a period of atleast 30 minutes.

c. Capacity requirements as above shall be calculated and indicated in the schedule of work.

d. The installation should be inspected at suitable intervals depending on the frequency of outages in the supply from the mains.

16.9 WIRING

16.9.1 Detector Circuits

i. The detector circuit wiring in a FAS may be either in "series or closed circuit" or in "parallel circuit".

ii. In series circuit design the circuit remains closed normally and there is a current flow in the detector circuit, only under healthy conditions. The circuit gets opened by the fire conditions. In this design the control panel cannot distinguish between an "open circuit fault" and a "fire". A

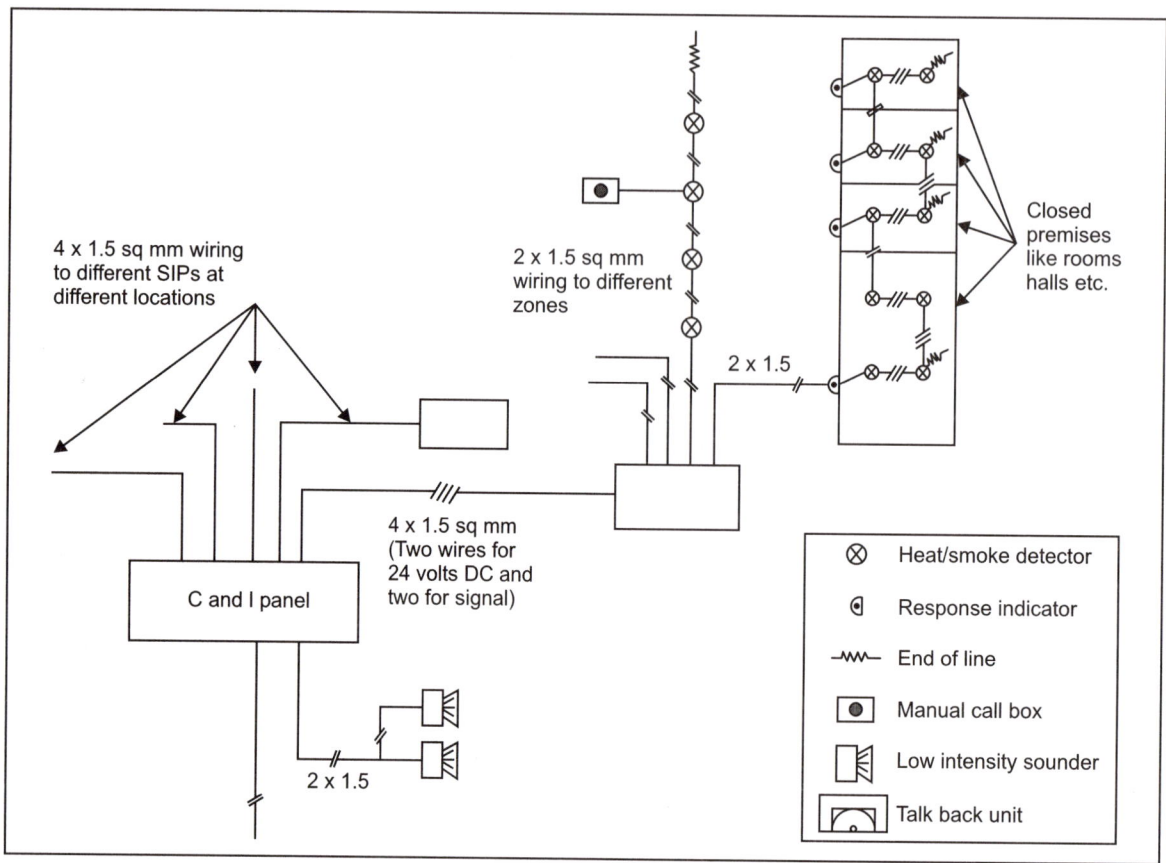

Fig. 16.15 Typical wiring schematic of fire alarm parallel arrangement

short circuited detector will not be shown off in the control panel.

iii. In a parallel circuit design, the detector circuit carries a negligibly small current (in milli-amperes) under fault or fire conditions.

iv. MOEFAS wiring may be circuited in either design.

Typical wiring circuit in parallel that is commonly used is shown in Fig. 16.15.

17

Automatic Sprinkler System

Perhaps no system has been found to be more efficient and reliable in protecting the lives and property from fire than the automatic sprinkler system.

In the beginning, four interdependent approaches led to the development of the early sprinkler system. The first approach was the invention of the automatic sprinkler by Henry S. Parmelee in 1874. The first practical, usable sprinkler was manufactured in 1875.

The second approach was the development of the original configuration and pipe size schedules. These were based on pipe arrangements by Parmelle and providence steam and gap pipe company (PSGP), which later became the Grinnell corporation.

The third approach was the development of alarm valves and dry valves needed to control and detect water flow and sound an alarm.

The fourth approach was the recognition by users like textile companies and later by insurance companies who recognized the benefits of reduction in losses and started offering more insurance coverage or insurance premium reduction where these system were installed.

Regardless of the importance of Parmelle, Buell and other's invention, it was Frederick Grinnell's sprinkler, patented in 1881, that was the first sensitive sprinkler.

The first manual on sprinkler standard came into being by National Fire Protection Association, at New York on November 6, 1896. This standard contained requirements for sprinkler positioning, distance of deflector from bottom of joists or ceilings and full sprinkler coverage throughout a building. It also required a sprinkler orifice to produce 12 GPM (45 LPM) at 5 PSI (0.34 bar) and a minimum static pressure of 25 PSI (1.7 bar) at the highest sprinklers. Two water supplies were deemed essential.

Most early sprinklers were in the form of a link and a lever. In a link and lever design, the sprinkler cap is held closed against the water pressure by the mechanical advantage of the sprinkler levers, which in turn, are held closed by the soldered link.

Hydraulic calculations had not yet been developed in 1896, so design was based on preferred piping arrangements with no more than six sprinklers on a branch line.

In 1926, UL completed a series of tests on the distribution of water by automatic sprinkler system. The purpose was to evaluate the adequacy of the existing pipe schedules and spacing requirements of the NFPA sprinkler standard.

Hydraulic calculation of sprinkler and water supply systems by Clyde Wood was published in 1940s by which wood translated mathematics to practice and published a complete set of tables to identify friction factors used in the flow calculations. A hydraulic design method was first included in NFPA in 1966. This method is now mandatory for most system designs.

Subsequent to the introduction of hydraulic calculations, it was felt that hydraulic design criteria were needed for all categories of hazard. The original area/density design curves were developed by NFPA. The curves were based on analysis of various piping configurations including piping variations such as center/center, side/center and end, all with long and short branch lines. The concept was that the existing pipe schedule system had performed well in most hazards, so they were evaluated to determine the comparable hydraulic design.

Continuing research refined the design concepts for sprinkler system design for high challenged fires and storage occupancies. The sprinkler manufacturers were manufacturing about 2000 varieties of sprinklers by 1996. This large number of designs resulted in an increased level of complexity. Thus in 1999, a new marking system was introduced based on one or two character system followed by up to four numbers to identify each variation in orifice size, shape, defector characteristic and thermal sensitivity. This marking system has become mandatory in January, 2001.

17.1 DEFINITIONS

For the purpose of this part of regulations, the following definitions shall apply.

Assumed maximum area of operation, hydraulically most favourable location. The location in a sprinkler array of an AMAO of specified shape at which the water flow is the maximum for a specific pressure.

Assumed maximum area of operation, hydraulically most unfavourable location. The location in a sprinkler array of an AMAO of specified shape at which the water supply pressure is the minimum needed to give the specified design density.

Cut-off sprinkler. A sprinkler protecting a door or window between two areas only one of which is protected by the sprinkler.

Design density. The minimum density of discharge, in mm/min of water, for which a sprinkler installation is designed, determined from the discharge of a specified group of sprinklers, in litre/min, divided by the area covered, in M^2.

Distribution pipe. A pipe feeding either a range pipe directly or a single sprinkler on a non-terminal range pipe more than 300 mm long.

Distribution pipe spur. A distribution pipe from a main distribution pipe to a terminal branched pipe array.

Drop. A vertical pipe feeding a distribution or range pipe.

End-side array. A pipe array with range pipes on one side only of a distribution pipe.

End-centre array. A pipe with range pipes on both sides of a distribution pipe.

Fire resistance. The ability of a component or the construction of a building to satisfy for a stated period of time the appropriate criteria specified in the relevant part of BS 476.

Fully hydraulically calculated. A term applied to pipe work in an installation in which all the pipe work downstream of the main installation control valve set is sized as per hydraulic calculations.

Sprinkler installation. Part of a sprinkler system comprising a set of installation main control vales, the associated downstream pipes and sprinklers.

Jockey pump. A small pump used to replenish minor water loss to avoid starting an automatic suction or booster pump unnecessarily.

Main distribution pipe. A pipe feeding a distribution pipe.

Node. A point in pipe work at which pressure and flow(s) are calculated; each node is a datum point for the purpose of hydraulic calculations in the installation.

Range pipe. A pipe feeding sprinkler directly or via arm pipes of restricted length.

Riser. A vertical pipe feeding a distribution or range pipe above.

Rosette sprinkler rosette. A plate covering the gap between shank or the body of a sprinkler projecting through a suspended ceiling, and the ceiling.

Sprinkler, ceiling or flush pattern. A pendent sprinkler for fitting partly above but with the temperature sensitive element below, the lower plane of the ceiling.

Sprinkler concealed. A recessed sprinkler with a cover plate that disengages when the heat is applied.

Sprinkler conventional pattern. A sprinkler that gives a spherical pattern of water discharge.

Sprinkler glass bulb. A sprinkler which opens when a liquid filled glass bulb bursts.

Sprinkler horizontal. A sprinkler in which the nozzle directs the water horizontally.

Sprinkler intermediate. A sprinkler installed below, and additional to the roof or ceiling sprinklers.

Sprinkler pendent. A sprinkler in which the nozzle directs water downwards.

Sprinkler, roof or ceiling. A sprinkler protecting the roof or ceiling.

Sprinkler side-wall pattern. A sprinkler that gives a downward parabolic pattern discharge.

Sprinkler upright. A sprinkler in which the nozzle directs the water upwards.

Standard sprinkler layout. A rectilinear layout with the sprinkler aligned perpendicular to the run of the ranges.

Suction pump. An automatic pump supplying water to a sprinkler system from a suction tank.

Supply pipe. A pipe connecting a water supply to a trunk main or the installation main control valve set(s); or a pipe supplying water to a private reservoir, suction tank or gravity tank.

Suspended open cell ceiling. A ceiling of regular open cell construction through which water from sprinkler can be discharged freely.

17.2 TYPE OF SYSTEMS

The various type of sprinklers in use today are shown in Fig. 17.1, while the configurations in which pipes are laid to create a network of sprinklers are shown in Figs. 17.2 to 17.5.

The two end side as used in Fig. 17.2 is to depict that there are two sprinklers on each branch line, while central feed means that water supply enters at the center of the network and gets divided into two halves, while sprinklers are only on one side of the supply pipe.

Similarly in Fig. 17.3, three end side defines that three sprinklers are mounted on each branch line while water enters network from one end and that the branch lines are only on one side of water supply.

Figure 17.5 is a bit complicated to define. In this case, water enters the network at the center and gets divided in two halves. The sprinklers are on both sides of the supply pipe and three in number on each branch pipe (three end center).

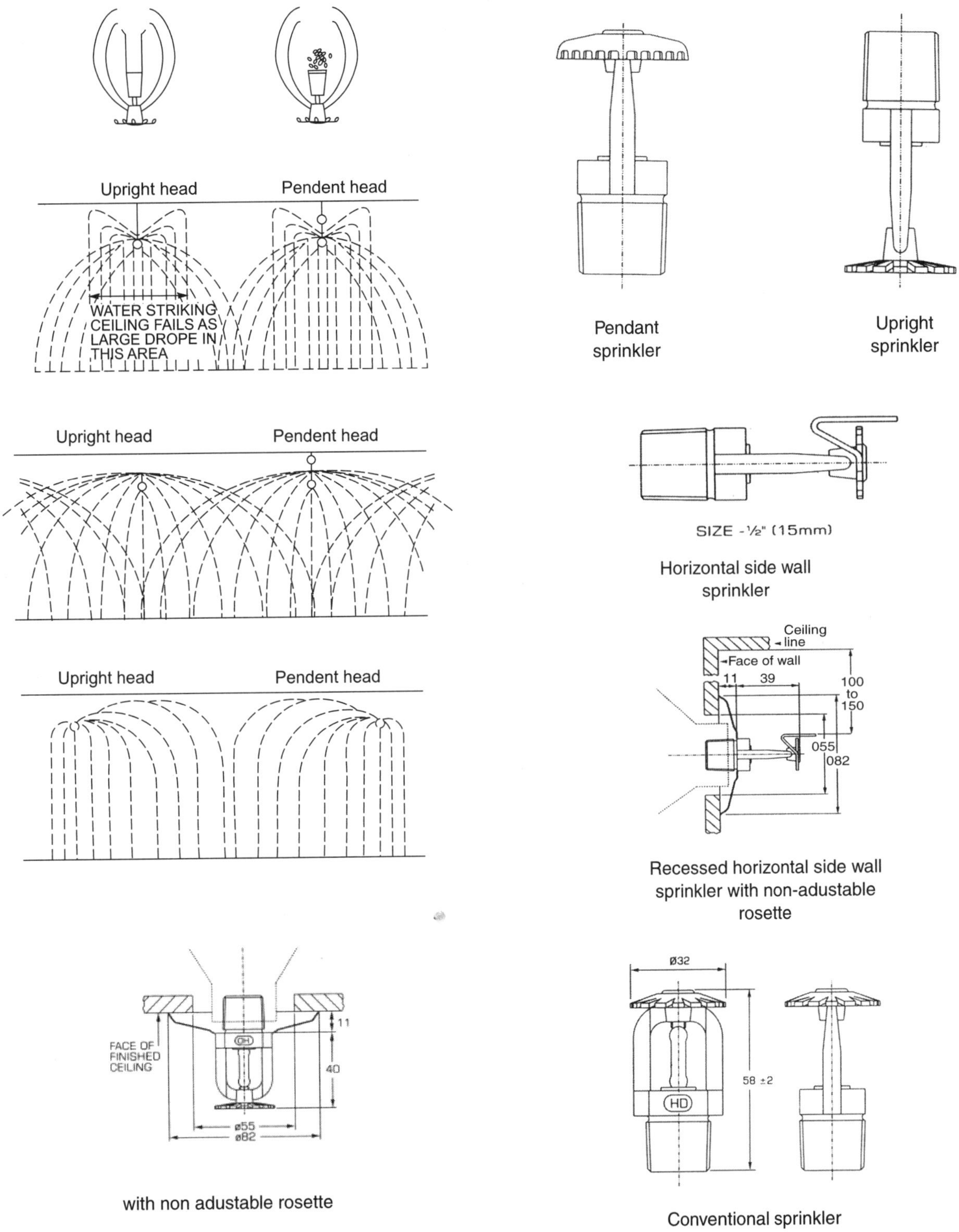

Upright head   Pendent head

WATER STRIKING CEILING FAILS AS LARGE DROPE IN THIS AREA

Pendant sprinkler

Upright sprinkler

Upright head   Pendent head

SIZE -½" (15mm)

Horizontal side wall sprinkler

Upright head   Pendent head

Ceiling line
Face of wall
11 39   100 to 150
055
082

Recessed horizontal side wall sprinkler with non-adustable rosette

FACE OF FINISHED CEILING
11
40
ø55
ø82

with non adustable rosette

Ø32
58 ±2
HD

Conventional sprinkler

Fig. 17.1 Various designs of sprinklers

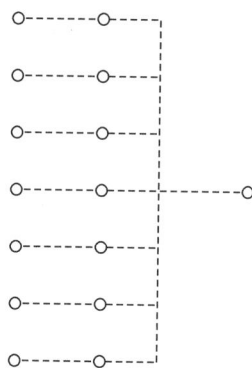

Fig. 17.2 Two end side with central feed

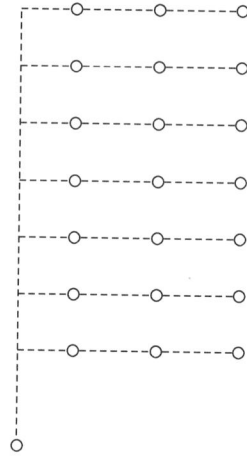

Fig. 17.3 Three end side with end feed

17.3 CLASSIFICATION OF OCCUPANCIES

Any designer who designs fire fighting measures for a building need to know the applications for which a building will be used, the kind of hazard fire may impose, the hydraulic parameters and thus calculate the number of sprinklers and pipe sizes.

17.3.1 Light Hazard Occupancies

Light hazard occupancies are those where the quantity and combustibility of contents is low and fires with relatively low rates of heat release are expected. These include.

   i. Clubs.
  ii. Educational institutions.
 iii. Hospitals.
 iv. Libraries except large stack rooms.
  v. Museums.
 vi. Offices including data processing.
 vii. Residential.
viii. Restaurants.
 ix. Theaters and auditoriums.

Sprinkler systems designed to protect against light hazard occupancies, therefore, have less demanding

water supply requirements. Additionally, more design flexibility is possible.

17.3.2 Ordinary Hazard Occupancies

These occupancies are classified in two groups, viz. Groups I and II.

Group I. These are the occupancies where combustibility is low, quantity of combustibles is moderate, stockpiles of combustibles do not exceed 2.4 m and fires with moderate rates of heat release are expected. These include.

   i. Automobile parking and showrooms.
  ii. Bakeries.
 iii. Beverage manufacturing.
 iv. Canneries.
  v. Dairy products manufacturing and processing.
 vi. Glass and glass products manufacturing.
 vii. Restaurant service area.

Group II. These are the occupancies where the quantity and combustibility of contents is moderate to high, stockpiles do not exceed 3.7 m and fires with moderate to high rates of heat release are expected. These include.

   i. Cereal mills.

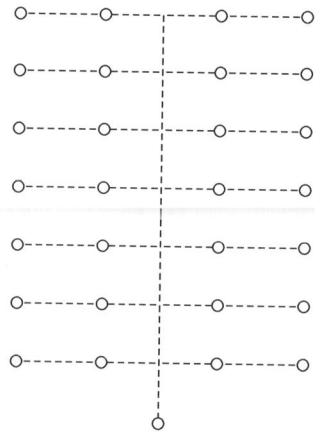

Fig. 17.4 Two end with end feed

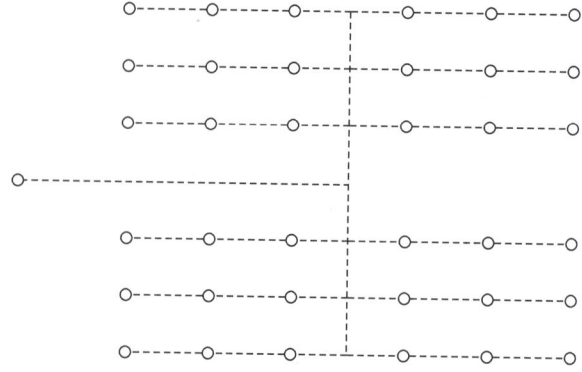

Fig. 17.5 Three end centre with centre feed

ii. Chemical plants ordinary.
iii. Distilleries.
iv. Drycleaners.
v. Leather goods manufacturing.
vi. Libraries with large stack room areas.
vii. Wood machining.
viii. Tyre manufacturing.

17.3.3 Extra Hazard Occupancies

These occupancies are classified into two groups again, i.e. Groups I and II.

Group I. These occupancies include those with hydraulic machinery or systems with flammable or combustible hydraulic fluids under pressure so that ruptures or leaks in piping or fittings would result in fine spray discharge of such liquids, causing intense fires.

These are the occupancies where quantity and combustibility of contents is very high and dust, lint, or other materials are present causing rapid fires with high rates of heat release but with little or no combustible or flammable liquids. These include.

i. Aircraft hangers.
ii. Die casting.
iii. Metal extruding.
iv. Printing.
v. Plywood and particle wood manufacturing.
vi. Saw mills.
vii. Upholstering with plastic foams.

Group II. These include occupancies with moderate to substantial amounts of flammable or combustible liquids, usually in open systems where rapid evaporation can occur when these liquids are subjected to high temperatures. These include.

i. Asphalt saturating.
ii. Flammable liquid spraying.
iii. Open oil quenching.
iv. Plastics processing.
v. Varnish and paint dipping.

The scope of this book is limited to design of sprinkler systems for building particularly commercial building. Thus, this chapter will be limiting itself to design of light and moderate hazard applications.

17.4 SPRINKLER TYPES

Numerous factors such as the available water supply, type of sprinklers, building construction features, and the anticipated fire hazards must be considered in the design of automatic sprinkler systems.

The selection of sprinkler type will vary with occupancy. Where more than one type of sprinkler is used within a compartment, sprinklers with similar response characteristics should be used (i.e. standard or quick response). However, some hazards might benefit from designs that include the use of both standard and quick response sprinklers. Examples include rack storage protected by standard response sprinklers and quick response in–rack sprinklers.

Today, many varieties of sprinklers are available in the market. As said earlier, there may be 2000 varieties of sprinklers available in international market. Thus, certain criteria have to be fixed to enable a designer to decide which sprinkler to be used in a particular application..

17.4.1 Selection on Type of Use

This section attempts to categorize sprinklers into various types and the application where each type can be used. These include.

i. Upright and pendant spray.
ii. Sidewall spray.
iii. Extended coverage.
iv. Open type.
v. Residential type.
vi. Early suppression fast response.
vii. Large drop type.
viii. Special sprinklers.

i. Upright and Pendant Spray Sprinklers

These type of sprinklers are the most common type and can be used in all occupancies and construction types (Fig. 17.6). The application of this type of sprinkler shall be decided in relation to K-factor (explained in articles to follow). General storage, rack storage, rubber storage, baled cotton storage which requires spray density of 13.9 mm/min, standard response sprinkler with K-factor of 115 or larger shall be used.

While for spray densities greater than 13.9 mm/min, the K-factor of 160 or larger shall be used.

For densities of 8.2 mm/min or less, standard response sprinklers with a K-factor of 80 shall be permitted.

Experiments have shown that sprinklers with large orifice, i.e. K-factor of 115 or large perform better during storage fires because due to less velocity, water droplets become large and penetrates more readily in the fire core and allow more water to reach fire seat. The distribution pattern of these kind of sprinklers is shown in Fig. 17.7.

ii. Sidewall Spray Sprinklers

These sprinklers shall be installed only in light hazard occupancies with smooth, flat ceilings. These sprinklers can be used to protect areas below overhead doors. The discharge characteristics of sidewall sprinklers are not as effective as those of upright and pendant sprinklers for all applications. Thus, their use is limited to light hazard occupancies. In addition to this limitation,

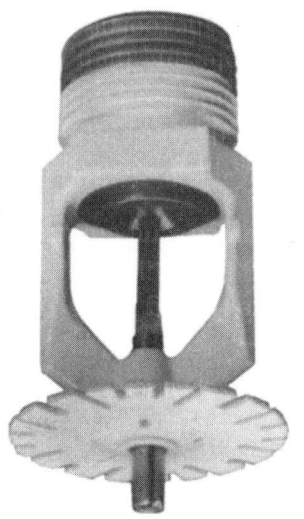

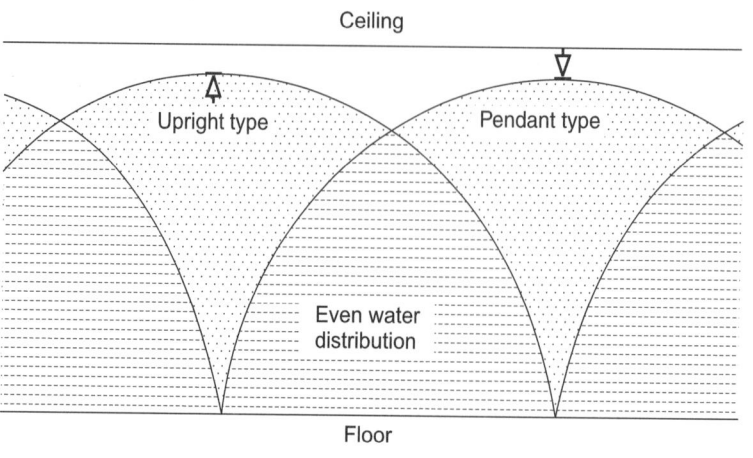

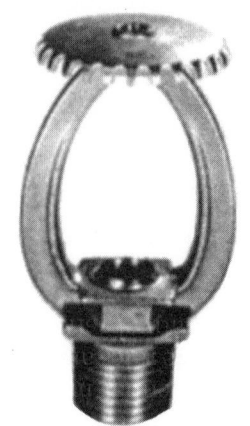

Fig. 17.6 Pendant type extended coverage sprinkler

Fig. 17.7 Distribution pattern of water from standard sprinkler

Fig. 17.8 Open sprinkler

sidewall sprinklers shall not be installed below sloped ceilings.

These sprinklers are generally used in retrofit situations where access to ceilings cannot be permitted.

iii. Extended Coverage Sprinklers

These sprinklers are limited to a construction consisting of flat, smooth ceilings with a slope not exceeding 1:6. These sprinklers have to be specifically mentioned for such constructions. The area coverage with EC sprinklers is larger than the area permitted for other types. The extended coverage allows fewer sprinklers to be used.

However, discharge from an EC sprinkler has a longer, flatter throw than a standard upright or sidewall sprinkler. This pattern is affected by obstructions and slope of ceiling more than other types. That is why, EC sprinklers are limited for use to smooth flat ceiling with relatively small slopes.

iv. Open Sprinklers

These are used in deluge system to protect special hazards or exposures (Fig. 17.8). These are spray sprinklers with their operating element removed. Aircraft hangers require foam water deluge systems which utilize open sprinklers as the predominant means of fire protection.

v. Residential Sprinklers

These sprinklers are tested using a residential fire scenario and hence are to be used in residential portion only (Fig. 17.9). These sprinklers can be used in dwelling units and their adjoining corridors. Residential sprinkles are intended to prevent flashover in the room of fire origin, thus improving the chance for occupants to escape.

The residential sprinklers are so selected that its area of coverage is equal to or greater than both the length and width of the hazard area. For example, if hazard area to be protected is 4.4 × 5.6 m (13′ 6″ × 17′ 6″), sprinkler to be selected shall have a coverage area of 4.5 × 5.8 m (14′ × 18′) or 5.8 × 5.8 m (18′ × 18′).

Fig. 17.9 Residential sprinkler

vi. Early Suppression Fast Response Sprinklers (ESFR)

ESFR sprinklers use fast response operating elements and are designed to respond very quickly to a fire. Occupancies where fire can grow extremely fast with large increase in heat release rates over a short period of time, the quickness with which water is applied is critical. ESFR sprinklers are used in such type of applications (Fig. 17.10).

Sloped roofs tend to cause a skewed distribution of heat from a fire burning beneath the roof. Thus, ESFR sprinklers shall not be used in such places.

Fig. 17.10 Pendant type ESFR sprinklers

vii. Large Drop Sprinklers

These sprinklers are used where fire grow in an exponential manner. A substantial escalation in heat release rates and thermal velocities would be expected in relatively short time periods (Fig. 17.11).

Fig. 17.11 Large drop sprinkler

viii. Special Sprinklers

As the name indicates, these are the sprinklers which are designed for specific applications. But these sprinklers shall nevertheless maintain orifice size, temperature rating, protection area of coverage as defined in various standards.

Nowadays, to provide more aesthetic look, sprinklers are mounted concealed inside the false ceiling. In these kind of sprinklers, all or part of the body, other than the shank thread, is mounted within a recessed housing (Fig. 17.12).

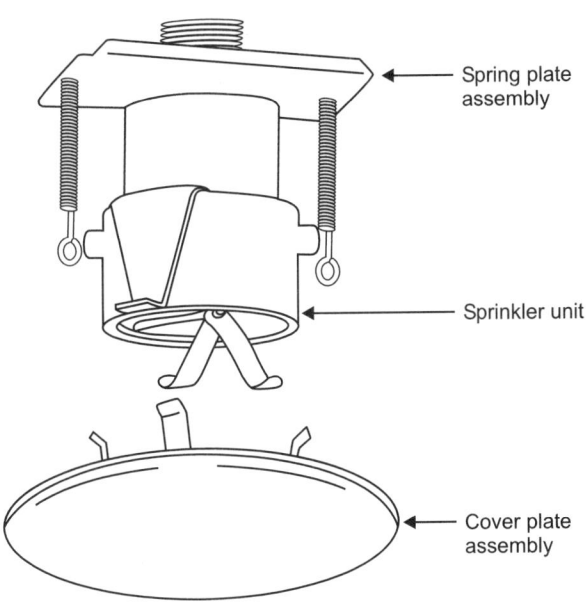

Fig. 17.12 Concealed ceiling sprinkler

17.4.2 Selection on Temperature Rating

The temperature rating of a sprinkler shall not be less than 30 °C than the highest anticipated temperature of the location of the installation.

Note: If the process conditions in a risk calls for continuous air-conditioning round the clock throughout the year, relaxation may be considered.

In high hazard installations, protecting high piled storage with intermediate sprinklers, the roof or ceiling sprinklers shall have a temperature rating of 141 °C.

Under glazed roof or where there are roof sheets of PVC or similar plastic material, the sprinkler rating shall be either 79 to 100 °C, or 141 °C for high storage.

The temperature rating of the roof or ceiling sprinklers within 3 m of the plan area of the boundary of either an oven or a hot process ventilating hood, fitted with sprinklers shall be the same as the oven or hood sprinklers, or 141 °C, whichever is lower. The following colour codes are used to distinguish sprinklers of different nominal temperature ratings (Tables 17.1 and 17.2).

Table 17.1

Fusible link type	
Temperature rating °C	Colour of yoke arms
68/74	Natural
93/100	White
141	Blue
182	Yellow
227	Red

Table 17.2

Glass bulb type	
Temperature rating °C	Colour of filled liquid
57	Orange
68	Red
79	Yellow
93	Green
141	Blue
182	Blue
204/260	Black

17.5 SPACING AND AREA OF COVERAGE OF SPRINKLERS

Sprinklers shall be located, spaced and positioned in accordance with the requirements of this section. The general requirements with which all sprinklers must comply are protection area per sprinkler, sprinkler spacing, deflector position, and obstruction to discharge and clearance to storage.

17.5.1 Upright Sprinklers

17.5.1.1 Area of Coverage

a. **Along branch lines.** Determine distance between sprinklers (or to wall in case of end sprinkler) upstream and down stream. Compare the value of distance between sprinklers with the double of distance from sprinkler to wall. Higher of the two values will be defined as *S*.

b. **Between branch lines.** Determine perpendicular distance to the sprinklers on the adjacent branch lines (or to a wall in case of last branch line) on each side of the branch line on which the subject sprinkler is located. Choose the larger of either twice the distance to the wall or distance to the next sprinkler. This dimension will be defined as *L*.

The protection area of coverage of a sprinkler will be $S \times L$. Thus, $As = S \times L$.

The maximum area of coverage for a sprinkler will vary with the type of sprinkler. However, in any case this value shall not exceed 36 m² for any type of sprinkler.

In a small room, the protection area of coverage for each sprinkler shall be the area of the room divided by the number of sprinklers in the room.

The maximum allowable protection area of coverage for a sprinkler (*As*) shall be in accordance with values indicated in Table 17.3 and Fig. 17.13.

17.5.1.2 Distance between Sprinklers

The maximum distance permitted between sprinklers shall be based on the centerline distance between sprinklers on the branch line or on adjacent branch lines. The maximum distance shall be measured along the slope of the ceiling. The maximum distance permitted shall be as per values given in Table 17.3. To minimize the amount of piping used, branch lines are usually spaced as far apart as possible while maintaining even spacing within a room. In case of sloped roof, the spacing of sprinkler is to be measured on the slope of the roof and the maximum distance and areas given in Table 17.3 shall apply. The distance in such cases should be measured along the slope projected between two sprinklers as shown in Fig. 17.14.

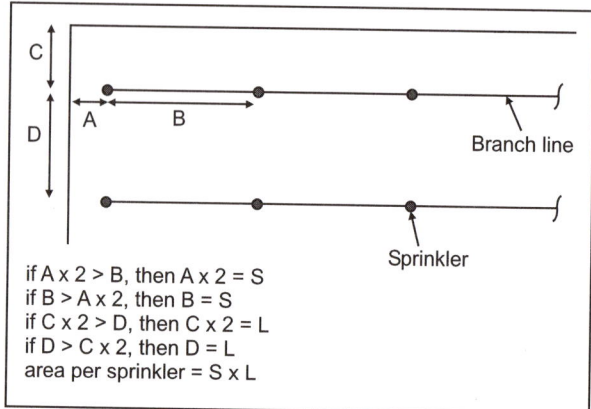

if A x 2 > B, then A x 2 = S
if B > A x 2, then B = S
if C x 2 > D, then C x 2 = L
if D > C x 2, then D = L
area per sprinkler = S x L

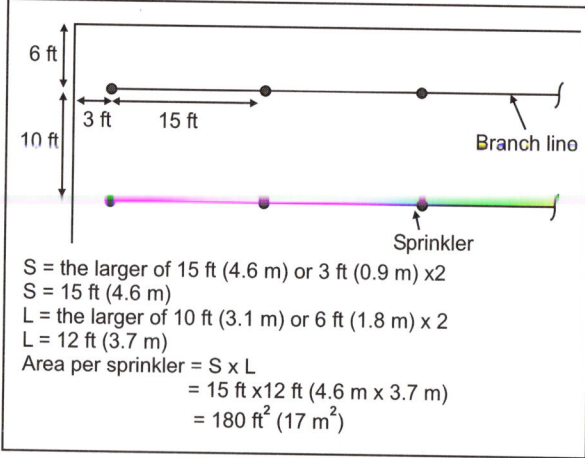

S = the larger of 15 ft (4.6 m) or 3 ft (0.9 m) x2
S = 15 ft (4.6 m)
L = the larger of 10 ft (3.1 m) or 6 ft (1.8 m) x 2
L = 12 ft (3.7 m)
Area per sprinkler = S x L
= 15 ft x 12 ft (4.6 m x 3.7 m)
= 180 ft² (17 m²)

Fig. 17.13 Area of coverage for a sprinkler

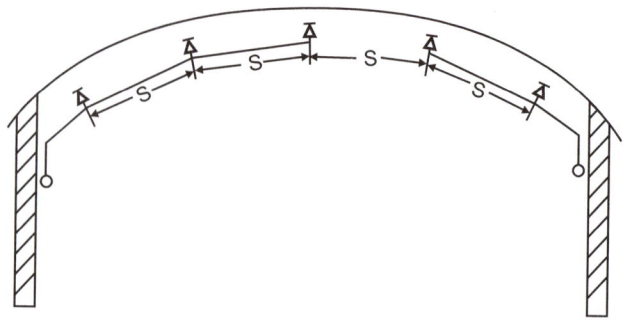

Fig. 17.14 Sprinklers installed along a curved ceiling

17.5.1.3 Maximum Distance from Walls

The distance from sprinklers to walls shall not exceed one half of the allowable maximum distance between sprinklers in Table 17.3.

The distance from the wall to the sprinkler shall be measured perpendicular to the wall.

Table 17.3 Protection area and maximum spacing of sprinklers

Hazard type	Protection area (m^2)	Spacing, m (Maximum)
Light hazard	21	4.6
Ordinary hazard	12	4.6

17.5.1.4 Minimum Distance between Sprinklers

The minimum distance shall be maintained between sprinklers to prevent the operating sprinkler from wetting the adjacent sprinkler and to prevent skipping of sprinkler. The sprinklers shall not be spaced less than 1.8 m on center unless baffles are present.

The sprinkler shall be located at a minimum distance of 10 cm from the wall.

An example is shown in Fig. 17.15 where the method of calculation of sprinkler distance is explained. If the distance of the last branch line from wall is 2.1 m and inter branch line distance is 4.0 m, then twice the distance of last branch line from wall being higher than the inter-branch line distance will be used in determining appropriate distance between sprinklers on branch line. For an ordinary hazard application.

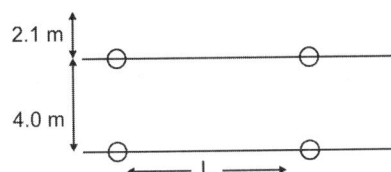

Fig. 17.15 Calculate distance between sprinklers

$$As = 12 \text{ m}^2$$
$$S \times L = 12 \text{ m}^2$$
$$L = \frac{12}{4.2} = 2.85 \text{ m}$$

Angled and other irregular walls and corners with less than 90° angles can cause situations where additional sprinklers are required to protect within a large area. This situation is countered by allowing the maximum distance from a sprinkler to a corner or wall to be slightly increased eliminating the need to install an additional sprinkler.

Figure 17.16a illustrates the arrangement for locating sprinklers when there are columns while Fig. 17.16b is the arrangement of sprinklers when there are continuous beams.

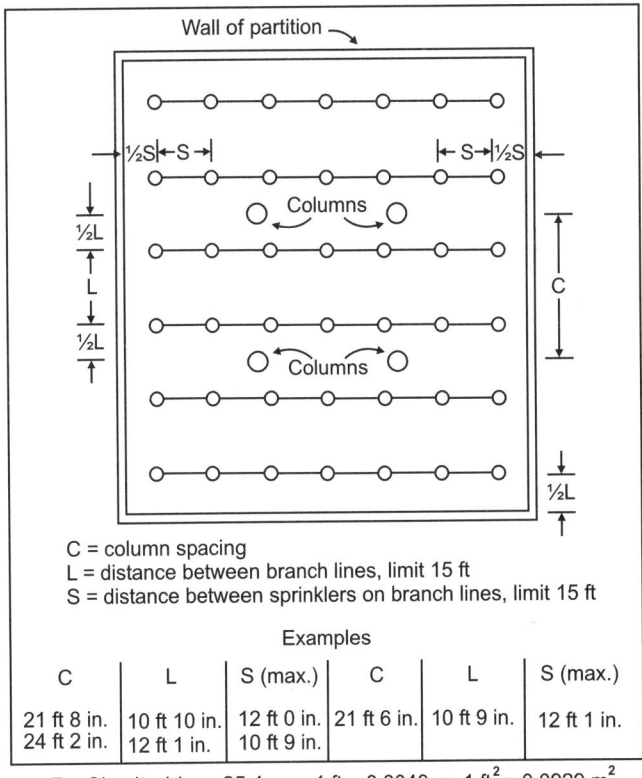

C = column spacing
L = distance between branch lines, limit 15 ft
S = distance between sprinklers on branch lines, limit 15 ft

Examples

C	L	S (max.)	C	L	S (max.)
21 ft 8 in.	10 ft 10 in.	12 ft 0 in.	21 ft 6 in.	10 ft 9 in.	12 ft 1 in.
24 ft 2 in.	12 ft 1 in.	10 ft 9 in.			

For SI units 1 in. = 25.4 mm; 1 ft = 0.3048 m; 1 ft^2 = 0.0929 m^2

Fig. 17.16a Positioning of sprinkler with columns of the building

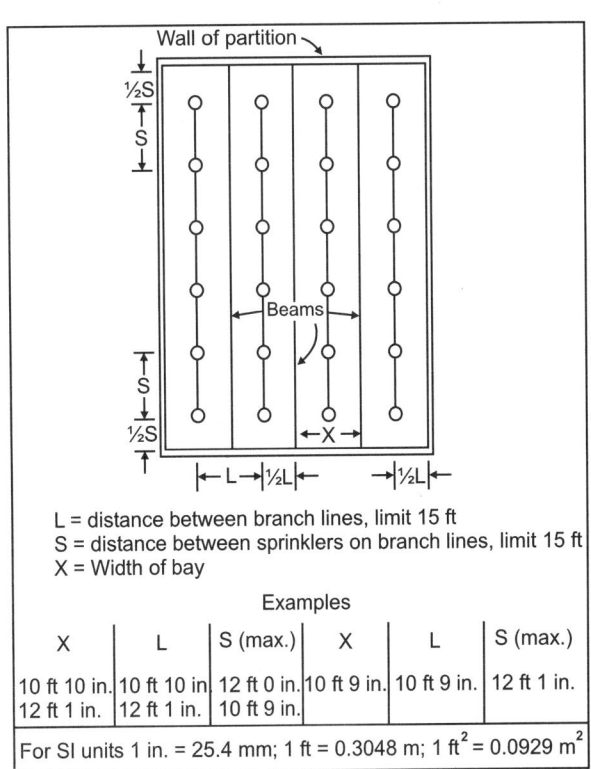

L = distance between branch lines, limit 15 ft
S = distance between sprinklers on branch lines, limit 15 ft
X = Width of bay

Examples

X	L	S (max.)	X	L	S (max.)
10 ft 10 in.	10 ft 10 in.	12 ft 0 in.	10 ft 9 in.	10 ft 9 in.	12 ft 1 in.
12 ft 1 in.	12 ft 1 in.	10 ft 9 in.			

For SI units 1 in. = 25.4 mm; 1 ft = 0.3048 m; 1 ft^2 = 0.0929 m^2

Fig. 17.16b Positioning of sprinkler with continuous beams of the building

17.5.1.5 Deflector Position

In all cases, a minimum of 25 mm clearance is required to be maintained between ceiling and defector to allow for the replacement of upright sprinklers. This distance is measured from ceiling to the sprinkler defector.

Under unobstructed construction, this distance shall be a minimum of 25 mm and a maximum of 305 mm.

Ceiling type sprinklers (concealed, recessed and flush type) shall be permitted to have the operating element above the ceiling and the deflector located nearer to the ceiling. The concern to locate the sprinklers as above is that hot gases from a fire can collect in these spaces and delay the activation of the sprinkler system.

Under obstructed construction, the sprinkler deflector shall be located within the horizontal planes of 25 to 152 mm (1 to 6 inch) below the structural member and a maximum distance of 559 mm (22 inch) below the ceiling/roof deck.

Figure 17.17a, b, the rule of placing sprinkler as per above paragraph has been strictly followed.

However, as an exception to this rule, sprinkler can be positioned at or above the bottom of the structural member, provided the distance from the ceiling or roof deck to the deflector does not exceed 559 mm (22 inch). Figure 17.18 depicts one such arrangement and illustrates that the sprinkler is positioned within 559 mm (22 inch) of the ceiling so as to not negatively impact sprinkler activation and so that the sprinkler discharge will adequately clear the structural members. Table 17.4 highlights the positioning of sprinkler's deflector above bottom of obstruction so as to avoid obstruction to discharge.

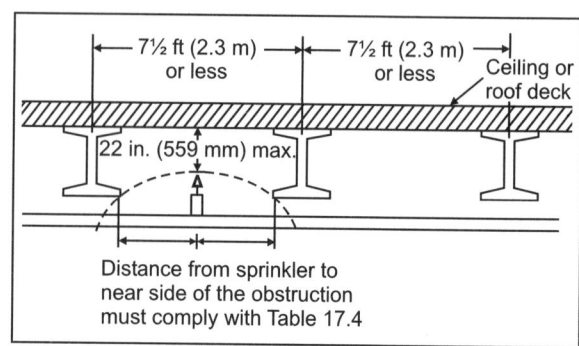

Fig. 17.18 Placement of sprinkler underneath obstructed construction

Table 17.4 Positioning of sprinklers to avoid obstruction to discharge

S. no.	Distance from sprinkler to side of construction	Maximum allowable distance of deflector above bottom of obstruction (mm)
1.	Less than 300 mm	0
2.	300 mm to less than 450 mm	00.5
3.	450 mm to less than 600 mm	89.0
4.	600 mm to less than 750 mm	140.0
5.	750 mm to less than 900 mm	190.5
6.	900 mm to less than 1050 mm	241.3
7.	1050 mm to less than 1200 mm	304.8
8.	1200 mm to less than 1350 mm	355.6
9.	1350 mm to less than 1500 mm	419.1
10.	1500 mm and greater	457.2

But where the depths of structural members are such that compliance with the obstruction rules of Table 17.4 is not possible, sprinkler would need to be installed in each bay. In these cases, sprinklers are located within 25 to 305 mm (1 to 12 inch) below the ceiling, as they are for unobstructed construction, but they can be located up to 559 mm (22 inch) below the ceiling (Fig. 17.19).

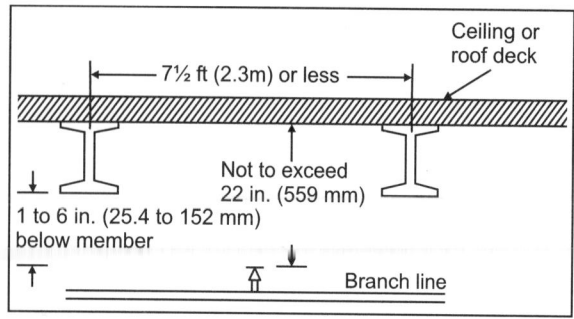

Fig. 17.17a Placement of sprinkler under obstructed construction

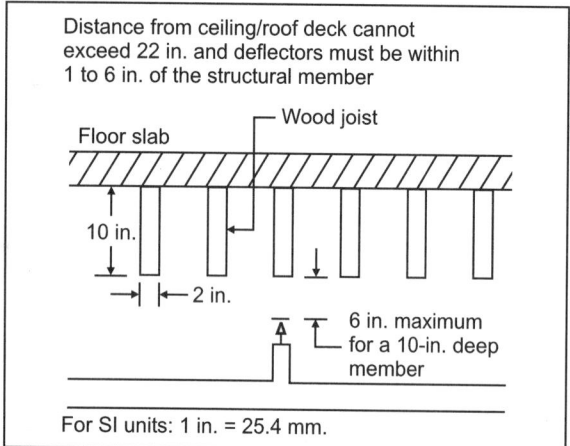

Fig. 17.17b Placement of sprinkler underneath solid wood joist construction

17.5.2 Sidewall Sprinklers

17.5.2.1 Area of Coverage

The protection area of coverage per sprinkler shall be determined as follows.

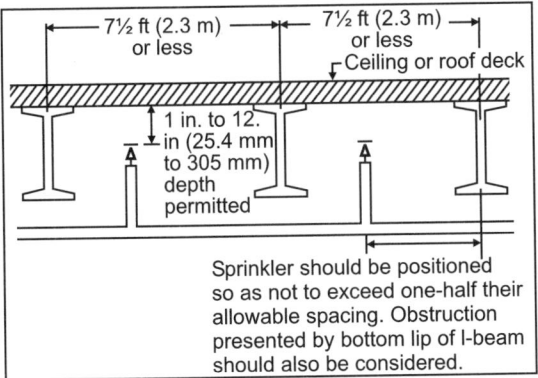

Fig. 17.19 Placement of sprinkler in each bay of obstructed construction

a. **Along the wall.** Determine the distance between sprinklers along the wall upstream and down stream (or to the end wall or obstruction in case of the end sprinkler on branch line). Choose the larger of either twice the distance to the end wall or the distance to the next sprinkler. This dimension will be defined as S.

b. **Across the room.** Determine the distance from the sprinkler to the wall opposite the sprinklers or to the mid-point of the room where sprinklers are installed on two opposite walls. This dimension will be defined as L.

The protection area coverage of sprinkler will be $As = S \times L$.

The maximum allowable protection area of coverage for a sprinkler (sidewall type) shall be as shown in Table 17.5.

17.5.2.2 Maximum Distance between Sprinklers

The maximum distance shall be based on the center line distance between sprinklers on the branch line. The maximum distance shall be measured along the slope of the ceiling where sloped ceilings are provided.

Table 17.5 Protection areas and maximum spacing of sidewall sprinklers

	Light hazard	Ordinary hazard
1. Maximum distance along the wall	4.6 m	3.4 m
2. Maximum room width	3.6 m	3.0 m
3. Maximum protection area	12.0 m²	9.0 m²

Sidewall sprinklers shall be installed along the length of a single wall of rooms or bays in accordance with Table 17.5.

17.5.2.3 Maximum Distance from Walls

The distance from sprinkler to the end walls shall not exceed one half of the allowable distance permitted between sprinklers as per Table 17.5 while sprinklers shall be located a minimum of 102 mm from an end wall. The distance from wall to sprinkler shall be measured perpendicular to wall.

17.5.2.4 Minimum Distance between Sprinklers

Inter-sprinkler distance shall not be less than 1.8 m on center.

17.5.2.5 Deflector Position

The sidewall sprinklers shall be located not more than 152 mm (6 inch) or less than 102 mm (4 inch) from ceiling. Moving beyond this range, the sprinkler will take longer time to operate.

The sidewall sprinklers shall be located not more than 152 mm (6 inch) or less than 102 mm (4 inch) from walls to which they are mounted.

Sprinkler shall be located so as to minimize obstruction to discharge or additional sprinklers shall be provided. Sidewall sprinklers shall be installed no closer than 1.2 m (4 feet) from light fixture or similar obstruction. But where distance between light fixtures or similar obstructions are more than 1.2 m (4 feet), the

Table 17.6 Positioning of sprinklers to avoid obstructions (sidewall sprinklers)

S. no.	Distance from sidewall sprinkler to side of obstruction (A)	Maximum allowable distance of deflector above bottom of obstruction (B)
1.	Less than 1200 mm	0
2.	1200 mm to less than 1500 mm	25.4 mm
3.	1500 mm to less than 1650 mm	50.8 mm
4.	1650 mm to less than 1800 mm	76.2 mm
5.	1800 mm to less than 1950 mm	102 mm
6.	1950 mm to less than 2100 mm	152.4 mm
7.	2100 mm to less than 2250 mm	177.8 mm
8.	2250 mm to less than 2400 mm	228.6 mm
9.	2400 mm to less than 2550 mm	279.4 mm
10.	2550 mm and greater	355.6 mm

location of sprinkler shall be decided as per Table 17.6 and Fig. 17.20.

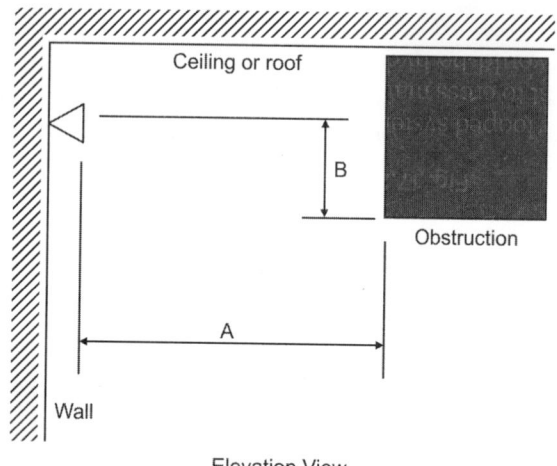

Fig. 17.20 Positioning of sprinklers to avoid obstructions (sidewall sprinklers)

17.6 WATER REQUIREMENT

This section illustrates the approach to calculate amount of water required to suppress a fire. The water quantity requirement can be calculated considering pipe schedule method or hydraulic calculation method.

Pipe schedule method is an older but still an acceptable method, although its application is not common. Hydraulic method results in a more efficient and economic design.

17.6.1 Pipe Schedule Method

Table 17.7 lists the minimum pressure flow requirements in determining the minimum water supply requirements for light and ordinary hazard categories when pipe is sized according to pipe schedule method. The pipe schedule method is normally used only for new installations of 465 m² (5000 ft²) or less or for additions or modifications to existing pipe schedule system.

But pipe schedule method can also be used for systems exceeding 465 m² (5000 ft²) where the flow requirement as in Table 17.7 are available at a minimum residual pressure of 3.50 kg/cm² at the highest elevation of sprinkler.

The residual pressure as defined in Table 17.7 shall be available at the highest sprinkler in the system.

The additional pressure to account for static pressure related to elevation of sprinkler above pump level and also friction losses in valves, bends, fittings and pipe friction losses shall be added to the residual pressure required at the highest sprinkler.

The static pressure at the rate of 0.0979 bar/m (0.1 kg/cm²/m) of elevation above water supply (pump) shall be added.

The lower value of flow requirements in Table 17.7 shall be considered where the potential areas of fire are limited by building size.

17.6.2 Hydraulic Calculation Method

There are two approaches to determine the water flow requirements by hydraulic calculation method. These are:

1. The area/density method.
2. The room design method.

The water supply requirement for sprinklers can be determined from the area/density curves shown in Fig. 17.21. These curves can be used both for area/density method or room design method.

Regardless of which method is used, if the area of operation is less than 139 m² (1500 ft²) for light and ordinary hazard areas, the density for 139 m² (1500 ft²) shall be used.

In the room design method, the water requirements are based upon the room that creates the greatest demand. The density can be selected from Fig. 17.21 corresponding to room size.

When opting for this method, designer must use the room that is also the most hydraulically demanding in terms of water supply and pressure.

17.7 HYDRAULIC CALCULATION

Hydraulic calculation method is the most reliable method of calculation of water flow, number of sprinklers and pipe sizes. The pipe schedule method shall not be used for designing a new installation except where the extension of existing system is required.

Hydraulic calculations shall be prepared on form sheets that include a summary sheet, detailed work sheets and graph sheets.

These sheets help monitor the water flow through the system and allow for a quick evaluation of the system. Water flow through the system is tracked in a direction opposite to that of actual flow. The first

Table 17.7 Water supply requirements for pipe schedule sprinkler systems

Hazard category	Minimum residual pressure required Kgf/cm²	Acceptable flow at the base of riser (LPM)	Duration of flow (Minutes)
Light	1.05	1892–2840	60
Ordinary	1.40	3217–5677	90

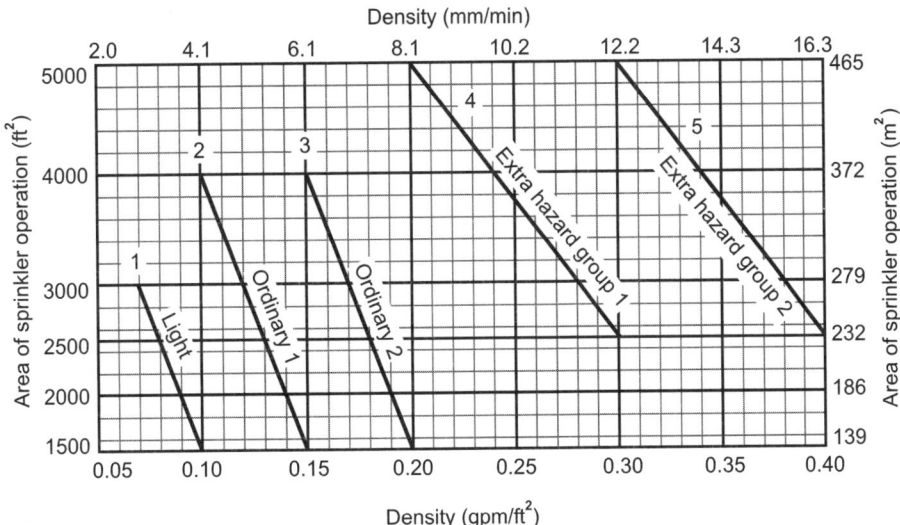

Fig. 17.21 Area/density curves

evaluation begins at the hydraulically most demanding sprinkler, which many time is the furthest sprinkler from the system riser and ends at the point of connection to the system water supply. The furthest point is chosen because if the sprinkler system is suitable to supply specified density and pressure at this point, then it will supply the required density at all other points. The fire is not expected to take place simultaneously at all places in a building. Designing a system to protect the total area of the building will thus be a futile exercise and will not only result in a heavy water storage requirements, but also heavy pumping needs.

These forms permit the designer or other persons reviewing the plans to account for individual pipe lengths and to keep track of changes in pressure and flow that results from friction loss and elevation changes.

The hydraulic calculation sheet is shown in Table 17.8 for a typical floor area to be protected shown in Fig. 17.22.

Let's start to fill the columns of hydraulic calculation sheet.

The floor area of the building is 39×60 m. The design density of 6.1 mm/min is selected from Fig. 17.21 for an area of operation of 139 m² (Refer Para 17.6.2) in ordinary hazard curves. Though a design density of 4.5 mm/min operating over an area of 279 m² is also acceptable for ordinary hazard occupancy. (One can assume 372 m² maximum area of operation also depending on the type of building and the protective coverage one intends to provide).

Number of sprinklers

$$= \frac{\text{Area of operation}}{\text{Area of coverage per sprinkler}}$$

$$= \frac{139}{12} = 11.54 \text{ (say } 12)$$

As the fractional number is not possible, it is rounded to next higher value, i.e. 12 sprinklers.

Number of sprinklers on a branch line

$$= \frac{1.2 \sqrt{A}}{L}$$

where, L = Spacing between sprinklers on a branch line.

A = Area of operation.

The multiplication by 1.2 ensures that the area of operation takes the shape of a rectangle with the longer dimension parallel to the direction of the branch line.

Note: Under certain conditions, where sprinklers were spaced 3.96 m apart on a branch line and the branch lines were spaced 3.05 m apart and calculation resulted in 12 sprinklers, some designers would calculate 4 branch lines ($4 \times 3.05 = 12.20$ m) with 3 sprinklers on each branch ($3 \times 3.96 = 11.88$ m).

This approach would result in a design area that was almost a perfect square. The square shape would be inadequate if four sprinklers operated on the branch line. Therefore, the requirement that the area of operation take the shape of a rectangle with the length of longer side having a dimension of $1.2 \sqrt{A}$ shall be used.

$L = 3.9$ m (by dividing total width of room equally in 10 sprinklers so as to limit the inter sprinkler distance within limit imposed by Table 17.3).

Number of sprinklers on a branch line

$$= \frac{1.2 \times \sqrt{139}}{3.9} = 3.58 \text{ sprinklers} = 4 \text{ sprinklers.}$$

Table 17.8 Hydraulic calculation sheet

Step no.	Nozzle identification and location	Flow in LPM		Pipe size mm	Pipe fittings and devices	Equiv pipe length m		Friction loss bar per meter C = 120	Pressure summary bar		Notes
1.	1 BL-1	q		25		L	3.9	0.028	Pt	0.837	
		Q	73.2			f			Pe		
						t	3.9		Pf	0.11	
2.	2	q	77.85	32		L	3.9	0.029	Pt	0.947	
		Q	151.05			f			Pe		
						t	3.9		Pf	0.11	
3.	3	q	82.9	40		L	3.9	0.030	Pt	1.057	
		Q	235.04			f			Pe		
						t	3.9		Pf	0.12	
4.	4	q	87.43			L	6.30	0.054	Pt	1.17	
		Q	322.50			f	4.80		Pe		
						t	10.95		Pf	0.59	
5.	Common main to BL 2	q				L	3.0	0.016	Pt	1.79	
		Q	322.5			f			Pe		
						t	3.0		Pf	0.048	
6.	BL 2 common main to BL-3	q	326.6			L	3.0	0.0245	Pt	1.84	
		Q	649.1	62		f			Pe		
						t	3.0		Pf	0.075	
7.	BL 3 to common main	q	333.45	62		L	21.00	0.053	Pt	1.915	
		Q	982.58			f			Pe		
						t	21.00		Pf	1.116	
8.	Common main to pump station	q		75		L	35.7	0.0186	Pt	3.03	
		Q	982.6			f	6.3		Pe	0.45	
						t	42.0		Pf	0.77	

Though the number of sprinklers on the branch line are 5, but only 4 sprinklers of this branch will form part of assumed maximum area of operation.

The values shall be rounded to next number even if the fractional part is less than 0.5.

The operating area thus becomes 4 sprinklers on each of 3 branch lines, which provides the required total of 12 sprinklers.

Inter branch line distance

$$= \frac{\text{Area of coverage of one sprinkler}}{\text{Inter sprinkler distance}}$$

$$= \frac{12}{3.9} = 3.0 \text{ m}$$

Coming back to hydraulic sheet, sprinkler 1 and branch line 1 (BL 1) are filled in the sheet. The flow for the first sprinkler is given by the formula.

$$Q = D \times A = 6.1 \text{ mm/min} \times 12 \text{ m}^2 = 73.2 \text{ LPM.}$$

where, D = Design density already selected,

A = Area of coverage per sprinkler.

The final pipe size is determined by trial and error and later modified through hydraulic analysis to determine the most ideal pipe size. In this case, 25 mm pipe has been selected.

The fitting directly connected to a sprinkler is not usually included in the calculation because it is accounted for in the sprinkler's K-factor. As a result, no values are shown under pipe fittings and devices column.

The equivalent pipe length is the total center to center distance between sprinklers, which in this case is the actual pipe length of 3.9 m. The friction loss is determined by using the Hazen-Williams formula.

$$P = \frac{6.05 \, Q^{1.85} \times 10^5}{C^{1.85} \times d^{4.87}}.$$

P = Frictional resistance in bar per m of pipe.

Q = Flow in litre per minute.

C = Friction loss coefficient.

d = Actual internal dia. of pipe in mm.

This is the most common of empirical formulas used to determine relationship between flow, friction loss and available pressure.

But this is dependent on the relationship between pipe type (C-factor).

The values of C-factor for various types of pipe is given in Table 17.9.

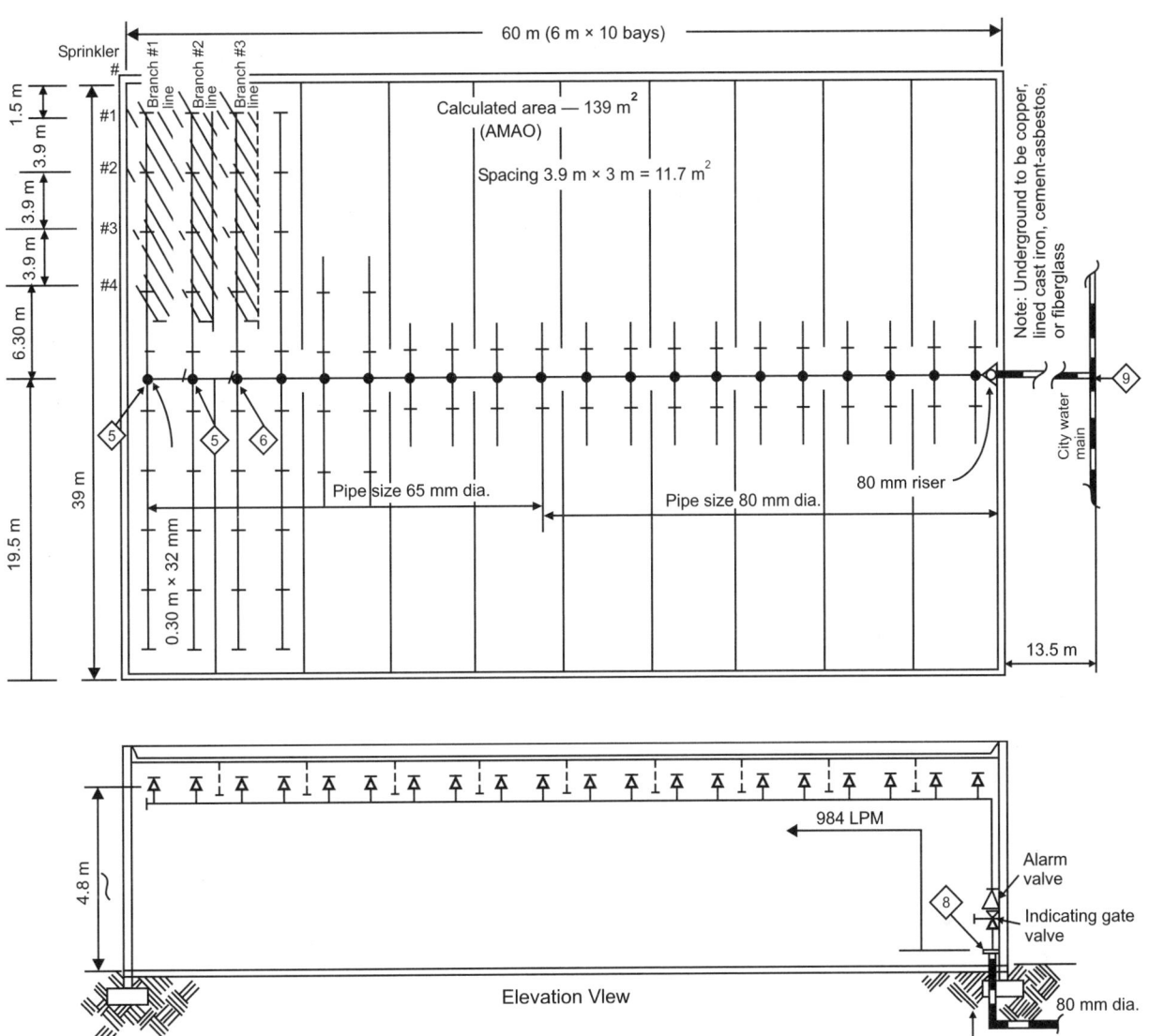

Fig. 17.22 A typical hydraulic calculation example

Table 17.9 Hazen-Williams *C*-values

S. no.	Pipe type	C-values
1.	Unlined cast or ductile iron	100
2.	Black steel	120
3.	Galvanized	120
4.	Plastic	150
5.	Copper tube or stainless steel	150
6.	Concrete	140

Thus, $p = \dfrac{6.05 \times 73.2^{1.85} \times 10^5}{120^{1.85} \times 25^{4.87}} = 0.0282 \text{ bar/m.}$

By multiplying with pipe length between sprinklers, i.e. 3.9 m,

$$p = 0.0282 \times 3.9 = 0.11 \text{ bar.}$$

The total pressure at sprinkler 1 is determined by the following formula.

$$P_t = \left(\frac{Q^2}{K^2} \right)$$

P_t = Pressure in bars.

Q = Flow in LPM.

K = *K*-factor of sprinkler.

Where *K*-factor of sprinkler depends upon the nominal orifice size of sprinkler. The sprinkler shall have a nominal orifice size of 10 mm, 15 mm and 20 mm. Table 17.10 lists the *K*-factors for various nominal orifice sizes.

Table 17.10 *K*-factor Values for sprinklers

Nominal orifice size (mm)	Mean value of K-factor	Limiting values	
		Minimum	Maximum
10	57	54	60
15	80	76	84
20	115	109	121

Thus, $P_t = \dfrac{(73.2)^2}{(80)^2} = 0.837$ bar.

Similarly, pressure at sprinkler 2 is determined by adding the pressure at sprinkler 1 plus the pressure drop caused by friction loss.

$$P_2 = 0.837 + 0.11 = 0.947 \text{ bar.}$$

The flow Q from sprinkler 2 is

$$Q_2 = K\sqrt{P} = 80\sqrt{0.947} = 77.85 \text{ LPM.}$$

This flow is added to sprinkler 1 and the same procedure for determining the friction loss and flow

In step 8, the static pressure due to elevation of sprinkler is added.

$P_e = 4.5$ m × 0.0979 (Building height 4.5 m) = 0.44 bar.

Thus, the results of step 8 will help the designer to select a pump of 982 LPM (rounded to 1000 LPM) capacity discharging at a pressure of 3.03 bar.

In addition, the friction loss up to distribution pipe is to be added. The fittings and devices equivalent lengths are totalled and friction loss is calculated.

This completes the hydraulic calculation sheet.

When additional sprinkler piping is added to an existing system, the existing piping does not have to be increased in size to compensate for the additional sprinklers, provided the new work is calculated and the calculations include that portion of the existing system that can be required to carry water to the new work.

Equivalent pipe lengths of valves and fittings is given in Tables 17.11 and 17.12.

Table 17.11   Equivalent length of fittings and valves

Fittings and valves	Equivalent length of medium grade steel pipe in m according to IS 1239 (Part I) (C = 120) for dia. in mm equal to						
	50	65	80	100	150	200	250
Screwed elbow 90°	1.46	1.89	2.37	3.04	4.30	5.67	7.42
Welded elbow 90°	0.69	0.88	1.10	1.43	2.00	2.64	3.35
Screwed elbow 45°	0.76	1.02	1.27	1.61	2.30	3.05	3.89
All other fittings	2.91	3.81	4.75	6.10	8.61	11.34	13.85
Gate valve	0.38	0.51	0.63	0.81	1.13	1.50	1.97
Alarm valve NR valve	–	–	3.94	5.07	7.17	9.40	12.30
Butterfly valve	2.19	2.86	3.55	4.56	6.38	8.62	9.90
Globe valve	6.43	21.64	126.80	34.48	48.79	64.29	84.11

for the pipe between sprinkler 1 and 2 are followed for determining the information between sprinkler 2 and 3, 3 and 4.

To determine pressure loss between sprinkler 4 and cross main.

Total length of pipe = Distance between sprinkler 4 and 5 + 2.0 m between sprinkler 5 and junction + 0.3 m for riser nipple = 4 + 2 + 0.3 = 6.3 m.

K-factor for branch line can now be determined

$$K = \frac{Q}{\sqrt{P}} = \frac{322.48}{\sqrt{1.8}} = 240$$

This *K*-factor is used in predicting the flow in subsequent branch lines which are identical to BL 1, i.e the *K*-factor for BL 1 describes the physical characteristics of the pipe opening in the cross main for the branch line and identifies the constant relationship between the flow and the pressure at this point.

Table 17.12   Equivalent length of fittings and valves

Fittings	Equivalent length of medium grade steel pipe in m according to IS 1239 (Part I) (C = 120) for dia. in mm equal to		
	25	32	40
Screwed elbow 90°	0.77	1.04	1.22
Welded elbow 90°	0.36	0.49	0.56
Screwed elbow 45°	0.40	0.50	0.66
All other fittings	1.54	2.13	2.44

17.8 HYDRAULICALLY MOST DEMANDING AREA

In the section 17.6, the concepts to determine amount of water each sprinkler will need to discharge, the pressure necessary to maintain this flow, the number of sprinklers needed, etc. were explained. In section 17.7, hydraulic calculations were done to determine flow, pressure demand at the most hydraulically demanding single point in the system, which is usually the most distant sprinkler from the water supply.

Now, we must determine what will happen when the water flow splits in two directions so that pressure and flow calculated at a single point may be sufficient to feed two branches. Under such cases, we must adjust the flow of the line demanding less pressure equivalent to high pressure. This will result in greater water flow in this line because greater the amount of pressure, greater the amount of water flow when all other variables remain unchanged. Consider Fig. 17.23, wherein at the point A, water flow splits in two branches, one feeding two sprinkler and another

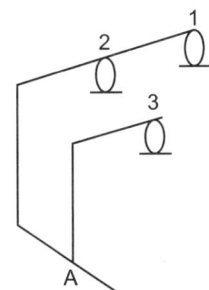

Fig. 17.23 Branch line water flow adjustment

feeding one sprinkler. Let's say, hydraulic calculations for first branch at point A indicate a demand of 140 LPM at 1.033 bar. Similarly, for branch line with sprinkler 3 operating indicates a demand of 68 LPM at 0.8 bar. At point A, only one pressure can exist, we need to provide 1.033 bar pressure at A so that proper discharge can be maintained from the first line. But this will create an overflow in the second line. Let's calculate by determining the K-factor at point A.

$$Q = K\sqrt{P}, \ 68 = K\sqrt{0.8}, \ K = 76.$$

This value of K-factor is now used to determine the flow in second branch line while pressure now becomes 1.033 bar.

$$Q = K\sqrt{P} = 76\sqrt{1.033} = 77.24 \text{ LPM}.$$

The total flow demand at A.

$$Q_A = 140 + 77.24 = 217.24 \text{ LPM}.$$

Figure 17.24 illustrates how to calculate the most demanding area or *AMAO*.

Let's assume a design area of 135 m² with sprinkler coverage of 10.8 m².

Number of sprinklers

$$= \frac{135}{10.8} = 12.5. \text{ Rounded to} = 13 \text{ sprinklers.}$$

Number of sprinklers on branch line

$$= \frac{1.2\sqrt{A}}{S}$$

where, A = Design area,

S = Distance between sprinklers on branch line.

Number of sprinklers on branch line

$$= \frac{1.2\sqrt{135}}{3.6} = 3.87. \text{ Rounded to}$$

$$= 4 \text{ sprinklers.}$$

As shown in the Fig. 17.24, the extra sprinkler after adjusting 4 sprinklers in three branch lines is adjusted in the fourth line from B to E at designer's option.

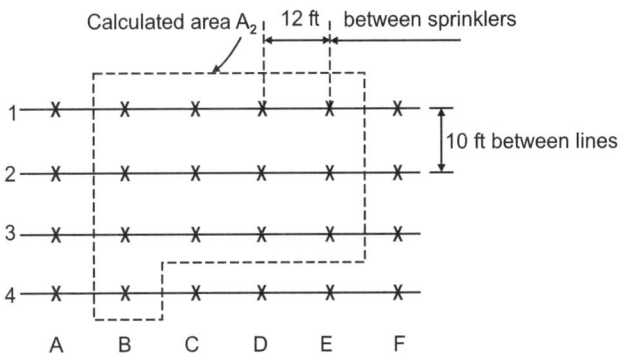

Fig. 17.24 Determination of *AMAO*

For looped system, this extra sprinkler shall be placed closest to cross mains because it is closest to the supply and would be hydraulically most demanding on the branch line.

Where the design is based on area/density method, the design area shall be rectangular having a dimension parallel to the branch lines at least 1.2 times the square root of the area of sprinkler operation used, which shall permit the inclusion of sprinklers on both sides of the cross main. Any fractional sprinkler shall be carried to the next higher whole sprinkler.

Figure 17.25 shows some exceptions, where design area may not be rectangular (A in this figure). System B and C shows location of extra sprinkler on the fourth branch line.

Figure 17.26 shows the design area for looped systems with various riser location and branch line configuration.

In systems having branch lines with an insufficient number of sprinklers to fulfill the $1.2\sqrt{A}$ requirement, the design area shall be extended to include sprinklers on adjacent branch lines supplied by same cross mains.

17.9 PUMP CAPACITY AND WATER STORAGE

Sufficient quantity of water at an adequate pressure shall be available for any sprinkler system. The effective capacity of the reservoir (above the level of foot valve seat or the level of top of pump) shall be as indicated in Table 17.13a and b.

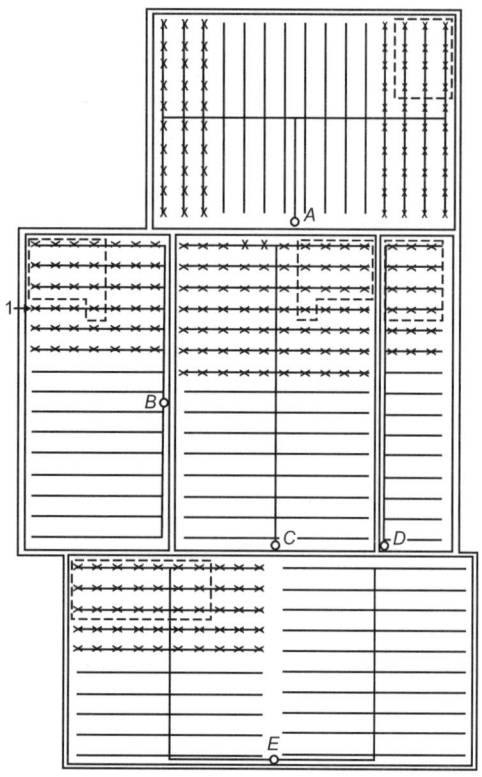

Fig. 17.25 Example of *AMAO* (non-rectangular)

The equilibrium water velocity shall not exceed 6 m/sec at any valve or flow monitoring device or 10 m/sec at any other point in the system for the stabilized flow condition at the demand point involving an *AMAO*.

17.10 PIPING

The water piping used in the sprinkler system shall be laid normally underground or in masonry culverts with removable covers of incombustible construction and shall be of any one of the following types.

a. Cast iron double flanged conforming to IS-718, IS-1537, IS-1536.

b. Centrifugally cast spun iron conforming to IS-1536.

c. Wrought or mild steel pipe (with or without galvanising of heavy grade conforming to IS-1239

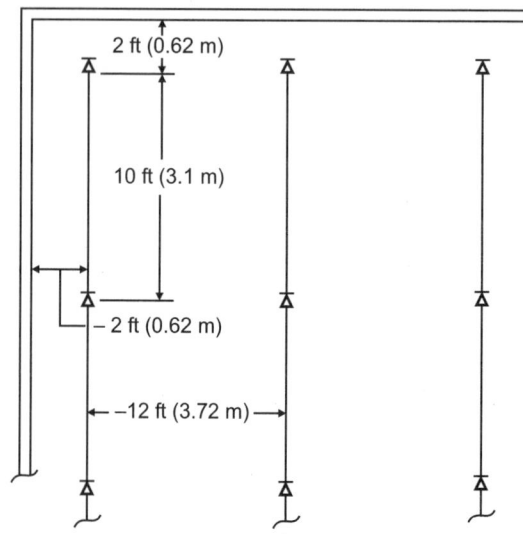

Fig. 17.26 Design area for looped system

Table 17.13a Pump capacity and storage requirement

Class of hazard	Exclusive water storage	Water pressure at installation valve	Pump capacity LPS (m^3/hr)	Delivery Kg/m^2
Light hazard	Not less than 20 minutes run for the pumping capacity or 35 m^3 whichever is greater.	2.2 bar plus static pressure* at 225 LPM	27(96) 30 (110)	5.6 5.6
Ordinary hazard	Not less than 1 hour run for the aggregate pumping capacity or 200 m^3 whichever is greater.	2.0 bar plus static pressure* at 1800 LPM or 1.5 bar plus static pressure at 2100 LPM	38 (137) 47 (171) 76 (273)	5.6/7.0 7.0 7.0
High hazard	Not less than 2 hours run for the aggregate pumping capacity.	—	47 (171) 76 (273) 114 (410)	7.0 7.0/8.8 7.0/8.8

* Static pressure is the pressure equivalent to the height of the highest sprinkler in the installation above pump outlet.

Table 17.13b *AMAO* and flow rate

Class of hazard	Minimum design density LPM/m^2	AMAO m^2	Minimum pressure sprinkler (bar) at discharging	Flow rate at LPM
Light hazard	2.25	84	0.70	225
Ordinary hazard	5.00	360	0.35	1800
High hazard	9.00	260	0.50	—

and IS-1978 or electric resistance welded steel pipes confirming to IS-3589.

d. Welded and seamless pipes as per ASTM-A 53.

e. Electric resistance welded steel confirming to ASTM-A 135.

f. Wrought steel pipe confirming to ASTM-B 36.10.

Important points related to pipe selection and laying.

a. The installation piping (from pump house up to installation valve and also the installation piping with sprinklers) shall be capable of with standing for two hours a pressure equivalent to 150% of maximum working pressure.

b. All installation pipe work above ground shall be installed at a slope not less than 1 : 500 for horizontal run of pipes.

c. Sprinkler pipes shall be supported from the building structure which itself shall be capable of supporting the water filled pipe work.

d. The thickness of all parts of pipe supports shall not be less than 3 mm.

e. The distance between pipe supports measured along the line of connected pipes shall not be less than as below.

   i. Up to 65 mm dia. 4.0 m.

   ii. 65 to 100 mm dia. 6.0 m.

   iii. 100 to 250 mm dia. 6.5 m.

f. Distribution pipes.

   i. The first support on a nominally horizontal distribution pipe shall not be at more than 2 m from the main distribution pipe.

   ii. The last support on a nominally horizontal distribution pipe shall not be more than 450 mm from the end.

iii. Drop or rise pipes shall be secured to the building structure either directly or indirectly at the adjacent nominally horizontal part of the pipe within 300 mm of the drop or rise.

g. Range pipes

   i. At least one support shall be provided for.

      a. Each pipe run connecting adjacent sprinkler.

      b. The pipe run connecting the distribution pipe and the first sprinkler on the range pipe.

   ii. Pipe supports shall not be closer than 150 mm to any sprinkler axial central line.

   iii. The first support on a range pipe shall not be more than 2 m from the distribution pipe.

   iv. The last support on a range pipe shall not be at more than 1.5 m from

      a. The range pipe end.

      b. Where there is a horizontal arm pipe of 450 mm or longer, the arm pipe end.

      c. Where there is a drop or rise exceeding 600 mm, the drop or rise pipe end.

h. Outgoing mains from the installation valve to the system shall be supported at every 3.5 m of its run.

i. The thickness of all components used in pipe supports shall not be less than 3 mm anywhere.

Remarks: *For a light hazard building of 10,000 sq m area, the pump capacity shall be designed on the basis of flow requirement calculations for the AMAO, hydraulically most favourable location i.e. for 84 m² area only [Table 17.13b]. This is because, fire is not expected to take place at all the places simultaneously.*

18

Lifts Design: Important Factors and Traffic Constraints

Lifts are installed into building to satisfy the vertical transportation needs of its occupants and visitors and are necessary by virtue of human comfort and convenience, or by statutory regulations. In offices and other commercial buildings, lifts are installed to aid efficiency by saving occupants time, and hence money. In residential towers, it is recommended that a lift be installed in all residences where there are four or more storeys, and that two lifts be installed where a building contains more than six storeys.

The increase in the numbers of high and medium rise building since the Second World War has been a challenge to the lift industry. The last three decades have seen the acceptance of automatic cars and the introduction of better control systems. Similar improvements have occurred in the engineering of lift systems. However, little change has occurred in the methods of traffic analysis and design.

18.1 TYPES OF LIFTS

There are four main type of lifts. Namely passenger, goods, hospital, and service lifts.

a. Passenger lifts are those designed primarily for passengers. For designing these lifts, human comfort and convenience are given primary considerations. These are available in car sizes to accommodate 4, 6, 8, 10, 13, 16 and 20 passengers. Weight of each passenger is taken as 68 kg. The standard speeds are 0.7, 1.00, 1.50, 1.75, 2.50, 3.00 m/s at present in India.

b. Goods lifts are mainly for the transport of materials, but may be required occasionally to carry passengers. These are available for standard loading of 500, 1000, 1500, 2000, 3000, 4000 and 5000 kg and speed of 0.25 to 1.00 m sec.

c. Hospital lifts are installed in hospital/dispensary/clinic and designed to accommodate one number bed/stretcher along its depth, with sufficient space all round to carry a minimum of three attendants in addition to lift operator. These are available in standard loading of 15, 20 and 26 persons and speed of 0.5 to 1.50 m/s. for high speed micro self levelling is preferable as this ensures safe movement of beds, stretchers, x-ray and other heavy equipment.

d. Service lifts (dumb-waiter) are those lifts, which are exclusively used for carrying materials, and shall not carry any person. Its car area does not exceed 1 m². Total inside height and capacity also does not exceed 1.25 m and 250 kg respectively. These are available in speeds from 0.25 to 0.5 m/s. These are mainly used in hotels and restaurants for service from kitchen to the dining rooms, in banks for transport of bullion and in libraries for transport of books.

18.2 CLASSIFICATION OF LIFTS

The lifts can be classified in two categories: (a) Rope lift, and (b) Hydraulic lift.

a. Rope lift is the most commonly used because of its low cost, availability from different manufacture's of reputed class, proven technology, availability of high speed and scope for future research and development and easy maintenance.

b. Hydraulic lift is generally used where accuracy is more important. The hydraulic lift is advantageous where the machine room cannot be mounted on roof due to aesthetic reasons but it shall be provided in the lift pit. The advantage with hydraulic lift is smooth and jerk free movement. However, the disadvantage with hydraulic lifts is that these are not suitable for high rise buildings as the horse power consumed in pumping the hydraulic fluid increases and become less efficient. The chances of bursting of the hydraulic piping and cylinder also increases.

To begin with, we shall concentrate on design and selection of various components of rope lift.

18.3 ESSENTIAL COMPONENTS OF LIFTS/ELEVATORS

Following are the essential components of lift utilization.

1. Machines.
2. Brakes.
3. Traction ropes.
4. Sheaves.
5. Divertor pulley.
6. Counter weights.
7. Governor.
8. Guides.
9. Buffers.
10. Door and door operator.
11. Selector.
12. Travelling cable.
13. Hoisting motor.
14. Controller.
15. Car.
16. Safety features.

18.3.1 Machines

Two type of machines are used in lifts. One is geared and the second is gear-less machine.

Geared machine consists of a worm gear reduction unit. Shaft of the hoisting motor is coupled with the shaft of the reduction gear with the help of pulleys. The pulley arrangement is used for braking. The system permits the use of a small but high speed motor. Geared machines are rarely used for speed beyond 0.5 m/s because of excessive problems with noise, vibration and difficulties with wear of the gear. For high rise buildings, to reduce the trip time (for reducing waiting interval) one has to go for gear-less machine.

In gear less machine, a grooved driving sheave and a brake pulley are directly mounted on the shaft of driving motor.

18.3.2 Brakes

Magnet operated brakes are generally used. The brakes are generally operated with DC supply. The brake shoes are provided with brake lining, which is of copper woven type as used in automobiles.

In the case of gear less machine, the function of the brake is also to hold and not only to slow down the lift.

Some manufacturers use small three phase motor for the opening and closing of brake.

18.3.3 Ropes

Lift car is raised and lowered by hoisting ropes which pass over the machine driving sheave. Necessary traction is obtained by way of friction between the ropes and the grooved surface of the sheave. Traction types of lifts have inherent safety feature that, when either the car or the counter weight bottoms, the tension in the hoisting ropes is relieved and the driving sheave may rotate without moving the elevator (lift) owing to loss in traction.

At least three ropes are used in parallel for traction purpose. The dia ranges from 1/2 to 1″. The ropes are of stranded type, each strand consists of number of steel wires. The core of rope consists of jute, hamp or manilla, impregnated with suitable lubricant. This acts as lubricating medium for reducing the wear of strands due to friction between them.

Apart from traction purpose, ropes are used for speed governor, selector, terminal limit switch, etc. The size of these ropes varies from 1/4 to 3/8″.

Factor of safety for the ropes as per IS: 4666 is given below.

Rope speed in m/s	Factor of safety
Up to 2.0	10
3.0	11
7.0	12

18.3.4 Sheave

Sheave is a pulley over which wire ropes pass for traction purpose. In geared machine, it is fitted to the output shaft of the gear. In case of gear less machine it is fitted on the shaft of the hoisting motor. The sheave is of V grooved shape. The ratio of sheave diameter to the rope diameter should not be less than that shown below.

Class of rope	Dia of sheave/pulley
6 × 19 (12/6/1)	
6 × 19 (12/6/1) Plus 6 filler wires	D (2.95 S + 37) with a minimum of 40 D
8 × 19 (12/6/1) Plus 6 filler wires	
8 × 19 (9/9/1) Scale	

where, D = Dia. of rope in cm

S = Rope speed in m/s.

For service lift, the ratio of sheave diameter to rope diameter should not be less than 30.

18.3.5 Divertor Pulley

It is an idler pulley to change direction of ropes or we can precisely say divert the ropes.

18.3.6 Counterweights

Counterweights are necessary to provide traction and balance the weight of the car plus predetermined additional load, usually 40 to 50% of the maximum car load so as to reduce the size of the motor. Counterweights are hanged with the help of wire ropes passing over the driving sheave. On the other side is car. The weights are in the form of cast iron slabs, which are fixed in the frame. The frame is provided with four guide shoes so that the counterweights move vertically within the guide rails.

18.3.7 Governor

The device is provided in the lift machine room to stop the lift when the speed increases beyond the predetermined value. It works on the principles of centrifugal force. It is driven by rope known as governor rope. This rope is attached to the safety gear provided below the car frame. An electrical switch is also provided with the governor. When the speed of the lift car increases beyond the rated speed, this switch gets actuated with the help of a lever fixed to the governor. Actuation of the switch causes the stoppage of electric supply to the controller. These speed governors are set to cause the application of safety gear at a speed not less than 115 percent of the rated/contract speed.

18.3.8 Guide

In lift system four guides are provided. Two are for car and other two for the counterweights. Earlier round guide rails were also used. But these days only *T* guide rails are universally used.

18.3.9 Buffers

Electric lifts are provided with buffers in the pit under the car and counterweight. Spring buffers are used for slow speed up to 1.5 m/s and oil buffers for high speed lifts. These buffers are symmetrically located with reference to the vertical center line of the car frame with a tolerance of 50 mm. Lift pit should not be punctured for providing buffers so as to avoid seepage of water. These are provided on the channel which is fixed at the bottom of the guide rails.

The use of buffers are to stop the car in case lift over travels *beyond terminal limits*. To restrict the car from over travelling, terminal limit switch and final terminal limit switch are provided.

Even so the lift car sometimes over travels. These buffers therefore act as final emergency device.

18.3.10 Door and Door Operators

Passenger lifts have horizontal sliding doors. They are either single sliding or center opening types. Central opening types are preferable as they reduce the time for operation of doors. This then reduces the round trip time and subsequently decreases waiting internal. The doors have rollers at the top which ride on a steel track for support and guidance. The doors are guided at the bottom by shoes sliding in a machined self cleaning slotted metal channel. The door operator is mounted on the lift car and is driven by electric motor and coupled to the car door by belts, chain or levers. The hoist way doors are automatically coupled to the car doors when the lift is at the landing and operate in synchronism with them.

The power operated doors are provided with safety shoes which when comes in contact with a person reverses the door operation. The operating mechanism shall operate with a force not exceeding 123 *N*. The leading edges of doors are provided with soft and fire resistance material.

18.3.11 Selector

The function of the selector is to give car position information to the controller and operating system so that automatic stops may be made at the selected landing. Selectors are coupled to the car with the help of wire rope or perforated steel tape.

For slow speed lifts, switches in the hoist way handle the whole positioning of the car.

The selector consists of many switches which are operated with the help of cams. It has slowing switches, stopping switches, car direction switches, car position switches, levelling switches, etc.

18.3.12 Travelling Cables

All electrical connections to the car are made by means of multi core hanging flexible cables, one end of which are connected to a terminal box fitted under the car floor or above the car top, the others to a terminal box fitted in the well at approximately in the mid position. The cable of 10 cores to 22 cores construction are generally used for higher speeds as the heavier cables give a better running performance than lighter cables.

18.3.13 Hoisting Motor

Different types of motors are used for lifts and the selection depend upon.

a. Supply characteristic.
b. Car speed.
c. Quality of service to be provided.

Usually a speed of 600–900 rpm is preferred, whilst at 1000 rpm, there is difficulty in obtaining the necessary degree of silence, even if special precautions are taken in the design of motor room and the mechanical equipment. Further higher the speed, greater the kinetic energy and more powerful braking effort required, but on the other hand, the price increases as the speed increases.

The main requirement of a lift motor are a starting torque equal to at least twice the full load torque, quietness and low kinetic energy. The last feature is necessary to obtain rapid acceleration and deceleration, together with a minimum amount of brake lining wear.

The theoretical power of the motor to drive any lift is calculated as follows.

$$\text{HP} = \frac{\text{Out of balance load} \times \text{speed in m/s}}{75}$$

Assume that maximum car load is 10 persons (680 kg), the maximum speed of 1 m/sec.

Since counter weight is 50% of car plus contract load, out of balance load will be 1/2 of the contract load.

$$\text{HP} = \frac{680/2 \times 1}{75} = \frac{340}{75} = 4.53$$

In practice actual HP is considerably more than the theoretical HP. It is essential to ensure that the motor is capable of providing necessary torque to accelerate the total moving mass of the lift system at the desired rate.

Following two type of motors are used in lifts.

a. Alternating current (AC) motors.
b. Direct current (DC) motors.

a. AC Motors

AC motors are almost exclusively squirrel cage induction motors, either single speed or two speeds. Squirrel cage motors generally have high resistance rotors to provide high starting torque and limited starting current, because a large proportion of the duty cycle consists of starting from rest. Usually the full load slip of these motors is about 20%. Thus, a motor having a synchronous speed of 900 rpm may run at 720 rpm when the lift is carrying full load in the up direction.

Double squirrel cage rotors are coming into use as well. They enable maintaining high starting torque characteristic, but the full load full speed slip can be reduced to approximately 9% giving better control and efficiency.

Better speed regulation, lower starting current and smooth acceleration are obtained by using a wound rotor motor which is accelerated by cutting out the rotor resistance in one or more steps. The following classification according to speed requirement can be made.

(a_1) *Motor for the single speed* (0.5 – 0.75 m/s) *squirrel cage induction motor, resistance start.*

(a_2) *Motor for the speed* (0.75 – 1 m/s).

i. **Squirrel cage induction motor.** Two speeds are obtained in the ratio of 2:1 by changing the number of poles. Another method of obtaining two speeds is to have two separate starter windings wound in the same slots. Ratios up to 6:1 are possible with this double wound motor. The high and low speed windings being cut in and out of circuit by means of contactor. These motors are also resistance start.

ii. **Slip ring induction motors.** Speed changes similar to squirrel cage induction motor, and involves use of two separate rotor windings. The use of two rotor windings may be avoided by carrying the rotor current through internal short circuit path during one of the speeds.

iii. **Tandem motors**
(a_3) *Motor for car speed above* 1 m/s
AC servo drive.
AC variable frequency geared machine.
AC variable frequency variable voltage.

b. DC Motors

(b_1) *Motor for speed of* 0.5 – 0.75 m/s. Single speed shunt or compound motor is used.

(b_2) *Motor for car speed of* 0.75 to 1 m/sec. Two speeds motor having speed ratios of 2:1, 3:1 or 4:1 are used. The motor is shunt regulated.

(b_3) *Motor for car speed* 1-1.75 m/s. Compound motors employing variable voltage or Wardleonard method of speed control is used with geared machines.

(b_4) *Motor for car speed* 2.5 m/sec and above. Shunt gear less motors are used for speed control.

Gearless motors. The shaft of this type of motor is directly coupled to the driving sheave without the use of reduction gear. Therefore, the speed of the motor is low. The motor, brake and sheave are mounted on common bed plate to form a single unit. These days variable voltage or Wardleonard principle is used for speed control. The major part of speed variation is accomplished by varying the voltage to the motor, the remaining small change being effected by field control.

18.3.14 Controller

This is located in the machine room. Its exact position is decided to ensure sufficient clearance between the controller, walls and other equipments so as avoid contact of the maintenance persons with the moving parts of the lifts.

The older design of controller consisting of relays, contactors, etc. have given way to microprocessor controller wherein the complete circuitry is on various printed circuit boards. The program logic is stored in the microprocessor which senses various parameters like speed of the lift, over-travel, position of limit switches, group control logic and gives required commands to control the operation of the lifts.

The controller in the machine room basically performs following three functions.

a. Motion control.
b. Operation control.
c. Door control.

18.3.15 Car

The car is the load carrying unit, including its platform, enclosure, car frame and car door. The car frame is the supporting structural frame to which the hoisting ropes or hoisting rope sheaves, car guides, car safety, platform and generally the door operating mechanism are attached. An average passenger requires about 2 sq ft of floor area to feel comfortable. On its basis, car dimensions are worked out. The standard dimension for various types of lifts are given in IS-3534 and have already been depicted in Chapter 2 in Figs. 2.8 to 2.11.

18.4 LIFT DESIGN: IMPORTANT FACTORS AND TRAFFIC CONSTRAINTS

While a lift designer designs the lift's dimension, speed, location and number thereof; the architect on the other hand is responsible to make all the necessary provisions in the building plans.

Efficient passenger transportation, both horizontal and vertical is the life line of any building, and hence, it is necessary that the architect take expert advice at conception stage. With a team approach, various aesthetic and conceptual ideas can be incorporated and optional solutions offered.

The net effect should be a building properly designed for good access and efficient transportation of the people using it or otherwise chaos prevails and the bad name reflects on the team and not on one individual.

When considering the traffic design of a new building the major building dimensions should be known. Unfortunately it is often the case that the architect responsible for the building conception will have fixed the building core limiting the space available for the lift system or even will have defined the number of shafts, their dimensions and travel. This removes one very important degree of freedom for the lift traffic designer.

Of course, at the low end of the market there may be only one lift or its dimensions may be fixed to conform with statutory regulations or to accommodate furniture, etc. But as the lift system moves 'up-market', initial design decisions become more important.

These fundamental constraints cannot be altered (at least not very much) when redesigning for modernisation as of course the building actually exists. However, there is one advantage that the building population to be served is known.

18.4.1 Human Constraints

A lift system has to be acceptable to the passengers using it. The passengers shall feel confident about the way they are handled, i.e. taking care of their physiological and psychological barriers.

i. Physiological Constraint

The physiological constraints limit the manner in which a passenger may be moved in the vertical plane. The human body is uncomfortable if its internal organs are caused to move in the body frame. This occurs when the body is subjected to acceleration or deceleration, i.e. the g effect. The magnitude of the effect depends on an individual's age, physical and mental health, and whether a sudden movement is expected. It is not clearly established the level of acceleration at which harm may be caused to the human body, but it is known by experience the levels of acceleration or deceleration, which have been found to be generally acceptable. These are shown in Fig. 18.1. Note there is no limit to the velocity at which a passenger may travel in an enclosed lift car, but that acceleration/deceleration should be limited to about $(1-1.5 \text{ m/s}^2)$ and jerk (rate of change of acceleration) to 2 m/s^3. It is the latter effect-jerk-which causes the most discomfort. If jerk is not

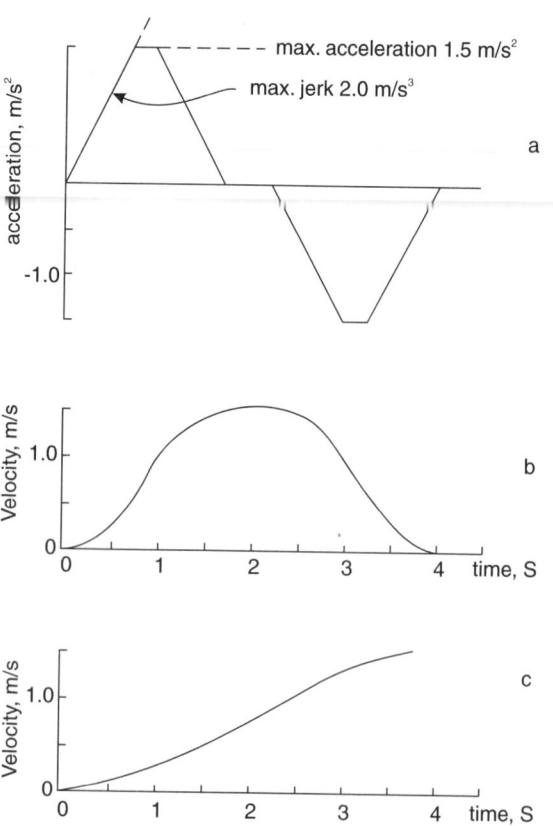

Fig. 18.1 Ideal acceleration, velocity and distance-travelled curves for a single floor jump lift systems

allowed to exceed 2 m/s^3 and is maintained constant the discomfort may be minimised.

ii. Psychological Constraint

As would be expected, psychological constraints are more subtle. A passenger expects a grade of service from a lift system. The same passenger expects a different grade of service at different times of the day and at different locations. For example, an office worker will not be too fussy when travelling up a building to work, but will become annoyed if he cannot quickly leave at night. In contrast the same office worker would not expect the same grade of service from a lift in a residential block. This constraint can be categorised as the passenger waiting time constraint. In general the maximum waiting time in an office block should not exceed 30 sec. and in the residential block should not exceed 60 sec. Waiting time is the prime psychological constraint. Tolerance will lengthen to about 120 sec. only if a few passengers are being served at each floor. Finally if monotony is relieved by a changing scene our passengers may tolerate a ride as long as 150 sec.

A secondary psychological constraint is transit time in the car after the passenger boards. Here the passenger is dependent on his fellow passengers in the car and other passengers making landing calls. A passenger travelling high up a building becomes intolerant of stops after about 90s of travel. Again the tolerance level depends on whether he is travelling in company of and on other passengers behaviour. For instance, one passenger boarding or alighting is obviously more 'selfish' than two or three transferrings at a time.

There are other psychological effects such as aesthetic appearance and 'gentle' doors, which add to a passenger's confidence in a lift system and overcome the fears of some persons who are afraid of such machines.

18.4.2 Traffic Constraints

The users of lift system, i.e. the passengers impose on the lift system, the need for it to respond to different traffic patterns. Consider Fig. 18.2; this shows the passenger demand in an office building as represented by the number of individual calls, aggregated for up and down call directions. This office building is subject to a strict time regime of fixed starting, break and leaving times. It illustrates clearly the different traffic patterns of morning up-peak, evening down-peak, mid-day four-way traffic and random (balanced) inter-floor traffic.

i. Up-peak Traffic

An up-peak traffic condition exists when the dominant or only traffic flow is in an upward direction with all or the majority of passengers entering the lift system at the main terminal of the building.

Up-peak occurs in considerable strength in the morning when prospective lift passengers enter a

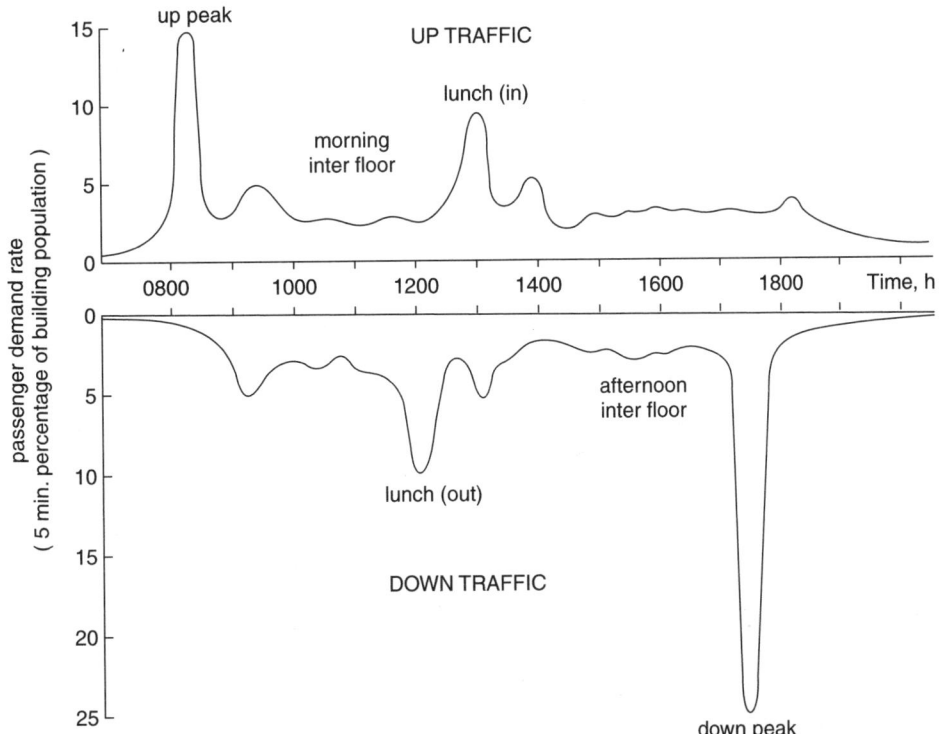

Fig. 18.2 Passenger demand rate for an office building. The total demand on the lift system is the sum of up and down direction demands

building, intent on travelling to destination on the upper floors of the building. To a lesser extent an up-peak occurs at the end of the mid-day. It is considered that if a lift system can cope efficiently with the morning up-peak, then it will cope with other patterns of traffic, such as down-peak and random inter-floor traffic. Figure 18.3 shows the arrival rate profile for morning up-peak.

The up-peak condition results from employers requiring their employees to arrive at work by a specific starting time. Human nature then exacerbates the condition as the majority of employees feel that in conscience all they must do is to be in a building before the defined starting time and that the employer then has the responsibility to transport them to their work station. The arrival rate profile for the morning up-peak thus takes a from as shown in Fig. 18.3. Here the envelope of the curve describes the arrival profile, in terms of the instantaneous passenger arrival rate, in calls per hour for a period of one hour. Note the gradual build-up prior to the official starting time and the more rapid decay afterwards.

The modern trend to flexi-time working will give way to alleviate the up-peak situation, but unfortunately it cannot be applied to all classes of employment.

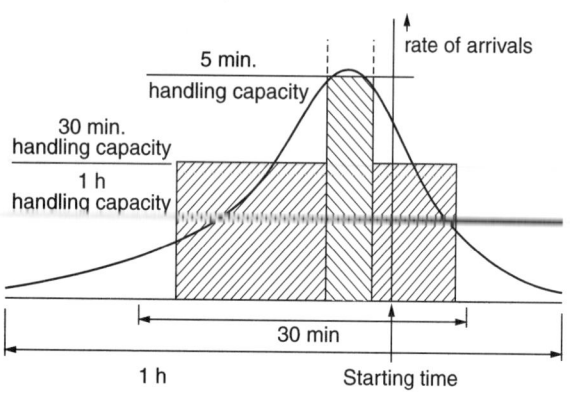

Fig. 18.3 Arrival rate profile for morning up-peak

The profile of Fig. 18.3 is often idealised by designers in terms of a 5 min peak rate taken as a percentage of the building population (the hatched area of Fig. 18.3). A second definition commonly used for the up-peak profile is to state the percentage of the building population that is likely to arrive over 30 min of peak activity (the large rectangular area of Fig. 18.3).

ii. Down-peak Traffic

A down-peak traffic condition exists when the dominant or only traffic flow is in a downward direction when with all or the majority of passengers leave the lift system at the main terminal of the building.

To some extent, down-peak is the reverse of the morning up-peak occurring at the end of the working

day, and to a lesser extent at the start of the mid-day break. The evening down peak is usually more intense than the morning up-peak with up to 50% higher demands with duration of up to 10 min. Figure 18.4 illustrates these effects.

iii. Other Traffic Situations

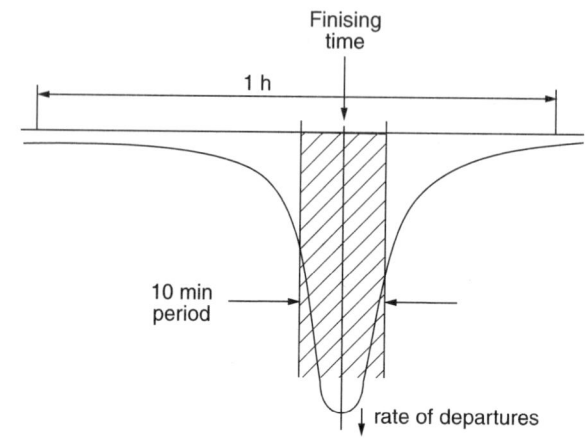

Fig. 18.4 Departure rate profile for evening down-peak

It is possible to find office buildings where no dominant traffic flows occur, especially where flexi-time working is used. Sometimes the up-peak situation occurs twice, as in Fig. 18.5, but at a lower intensity, and obviously traffic patterns are different in institutional and residential building; but often dominant pattern similar to those defined above do emerge and hence case design procedures.

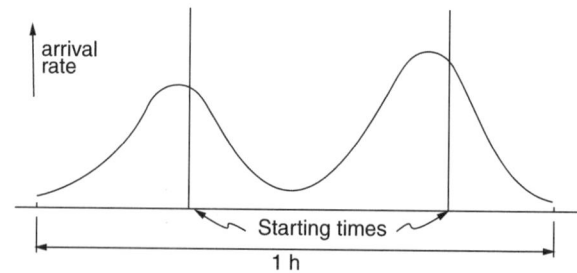

Fig. 18.5 Another arrival rate profile for morning up-peak times

iv. Requirements for Traffic Design and Control

It is extremely difficult to compare competitive tenders, where no standardised methods of specification or common design procedures are used. Each manufacture and lift consultant appear to use different methods, and are often not keen to explain their approach. Those methods that are published are often sketchy and some are inaccurate. Such a situation is confusing and thus a need exists to provide an easy to use, acceptable and standard design method. In addition, the use of modern control systems radically alters some of the design assumptions.

18.5 TRAFFIC ANALYSIS

The need of traffic analysis has gained importance with the increase in the number of storeys in the building, increase in the plan area of the building. The design of lift system thus becomes very complex when compared to the lift system requiring a single lift. The ill fated World Trade Center had 110 floors and consisted of 58 lifts. The complexity with which the lift system must have been designed can be understood from the fact that the passenger who had to attend his office on 108th or 110th floor must feel confident both physiologically and psychologically.

The traffic analysis is a tool to determine the speed, numbers and size of lift required to take care of all these factors.

The importance of close liaison between lift designer, architect and building contractor is the key to a successful design of the system and hence building.

18.5.1 Positioning of Lift

The first matter for consideration is the position to be occupied by the lift and lift-well. The main requirement being that users will be able to pass quickly from the building entrance to the lifts and that the lift exiting on various floors will be as near as possible to the centers of population of the floors.

Particular attention must be given to the traffic due to basement parking area, canteen and conference rooms. If the building has one main street entrance, the passenger lifts should be arranged adjacent to each other in a single bank and conveniently situated with regard to the entrance.

From the lift service aspect, the bank should not be arranged in two sections, i.e. one on each side of a central staircase. This practice is sometimes adopted for appearance or to satisfy architectural requirements. With a single bank, a common machine room can be used and so simplify maintenance. It is easier to arrange lift interconnection facilities and group control and the resultant service is better than if the lifts are separated.

If the building has two main entrances, two banks of lifts are necessary, the number in each bank being governed by the number of passengers that will be expected to use each entrance. Although the lift entrances are usually near the stair-way, care should be taken in the design to ensure that persons intending to use the lifts are kept clear of those who wish to use the stairs.

The design of lift installation is governed very largely by the type of building which it has to serve.

18.5.2 Population in a Building

To know the requirement of elevators it is essential to assess the population in a building. In national building code, occupant level for various types of buildings is given as follow.

If we know how many people will require elevator service within a given period of time, the task of providing that service is one of **time and motion study**.

Table 18.1 Occupancy in a building

S. no.	Occupancy	Occupant load gross area in m^2/person
i.	Residential	12.5
ii.	Educational	4.0
iii.	Institutional	15
iv.	Assembly	
	a. With fixed or loose seats and dance floor.	0.6
	b. Without seating facilities i/c dining rooms	1.5
v.	Mechantile	
	a. Street floor and sales basement	3
	b. Upper sale floors	6
vi.	Business and industrial	10
vii.	Storage	30
viii.	Hazardous	10

It embraces many variables, one of the most important being passenger reactions to elevators.

18.5.3 Calculation of Time Factors

To calculate the total time for an elevator trip, a practical procedure is to break the trip down into its components. A two stop elevator travelling one floor of height 3.05 m is considered here. As soon as a passenger arrives inside a building and presses call button, the following time factors come into play till the passenger leaves the elevator.

A. Passenger going up

Door opening time   : 0.5 sec.
Passenger entry time   : 2 sec.
Door closing time   : 2 sec.
Car travel one floor   : 7.5 sec.
Door opening time   : 0.5 sec.
(As the doors can start opening before the elevator has finally stopped, hence time of opening is taken 0.5 sec. here).
Passenger leaving time : 1.2 sec.

B. Elevator going down to pickup second passenger.

Door closing time   : 2 sec.
Car travel one floor   : 7.5 sec.

18.5.4 Handling Capacity and Number of Lifts

The handling capacity of the lift system is the total number of passenger that it can transport in a period of 5 minute during the up-peak traffic condition with a specified average car loading.

If the average arrival rate is low (Fig. 18.3), it can be distributed over a period of time (say one hour) and it can be handled by a small sized or less number of elevators. But if the arrival rate increases from the one hour handling capacity, queues start building up and so on.

Thus, it is not sufficient to size a lift system to handle 1 hour average rate of arrival because in practical situations, the passengers arrive suddenly in 5 minute up-peak interval and leave suddenly in 10 minutes down-peak interval. Thus, if it is possible to equate the passenger demand as expressed by the 5 minute passenger up-peak arrival rate with the handling capacity of a lift system, then a suitable configuration has been designed.

Round Trip Time

The round trip time (RTT) is the time in seconds for a single car trip around a building from the time the car doors open at the main terminal until the doors reopen when the car has returned to the main terminal floor after its trip around the building.

A round trip time should not usually exceed two to three minutes (except in very tall buildings) as the majority of this time can represent the journey time for some passengers with destinations on the top floors of a building, which is undesirable.

Now if it is known how many round trips a single lift car can complete during the peak 5 minutes period, then the handling capacity can be defined.

Number of round trip for a single car

$$= \frac{5 \text{ min}}{\text{RTT}} = \frac{300 \text{ sec.}}{\text{RTT}} \text{ (RTT in seconds)}.$$

Therefore, the 5 min handling capacity for a single car

$$HC_5 = \text{Number of passenger per trip} \times \frac{300}{RTT} \quad \dots \text{(i)}$$

The lift manufactures and designers over a period of time have found that if the car is loaded up to 80% of the rated passenger capacity, this is the most comfortable option for passengers. This gives comfortable breathing space as also the space to the passenger standing in the last to leave the door comfortably when the elevator stops at his landing.

Thus, 5 min handling capacity of a single car becomes.

$$HC_5 = \frac{80 \times CC \times 300}{100 \times RTT}$$

where, CC = Contract capacity or rated capacity of the car.

or $$HC_5 = \frac{300 \times Q}{RTT}$$

where, Q = 80% of contract capacity.

If there are L number of elevators, then the system handling capacity becomes.

$$Or \ HC = \frac{300 \times Q \times L}{RTT}$$

Handling capacity indicates the quantity of service, but the passenger is more concerned with the quality of service he receives in terms of waiting interval.

$$Interval = \frac{RTT}{L}$$

Interval is the average time between successive lift car arrivals at the main terminal floor.

$$Thus, \ HC = \frac{300 \times Q}{Interval}$$

If the population to be served in peak morning period is known or derived from Table 18.1 or otherwise, then.

HC as percentage of population

$$= \frac{300 \times Q \times 100}{Interval \times P}$$

where, P = population to be served in the peak morning period.

National building code defines the handling capacity as below.

Table 18.2 Handling capacity

Class of occupancy	Handling capacity
a. Diversified (mixed) office occupancy	10–15%
b. Single purpose office occupancy	15–25%
c. Residential	7.5%

The quality of service as already defined is measured in terms of interval a passenger has to wait. The quality of service in terms of waiting interval can be rated as below.

Table 18.3 Waiting interval

Quality of service or waiting interval	Rating
20–25 sec.	Excellent
30–35 sec.	Good
35–40 sec.	Fair
40–45 sec.	Poor
Over 45 sec.	Unsatisfactory

For residential buildings, greater intervals up to 90 sec. should be permissible.

Example 1

Consider an office building of 11 floors (G + 10) with inter floor distance of 3.00 m. Design a suitable lift system. The area of each floor being 750 sq m.

Analysis. From the Table 18.1, let's calculate the total population of the building considering the building to be under institutional category.

$$\text{Population} = \frac{750}{15} = 50 \text{ persons/floor.}$$

From Table 18.2, the handling capacity for diversified (mixed) office occupancy is 10–15% let's take the HC as 15%.

From Table 18.3, the waiting interval may be selected as 30 sec.

From Table 18.4, the lift speed for 10–11 floors may be taken to be 1.0 m/s.

Let's now calculate the number of lifts, car capacity and round trip time.

Number of car trips in 5 minutes

$$= \frac{300 \times L}{\text{RTT}} = \frac{300}{\text{Interval}} = \frac{300}{30} = 10 \text{ trips.}$$

Persons to be handled in 5 minutes.

$$\text{HC} = \text{Population} \times 15\% = 50 \times 10 \times 15\% = 75.$$

Car occupancy

$$= \frac{\text{Number of persons to be carried in 5 minutes}}{\text{Car trips}}$$

$$= \frac{75}{10} = 7.5.$$

$$\text{Car capacity} = \frac{7.5}{0.8} \text{ (assuming 80\% occupancy)} = 9.4.$$

This nearest car size shall be of 10 passenger capacity.

In the above example, the only parameter that remains to be determined is the number of lifts and its division into number of banks. Moreover, the percentage handling capacity, interval have been picked up from the tables. Though these will give a solution which will be very nearly accurate. But in many cases depending on traffic pattern, the calculated values had to be revised. Thus, we should have an accurate mathematical model to check the validity of the values so calculated.

Mathematical Model

Since the population on all floors is not same at one point of time, so is the arrival and departure pattern of passengers. Similarly, a lift may stop at all the floors or it may stop on some floors and skip others.

Thus, a lift designer must know the expected number of stops to calculate round trip time (RTT). Here we use theory of probability. Assuming there are N floors above the main terminal in a building and each floor is equally likely, then the probability that at least one passenger will leave the car on each floor is

$$= \frac{1}{N}$$

Or the probability that one passenger will not leave the car is

$$\mu = 1 - \frac{1}{N} = \frac{N-1}{N}$$

Since each passenger is independent of all other, the product law of probability that no passenger from a car containing *P* passengers will leave the car at any particular floor is:

Table 18.4 The speed of elevators for different class of buildings

S. no. (1)	Type of lift (2)	Occupancy (3)	No. of floors served (4)	Car speed (m/s) (5)
i.	Passenger lifts	Low and medium class flats (residential) office building, hotels	–	0.5
			4–5	0.5–0.75
		Large flats (high)	6–12	0.75–1.5
		Hospital	13–20	Above 1.5
		Shops and departmental stores*	–	2.25
ii.	Goods Lifts	Normal**	–	0.25–0.5
		Serving many floors	–	1
iii.	Hospital bed lifts	Short travel lifts in small hospitals	–	0.25
		Normal	–	0.5
		Long travel lifts in general hospitals	–	1

* The high speed is desirable especially where restaurant or other facilities are provided on the top floor as the traffic would at times demand a lift or lifts to be used entirely between the ground and the top floors.

** Slower speeds may be used for heavier leads.

$$\mu = \left(\frac{N-1}{N}\right)\left(\frac{N-1}{N}\right) \ldots \left(\frac{N-1}{N}\right) \ldots P \text{ terms}$$

$$= \left(\frac{N-1}{N}\right)^{p}$$

Thus, the probability that a stop will be made at any particular floor is.

$$\rho = 1 - \left(\frac{N-1}{N}\right)^{p}$$

And the expected number of stops for N floors is.

$$S = N\left[1 - \left(\frac{N-1}{N}\right)^{p}\right] \qquad \ldots (1)$$

The designer must also know the probable highest reversal floor to calculate round trip time.

Let us calculate the probability of the car travelling no higher than the ith floor. This is similar to assuming that no passenger leaves the car above i^{th} floor, i.e. at N^{th}, $(N-1)^{th} \ldots (i+1)^{th}$ floor.

Simplifying, we get

$$= \left(\frac{1}{N}\right)^{p}$$

or probability that i is the highest floor is equal to the probability that the car travels no higher than the ith floor – probability that the car travels no higher than the $(i-1)^{th}$ floor.

$$= \left(\frac{1}{N}\right)^{p} - \left(\frac{i-1}{N}\right)^{p}$$

The mean (average) highest reversal floor H is.

$$\sum_{i=1}^{H=N} i\left[\left(\frac{i}{N}\right)^{p} - \left(\frac{i-1}{N}\right)^{p}\right]$$

Expanding and simplifying, we get

$$H = N - \sum_{i=1}^{N-1} i\left(\frac{i}{N}\right)^{p}$$

However, the calculations become complicated as the number of floors increase.

As a rule of thumb. Highest reversal floor = Number of floors up to 20 floors building = Number of floors less one above 20 floors building.

H and S have been tabulated in Table 18.5.

Once we have calculated, expected number of stops and the highest reversal floor, RTT can be calculated.

RTT comprises of

1. Standing time.
   a. Time to load passengers.
   b. Time to open and close doors.
   c. Time to open and close at S stops.
   d. Time to unload P passengers at S stops.
2. Running time.
   a. Time to start, accelerate, run at contract speed, decelerate and level car at $(S + 1)$ stops.
   b. Time to pass remaining floors at contract speed to top floor.
   c. Time for express run down from highest reversal floor to main terminal.

$$t_1 = \text{time to load passengers}$$
$$= 12\left(1 - e^{-0.11p}\right) \ldots \text{Empirical formula.}$$
$$t_u = \text{time to unload passengers}$$
$$= 0.5(S + P) + 0.67(P)^{1.5} \ldots \text{Empirical formula.}$$
$$P = \text{number of passengers.}$$

However, practical observations reveal that.

$$t_1 = t_u = 1.2 \text{ sec. per person.}$$
$$t_o = \text{door opening time.}$$
$$t_c = \text{door closing time.}$$
$$t_f = \text{single floor jump time (includes acceleration, deceleration and travel at contract speed).}$$
$$T_e = \text{time for express run to main terminal.}$$
$$t_v = \text{travelling time at velocity } v \text{ for skipping some floors.}$$
$$= \frac{d}{v}$$
$$d = \text{inter floor distance.}$$
$$v = \text{contract speed to be decided from Table 18.4.}$$
$$\text{RTT} = (t_1 + t_u)^p + (S + 1)(t_c + t_o) + (S+1)\,t_f + (H - S)(t_v) + (H - 1)(t_v).$$
$$= 2H\,t_v + (S + 1)(t_c + t_o + t_f - t_v) + P(t_1 + t_v).$$
$$= 2H\,t_v + (S + 1)\,t_s + P(t_1 + t_v).$$

Since calculation of t_f require knowledge of characteristics of motor to accelerate and decelerate which becomes a very complex task.

Hence, designers usually adopt $t_f = 3.5\,t_v$.

Once RTT is known, the number of lifts to be installed can be derived as follow.

$$L = \text{Number of lifts} = \frac{\text{RTT}}{\text{Interval}}$$

Summary to Traffic Analysis

a. Decide on passengers arrival rate over 5 minute
$$= \frac{\text{Population} \times \text{HC}}{100}$$

b. Estimate an appropriate waiting interval from Table 18.3.

Table 18.5 Highest reversal floor (*H*) and expected number of stops (*S*) for buildings with 5–24 floors above the main terminal. cars assumed to carry 80% of contract capacity

Number of floor above main terminal		Car contract capacity						
		6	8	10	13	16	20	24
5	H	4.6	7.0	4.8	4.9	4.9	5.0	5.0
	S	3.3	3.8	4.2	4.5	4.7	4.9	4.9
6	H	5.4	5.6	5.7	5.8	5.9	5.9	6.0
	S	3.5	4.1	4.6	5.1	5.4	5.7	5.8
7	H	6.2	6.5	6.6	6.8	6.8	6.9	6.9
	S	3.7	4.4	5.0	5.6	6.0	6.4	6.6
8	H	7.1	7.4	7.5	7.7	7.8	7.9	7.9
	S	3.8	4.6	5.3	6.0	6.6	7.1	7.4
9	H	7.9	8.2	8.4	8.6	8.7	8.8	8.9
	S	3.9	4.8	5.5	6.3	7.0	7.6	8.1
10	H	8.7	9.1	9.3	9.5	9.7	9.8	9.9
	S	4.0	4.9	5.7	6.7	7.4	8.1	8.7
11	H	9.6	10.0	10.2	10.5	10.6	10.7	10.8
	S	4.0	5.0	5.9	6.9	7.8	8.6	9.2
12	H	10.4	10.8	11.1	11.3	11.5	11.7	11.8
	S	4.1	5.1	6.0	7.2	8.1	9.0	9.7
13	H	11.2	11.7	12.0	12.3	12.5	12.6	12.7
	S	4.1	5.2	6.1	7.4	8.3	9.4	10.2
14	H	12.1	12.6	12.9	13.2	13.4	13.6	13.7
	S	4.2	5.3	6.3	7.6	8.6	9.7	10.6
15	H	12.9	13.4	13.8	14.0	14.3	14.5	14.7
	S	4.2	5.4	6.4	7.7	8.8	10.0	11.0
16	H	13.7	14.3	14.7	15.0	15.3	15.5	15.6
	S	4.3	5.4	6.5	7.8	9.0	10.3	11.4
17	H	14.5	15.2	15.6	15.9	16.2	16.4	16.6
	S	4.3	5.5	6.5	8.0	9.2	10.6	11.7
18	H	15.4	16.0	16.5	16.8	18.1	18.4	18.5
	S	4.3	5.5	6.6	8.1	9.3	10.8	12.0
19	H	16.2	16.9	18.4	18.8	18.1	18.3	18.5
	S	4.3	5.6	6.7	8.2	9.5	11.0	12.3
20	H	18.0	18.8	18.2	18.8	19.0	19.3	19.4
	S	4.4	5.6	6.7	8.3	9.6	11.2	12.5
21	H	18.9	18.6	19.1	19.5	19.9	20.2	20.4
	S	4.4	5.6	6.7	8.4	9.8	11.4	12.8
22	H	18.7	19.5	20.0	20.5	20.9	21.1	22.3
	S	4.4	5.7	6.8	8.4	9.9	11.5	13.0
23	H	19.5	20.4	20.9	21.3	21.8	22.1	21.3
	S	4.4	5.7	6.9	8.5	10.0	11.7	13.2
24	H	20.3	21.2	21.8	22.2	22.7	23.0	23.2
	S	4.4	5.7	6.9	8.6	10.1	11.9	13.4

c. Calculate highest reversal floor

$$H = N - \sum_{i-1}^{N-1} \left(\frac{i}{N}\right)^p$$

or by rule of thumb.
Expected number of stops

$$S = N\left[1 - \left(\frac{N-1}{N}\right)^p\right].$$

d. Calculate average car capacity as follow.
Number of car trips in 5 min

$$= \frac{300}{\text{Interval}}$$

Car occupancy

$$= \frac{\begin{array}{c}\text{Number of persons handled}\\ \text{in 5 min from (a) above}\end{array}}{\text{Number of trips in 5 min}}$$

Car capacity

$$= \frac{\text{Car occupancy}}{80\% \text{ loading}}$$

Round to the nearest available car size (on higher side).

e. Calculate RTT

$$\text{RTT} = 2 H t_v + (S + 1) t_s + P (t_1 + t_v).$$

f. Select number of lifts to provide an interval close to that in step (b) above.

$$\text{Interval} = \frac{\text{RTT}}{L}$$

L = number of lifts selected and shall be varied till a desirable waiting interval close to that assumed in step b. above is reached.

g. Find % loading and % handling capacity

$$\text{HC} = \frac{300 \times 80 \times \text{Car capacity} \times \text{Number of lifts}}{100 \times \text{RTT}}$$

HC so calculated shall be near to that assumed in step a. above.

Example 2

Consider an office building of 16 floors (G + 16) with population who arrive at a fixed time. The inter-floor distance is 3.0 m. Design a suitable lift system.

The built up area of the building may be taken as 10000 sq. m.

Determination of Capacity of Lift

Calculate population from Table 18.1 considering 15 m²/person occupancy

$$= \frac{10000}{15} = 667$$

To begin with, assume waiting interval = 35 sec.

Number of car trips in 5 minutes

$$= \frac{300}{\text{Interval}}$$

$$= \frac{300}{35}$$

$$= 8.57$$

Assume handling capacity $= 15\%$
Number of persons to be handled in 5 minutes $= \text{Population} \times \text{HC}$

$$= 667 \times \frac{15}{100}$$

$$= 100$$

$$\text{Car occupancy} = \frac{\text{Number of persons in 5 minutes}}{\text{Car trips}}$$

$$= \frac{100}{10}$$

$$= 10$$

Car contract capacity (80% loading) $= \dfrac{10}{0.8}$

$$= 12.5$$

Nearest car size $= 13$

Highest reversal floor $(H) = N - \sum_{i=1}^{N-1}\left(\frac{1}{N}\right)^p$

(where $p = 13 \times 0.8 = 10.4 \simeq 10$)

$$H = 16 - \sum_{i=1}^{15}\left(\frac{1}{16}\right)^{10}$$

$$= 15$$

Probable number of stops $(S) = N\left[1 - \left(1 - \frac{1}{N}\right)^p\right]$

$$= 16\left[1 - \left(1 - \frac{1}{16}\right)^{10}\right]$$

$$= 7.8$$

Assume

Door opening time	= 0.5 sec.
Door closing time	= 2.0 sec.
Time to load and unload passenger	= 1.2 sec.

Calculation of RTT

a. Standing time

a_1 = Time to load passengers at main terminal (MT).

$$= 1.2 \times 10$$

$$= 12$$

a_2 = Time to open and close doors at MT

$\quad$ = 0.5 + 2

$\quad$ = 2.5 sec.

a_3 = Time to open and close doors at S stops

$\quad$ = 2.5 × 7.8

$\quad$ = 19.5 sec.

a_4 = Time to unload passengers at S stops

$\quad$ = 1.2 × 7.8 × 10

$\quad$ = 93.6 sec.

Subtotal of standing time = $a_1 + a_2 + a_3 + a_4$

$\qquad$ = 12 + 2.5 + 19.5 + 93.6

$\qquad$ = 127.6 sec.

b. Running time

t_v = Time to run at contract speed per floor assuming car speed 1.5 m/s

$\quad = \dfrac{3.0}{1.5}$

$\quad$ = 2 sec.

b_1 = Time to accelerate, decelerate and run at contract speed per floor for S floors

$\quad = 3.5 \times t_v \times S$

$\quad$ = 3.5 × 2 × 7.8

$\quad$ = 54.6 sec.

b_2 = Time for express run from highest floor (H) to MT to pick up passengers

$\quad = \dfrac{\text{Floor height} \times H}{\text{Velocity}}$

$\quad = \dfrac{3 \times 15}{1.5}$

$\quad$ = 30 sec.

Subtotal of running time = $b_1 + b_2$

$\qquad$ = 54.6 + 30

$\qquad$ = 84.6 sec.

Round trip time = $a + b$

$\qquad$ = 127.6 + 84.6

$\qquad$ = 212.2 sec.

Consider providing 6 elevators

RTT for a single car $= \dfrac{212}{6}$

$\qquad$ = 35 sec.

Round trip time is same as assumed in the beginning

Handling capacity (HC) $= \dfrac{300 \times Q \times 100}{\text{Interval} \times \text{Population}}$

$\qquad = \dfrac{300 \times (13 \times 0.8) \times 100}{35 \times 667}$

$\qquad$ = 13.36%

The handling capacity is slightly less than that assumed in the beginning, but since waiting interval is within the range of 'good' category of Table 18.3, this configuration can be adopted.

18.6 POWER CONTROL

An important part of the lift design include selection of the method to apply power to the elevator. The following methods are available.

18.6.1 AC Resistance Control

It consists of starting the elevator machine by using induction motor directly across the AC line or through resistance steps.

Its disadvantage is that accuracy of floor levelling depends on effectiveness of braking system and on higher load, it varies from 50 to 75 mm. This kind of system may be accepted for apartments but not for office/hospitals.

This system is seldom used for speeds more than 0.5 m/s.

18.6.2 Variable Voltage Variable Frequency Drive

This system consists of providing a varying voltage to a DC elevator motor. The characteristics of the DC drive motor are that it has the torque to move the elevator load smoothly up to speed and can absorb the inertia of the moving load by regeneration to stop the elevator with smooth retardation. Stopping is independent of the brake with all the power being absorbed back through the electrical system. These systems are available nowadays up to elevator speed of 9 m/s.

18.7 OPERATING SYSTEM

The method of causing an elevator to move in response to demands for service is known as the operating system. With the increasing demand for an efficient elevator system as the number of storeys in a building increases, it became very essential to operate the system in close unison within and with external demands. The various types of operating systems are as follow.

18.7.1 Single Automatic Operation

This consists of single buttons at each landing and a button for each floor in a car operating panel. This system has one call button for each elevator on each floor. This type of system is used for exclusive use of elevators like for light traffic buildings.

18.7.2 Full Selective Collective

The collective operation, as the name implies, is a mean to collect and answer all the calls in one direction, reverse the elevator, and then collect and answer all the calls in opposite direction.

A variation of selective collective is "down collective" when only down hall buttons are provided at each upper floor. This can be acceptable in an apartment house, in which most people want service up to their apartments from the lobby (up car calls) and down to the lobby from their floors (down hall calls).

18.7.3 Duplex Collective

Two cars selective collective operation is accomplished by duplexing, hence the name duplex collective.

In this system, one of the two cars is designated as the home landing car. It can be either car and may change after each trip. The other car is called "Free" and may park at its last call or some designated landing (middle free car parking), or in a designated upper area of the building. The home-landing car is designated to respond to any hall call at the lobby or basement floor in the building. The free car is designated to respond to any call above the lobby or in a designated area. Thus, both the cars work in unison with each other and provide the most efficient service to the user.

18.7.4 Group Automatic Operation

The automatic operation of two or more more non-attendant elevators equipped with power operated car and hoist way doors.

The operation of the cars is coordinated by a supervisory control system including automatic despatching means whereby selected cars at designated despatching points automatically close their doors and proceed on their trips in a regulated manner. The passengers are served by one set of up and down buttons at each landing, for a group of elevators. The stops set up by the momentary actuation of the landing buttons may be accomplished by any elevator in the group, and are made automatically by the first available car that approaches the landing in the corresponding direction.

The supervisory control system is capable of changing the coordination of cars according to the different needs of traffic at different times of day, i.e. it includes scheduling feature.

18.8 HYDRAULIC ELEVATORS

These elevator are actuated through a single acting displacement type hydraulic cylinder which is connected to the passenger car through a sling assembly. The sling assembly slides on a pair of guide rails during upward and downward motion of passenger car. Controlled supply of fluid to the cylinder through a spirally designed power unit ensures precise motion control and floor levelling. Down motion is by gravity.

In this system, an electric pump is used to force the hydraulic fluid from the tank to the cylinder to move the car up. To ensure accuracy of levelling or to move the cylinder down, a control valve/bypass valve is used which allows the fluid to flow back into the cylinder, thus reducing the speed and levelling accuracy.

The hydraulic elevators have not found acceptance in the market perhaps due to high cost of maintenance once the system breaks down and due to less vendors available in the market. Also because of the fact that due to precision and accuracy now available in conventional lifts with the introduction of VVVF power drives and microprocessor control, the need to substitute the motorised rope elevator is practically not required.

18.9 DIMENSIONAL AND STRUCTURAL REQUIREMENTS

18.9.1 Dimensional Requirement

To have a standard in the construction aspects of lift manufacturing and installation, the national building code recommends the size of lift well vis-à-vis the car size and passenger capacity. The dimensional requirement of various categories of lift is given in Chapter 2 at Figs. 2.8 to 2.11.

18.9.2 Structural Requirement

Lift shafts should be plumb from all sides.

Tolerance allowed

For well up to 30 m, vertically < 25 mm.
For well up to 60 m, vertically < 35 mm.
For well up to 90 m, vertically < 50 mm.

A few important informations which may not be specifically asked for by the architect, but shall be passed on to him by the designer are.

1. **Machine room.** Floors shall be designed to carry a load of not less than 500 kg/m^2 over the whole area.
2. **Load on beam.** The total load on overhead beams shall be assumed as equal to all equipment resting on the beams plus twice the minimum load suspended from the beam.
3. **Deflection of beam.** Deflection of overhead beam under the minimum static load shall not exceed 1/1500 of the span.
4. In the case of large lift installation, the roof of the machine room also should be designed to take up the pulley which could be used for lifting up parts of the lift machinery for inspection and repair.

18.10 CLASSIFICATION ACCORDING TO INTERIORS

The lift is not considered to be a simple mean of transportation these days. The importance of the building and its constructional features, elegance all gets reflected to an outsider who is entering the building for the first time by the interiors of the lift.

Various concepts and materials used to define the interiors of a lift are as follow.

a. Mirror on all the faces except door face. As the car button panel is mounted on this face, it needs to be opened for repair and maintenance. Hence, a nice teak work or beautifully painted aluminum paneling can be provided. The hand rail in golden polish looks quite elegant with mirror work.

b. Lifts used in theme building can be provided with teak paneling with theme painting on the walls and elegantly painted teak guide rails, e.g. in a hotel with Mughal or Rajputana theme; the painting of a darbar with the king or any historical event will look amazing.

c. In a commercial building, sunmica, ply, stainless steel paneling will suffice. Stainless steel paneling with moon rock finish (pot grooves) and sunmica are the most preferred ones as no rubbish can be written over it.

18.11 LIST OF IS CODES

1. IS 4666–1980 specification for electric passenger and goods lifts (first revision).
2. IS 6383–1971 specification for electric service lifts.
3. IS 2332–1963 nomenclature of floors and stories.
4. IS 4289–1967 specification for lift cables.
5. IS 3043–1966 code of practice for earthing.
6. IS 1860–1980 code of practice for installation, operation and maintenance of electric passenger and goods lifts (second revision).
7. IS 6620–1972 code of practice for installation, operation and maintenance of electric service lifts.
8. IS 2309–1969 code of practice for the protection of building and allied structures against lightning (first revision).
9. IS 4591–1968 code of practice for installation and maintenance of escalators.

Also the provision of national building code part VII. Installation of lifts and escalators shall be observed carefully. Similarly, special considerations have been given in part IV of NBC which shall be taken care of before floating tenders.

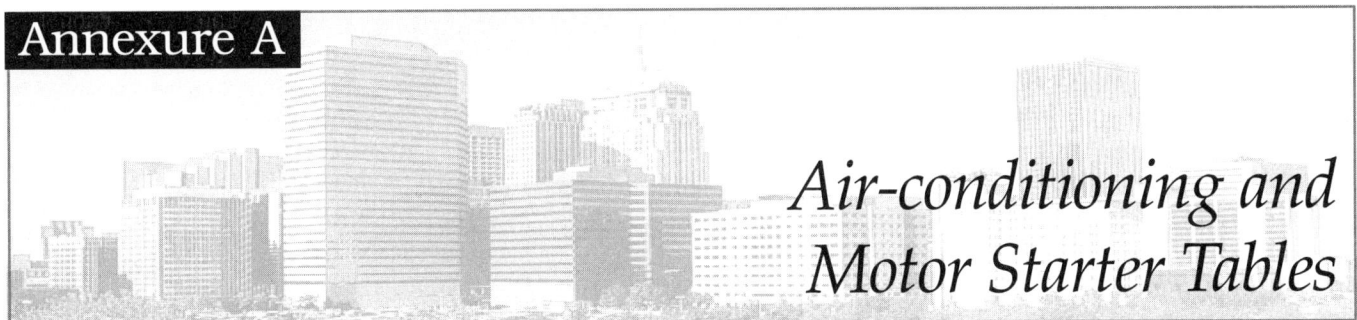

Annexure A

Air-conditioning and Motor Starter Tables

Table A.1 Relative density of air at various temperatures (70 °F = 1.00)

Temperature (°F)	Relative density	Temperature (°F)	Relative density	Temperature (°F)	Relative density	Temperature (°F)	Relative density	Temperature (°F)	Relative density
−40	1.261	110	0.93	260	0.736	410	0.609	560	0.52
−30	1.233	120	0.914	270	0.726	420	0.603	570	0.515
−20	1.205	130	0.898	280	0.716	430	0.596	580	0.510
−10	1.178	140	0.883	290	0.707	440	0.589	590	0.505
0	1.152	150	0.869	300	0.697	450	0.582	600	0.500
10	1.128	160	0.855	310	0.688	460	0.576	650	0.477
20	1.104	170	0.841	320	0.680	470	0.57	700	0.457
30	1.082	180	0.828	330	0.671	480	0.565	750	0.438
40	1.06	190	0.815	340	0.622	490	0.559	800	0.421
50	1.039	200	0.803	350	0.654	500	0.553	850	0.404
60	1.019	210	0.791	360	0.646	510	0.547	900	0.39
70	1.000	220	0.779	370	0.638	520	0.541	950	0.376
80	0.982	230	0.768	380	0.631	530	0.535	1000	0.364
90	0.964	240	0.757	390	0.624	540	0.530	1050	0.352
100	0.946	250	0.747	400	0.616	550	0.525	1100	0.340

Table A.2 Typical bypass factors (for finned coils)

Depth of coils (rows)	Without sprays		With sprays*	
	8 fins/in.	14 fins/in.	8 fins/in.	14 fins/in.
	Velocity (fpm)			
	300−700	300−700	300−700	300−700
2	.42−.55	.22−.38	—	—
3	.27−.40	.10−.23	—	—
4	.19−.30	.05−.14	.12−.22	.03−.10
5	.12−.23	02−.09	.08−.14	.01−.08
6	.08−.18	.01−.06	.06−.11	.01−.05
8	.03−.08		.02−.05	

* The bypass factor with spray-coils is decreased because the spray provides more surface for contacting the air.

Table A.3 Typical bypass factors (for various applications)

Coil bypass factor	Type of application	Example
0.30 to 0.50	A small total load or a load with a low sensible heat factor (high latent load)	Residence
0.20 to 0.30	Typical comfort application with a relatively small total load or a low sensible heat factor with a somewhat larger load	Residence, Small retail Shop, Factory
0.10 to 0.20	Typical comfort application	Dept. Store, Bank, Factory
0.05 to 0.10	Applications with high internal sensible loads or requiring a large amount of outdoor air for ventilation	Dept. Store, Restaurant, Factory
0.00 to 0.10	All outdoors air applications	Hospital Operating room, Factory

Table A.4 Relative density of air at various altitudes (sea level @ 70 °F = 1.00)

Altitude (ft)	Relative density	Corresponding barometric pressure	Altitude (ft)	Relative density	Corresponding barometric pressure
Sea Level	1.00	29.92	2500	0.913	27.31
200	0.996	29.7	3000	0.896	26.81
400	0.985	29.49	3500	0.873	26.32
600	0.978	29.28	4000	0.864	25.84
800	0.971	29.06	5000	0.832	24.89
1000	0.964	28.85	6000	0.799	23.98
1200	0.957	28.65	7000	0.774	23.09
1400	0.951	28.44	8000	0.739	22.12
1600	0.944	28.33	9000	0.715	21.38
1800	0.937	28.02	10000	0.687	20.54
2000	0.93	27.82			

Table A.5 Standard atmospheric data for altitudes to 60,000 ft

Altitude	Temperature, °F	Pressure		Altitude	Temperature, °F	Pressure	
		Inch Hg	Psia			Inch Hg	Psia
−1000	62.6	31.02	15.236	7000	34.0	23.09	11.341
−500	60.8	30.47	14.966	8000	30.5	22.22	10.914
0	59.0	29.921	14.696	9000	26.9	21.39	10.506
500	57.2	29.38	14.430	10,000	23.4	20.58	10.108
1000	55.4	28.86	14.175	15,000	5.5	16.89	8.296
2000	51.9	27.82	13.664	20,000	−12.3	13.76	6.758
3000	48.3	26.82	13.173	30,000	−47.8	8.90	4.371
4000	44.7	25.82	12.682	40,000	−69.7	5.56	2.731
5000	41.2	24.90	12.230	50,000	−69.7	3.44	1.690
6000	37.6	23.98	11.778	60,000	−69.7	2.14	1.051

Table A.6 Dew points in °C

Dry-bulb temperature °C	Relative humidity percent									
	10	20	30	40	50	60	70	80	90	100
20	−10	−3	2	6	9	12	14	16	18	20
22	−9	−2	4	8	11	14	16	18	20	22
24	−8	0	6	10	13	16	18	20	22	24
26	−7	1	7	11	15	18	20	22	24	26
28	−6	3	9	13	17	20	22	24	26	28
30	−4	5	11	15	18	21	24	26	28	30
32	−3	6	12	17	20	23	26	28	30	32
34	−2	8	14	19	22	25	28	30	32	34
36	0	9	16	20	24	27	30	32	34	36
38	1	11	18	22	26	29	32	34	36	38
40	3	13	19	24	28	31	34	36	38	40
42	4	14	21	26	30	33	36	38	40	42
44	6	16	22	27	31	35	38	40	42	44
46	8	18	24	29	33	37	40	42	44	46
48	9	19	26	31	35	38	41	44	46	48
50	11	21	28	33	37	40	43	46	48	50

Table A.7 Temperature converter

To °C	From	To °F	To °C	From	To °F	To °C	From	To °F
−40.00	−40	−40.0	−9.44	+15	+59.0	+21.11	+70	+158.0
−39.44	−39	−38.2	−8.89	+16	+60.8	+21.67	+71	+159.8
−38.89	−38	−36.4	−8.33	+17	+62.6	+22.22	+72	+161.6
−38.33	−37	−34.6	−7.78	+18	+64.4	+22.78	+73	+163.4
37.78	−36	−32.8	−7.22	+19	+66.2	+23.33	+74	+165.2
−37.22	−35	−31.0	−6.67	+20	+68.0	+23.89	+75	+167.0
−36.67	−34	−29.2	−6.11	+21	+69.8	+24.44	+76	+168.8
−36.11	−33	−27.4	−5.56	+22	+71.6	+25.00	+77	+170.6
−35.56	−32	−25.6	−5.00	+23	+73.4	+25.56	+78	+172.4
−35.00	−31	−23.8	−4.44	+24	+75.2	+26.11	+79	+174.2
−34.44	−30	−22.0	−3.89	+25	+77.0	+26.67	+80	+176.0
−33.89	−29	−20.2	−3.33	+26	+78.8	+27.22	+81	+177.8
−33.33	−28	−18.4	−2.78	+27	+80.6	+27.89	+82	+179.6
−32.78	−27	−16.6	−2.22	+28	+82.4	+28.33	+83	+181.4
32.22	−26	−14.8	−1.67	+29	+84.2	+28.89	+84	+183.2
−31.67	−25	−13.0	−1.11	+30	+86.0	+29.44	+85	+185.0
−31.11	−24	−11.2	−0.56	+31	+87.8	+30.00	+86	+186.8
−30.56	−23	−9.4	0.00	+32	+89.6	+30.56	+87	+188.6
−30.00	−22	−7.6	+0.56	+33	+91.4	+31.11	+88	+190.4
−29.44	−21	−5.8	+1.11	+34	+93.2	+31.67	+89	+192.2
−28.89	−20	−4.0	+1.67	+35	+95.0	+32.22	+90	+194.0
−28.33	−19	−2.2	+2.22	+36	+96.8	+32.78	+91	+195.8
−27.78	−18	−0.4	+2.78	+37	+98.6	+33.33	+92	+187.6
−27.22	−17	+1.4	+3.33	+38	+100.4	+33.89	+93	+199.4
−26.67	−16	+3.2	+3.89	+39	+102.2	+34.44	+94	+201.2
−26.11	−15	+5.0	+4.44	+40	+104.0	+35.00	+95	+203.0
−25.56	−14	+6.8	+5.00	+41	+105.8	+35.56	+96	+204.8
−25.00	−13	+8.6	+5.56	+42	+107.6	+36.11	+97	+206.6
−24.44	−12	+10.4	+6.11	+43	+109.4	+36.67	+98	+208.4
−23.89	−11	+12.2	+6.67	+44	+111.2	+37.22	+99	+210.2
−23.33	−10	+14.0	+7.22	+45	+113.0	+37.78	+100	+212.0
−22.78	−9	+15.8	+7.78	+46	+114.8	+38.33	+101	+213.8
−22.22	−8	+17.6	+8.33	+47	+116.6	+38.89	+102	+215.6
−21.67	−7	+19.4	+8.89	+48	+118.4	+39.44	+103	+217.4
−21.11	−6	+21.2	+9.44	+49	+120.2	+40.00	+104	+219.2
−20.56	−5	+23.0	+10.00	+50	+122.0	+40.56	+105	+221.0
−20.00	−4	+24.8	+10.56	+51	+123.8	+41.11	+106	+222.8
−19.44	−3	+26.6	+11.11	+52	+125.6	+41.67	+107	+224.6
−18.89	−2	+28.4	+11.67	+53	+127.4	+42.22	+108	+226.4
−18.33	−1	+30.2	+12.22	+54	+129.2	+42.78	+109	+228.2
−17.78	0	+32.0	+12.78	+55	+131.0	+43.33	+110	+230.0
−17.22	+1	+33.8	+13.33	+56	+132.8	+43.89	+111	+231.8
−16.67	+2	+35.6	+13.89	+57	+134.6	+44.44	+112	+233.6
−16.11	+3	+37.4	+14.44	+58	+136.4	+45.00	+113	+235.4
−15.56	+4	+39.2	+15.00	+59	+138.2	+45.56	+114	+237.2
−15.00	+5	+41.0	+15.56	+60	+140.0	+46.11	+115	+239.0
−14.44	+6	+42.8	+16.11	+61	+141.8	+46.67	+116	+240.8
−13.89	+7	+44.6	+16.67	+62	+143.6	+47.22	+117	+242.6
−13.33	+8	+46.4	+17.22	+63	+145.4	+47.78	+118	+244.4
−12.78	+9	+48.2	+17.78	+64	+147.2	+48.33	+119	+246.2
−12.22	+10	+50.0	+18.33	+65	+149.0	+48.89	+120	+248.0
−11.67	+11	+51.8	+18.89	+66	+150.8	+49.44	+121	+249.8
−11.11	+12	+53.6	+19.44	+67	+152.6	+50.00	+122	+251.6
−10.56	+13	+55.4	+20.00	+68	+154.4	+50.56	+123	+253.4
−10.00	+14	+57.2	+20.56	+69	+156.2	+51.11	+124	+255.2

C = 5/9 (F − 32)   F = 9/5 C + 32

Table A.8 Desirable wind speeds (m/s) for thermal comfort conditions

Dry-bulb temperature °C	Relative humidity (percentage)						
	30	40	50	60	70	80	90
28	*	*	*	*	*	*	*
29	*	*	*	*	*	0.06	0.19
30	*	*	*	0.06	0.24	0.53	0.85
31	*	0.06	0.24	0.53	1.04	1.47	2.10
32	0.20	0.46	0.94	1.59	2.26	3.04	+
33	0.77	1.36	2.12	3.00	+	+	+
34	1.85	2.72	+	+	+	+	+
35	3.20	+	+	+	+	+	+

Minimum wind speeds (m/s) for just acceptable warm conditions

Dry-bulb temperature °C	Relative humidity (percentage)						
	30	40	50	60	70	80	90
28	*	*	*	*	*	*	*
29	*	*	*	*	*	*	*
30	*	*	*	*	*	*	*
31	*	*	*	*	*	0.06	0.23
32	*	*	*	0.09	0.29	0.60	0.94
33	*	4.04	0.24	0.60	1.04	1.85	2.10
34	0.15	0.46	0.94	1.60	2.26	3.05	+
35	0.68	1.36	2.10	3.05	+	+	+
36	1.72	2.70	+	+	+	+	+

* None + Higher than those acceptable in practice

Table A.9 Relative humidity in percent

For various room temperatures and various differences between wet and dry-bulb temperatures

Dry-bulb temperature °F	Differences between dry-bulb and wet-bulb temperatures (°F)										
	0	2	4	6	8	10	12	14	16	18	20
50	100	87	74	62	50	39	28	17	7	0	0
52	100	88	75	63	52	41	30	20	10	0	0
54	100	88	76	65	54	43	33	23	14	5	0
56	100	88	77	66	55	45	35	26	17	8	0
58	100	88	77	67	57	47	38	28	20	11	3
60	100	89	78	68	58	49	40	31	22	14	6
62	100	89	79	69	60	50	41	33	25	17	9
64	100	90	79	70	61	52	43	35	27	20	12
66	100	90	80	71	62	53	45	37	29	22	15
68	100	90	81	72	63	55	47	39	31	24	17
70	100	90	81	72	64	56	48	40	33	26	20
72	100	91	82	73	65	57	49	42	35	28	22
74	100	91	82	74	66	58	51	44	37	30	24
76	100	91	83	74	67	59	52	45	38	32	26
78	100	91	83	75	67	60	53	46	40	34	28
80	100	91	83	76	68	61	54	47	41	35	29
82	100	92	84	76	69	62	55	49	43	37	31
84	100	92	84	77	70	63	56	50	44	38	32
86	100	92	85	77	70	63	57	51	45	39	34
88	100	92	85	78	71	64	58	52	46	41	35
90	100	92	85	78	71	65	59	53	47	42	37
92	100	92	85	78	72	65	59	54	48	43	38
94	100	93	86	79	72	66	60	54	49	44	39
96	100	93	86	79	73	67	61	55	50	45	40
98	100	93	86	79	73	67	61	56	51	46	41
100	100	93	86	80	74	68	62	57	52	47	42

Table A.10 The Beaufort scale of wind force with specifications and velocity equivalents

Beaufort number	General description	Specifications	Limits of velocity 20 ft above level ground			
			m/sec	km/hr	mph	knots
0	Calm	Smoke rises vertically	Under 0.6	Under 1	Under 1	Under 1
1	Light air	Wind direction shown by smoke drift but not by vanes	0.6–1.7	1–6	1–3	1–3
2	Slight breeze	Wind felt on face; leaves rustle; ordinary vane moved by wind	1.8–3.3	7–12	4–7	4–6
3	Gentle breeze	Leaves and twigs in constant motion; wind extends light flage	3.4–5.2	13–18	8–11	7–10
4	Moderate breeze	Dust and loose paper; small branches are moved	5.3–7.4	19–26	12–16	10–14
5	Fresh breeze	Small trees in leaf begin to sway	7.5–9.8	27–35	17–22	15–19
6	Strong breeze	Large branches in motion; whistling in telegraph wires	9.9–12.4	36–44	23–27	19–24
7	Moderate gale	Whole trees in motion	12.5–15.2	45–55	28–34	24–30
8	Fresh gale	Twigs broken off trees; progress generally impeded	15.3–18.2	56–66	35–41	30–35
9	Strong gale	Slight structural damage occurs; chimney pots removed	18.3–21.5	67–77	42–48	36–42
10	Whole gale	Trees uprooted; considerable structural damage	21.6–25.4	78–90	49–56	42–49
11	Storm	Very rarely experienced; widespread damage	25.5–29.0	91–104	57–67	49–56
12	Hurricane	....	Above 29.0	Above 104	Above 67	Above 56

Table A.11 Equivalent wind chill temperatures of cold environment[a]

Wind speed (in km/h)	Actual thermometer reading (°C)												
	10	5	0	−5	−10	−15	−20	−25	−30	−35	−40	−45	−50
	Equivalent chill temperature (°C)												
Calm	10	5	0	−5	−10	−15	−20	−25	−30	−35	−40	−45	−50
10	8	2	−3	−9	−14	−20	−25	−31	−37	−42	−48	−53	−59
20	3	−3	−10	−16	−23	−29	−35	−42	−48	−55	−61	−68	−74
30	1	−6	−13	−20	−27	−34	−42	−49	−56	−63	−70	−77	−84
40	1	−8	−16	−23	−31	−38	−46	−53	−60	−68	−75	−83	−90
50	2	−10	−18	−25	−33	−41	−48	−56	−64	−71	−79	−87	−94
60	3	−11	−19	−27	−35	−42	−50	−58	−66	−74	−82	−90	−97
70[b]	4	−12	−20	−28	−35	−43	−51	−59	−67	−75	−83	−91	−99

Little danger: In less than 5 hr, with dry skin. Maximum danger from false sense of security. (WCI less than 1400)

Increasing Danger: Danger of freezing exposed flesh within one minute (WCI between 1400 and 2000)

Great Danger: Flesh may freeze within 30 seconds (WCI greater than 2000)

[a] Cooling power of environment expressed as an equivalent temperature under calm conditions.

[b] Winds greater than 70 km/h have little added chilling effect.

Table A.12 General emission standards

Concentration based standards.

Sl. no.	Parameter	Standard concentration not to exceed (in mg/Nm^3)
1.	Suspended particulate matter (SPM)	150
2.	Fluoride	10
3.	Asbestos	4 fibre/cc
4.	Mercury	0.2
5.	Chlorine	15
6.	Hydrochloric acid vapour and mist	35
7.	Hydrogen sulphide	10
8.	Sulphuric acid mist	50
9.	Carbon monoxide	1 % max
10.	Chloride	100
11.	Lead	20
12.	Sulphur dioxide	50

Table A.13 Fan laws

Dependant variables				Independent variables
Q_1	=	Q	×	$(D_1/D_2)^3 (N_1/N_2)$
p_1	=	p_2	×	$(D_1/D_2)^2 (N_1/N_2)^2 d_1/d_2$
W_1	=	W_2	×	$(D_1/D_2)^5 (N_1/N_2)^3 d_1/d_2$
Q_1	=	Q_2	×	$(D_1/D_2)^2 (p_1/p_2)^{1/2} (d_1/d_2)^{1/2}$
N_1	=	N_2	×	$(D_2/D_1) (p_1/p_2)^{1/2} (p_2/p_1)^{1/2}$
W_1	=	W_2	×	$(D_1/D_2)^2 (p_1/p_2)^{3/2} (p_2/p_1)^{1/2}$
N_1	=	N_2	×	$(D_2/D_1)^3 (Q_1/Q_2)$
W_1	=	W_2	×	$(D_2/D_1)^4 (Q_1/Q_2)^3 (p_1/p_2)$

Legend

D =	Fan size	N =	Rotational speed	d =	Gas density
Q =	Volume flow rate	p =	Pressure	W =	Power

Table A.14 Weight of aluminum sheet

Size (ft)	Size (mm)	Thickness SWG	Thickness (mm)	Weight per sheet (kg)
8 × 3	2438 × 914	1/4	6.4	51.600
8 × 4	2438 × 1219	1/4	6.4	39.000
8 × 3	2438 × 914	3/16	4.877	29.500
8 × 4	2438 × 1219	3/16	4.877	38.500
8 × 3	2438 × 914	1/8	3.251	20.000
8 × 4	2438 × 1219	1/8	3.251	26.200
8 × 3	2438 × 914	12	2.642	16.000
8 × 4	2438 × 1219	12	2.642	21.500
8 × 3	2438 × 914	14	2.032	13.000
8 × 4	2438 × 1219	14	2.032	16.000
8 × 3	2438 × 914	16	1.626	10.000
8 × 4	2438 × 1219	16	1.626	13.000
8 × 3	2438 × 914	18	1.219	7.500
8 × 4	2438 × 1219	18	1.219	10.000
8 × 3	2438 × 914	20	0.914	5.300
8 × 4	2438 × 1219	20	0.914	7.000
8 × 3	2438 × 914	22	0.711	4.200
8 × 4	2438 × 1219	22	0.711	6.000
8 × 3	2438 × 914	24	0.559	3.400
8 × 4	2438 × 1219	24	0.559	4.700
8 × 3	2438 × 914	26	0.457	2.800
8 × 4	2438 × 1219	26	0.457	4.000

Table A.15 Properties of rigid PVC sheets

Types	Softening point °C, min	Tensile strength kgf / cm², min	Dimensional change at 120 °C		Application
			Extruded or calendared %	Calendared laminated %	
Type 1	75	450	20	5	General purpose
Type 2	75	450	20	5	Parts requiring chemical resistance
Type 3	65	380	20	5	Parts requiring high impact strength
Type 4	50	380	–	–	For deep draw vacuum forming

Table A.16 Refrigerant properties

Refrigerant structure		Boiling point °C	Atmospheric life (years*)	ODP[a]	HGWP[b]	Flammable
R-32	CH_2F_2	−52.5	7.3	0	0.11	Yes
R-125	CF_3—CHF_2	−47.8	40.5	0	0.84	No
R-143a	CF_3—CH_3	−47.8	64.2	0	1.1	No
R-502	R-22/115	−45.4	15.8/400	0.22	3.7	No
R-22	$CHClF_2$	−40.8	15.8	0.05	0.34	No
E-125	CF_3—O—CHF_2	−34.6	21.0	0		No
R-12	CCl_2F_2	−29.8	130.0	1.0	3.1	No
R-134a	CF_3—CH_2F	−26.5	15.6	0	0.28	No
R-152a	CHF_2—CH_3	−25.0	1.8	0	0.03	Yes
E-143a	CF_3—O—CH_3	−24.1	3.4	0		Yes
R-134	CHF_2—CHF_2	−20.0		0		No
R-227ea	CF_3—CHF—CHF_3	−18.3	30.0*	0		No
R-227ca	CF_3—CF_2—CHF_2	−16.3	15.0*			No
E-134	CHF_2—O—CHF_2	4.7	2.8			
R-143	CHF_2—CHF_2	5.0	41.0	0	1.1	Yes
R-236ea	CF_3—CHF—CHF_2	6.5		0		No
R-11	CCl_3F	23.6	55.0	1.0	1.0	No
R-245ca	CHF_2—CF_2—CH_2F	25.0	6.4	0		Yes
R-123	CF_2—$CHCl_2$	27.9	1.8	0.016	0.02	No
E-143	CHF_2—O—CH_2F	29.9		0		Yes
R-152	CH_2F—CH_2F	30.7		0		Yes

* = Estimated

a = Ozone depletion potential

b = Halocarbon global warming potential

Table A.17 Weight of copper tubes in various sizes and gauges

ISWG	6	7	8	9	10	11	12	13	14	15	16	17	18	19	20	21	22	23	24	25	26	27	28	30
mm.	4.877	4.470	4.064	3.658	3.251	2.946	2.642	2.337	2.032	1.829	1.626	1.422	1.219	1.016	0.914	0.813	0.711	0.610	0.559	0.508	0.457	0.4166	0.3759	0.315
In.	0.192	0.176	0.160	0.144	0.128	0.116	0.104	0.092	0.080	0.072	0.064	0.056	0.048	0.040	0.036	0.032	0.028	0.024	0.022	0.020	0.018	0.0164	0.0148	0.0124

Weight In Kilograms

Ext. Dia Inch	6	7	8	9	10	11	12	13	14	15	16	17	18	19	20	21	22	23	24	25	26	27	28	30
1/4	–	–	–	–	–	–	0.273	0.262	0.245	0.231	0.214	0.196	0.175	0.151	0.139	0.126	0.112	0.098	0.090	0.083	0.076	0.069	0.063	0.053
5/16	–	–	–	–	–	–	0.390	0.365	0.335	0.312	0.286	0.259	0.229	0.196	0.179	0.162	0.143	0.125	0.115	0.106	0.096	0.088	0.080	0.067
3/8	–	0.631	0.619	0.599	0.569	0.541	0.507	0.469	0.425	0.393	0.358	0.322	0.283	0.241	0.220	0.198	0.175	0.152	0.140	0.128	0.116	0.106	0.096	0.081
7/16	–	0.829	0.799	0.761	0.713	0.671	0.624	0.572	0.515	0.474	0.430	0.385	0.337	0.286	0.260	0.234	0.206	0.179	0.165	0.151	0.137	0.125	0.113	0.095
1/2	1.065	1.027	0.979	0.923	0.857	0.802	0.742	0.676	0.605	0.555	0.501	0.448	0.391	0.331	0.301	0.270	0.238	0.206	0.189	0.173	0.157	0.144	0.130	0.109
9/16	1.281	1.225	1.160	1.085	1.001	0.933	0.859	0.779	0.695	0.636	0.574	0.511	0.445	0.376	0.341	0.306	0.269	0.233	0.214	0.196	0.177	0.162	0.147	0.123
5/8	1.497	1.423	1.340	1.247	1.145	1.063	0.976	0.883	0.785	0.717	0.646	0.574	0.499	0.421	0.382	0.342	0.301	0.260	0.239	0.219	0.198	0.181	0.163	0.138
11/16	1.713	1.621	1.520	1.409	1.289	1.194	1.093	0.986	0.875	0.798	0.718	0.637	0.553	0.466	0.422	0.378	0.330	0.287	0.264	0.242	0.218	0.199	0.180	0.152
3/4	1.929	1.819	1.700	1.571	1.433	1.324	1.210	1.090	0.965	0.879	0.790	0.700	0.607	0.511	0.463	0.414	0.364	0.314	0.288	0.264	0.238	0.218	0.197	0.160
13/16	2.145	2.017	1.880	1.733	1.578	1.455	1.327	1.193	1.055	0.960	0.863	0.763	0.661	0.556	0.503	0.450	0.395	0.341	0.313	0.287	0.259	0.236	0.214	0.180
7/8	2.361	2.215	2.060	1.895	1.722	1.585	1.444	1.297	1.145	1.041	0.935	0.826	0.715	0.601	0.544	0.486	0.427	0.368	0.338	0.310	0.279	0.255	0.230	0.194
15/16	2.577	2.413	2.240	2.057	1.866	1.716	1.561	1.401	1.235	1.122	1.007	0.889	0.769	0.646	0.584	0.522	0.459	0.395	0.363	0.332	0.299	0.273	0.247	0.208
1	2.793	2.611	2.420	2.219	2.010	1.846	1.678	1.504	1.325	1.203	1.079	0.952	0.823	0.691	0.625	0.558	0.490	0.422	0.387	0.355	0.320	0.292	0.264	0.222
1–1/8	3.225	3.007	2.780	2.543	2.298	2.107	1.912	1.711	1.505	1.365	1.223	1.078	0.931	0.781	0.706	0.630	0.553	0.476	0.437	0.400	0.361	0.329	0.297	0.250
1–1/4	3.657	3.403	3.140	2.868	2.586	2.368	2.146	1.918	1.685	1.527	1.367	1.204	1.039	0.871	0.787	0.702	0.616	0.530	0.486	0.445	0.401	0.366	0.331	0.278
1–3/8	4.090	3.799	3.500	3.192	2.874	2.630	2.380	2.125	1.865	1.689	1.511	1.330	1.147	0.961	0.868	0.774	0.679	0.584	0.536	0.490	0.442	0.403	0.364	0.306
1–1/2	4.522	4.220	3.840	3.516	3.162	2.891	2.614	2.332	2.045	1.851	1.653	1.471	1.255	1.051	0.949	0.846	0.742	0.638	0.585	0.536	0.483	0.440	0.398	0.334
1–5/8	4.954	4.592	4.220	3.840	3.450	3.152	2.848	2.539	2.225	2.013	1.799	1.582	1.363	1.142	1.030	0.918	0.805	0.692	0.635	0.581	0.525	0.477	0.431	0.362
1–3/4	5.386	4.988	4.580	4.164	3.738	3.413	3.082	2.746	2.405	2.175	1.943	1.708	1.471	1.232	1.111	0.990	0.868	0.746	0.684	0.626	0.564	0.515	0.465	0.390
1–7/8	5.818	5.384	4.941	4.488	4.026	3.674	3.316	2.953	2.586	2.337	2.087	1.834	1.579	1.322	1.192	1.062	0.931	0.800	0.734	0.671	0.605	0.552	0.498	0.418
2	6.250	5.780	5.301	4.812	4.314	3.935	3.550	3.161	2.766	2.499	2.231	1.960	1.687	1.412	1.273	1.134	0.994	0.854	0.784	0.717	0.646	0.589	0.532	0.446
2–1/8	6.682	6.176	5.661	5.136	4.602	4.196	3.784	3.368	2.946	2.661	2.375	2.086	1.795	1.502	1.354	1.206	1.057	0.908	0.833	0.762	0.687	0.626	0.565	0.474
2–1/4	7.114	6.572	6.021	5.460	4.890	4.457	4.018	3.575	3.126	2.823	2.519	2.212	1.903	1.592	1.435	1.278	1.120	0.962	0.883	0.807	0.727	0.663	0.599	0.502
2–3/8	7.547	6.968	6.381	5.784	5.179	4.718	4.252	3.782	3.306	2.986	2.663	2.338	2.001	1.682	1.516	1.350	1.183	1.016	0.932	0.852	0.768	0.700	0.632	0.530
2–1/2	7.979	7.364	6.741	6.108	5.467	4.979	4.487	3.989	3.486	3.148	2.807	2.464	2.119	1.772	1.597	1.422	1.246	1.070	0.982	0.898	0.809	0.737	0.666	0.558
1.327	1.116	0.922	0.747	0.590	0.485	0.389	0.305	0.230	0.187	0.148	0.113	0.083	0.057	0.047	0.037	0.028	0.021	0.017	0.015	0.012	0.010	0.008	0.006	

To calculate the weight when internal diameter is given, add figure at the bottom of column to the corresponding figure for the weight of external diameter of size required.

Table A.18 Recommended insulation thickness of phenolic foam

Ambient air temperature : 40 °C

Relative humidity : 85%

Weather barrier : Aluminium cladding

Operating temperature range °C	Pipe diameters (NB) valves and fittings (mm)															Vessels columns and flat surfaces
	15	20	25	40	50	80	100	150	200	250	300	350	400	450	500	
4 and above	25	25	30	30	40	40	40	40	50	50	50	60	60	60	60	60
4 to −3	40	40	40	40	50	50	50	50	60	60	60	60	75	75	75	75
−4 to −18	40	40	40	50	50	65	75	75	75	75	75	75	90	90	90	90
−19 to −32	50	50	50	65	65	70	75	75	90	90	90	90	90	100	100	100
−33 to −46	50	65	75	75	80	80	90	90	100	100	100	110	110	115	115	110
−47 to −60	65	75	90	90	90	90	100	100	110	110	110	120	120	130	130	120
−61 to −74	75	75	90	90	90	100	110	120	120	120	120	130	130	140	140	145
−75 to −87	75	75	90	90	90	100	110	130	130	130	130	140	150	150	150	155
−88 to −100	90	90	100	100	100	110	120	130	130	140	130	150	160	160	160	165
−101 to −115	90	90	100	105	110	115	130	130	140	140	150	160	160	160	160	180
−116 to −129	100	100	110	110	120	120	130	140	140	140	150	160	165	165	165	180
−130 to −143	100	100	110	115	120	130	135	140	150	150	160	165	165	165	165	190
−144 to −157	100	110	110	120	120	130	140	150	150	150	160	165	165	175	175	200
−158 to −170	110	120	125	130	140	140	140	160	160	160	175	175	175	190	190	210
−171 to −184	120	125	130	140	150	150	160	170	170	170	180	200	200	200	210	220

Manufacturer's data.

Based on average conditions. Should be modified to suit individual technical requirements.

Table A.19 Recommended thickness of expanded polystyrene in mm for pipe insulation at various operating temperatures

Temperature °C	15 mm	20 mm	25 mm	32 mm	40 mm	50 mm	65 mm	80 mm	100 mm	125 mm	150 mm	Temperature °F
20	25	25	25	25	25	25	40	40	40	50	50	68
10	25	25	25	40	40	40	40	40	50	50	50	50
0	40	40	40	50	50	50	50	50	50	50	75	32
−10	50	50	50	50	65	65	75	75	75	75	75	14
−20	50	50	65	75	75	75	75	75	100	100	100	−4
−30	65	75	75	75	100	100	100	100	100	100	125	−22
−40	75	75	100	100	100	100	100	100	125	125	125	−40
−50	100	100	100	100	100	100	125	125	125	150	150	−58
−60	100	100	125	125	125	125	125	125	150	150	150	−76
−70	125	125	125	125	125	150	150	150	150	175	175	−94
−	1/2"	3/4"	1"	1 − 1/4"	1 − 1/2"	2"	2 − 1/2"	3"	4"	5"	6"	

Manufacturer's data.

Based on average conditions. Should be modified to suit individual technical requirements.

Table A.20 Recommended thickness of high density polyethylene in mm for pipe insulation at various operating temperatures

Temperature °C	Nominal pipe dia. in mm						Temperature °F
	15	25	40	50	80	100	
20	8	10	10	10	10	10	68
10	10	12	15	15	15	15	50
0	15	15	20	20	20	20	32
−10	20	20	25	25	25	25	14
−20	25	25	30	30	30	30	−4
−30	25	30	30	30	35	35	−22
−40	30	30	35	35	40	40	−40
−50	30	35	40	40	40	50	−58

Manufacturer's data.
Based on average conditions. Should be modified to suit individual technical requirements.

Table A.21 Recommended thickness of polyurethane foam in mm for pipe insulation at various operating temperatures

Bore of pipe (inch)	50 °F to 32 °F (10 °C to 0 °C)	32 °F to 14 °F (0 °C to −10 °C)	14 °F to −4 °F (−10 °C to −20 °C)	−4 °F to 22 °F (−20 °C to −30 °C)
1/2	25 mm	30 mm	50 mm	50 mm
3/4	25 mm	30 mm	50 mm	50 mm
1	25 mm	30 mm	50 mm	50 mm
1–1/2	30 mm	30 mm	50 mm	75 mm
2	30 mm	30 mm	50 mm	75 mm
2–1/2	30 mm	30 mm	50 mm	75 mm
3	30 mm	50 mm	50 mm	75 mm
4	30 mm	50 mm	50 mm	75 mm
5	50 mm	50 mm	50 mm	75 mm
6	50 mm	50 mm	50 mm	75 mm
Flat surface/Equipments	30 mm	50 mm	75 mm	100 mm

Manufacturer's data.
Based on average conditions. Should be modified to suit individual technical requirements.

Table A.22 Thickness recommendation to control condensation in pipe. Insulation of cooling lines–nitrile foam rubber

Design conditions	Pipe size	Pipeline temperature			
		14.4 °C (58 °F)	7 °C (44.6 °F)	2.5 °C (36.5 °F)	−18 °C (0 °F)
Mild condition 26.7 °C (80 °F) 50% RH	10 mm ID to 76 mm IPS	9 mm	9 mm	9 mm	19 mm
	Above 76 mm IPS to 127 mm IPS	9 mm	13 mm	13 mm	25 mm
	Above 127 mm IPS to 254 mm IPS	13 mm	13 mm	13 mm	25 mm
Normal condition 29.4 °C (85 °F) 70% RH	10 mm ID to 76 mm IPS	9 mm	13 mm	13 mm	25 mm
	Above 76 mm IPS to 127 mm IPS	13 mm	13 mm	13 mm	31 mm
	Above 127 mm IPS to 254 mm IPS	13 mm	13 mm	19 mm	31 mm
Severe condition 32.2 °C (90 °F) 80% RH	10 mm ID to 76 mm IPS	13 mm	19 mm	19 mm	38 mm
	Above 76 mm IPS to 127 mm IPS	13 mm	25 mm	25 mm	38 mm
	Above 127 mm IPS to 254 mm IPS	13 mm	25 mm	31 mm	50 mm
Extremely severe condition	10 mm ID to 76 mm IPS	13 mm	25 mm	25 mm	38 mm
32.2 °C (90 °F)	Above 76 mm IPS to 127 mm IPS	19 mm	31 mm	31 mm	50 mm
85% RH	Above 127 mm IPS to 254 mm IPS	25 mm	31 mm	38 mm	50 mm

Table A.23 Thickness recommendation for insulation of ducting, tanks and equipment of cooling systems–nitrile foam rubber

Design conditions	Metal surface temperature				
	15 °C (59 °F)	12 °C (53.6 °F)	7 °C (44.6 °F)	2.5 °C (36.5 °F)	−18 °C (0 °F)
26.7 °C (80 °F) 50% RH	9 mm	9 mm	13 mm	19 mm	25 mm
29.4 °C (85 °F) 70% RH	13 mm	13 mm	19 mm	25 mm	31 mm
32.2 °C (90 °F) 80% RH	13 mm	19 mm	25 mm	31 mm	50 mm
32.2 °C (90 °F) 85% RH	25 mm	25 mm	31 mm	38 mm	50 mm

Average physical properties	Rating	Average physical properties	Rating
Density (ASTM D 1667)	76.6 kg/m³	Thermal Stability (% shrinkage)	
		7 days-200 °F	4.5
		7 days-220 °F	5.5
Thermal conductivity (ASTM C 5 18) at 10 °C mean temperature	0.04 W/m °C (0.0346 K cal/m. °Ch)	Spread of flame (UL 94)	Self extinguishing
Temperature limits °C	−40 °C to +105 °C	Flexibility	Excellent
Water absorption [(% W/W) ASTM C 272]	4.37	Weather and ultraviolet resistance	Good
Water vapour permeability (ASTM E 96) Perm-in. Max g/Pa.s.m²	9.33×10^{-8}	Chemical resistance	Good
Ozone resistance	Excellent	Order	Negligible
		Mildew resistance	No fungal growth

Manufacturer's data.

Table A.24 Recommended indoor design goals for air-conditioning system: sound control

Type of area	Recommended RC or NC criteria range	Type of area	Recommended RC or NC criteria range
1. Private residences	25 to 30	5. Hospitals and clinics	
2. Apartments	25 to 30	a. Private rooms	25 to 30
3. Hotels/motels		b. Wards	30 to 40
a. Individual rooms or suites	30 to 35	c. Operating rooms	35 to 40
b. Meeting/banquet rooms	25 to 30	d. Corridors	35 to 40
c. Halls, corridors, lobbies	35 to 40	e. Public areas	35 to 40
d. Service/support areas	40 to 45	6. Churches	25 to 30
4. Offices		7. Schools	
a. Executive	25 to 30	a. Lecture and classrooms	25 to 30
b. Conference rooms	25 to 30	b. Open-plan classrooms	30 to 35
c. Private	30 to 35	8. Libraries	35 to 40
d. Open-plan areas	35 to 40		
e. Computer equipments rooms	40 to 45		
f. Public circulation	40 to 45		

Note: These are for unoccupied spaces, with all systems operating.

Table A.25 Quality of condenser and chiller water for air-conditioning

1. *Quality of circulating water:* "Normal" circulating water chemistry should fall within the following limits.		2. *Quality of make up water:* "Acceptable" make up water chemistry should fall within the following limits.	
1. pH	Between 6.4 to 8	1. pH	Between 6.5 to 8
2. Chlorides	< 750 ppm	2. Chlorides	< 50 ppm
3. Hardness as CaCO3	< 1200 ppm and < 800 ppm for arid areas	3. Hardness as CaCo3	< 50 ppm
4. Sulphates	< 5000 ppm	4. Sulphates	< 50 ppm
5. Sulphides	< 1 ppm	5. Sulphides	< 1 ppm
6. Silica	< 150 ppm	6. Silica	< 30 ppm
7. Iron	< 3 ppm	7. Iron	< 1 ppm
8. Manganese	< 0.1 ppm	8. Manganese	< 0.1 ppm
9. Langlier saturation index	Between −0.5 and +0.5	9. Suspended solids (Non abrasive)	< 50 ppm
10. Suspended solids (Non abrasive)	< 150 ppm	10. Oil and grease	NIL
11. Oil and grease	< 10 ppm	11. Organic solvents	NIL
12. Organic solvents	NIL	12. Organic nutrients	NIL
13. Organic nutrients	NIL	13. Free chlorine	< 1 ppm
14. Free chlorine	< 1 ppm		

Table A.26 Global warming potential (GWP) of some refrigerants

S. no.	Reference	GWP values	S. no.	Reference	GWP values
1.	R-12	10600	5	R-407	1980
2.	R-22	1900	6	R-407	2340
3.	R-134	1600	7	R-410	2340
4.	R-123	120			

Ozone depletion potential

S. no.	Reference	GWP values	S. no.	Reference	GWP values
1.	R-11	1.000	5	R-134	0.000
2.	R-12	0.820	6	R-407	0.000
3.	R-22	0.050	7	R-410	0.000
4.	R-123	0.012	8	R-410	0.000

Table A.27 Ready-reckoner for DOL start motors 415 V 3 Phase, 50 Hz

Motor rating		Full load current amps	Over load relay range amps.	Recommended backupHRC fuse amps	Recommended switchgear rating–amps	Recommended cable size sq mm	
HP	kW					Al.	Cu.
0.5	0.37	1	1.0−1.6	4	16	4	2.5
0.75	0.55	1.4	1−2	6	16	4	2.5
1	0.75	2	1.5−2.5	6	16	4	2.5
1.5	1.1	2.6	2.5−4	10	16	4	2.5
2	1.5	3.5	2.5−4	16	16	4	2.5
3	2.2	4.8	4−6.5	16	16	4	2.5
5	3.7	7.5	6−10	25	32	6	4
7.5	5.5	11	9−14	25	32	6	4
10	7.5	14	10−16	35	63	6	4
12.5	9.3	18	13−21	35	63	10	6
15	11	22	18−24	50	63	10	6
20	15	28	20−32	63	63	10	6
25	18.5	35	28−42	80	100	16	10
30	22	40	28−42	100	100	16	10

Table A.28 Ready-reckoner for start delta start motors 415 V 3 Phase, 50 Hz

Motor rating		Full load current		Over load relay range (amps)	Recommended backup HRC fuse (amps)	Recommended switchgear rating (amps)	Recommended cable size (sq mm)	
HP	kW	Line amps	Phase amps				Supply side	Motor side
10	7.5	15	9	6–10	25	32	4	2.5
12.5	9.3	18	10.4	9–14	35	63	6	4
15	11	22	12.7	10–16	35	63	6	4
20	15	29	16.8	13–21	63	63	10	6
25	18.5	35	20.2	18–24	50	63	16	6
30	22.5	40	23	20–32	63	125	16	6
35	26	47	27	20–32	63	125	25	10
40	30	53	30.6	20–32	63	125	25	10
45	33.5	60	35	28–42	100	250	35	16
50	37	66	38	28–42	100	250	35	16
60	45	80	46	45–70	125	250	50	25
75	55	100	57	45–70	160	250	70	35
90	67.5	120	69	60–100	200	250	95	50
100	75	135	78	60–100	200	250	95	50
125	90	165	95	90–150	250	300	150	70
150	110	200	115	90–150	250	400	185	95
175	130	230	133	120–200	320	400	240	120
200	150	275	159	120–200	350	400	240	120

Table A.29 Comparision of starters

Type of starter	Percent tap	Starting charcteristics (Percent of rated value)					Advantages	Limitations
		Voltage at motor	Motor current	Line current	Torque	Torque efficiency		
					Reduced – Voltage			
Auto-transformer (Closed–transition standard)	80	80	80	64	64	100	1. Starting characteristics easily adjusted.	1. Additional external voltage reducing component impose additional limits to duty cycle.
	65	65	65	42	42	100	2. Provides maximum torque per line ampere for reduced voltage starters (high torque efficiency).	
	50	50	50	25	25	100	3. Closed-transition starting.	
Star-delta (Open or closed transition)	–	100	33	33	33	100	1. Starting duty cycle usually limited by motor heating only.	1. Starting characteristics not adjustable.
							2. High torque efficiency for all speeds.	2. Requires motor with a normally delta-connected winding with all leads brought out for connection to control.
							3. No torque dips or unusual stresses because full winding energized.	
							4. Closed–circuit transition type eliminates line surges during transition from start to run.	

Table A.30 Comparison of starting methods

Method of starting	Inrush current (Percent full voltage locked rotor current)	Starting torque (Percent full voltage locked rotor torque)
Across-the-line	100	100
Auto-transformer		
80% tap	71	64
65% tap	48	42
50% tap	28	25
Primary resistor or reactor		
80% applied voltage	80	64
65% applied voltage	65	42
58% applied voltage	58	33
50% applied voltage	50	25
Star-Delta	33	33
Part-Winding	60	48
Part-Winding with Resistors	60–30	48–12
Wound Rotor (approximate)	25	150

Table A.31 Solar heat gain through ordinary glass btu/(hr) (sq ft)

10° North latitude				AM				Sun time			PM				10° South latitude	
Time of year	Exposure	6	7	8	9	10	11	Noon	1	2	3	4	5	6	Exposure	Time of year
	North	19	44	50	45	44	43	41	43	44	45	50	44	2	South	
	Northeast	55	131	153	140	106	65	28	14	14	13	11	8	2	Southeast	
	East	54	134	155	139	98	41	14	14	14	13	11	8	2	East	
June 21	Southeast	18	49	55	43	25	14	14	14	14	13	11	8	2	Northeast	Dec. 22
	South	2	8	11	13	14	14	14	14	14	13	11	8	2	North	
	Southwest	2	8	8	13	14	14	14	14	25	43	55	49	18	Northwest	
	West	2	8	8	13	14	14	14	41	98	139	155	134	54	West	
	Northwest	2	8	8	13	14	18	28	65	106	140	153	131	55	Southwest	
	Horizontal	4	44	107	166	205	233	243	233	205	166	107	44	4	Horizontal	
	North	5	34	39	35	33	31	30	31	33	35	39	34	5	South	
	Northeast	42	127	148	133	109	56	22	14	14	13	11	7	1	Southeast	
	East	50	135	158	142	98	43	14	14	14	13	11	7	1	East	
July 23 and May 21	Southeast	26	57	66	56	32	14	14	14	14	13	11	7	1	Northeast	Jan 21 and Nov. 21
	South	1	7	11	13	14	14	14	14	14	13	11	7	1	North	
	Southwest	1	7	11	13	14	14	14	14	32	56	66	57	26	Northwest	
	West	1	7	11	13	14	14	14	43	98	142	158	135	50	West	
	Northwest	1	7	11	13	14	14	22	56	109	133	148	127	42	Southwest	
	Horizontal	3	42	107	166	210	236	247	236	210	166	107	42	3	Horizontal	
	North	1	15	16	15	15	15	14	14	15	15	16	15	1	South	
	Northeast	17	113	130	111	80	34	14	14	14	13	11	7	1	Southeast	
	East	25	138	163	149	104	46	14	14	14	13	11	7	1	East	
Aug. 24 and April 20	Southeast	18	79	94	85	60	27	14	14	14	13	11	7	1	Northeast	Feb. 20 and Oct. 23
	South	1	7	11	13	14	14	14	14	14	13	11	7	1	North	
	Southwest	1	7	11	13	14	14	14	27	60	85	94	79	18	Northwest	
	West	1	7	11	13	14	14	14	46	80	149	163	138	25	West	
	Northwest	1	7	11	13	14	14	14	34	15	111	130	113	17	Southwest	
	Horizontal	2	38	105	167	213	242	250	242	213	167	105	38	2	Horizontal	
	North	1	6	11	13	14	14	14	14	14	13	11	6	1	South	
	Northeast	1	89	103	80	45	17	14	14	14	13	11	6	1	Southeast	
	East	1	130	164	151	106	47	14	14	14	13	11	6	1	East	
Sept. 22 and Mar. 22	Southeast	1	97	127	122	94	56	21	14	14	13	11	6	1	Northeast	Mar. 22 and Sept. 22
	South	1	6	13	19	24	27	28	27	24	19	13	6	1	North	
	Southwest	1	6	11	13	14	14	21	56	94	122	127	97	1	Northwest	
	West	1	6	11	13	14	14	14	47	106	151	164	130	1	West	
	Northwest	1	6	11	13	14	14	14	17	18	88	103	89	1	Southwest	
	Horizontal	1	31	97	160	207	235	247	235	207	160	97	31	1	Horizontal	
	North	0	5	10	13	14	14	14	14	14	13	10	5	0	South	
	Northeast	0	58	66	44	28	14	14	14	14	13	10	5	0	Southeast	
	East	0	118	155	145	100	40	14	14	14	13	10	5	0	East	
Oct. 23 and Feb. 20	Southeast	0	103	147	149	123	81	46	18	14	13	10	5	0	Northeast	April 20 and Aug. 24
	South	0	18	40	55	65	71	73	71	65	55	40	18	0	North	
	Southwest	0	5	10	13	14	18	46	81	123	149	147	103	0	Northwest	
	West	0	5	10	13	14	14	14	40	100	145	155	118	0	West	
	Northwest	0	5	10	13	14	14	14	14	28	44	66	58	0	Southwest	
	Horizontal	0	22	85	139	193	220	230	220	193	139	85	22	0	Horizontal	
	North	0	4	9	12	13	14	14	14	13	12	9	4	0	South	
	Northeast	0	27	37	17	13	14	14	14	13	12	9	4	0	Southeast	
	East	0	99	143	132	93	39	14	14	13	12	9	4	0	East	
Nov. 21 and Jan 21	Southeast	0	99	153	161	146	109	70	31	17	12	9	4	0	Northeast	May 21 and July 23
	South	0	35	65	91	96	104	106	104	96	91	65	35	0	North	
	Southwest	0	4	9	12	17	31	70	109	146	161	153	99	0	Northwest	
	West	0	4	9	12	13	14	14	39	93	132	143	99	0	West	
	Northwest	0	4	9	12	13	14	14	14	13	17	37	27	0	Southwest	
	Horizontal	0	17	62	131	175	202	210	202	175	131	62	17	0	Horizontal	
	North	0	4	9	12	13	14	14	14	13	12	9	4	0	South	
	Northeast	0	15	28	17	13	14	14	14	13	12	9	4	0	Southeast	
	East	0	86	137	130	91	42	14	14	13	12	9	4	0	East	
Dec. 22	Southeast	0	99	154	163	149	121	79	36	23	12	9	4	0	Northeast	June 21
	South	0	50	74	94	109	116	120	116	109	94	74	50	0	North	
	Southwest	0	4	9	12	23	36	79	121	149	163	154	99	0	Northwest	
	West	0	4	9	12	13	14	14	42	91	130	137	86	0	West	
	Northwest	0	4	9	12	13	14	14	14	13	17	28	15	0	Southwest	
	Horizontal	0	14	66	120	167	193	202	193	167	120	66	14	0	Horizontal	

Courtesy: Reprinted with permission of M/S Carrier Corporation from *Handbook of Air-conditioning System Design.*

Table A.32 Solar heat gain through ordinary glass btu/(hr) (sq ft)

20° North latitude Time of year	Exposure	6	7	8	9	10	11	Noon	1	2	3	4	5	6	Exposure	20° South latitude Time of year
		AM						Sun time	PM							
June 21	North	28	41	33	25	19	17	15	17	19	25	33	41	28	South	Dec. 22
	Northeast	81	154	144	122	83	38	15	14	14	14	12	9	3	Southeast	
	East	81	148	160	143	96	41	14	14	14	14	12	9	3	East	
	Southeast	28	62	73	66	44	21	14	14	14	14	12	9	3	Northeast	
	South	3	9	12	14	14	14	14	14	14	14	12	9	3	North	
	Southwest	3	9	12	14	14	14	14	21	44	66	73	62	28	Northwest	
	West	3	9	12	14	14	14	14	41	96	143	160	148	81	West	
	Northwest	3	9	12	14	14	14	15	18	83	122	144	154	81	Southwest	
	Horizontal	11	60	121	176	216	232	250	232	216	176	121	60	11	Horizontal	
July 23 and May 21	North	20	28	23	17	15	14	14	14	15	17	23	28	20	South	Jan. 21 and Nov. 21
	Northeast	71	132	138	111	73	31	14	14	14	13	12	8	3	Southeast	
	East	75	148	163	145	99	46	14	14	14	13	12	8	3	East	
	Southeast	31	70	85	79	57	29	14	14	14	13	12	8	3	Northeast	
	South	3	8	12	13	14	14	14	14	14	13	12	8	3	North	
	Southwest	3	8	12	13	14	14	14	29	57	79	85	70	31	Northwest	
	West	3	8	12	13	14	14	14	46	99	145	163	148	75	West	
	Northwest	3	8	12	13	14	14	14	31	73	111	138	132	71	Southwest	
	Horizontal	8	55	118	175	216	240	251	240	216	175	118	55	8	Horizontal	
Aug. 24 and April 20	North	6	10	11	13	14	14	14	14	14	13	11	10	6	South	Feb. 20 and Oct. 23
	Northeast	45	111	118	89	50	18	14	14	14	13	11	7	2	Southeast	
	East	53	142	165	149	106	51	14	14	14	13	11	7	2	East	
	Southeast	29	89	113	108	98	55	20	14	14	13	11	7	2	Northeast	
	South	2	7	11	14	20	24	26	24	20	14	11	7	2	North	
	Southwest	2	7	11	13	14	14	20	55	98	108	113	89	29	Northwest	
	West	2	7	11	13	14	14	14	51	106	149	165	142	53	West	
	Northwest	2	7	11	13	14	14	14	18	50	89	118	111	45	Southwest	
	Horizontal	5	48	107	167	210	235	247	235	210	167	107	48	5	Horizontal	
Sept. 22 and Mar. 22	North	0	6	11	13	14	14	14	14	14	13	11	6	0	South	Mar. 22 and Sept. 22
	Northeast	0	81	67	59	22	14	14	14	14	13	11	6	0	Southeast	
	East	0	130	163	147	104	45	14	14	14	13	11	6	0	East	
	Southeast	0	99	136	140	120	84	41	15	14	13	11	6	0	Northeast	
	South	0	8	22	38	52	63	65	63	52	38	22	8	0	North	
	Southwest	0	6	11	13	14	15	41	84	120	140	136	99	0	Northwest	
	West	0	6	11	13	14	14	14	45	104	149	163	130	0	West	
	Northwest	0	6	11	13	14	14	14	22	59	87	83	30	0	Southwest	
	Horizontal	0	30	93	153	198	225	233	225	198	153	93	30	0	Horizontal	
Oct. 23 and Feb. 20	North	0	4	9	12	13	14	14	14	13	12	9	4	0	South	April 20 and Aug. 24
	Northeast	0	44	52	29	13	14	14	14	13	12	9	4	0	Southeast	
	East	0	99	147	141	100	49	14	14	13	12	9	4	0	East	
	Southeast	0	91	146	160	149	119	74	27	13	12	9	4	0	Northeast	
	South	0	21	50	76	93	106	111	106	93	76	50	21	0	North	
	Southwest	0	4	9	12	13	27	74	119	149	160	146	91	0	Northwest	
	West	0	4	9	12	13	14	14	49	100	141	147	99	0	West	
	Northwest	0	4	9	12	13	14	14	14	13	29	52	44	0	Southwest	
	Horizontal	0	18	68	127	171	196	208	196	171	127	68	18	0	Horizontal	
Nov. 21 and Jan 21	North	0	3	8	11	13	13	13	13	13	11	8	3	0	South	May 21 and July 23
	Northeast	0	24	26	14	13	13	13	13	13	11	8	3	0	Southeast	
	East	0	71	128	127	91	43	13	13	13	11	8	3	0	East	
	Southeast	0	73	144	164	158	135	91	46	16	11	8	3	0	Northeast	
	South	0	29	69	100	123	136	141	136	123	100	69	28	0	North	
	Southwest	0	3	8	11	16	46	91	135	158	164	144	73	0	Northwest	
	West	0	3	8	11	13	13	13	43	91	127	128	71	0	West	
	Northwest	0	3	8	11	13	13	13	13	13	14	26	24	0	Southwest	
	Horizontal	0	5	48	101	146	172	180	172	146	101	48	5	0	Horizontal	
Dec. 22	North	0	2	7	11	12	13	13	13	12	11	7	2	0	South	June 21
	Northeast	0	14	18	12	12	13	13	12	12	11	7	2	0	Southeast	
	East	0	56	118	121	85	34	13	12	12	11	7	2	0	East	
	Southeast	0	59	139	167	159	134	97	60	20	11	7	2	0	Northeast	
	South	0	25	74	111	132	146	149	146	132	111	74	25	0	North	
	Southwest	0	2	7	11	20	60	97	134	159	167	139	59	0	Northwest	
	West	0	2	7	11	12	13	13	34	85	121	118	56	0	West	
	Northwest	0	2	7	11	12	13	13	13	12	12	18	14	0	Southwest	
	Horizontal	0	4	36	92	135	161	170	161	135	92	36	4	0	Horizontal	

Courtesy: Reprinted with permission of M/S Carrier Corporation from *Handbook of Air-conditioning System Design.*

Table A.33 Solar heat gain through ordinary glass btu/(hr) (sq ft)

40° North latitude		AM					Sun time			PM				40° South latitude	
Time of year	Exposure	6	7	8	9	10	11	Noon	1	2	3	4	5	6	Exposure	Time of year
June 21	North	32	20	12	13	14	14	14	14	14	13	12	20	32	South	Dec. 22
	Northeast	118	133	112	73	30	14	14	14	14	13	12	10	6	Southeast	
	East	126	161	162	142	95	44	14	14	14	13	12	10	6	East	
	Southeast	51	88	109	111	99	71	34	14	14	13	12	10	6	Northeast	
	South	6	10	12	19	33	44	54	44	35	19	12	10	6	North	
	Southwest	6	10	12	13	14	14	14	71	99	111	109	88	51	Northwest	
	West	6	10	12	13	14	14	14	44	95	142	162	161	126	West	
	Northwest	6	10	12	13	14	14	14	14	30	73	112	133	118	Southwest	
	Horizontal	31	82	134	179	210	232	237	232	210	179	134	82	31	Horizontal	
July 23 and May 21	North	24	14	12	13	14	14	14	14	14	13	12	14	24	South	Jan 21 and Nov. 21
	Northeast	106	127	105	66	26	14	14	14	14	13	12	10	5	Southeast	
	East	118	116	164	144	98	43	14	14	14	13	12	10	5	East	
	Southeast	54	95	119	125	110	82	42	15	14	13	12	10	5	Northeast	
	South	5	10	13	26	44	63	69	63	44	26	13	10	5	North	
	Southwest	5	10	12	13	14	15	42	82	110	125	119	95	54	Northwest	
	West	5	10	12	13	14	14	14	43	98	144	164	161	118	West	
	Northwest	5	10	12	13	14	14	14	14	26	66	105	127	106	Southwest	
	Horizontal	24	73	126	171	203	225	233	225	203	171	126	73	24	Horizontal	
Aug. 24 and April 20	North	7	8	11	13	14	14	14	14	14	13	11	8	7	South	Feb. 20 and Oct. 23
	Northeast	68	102	82	46	16	14	14	14	14	13	11	8	3	Southeast	
	East	84	147	162	145	101	45	14	14	14	13	11	8	3	East	
	Southeast	48	105	138	146	139	107	66	25	14	13	11	8	3	Northeast	
	South	3	8	24	51	89	97	102	97	89	51	24	8	3	North	
	Southwest	3	8	11	13	14	25	66	107	139	146	138	105	48	Northwest	
	West	3	8	11	13	14	14	14	45	101	145	162	147	84	West	
	Northwest	3	8	11	13	14	14	14	14	16	46	82	102	68	Southwest	
	Horizontal	9	47	100	150	185	205	214	205	185	150	100	47	9	Horizontal	
Sept. 22 and Mar. 22	North	0	5	9	12	13	13	14	13	13	12	9	5	0	South	Mar. 22 and Sept. 22
	Northeast	0	51	58	26	13	13	14	13	13	12	9	5	0	Southeast	
	East	0	116	149	139	99	45	14	13	13	12	9	5	0	East	
	Southeast	0	95	144	162	157	133	90	41	14	12	9	5	0	Northeast	
	South	0	12	44	81	110	122	140	122	110	81	44	12	0	North	
	Southwest	0	5	9	12	14	41	90	133	157	162	144	95	0	Northwest	
	West	0	5	9	12	13	13	14	45	99	139	149	116	0	West	
	Northwest	0	5	9	12	13	13	14	13	13	26	58	51	0	Southwest	
	Horizontal	0	21	67	124	153	176	183	176	153	124	67	21	0	Horizontal	
Oct. 23 and Feb. 20	North	0	2	6	10	11	12	12	12	11	10	6	2	0	South	April 20 and Aug. 24
	Northeast	0	35	33	12	11	12	12	12	11	10	6	2	0	Southeast	
	East	0	85	117	122	88	39	12	12	11	10	6	2	0	East	
	Southeast	0	81	132	161	163	144	107	63	20	10	6	2	0	Northeast	
	South	0	21	59	104	137	154	162	154	137	104	59	21	0	North	
	Southwest	0	2	6	10	20	63	107	144	163	161	132	81	0	Northwest	
	West	0	2	6	10	11	12	12	39	88	122	117	85	0	West	
	Northwest	0	2	6	10	11	12	12	12	11	12	33	35	0	Southwest	
	Horizontal	0	8	29	64	101	123	129	123	101	64	29	8	0	Horizontal	
Nov. 21 and Jan 21	North	0	0	3	7	9	10	11	10	9	7	3	0	0	South	May 21 and July 23
	Northeast	0	0	12	7	9	10	11	10	9	7	3	0	0	Southeast	
	East	0	0	91	100	74	33	11	10	9	7	3	0	0	East	
	Southeast	0	0	109	144	156	144	116	70	27	7	3	0	0	Northeast	
	South	0	0	59	104	139	158	166	158	139	104	59	0	0	North	
	Southwest	0	0	3	7	27	70	116	144	156	144	109	0	0	Northwest	
	West	0	0	3	7	9	10	11	33	74	100	91	0	0	West	
	Northwest	0	0	3	7	9	10	11	10	9	7	12	0	0	Southwest	
	Horizontal	0	0	16	43	73	92	103	92	73	43	16	0	0	Horizontal	
Dec. 22	North	0	0	2	6	9	10	10	10	9	6	2	0	0	South	June 21
	Northeast	0	0	7	6	9	10	10	10	9	6	2	0	0	Southeast	
	East	0	0	72	86	68	31	10	10	9	6	2	0	0	East	
	Southeast	0	0	88	134	148	142	115	73	30	7	2	0	0	Northeast	
	South	0	0	51	99	134	158	165	158	134	99	51	0	0	North	
	Southwest	0	0	2	7	30	73	115	142	148	134	88	0	0	Northwest	
	West	0	0	2	6	9	10	10	31	68	86	72	0	0	West	
	Northwest	0	0	2	6	9	10	10	10	9	6	7	0	0	Southwest	
	Horizontal	0	0	8	32	55	76	85	76	55	32	8	0	0	Horizontal	

Courtesy: Reprinted with permission of M/S Carrier Corporation from *Handbook of Air-conditioning System Design.*

Table A.34 Storage load factors, solar heat gain through glass with internal shade (16 hour operation, constant space temperature)

Exposure (North lat)	Weight (lb per sq ft of floor area)	AM							PM									Exposure (South lat)
		6	7	8	9	10	11	12	1	2	3	4	5	6	7	8	9	
Northeast	150 and over	.53	.64	.59	.47	.31	.25	.24	.22	.18	.17	.16	.14	.12	.09	.08	.07	Southeast
	100	.53	.65	.61	.50	.33	.27	.22	.21	.17	.16	.15	.13	.11	.08	.07	.06	
	30	.56	.77	.73	.58	.36	.24	.19	.17	.15	.13	.12	.11	.07	.04	.02	.02	
East	150 and over	.47	.63	.68	.64	.54	.38	.27	.25	.20	.18	.17	.15	.12	.10	.09	.08	East
	100	.46	.63	.70	.67	.56	.38	.27	.24	.20	.18	.16	.14	.12	.09	.08	.07	
	30	.47	.71	.80	.79	.64	.42	.25	.19	.16	.14	.11	.09	.07	.04	.02	.02	
Southeast	150 and over	.14	.37	.55	.66	.70	.68	.58	.46	.27	.24	.21	.19	.16	.14	.12	.11	Northeast
	100	.11	.35	.53	.66	.72	.69	.61	.47	.29	.24	.21	.18	.15	.12	.10	.09	
	30	.02	.31	.57	.75	.84	.81	.69	.50	.30	.20	.17	.13	.09	.05	.04	.03	
South	150 and over	.19	.18	.34	.48	.60	.68	.73	.74	.64	.59	.42	.24	.22	.19	.17	.15	North
	100	.16	.14	.31	.46	.59	.69	.76	.70	.69	.59	.45	.26	.22	.18	.16	.13	
	30	.12	.23	.44	.64	.77	.86	.88	.82	.56	.50	.24	.16	.11	.08	.05	.04	
Southwest	150 and over	.22	.21	.20	.20	.20	.32	.47	.60	.63	.66	.61	.47	.23	.19	.18	.16	Northwest
	100	.20	.19	.18	.17	.18	.31	.46	.60	.66	.70	.64	.50	.26	.20	.17	.15	
	30	.08	.08	.09	.09	.10	.24	.47	.67	.81	.86	.79	.60	.26	.17	.12	.08	
West	150 and over	.23	.23	.21	.21	.20	.19	.18	.25	.36	.52	.63	.65	.55	.22	.19	.17	West
	100	.22	.21	.19	.19	.17	.16	.15	.23	.36	.54	.66	.68	.60	.25	.20	.17	
	30	.12	.10	.10	.10	.10	.10	.09	.19	.42	.65	.81	.85	.74	.30	.19	.13	
Northwest	130 and over	.21	.21	.20	.19	.18	.18	.17	.16	.16	.33	.49	.61	.60	.19	.17	.15	Southwest
	100	.19	.19	.18	.17	.17	.16	.16	.15	.16	.34	.52	.65	.23	.18	.15	.12	
	30	.12	.11	.11	.11	.11	.11	.11	.10	.17	.39	.63	.80	.79	.28	.18	.12	
North and shade	130 and over	.23	.58	.75	.79	.80	.80	.81	.82	.83	.84	.86	.87	.88	.39	.35	.31	South and shade
	100	.25	.46	.73	.78	.82	.82	.83	.84	.85	.87	.88	.89	.90	.40	.34	.29	
	30	.07	.22	.69	.80	.86	.93	.94	.95	.97	.98	.98	.99	.99	.35	.23	.16	

Courtesy: Reprinted with permission of M/S Carrier Corporation from *Handbook of Air-conditioning System Design.*

Table A.35 Storage load factors, solar heat gain through glass (12 hour operation, constant space temperature)

Exposure (North lat)	Weight (lb per sq ft of floor area)	AM							PM												AM					Exposure (South lat)
		6	7	8	9	10	11	12	1	2	3	4	5	6	7	8	9	10	11	12	1	2	3	4	5	
Northeast	150 and over	.59	.67	.62	.49	.33	.27	.25	.24	.22	.21	.20	.17	.34	.42	.47	.45	.42	.39	.36	.33	.30	.29	.26	.25	Southeast
	100	.59	.68	.64	.52	.35	.29	.24	.23	.20	.19	.17	.15	.35	.45	.50	.49	.45	.42	.34	.30	.27	.26	.23	.20	
	30	.62	.80	.75	.60	.37	.25	.19	.17	.15	.13	.12	.11	.40	.62	.69	.64	.48	.34	.27	.22	.18	.16	.14	.12	
East	150 and over	.51	.66	.71	.67	.57	.40	.29	.26	.25	.23	.21	.19	.36	.44	.50	.53	.53	.50	.44	.39	.36	.34	.30	.28	East
	100	.52	.67	.73	.70	.58	.40	.29	.26	.24	.21	.19	.16	.34	.44	.54	.58	.57	.51	.44	.39	.34	.31	.28	.24	
	30	.53	.74	.82	.81	.65	.43	.25	.19	.16	.14	.11	.09	.36	.56	.71	.76	.70	.54	.39	.28	.23	.18	.15	.12	
Southeast	150 and over	.20	.42	.59	.70	.74	.71	.61	.48	.33	.30	.26	.24	.34	.37	.43	.50	.54	.58	.57	.55	.50	.45	.41	.37	Northeast
	100	.18	.40	.57	.70	.75	.72	.63	.42	.34	.28	.25	.21	.29	.33	.41	.51	.58	.61	.61	.56	.49	.44	.37	.33	
	30	.09	.35	.61	.78	.86	.82	.69	.50	.30	.20	.17	.13	.14	.27	.47	.64	.75	.79	.73	.61	.45	.32	.23	.18	
South	150 and over	.28	.25	.40	.53	.64	.72	.77	.77	.73	.67	.49	.31	.47	.43	.42	.46	.51	.56	.61	.65	.66	.65	.61	.54	North
	100	.26	.22	.38	.51	.64	.73	.79	.79	.77	.65	.51	.31	.44	.37	.39	.43	.50	.57	.64	.68	.70	.68	.63	.53	
	30	.21	.29	.48	.67	.79	.88	.89	.83	.56	.50	.24	.16	.28	.19	.25	.38	.54	.68	.78	.84	.82	.76	.61	.42	
Southwest	150 and over	.31	.27	.27	.26	.25	.27	.50	.63	.72	.74	.69	.54	.51	.44	.40	.37	.34	.36	.41	.47	.54	.57	.60	.58	Northwest
	100	.33	.28	.25	.23	.23	.35	.50	.64	.74	.77	.70	.55	.53	.44	.37	.35	.31	.33	.39	.46	.55	.62	.64	.60	
	30	.29	.21	.18	.15	.14	.27	.50	.69	.82	.87	.79	.60	.48	.32	.25	.20	.17	.19	.39	.56	.70	.80	.79	.60	
West	150 and over	.63	.31	.28	.27	.25	.24	.22	.29	.46	.61	.71	.72	.56	.49	.44	.39	.36	.33	.31	.31	.35	.42	.49	.54	West
	100	.67	.33	.28	.26	.24	.22	.20	.28	.44	.61	.72	.73	.60	.52	.44	.39	.34	.31	.29	.28	.33	.43	.51	.57	
	30	.77	.34	.25	.20	.17	.14	.13	.22	.44	.67	.82	.85	.77	.56	.38	.28	.22	.18	.16	.19	.33	.52	.69	.77	
Northwest	130 and over	.68	.28	.27	.25	.23	.22	.20	.19	.24	.41	.56	.67	.49	.44	.39	.36	.33	.30	.28	.26	.26	.30	.37	.44	Southwest
	100	.71	.31	.27	.24	.22	.21	.19	.18	.23	.40	.58	.70	.54	.49	.41	.35	.31	.28	.25	.23	.24	.30	.39	.48	
	30	.82	.33	.25	.20	.18	.15	.14	.13	.19	.41	.64	.80	.75	.53	.36	.28	.24	.19	.17	.15	.17	.30	.50	.66	
North and shade	130 and over	.96	.96	.96	.96	.96	.96	.96	.96	.96	.96	.96	.96	.75	.75	.79	.83	.84	.86	.88	.88	.91	.92	.93	.93	South and shade
	100	.98	.98	.98	.98	.98	.98	.98	.98	.98	.98	.98	.98	.81	.84	.86	.89	.91	.93	.93	.94	.94	.95	.95	.95	
	30	1	1	1	1	1	1	1	1	1	1	1	1	1	1	1	1	1	1	1	1	1	1	1	1	

Courtesy: Reprinted with permission of M/S Carrier Corporation from *Handbook of Air-conditioning System Design.*

Table A.36 Performance charcteristics of typical airfoil blade centrifugal fans (27 in. wheel diameter)

CFM	OV	1/4" SP		3/8" SP		1/2" SP		5/8" SP		3/4" SP		1" SP		1¼" SP		1½" SP		1¾" SP		2" SP	
		RPM	BHP	RPM	BHP	RPM	BHP	RPM	BHP	RPM	BHP	RPM	BHP	RPM	BHP	RPM	BHP	RPM	BHP	RPM	BHP
2085	500	325	.10	376	.15	–	–	–	–	–	–	–	–	–	–	–	–	–	–	–	–
2502	600	351	.13	395	.18	438	.24	481	.30	–	–	–	–	–	–	–	–	–	–	–	–
2919	700	382	.16	421	.22	458	.28	495	.35	532	.42	–	–	–	–	–	–	–	–	–	–
3336	800	414	.19	450	.26	484	.33	516	.40	548	.47	613	.63	–	–	–	–	–	–	–	–
3753	900	447	.23	481	.31	513	.38	542	.46	572	.54	629	.70	686	.89	744	1.09	–	–	–	–
4170	1000	482	.28	514	.36	543	.45	571	.53	598	.62	650	.79	702	.98	753	1.18	805	1.40	–	–
4587	1100	518	.34	547	.42	575	.52	602	.61	627	.70	676	.89	723	1.08	770	1.29	816	1.51	863	1.74
5004	1200	555	.40	582	.50	609	.59	634	.69	658	.79	703	1.00	748	1.20	791	1.42	834	1.64	876	1.87
5421	1300	592	.47	618	.57	642	.68	667	.79	689	.89	733	1.11	775	1.33	815	1.56	855	1.79	895	2.03
5838	1400	629	.56	654	.66	677	.78	700	.89	722	1.00	765	1.23	804	1.47	842	1.71	880	1.96	917	2.21
6255	1500	668	.65	691	.76	714	.88	734	1.00	755	1.12	796	1.37	834	1.62	870	1.88	906	2.13	941	2.40
6672	1600	707	.76	728	.87	749	1.00	770	1.13	789	1.25	8282	1.51	866	1.78	900	2.05	934	2.32	967	2.60
7089	1700	745	.87	765	1.00	786	1.13	806	1.26	824	1.40	861	1.67	897	1.95	932	2.24	964	2.53	996	2.81
7506	1800	785	.99	804	1.14	823	1.27	842	1.41	860	1.55	895	1.84	930	2.13	963	2.43	995	2.73	1025	3.05
7923	1900	823	1.13	842	1.29	860	1.43	878	1.57	896	1.72	930	2.02	963	2.34	995	2.64	1027	2.96	1056	3.28
8340	2000	863	1.28	881	1.45	898	1.61	915	1.75	933	1.91	965	2.23	997	2.55	1028	2.87	1058	3.19	1088	3.53
8757	2100	903	1.45	920	1.62	936	1.80	952	1.94	969	2.11	1000	2.44	1031	2.77	1061	3.12	1090	3.45	1119	3.79
9174	2200	942	1.65	960	1.81	975	2.00	990	2.16	1006	2.31	1036	2.66	1066	3.01	1095	3.37	1123	3.72	1151	4.08
9591	2300	982	1.86	998	2.01	1014	2.21	1028	2.40	1043	2.55	1073	2.91	1102	3.28	1130	3.63	1157	4.02	1184	4.38
10008	2400	1023	2.09	1037	2.23	1053	2.43	1068	2.64	1080	2.81	1110	3.17	1137	3.54	1165	3.93	1191	4.32	1217	4.71
10425	2500	1062	2.33	1077	2.47	1092	2.68	1106	2.89	1119	3.09	1147	3.45	1173	3.84	1200	4.25	1226	4.63	1251	5.05
10842	2600	1102	2.59	1117	2.74	1131	2.94	1145	3.16	1158	3.38	1184	3.74	1210	4.16	1235	4.56	1261	4.98	1285	5.40
11259	2700	1142	2.85	1156	3.04	1170	3.22	1184	3.45	1196	3.68	1222	4.07	1247	4.49	1271	4.90	1296	5.35	1320	5.76
11676	2800	1183	3.13	1196	3.36	1209	3.51	1223	3.76	1236	4.00	1259	4.40	1284	4.83	1308	5.28	1331	5.72	1355	6.16
12093	2900	1223	3.44	1237	3.69	1249	3.84	1262	4.09	1274	4.33	1297	4.75	1321	5.20	1344	5.67	1367	6.11	1390	6.59
12510	3000	1264	3.78	1277	4.04	1289	4.21	1301	4.43	1313	4.69	1335	5.14	1359	5.60	1381	6.07	1403	6.54	1425	7.01
12927	3100	1304	4.14	1316	4.42	1328	4.60	1340	4.79	1353	5.07	1373	5.55	1397	6.02	1419	6.49	1440	6.99	1461	7.46
13344	3200	1345	4.52	1356	4.81	1368	5.02	1379	5.19	1392	5.47	1412	5.98	1434	6.44	1456	6.95	1477	7.45	1498	7.95
13761	3300	1385	4.92	1396	5.2	1408	5.46	1419	5.63	1430	5.88	1450	6.45	1472	6.92	1494	7.43	1514	7.94	1535	8.47
14178	3400	1426	5.35	1437	5.62	1449	5.92	1459	6.10	1469	6.32	1490	6.91	1510	7.41	1531	7.94	1552	8.45	1571	8.99
14595	3500	1466	5.80	1477	6.07	1489	6.40	1499	6.61	1506	6.79	1529	7.40	1549	7.94	1569	8.44	1590	8.99	1608	9.54

(Contd.)

CFM	OV	1/4" SP		3/8" SP		1/2" SP		5/8" SP		3/4" SP		1" SP		1¼" SP		1½" SP		1¾" SP		2" SP	
		RPM	BHP	RPM	BHP	RPM	BHP	RPM	BHP	RPM	BHP	RPM	BHP	RPM	BHP	RPM	BHP	RPM	BHP	RPM	BHP
								(30 in. wheel diameter)														
2575	500	292	.13	339	.19	–	–	–	–	–	–	–	–	–	–	–	–	–	–	–	–
3090	600	316	.16	356	.22	394	.29	433	.37	–	–	–	–	–	–	–	–	–	–	–	–
3605	700	344	.19	379	.27	412	.34	445	.43	478	.51	–	–	–	–	–	–	–	–	–	–
4120	800	373	.24	405	.32	435	.41	465	.49	494	.58	551	.78	–	–	–	–	–	–	–	–
4635	900	403	.29	433	.38	461	.47	488	.57	514	.67	566	.87	617	1.09	669	1.35	–	–	–	–
5150	1000	434	.35	463	.45	489	.55	514	.66	538	.76	585	.98	631	1.20	678	1.45	725	1.73	–	–
5665	1100	466	.42	792	.52	518	.64	542	.75	565	.86	608	1.10	651	1.34	693	1.59	735	1.86	777	2.15
6180	1200	499	.49	524	.61	548	.73	570	.85	592	.97	633	1.23	673	1.49	712	1.75	750	2.02	789	2.31
6695	1300	533	.58	556	.71	578	.84	600	.97	620	1.10	660	1.37	697	1.64	734	1.92	770	2.21	805	2.51
7210	1400	566	.69	589	.82	609	.96	630	1.09	650	1.24	688	1.52	723	1.82	758	2.11	792	2.42	825	2.72
7725	1500	601	.81	622	.94	942	1.09	661	1.23	680	1.39	716	1.69	751	2.00	783	2.32	815	2.63	847	2.96
8240	1600	636	.93	655	1.08	974	1.23	693	1.39	710	1.54	745	1.87	779	2.20	810	2.53	841	6.87	871	3.21
8755	1700	671	1.07	689	1.24	707	1.39	725	1.56	742	1.73	775	2.07	807	2.40	839	2.76	868	3.12	896	3.47
9270	1800	707	1.23	723	1.41	741	1.56	758	1.74	774	1.92	806	2.28	837	2.63	867	3.00	895	3.37	923	3.76
9785	1900	741	1.39	758	1.59	774	1.76	791	1.94	807	2.13	837	2.50	867	2.88	896	3.26	924	3.65	951	4.05
10300	2000	777	1.58	793	1.79	808	1.99	824	2.16	839	2.36	869	2.75	897	3.15	925	3.54	952	3.94	979	4.36
10815	2100	812	1.79	828	2.00	843	2.22	857	2.40	872	2.60	900	3.01	928	3.42	955	3.85	981	4.26	1007	4.69
11330	2200	848	2.03	864	2.24	878	2.47	891	2.67	906	2.86	933	3.29	960	3.72	986	4.16	1011	4.60	1036	5.04
11845	2300	884	2.30	899	2.48	913	2.73	926	2.96	939	3.15	966	3.59	992	4.05	1017	4.49	1041	4.96	1065	5.41
12360	2400	921	2.58	934	2.75	948	3.01	961	3.26	973	3.48	999	3.92	1023	4.38	1049	4.85	1072	5.33	1095	5.82
12875	2500	956	2.88	969	3.05	983	3.31	996	3.57	1007	3.82	1032	4.26	1056	4.74	1080	5.24	1103	5.72	1126	6.24
13390	2600	992	3.20	1005	3.38	1018	3.64	1030	3.91	1043	4.18	1066	4.62	1089	5.14	1112	5.63	1135	6.15	1157	6.66
13905	2700	1028	3.53	1041	3.75	1053	3.98	1066	4.26	1077	4.54	1100	5.02	1122	5.54	1144	6.06	1167	6.61	1188	7.12
14420	2800	1065	3.87	1077	4.15	1088	4.34	1101	4.65	1112	4.94	1134	5.44	1156	5.97	1177	6.52	1198	7.06	1220	7.61
14935	2900	1101	4.25	1113	4.56	1124	4.75	1136	5.05	1147	5.35	1168	5.87	1190	6.43	1210	7.00	1231	7.55	1251	8.14
15450	3000	1138	4.67	1149	5.00	1160	5.20	1171	5.47	1182	5.79	1202	6.35	1224	6.92	1243	7.50	1263	8.08	1283	8.66
15965	3100	1174	5.12	1185	5.46	1196	5.69	1206	5.92	1218	6.26	1236	6.86	1257	7.44	1277	8.02	1297	8.64	1315	9.2
16480	3200	1211	5.58	1221	5.95	1232	6.20	1242	6.41	1253	6.75	1271	7.39	1291	7.96	1311	8.58	1330	9.21	1348	9.83
16995	3300	1247	6.08	1257	6.44	1268	6.74	1278	6.95	1288	7.26	1306	7.96	1325	8.55	1345	9.18	1363	9.80	1382	10.46
17510	3400	1284	6.60	1294	6.95	1304	7.31	1313	7.54	1323	7.80	1341	8.53	1360	9.16	1379	9.81	1397	10.43	1415	11.11
18025	3500	1320	7.16	1330	7.50	1340	7.91	1349	8.16	1358	8.39	1377	9.14	1394	9.81	1413	10.42	1431	11.11	1448	11.78

(Contd.)

CFM	OV	1/4" SP		3/8" SP		1/2" SP		5/8" SP		3/4" SP		1" SP		1¼" SP		1½" SP		1¾" SP		2" SP	
		RPM	BHP	RPM	BHP	RPM	BHP	RPM	BHP	RPM	BHP	RPM	BHP	RPM	BHP	RPM	BHP	RPM	BHP	RPM	BHP

(33 in. wheel diameter)

CFM	OV	RPM	BHP	RPM	BHP	RPM	BHP	RPM	BHP	RPM	BHP	RPM	BHP	RPM	BHP	RPM	BHP	RPM	BHP	RPM	BHP
3130	500	261	.15	302	.22	–	–	–	–	–	–	–	–	–	–	–	–	–	–	–	–
3756	600	282	.19	317	.26	351	.35	385	.44	–	–	–	–	–	–	–	–	–	–	–	–
4382	700	306	.23	337	.32	367	.41	397	.50	426	.61	–	–	–	–	–	–	–	–	–	–
5008	800	332	.28	360	.38	388	.48	414	.58	440	.69	491	.93	–	–	–	–	–	–	–	–
5634	900	359	.35	386	.45	410	.56	435	.68	458	.79	504	1.03	550	1.30	595	1.60	–	–	–	–
6260	1000	388	.42	412	.54	436	.65	457	.77	479	.90	521	1.16	563	1.43	604	1.73	645	2.06	–	–
6886	1100	416	.50	440	.63	461	.76	482	.89	502	1.02	541	1.30	580	1.59	617	1.89	655	2.21	692	2.57
7512	1200	446	.59	468	.74	488	.87	508	1.02	527	1.16	564	1.46	599	1.76	634	2.08	669	2.40	703	5.75
8138	1300	476	.70	496	.86	516	1.01	535	1.16	553	1.31	587	1.62	621	1.95	653	2.28	686	2.63	718	2.98
8764	1400	506	.82	526	.99	544	1.15	562	1.31	579	1.48	612	1.81	644	2.15	675	2.51	705	2.87	735	3.23
9390	1500	536	.95	555	1.13	573	1.32	590	1.49	607	1.66	638	2.01	668	2.37	698	2.74	726	3.13	754	3.51
10016	1600	567	1.10	585	1.30	602	1.49	618	1.68	634	1.86	665	2.24	693	2.61	721	2.99	749	3.40	776	3.81
10642	1700	598	1.27	615	1.47	631	1.68	647	1.89	663	2.08	692	2.48	720	2.87	746	3.28	772	3.68	798	4.12
11268	1800	630	1.45	645	1.67	661	1.89	676	2.11	691	2.32	719	2.73	746	3.16	772	3.57	797	4.00	822	4.44
11894	1900	661	1.66	676	1.88	691	2.11	706	2.34	720	2.58	747	3.01	773	3.46	799	3.90	822	4.35	846	4.80
12520	2000	693	1.90	707	2.12	721	2.36	735	2.61	749	2.84	776	3.33	801	3.76	825	4.25	849	4.71	871	5.19
13146	2100	724	2.13	738	2.38	752	2.63	766	2.88	779	3.13	804	3.65	829	4.12	852	4.60	875	5.11	897	5.59
13772	2200	756	2.39	770	2.64	783	2.91	796	3.17	808	3.46	833	3.99	857	4.50	880	4.98	902	5.52	924	6.04
14398	2300	787	2.69	801	2.95	814	3.23	826	3.51	838	3.79	862	4.34	886	4.90	908	5.41	929	5.93	950	6.50
15024	2400	819	3.02	833	3.30	845	3.58	857	3.85	868	4.13	892	4.73	914	5.32	936	5.87	957	6.40	977	6.97
15650	2500	851	3.36	864	3.66	876	3.92	888	4.21	899	4.53	922	5.15	943	5.74	965	6.35	985	6.90	1005	7.45
16276	2600	884	3.73	896	4.01	908	4.29	919	4.64	930	4.94	951	5.59	973	6.20	993	6.85	1013	7.44	1033	8.01
16902	2700	917	4.13	928	4.40	939	4.74	950	5.07	960	5.36	980	6.01	1002	6.69	1022	7.35	1042	8.00	1061	8.60
17528	2800	950	4.57	959	4.85	971	5.22	981	5.50	992	5.84	1012	6.51	1032	7.22	1051	7.88	1071	8.58	1089	9.22
18154	2900	981	5.03	991	5.34	1002	5.69	1013	5.97	1023	6.37	1044	7.04	1061	7.77	1081	8.45	1099	9.16	1118	9.86
18780	3000	1013	5.51	1023	5.86	1034	6.16	1044	6.51	1054	6.89	1074	7.58	1090	8.30	1111	9.05	1129	9.77	1146	10.52
19406	3100	1046	6.02	1055	6.37	1066	6.68	1076	7.10	1085	7.40	1104	8.14	1122	8.89	1140	9.72	1158	10.43	1175	11.19
20032	3200	1078	6.55	1088	6.95	1097	7.26	1107	7.70	1117	7.98	1136	8.78	1154	9.55	1169	10.36	1188	11.11	1204	11.89
20658	3300	1111	7.13	1122	7.54	1129	7.89	1139	8.28	1148	8.65	1169	9.50	1184	10.23	1198	11.00	1217	11.86	1234	12.63
21284	3400	1144	7.74	1154	8.19	1161	8.57	1171	8.89	1180	9.37	1201	10.24	1214	10.90	1230	11.72	1247	12.64	1263	13.40
21910	3500	1177	8.42	1186	8.87	1193	9.27	1203	9.57	1211	10.08	1231	10.97	1245	11.63	1262	12.51	1275	13.37	1293	14.23

(Contd.)

(36½ in. wheel diameter)

CFM	OV	1/4" SP		3/8" SP		1/2" SP		5/8" SP		3/4" SP		1" SP		1¼" SP		1½" SP		1¾" SP		2" SP	
		RPM	BHP	RPM	BHP	RPM	BHP	RPM	BHP	RPM	BHP	RPM	BHP	RPM	BHP	RPM	BHP	RPM	BHP	RPM	BHP
3830	500	235	.18	273	.27	–	–	–	–	–	–	–	–	–	–	–	–	–	–	–	–
4596	600	255	.23	287	.32	317	.43	348	.54	–	–	–	–	–	–	–	–	–	–	–	–
5362	700	277	.28	305	.39	332	.50	359	.62	385	.75	–	–	–	–	–	–	–	–	–	–
6128	800	300	.35	326	.46	350	.59	374	.71	398	.84	444	1.13	–	–	–	–	–	–	–	–
6894	900	325	.42	349	.55	371	.68	393	.83	414	.96	456	1.26	491	1.59	538	1.96	–	–	–	–
7660	1000	350	.51	373	.65	394	.80	414	.94	433	1.10	471	1.42	509	1.75	546	2.12	583	2.52	–	–
8426	1100	377	.61	397	.77	417	.96	436	1.09	454	1.25	489	1.59	524	1.94	558	2.31	592	2.71	625	3.14
9192	1200	403	.73	423	.90	442	1.07	460	1.24	477	1.42	510	1.78	541	2.16	573	2.54	605	2.94	636	3.37
9958	1300	430	.86	449	1.05	467	1.23	484	1.42	500	1.60	531	1.98	562	2.39	590	2.79	620	3.21	649	3.64
10724	1400	457	1.00	475	1.21	492	1.41	508	1.60	524	1.81	553	2.21	582	2.63	610	3.07	637	3.51	665	3.96
11490	1500	485	1.17	502	1.39	518	1.61	534	1.82	548	2.03	577	2.46	604	2.90	631	3.36	657	3.82	682	4.30
12256	1600	513	1.34	529	1.59	544	1.82	559	2.05	574	2.28	601	2.74	627	3.19	652	3.66	677	4.16	701	4.66
13022	1700	541	1.56	556	1.80	571	2.05	585	2.31	599	2.55	625	3.03	651	3.51	674	4.01	698	4.51	722	5.04
13788	1800	569	1.77	584	2.04	598	2.31	612	2.58	625	2.84	651	3.33	675	3.87	698	4.37	720	4.90	743	5.43
14554	1900	598	2.03	611	2.30	625	2.59	638	2.87	651	3.16	676	3.69	699	4.23	722	4.78	743	5.32	765	5.87
15320	2000	626	2.32	640	2.59	652	2.88	665	3.19	678	3.48	701	4.07	724	4.61	746	5.21	767	5.76	787	6.34
16086	2100	655	2.61	668	2.92	680	3.22	692	3.53	704	3.83	727	4.47	749	5.04	771	5.63	791	6.26	811	6.84
16852	2200	684	2.92	696	3.23	708	3.56	719	3.88	731	4.23	753	4.88	775	5.51	796	6.10	815	6.76	835	7.2
17618	2300	712	3.29	725	3.61	736	3.95	747	4.29	758	4.64	780	5.31	801	6.00	821	6.62	840	7.25	859	7.96
18384	2400	741	3.69	753	4.04	764	4.38	775	4.71	785	5.06	807	5.78	827	6.51	846	7.18	865	7.83	884	8.53
19150	2500	770	4.12	782	4.47	792	4.80	803	5.15	813	5.55	834	6.30	853	7.03	872	7.77	890	8.44	909	9.11
19916	2600	799	4.57	810	4.91	821	5.26	831	5.67	841	6.05	860	6.84	880	7.59	898	8.38	916	9.10	934	9.80
20682	2700	830	5.06	839	5.39	849	5.80	859	6.21	868	6.56	886	7.36	906	8.18	924	8.99	942	9.79	959	10.52
21448	2800	859	5.59	867	5.94	878	6.38	887	6.73	897	7.15	915	7.96	933	8.84	951	9.64	968	10.50	985	11.28
22214	2900	887	6.16	896	6.53	906	6.96	916	7.30	925	7.79	944	8.62	959	9.51	977	10.34	994	11.21	1011	12.0
22980	3000	916	6.75	925	7.17	935	7.54	944	7.97	953	8.43	971	9.28	986	10.15	1004	11.07	1020	11.96	1037	12.88
23746	3100	946	7.37	954	7.80	964	8.17	973	8.69	981	9.06	998	9.96	1014	10.89	1031	11.89	1047	12.76	1063	13.70
24512	3200	975	8.02	984	8.50	992	8.89	1001	9.42	1010	9.77	1027	10.74	1043	11.69	1057	12.68	1074	13.60	1089	14.55
25278	3300	1005	8.73	1014	9.23	1021	9.66	1030	10.14	1038	10.59	1057	11.63	1071	12.52	1084	13.46	1101	14.51	1116	15.46
26044	3400	1034	9.48	1043	10.02	1050	10.49	1059	10.88	1067	11.46	1086	12.53	1098	13.35	1112	14.35	1127	15.47	1142	16.40
26810	3500	1064	10.30	1072	10.86	1078	11.34	1087	11.72	1095	12.34	1114	13.43	1126	14.24	1141	15.32	1153	16.36	1169	17.42

(Contd.)

CFM	OV	1/4" SP		3/8" SP		1/2" SP		5/8" SP		3/4" SP		1" SP		1¼" SP		1½" SP		1¾" SP		2" SP	
		RPM	BHP	RPM	BHP	RPM	BHP	RPM	BHP	RPM	BHP	RPM	BHP	RPM	BHP	RPM	BHP	RPM	BHP	RPM	BHP
								(40¼ in. wheel diameter)													
4655	500	213	.22	248	.33	–	–	–	–	–	–	–	–	–	–	–	–	–	–	–	–
5586	600	230	.28	259	.39	288	.52	318	.66	–	–	–	–	–	–	–	–	–	–	–	–
6517	700	248	.34	276	.47	301	.61	325	.75	350	.91	–	–	–	–	–	–	–	–	–	–
7448	800	268	.42	293	.56	317	.72	339	.87	360	1.03	404	1.38	–	–	–	–	–	–	–	–
8379	900	289	.50	313	.67	335	.84	355	1.01	375	1.17	413	1.54	452	1.94	492	2.37	–	–	–	–
9310	1000	312	.61	333	.78	354	.97	373	1.16	392	1.35	427	1.72	461	2.13	496	2.58	532	3.05	–	–
10241	1100	335	.73	354	.92	373	1.12	392	1.32	409	1.53	443	1.94	475	2.36	505	2.81	537	3.30	570	3.81
11172	1200	358	.87	376	1.07	394	1.28	412	1.50	428	1.72	460	2.18	490	2.62	519	3.09	547	3.58	577	4.11
12103	1300	382	1.03	399	1.25	416	1.47	432	1.70	448	1.94	479	2.43	507	2.92	935	3.40	561	3.91	587	4.43
13034	1400	406	1.20	422	1.45	438	1.68	453	1.92	468	2.17	497	2.70	525	3.22	552	3.75	577	4.26	602	4.81
13965	1500	431	1.40	446	1.67	460	1.91	475	2.17	489	2.43	517	2.99	544	3.54	569	4.10	594	4.67	617	5.22
14896	1600	455	1.63	470	1.91	484	2.18	497	2.44	511	2.72	537	3.29	563	3.89	588	4.48	611	5.08	634	5.69
15827	1700	480	1.87	494	2.17	508	2.47	520	2.74	533	3.04	558	3.62	583	4.26	606	4.89	630	5.52	652	6.16
16758	1800	504	2.13	519	2.46	532	2.78	544	3.08	555	3.37	580	3.99	603	4.63	626	5.32	648	5.99	670	6.65
17689	1900	529	2.43	543	2.78	555	3.11	568	3.44	579	3.75	602	4.40	624	5.05	646	5.76	667	6.48	688	7.18
18620	2000	554	2.76	567	3.13	580	3.47	591	3.83	602	4.17	624	4.83	645	5.52	666	6.22	687	6.99	707	7.74
19551	2100	579	3.13	592	3.51	604	3.87	615	4.24	626	4.61	647	5.28	667	6.02	687	6.74	707	7.51	727	8.31
20482	2200	605	3.52	616	3.90	629	4.31	639	4.68	650	5.08	670	5.81	689	6.56	709	7.30	728	8.07	747	8.90
21413	2300	630	3.95	641	4.34	653	4.77	664	5.16	674	5.57	693	6.36	712	7.09	731	7.91	749	8.69	767	9.52
22344	2400	656	4.43	666	4.81	677	5.27	688	5.69	698	6.09	717	6.94	735	7.71	753	8.56	771	9.37	788	10.19
23275	2500	682	4.94	691	5.34	702	5.78	713	6.25	722	6.66	741	7.57	758	8.40	776	9.19	793	10.09	810	10.92
24206	2600	707	5.50	716	5.91	727	6.34	737	6.85	747	7.29	765	8.21	782	9.09	798	9.91	815	10.84	832	11.72
25137	2700	733	6.09	742	6.50	751	6.95	761	7.47	771795		789	8.88	806	9.83	821	10.72	838	11.58	854	12.56
26068	2800	758	6.70	768	7.13	776	7.60	786	8.11	796	8.65	813	9.61	830	10.62	845	11.54	860	12.41	876	13.41
26999	2900	784	7.35	793	7.83	802	8.31	811	8.80	820	9.38	838	10.37	854	11.42	869	12.39	883	13.35	899	14.27
27930	3000	809	8.06	819	8.58	827	9.07	836	9.56	845	10.15	861	11.19	878	12.25	893	13.23	907	14.31	921	15.23
28861	3100	835	8.85	845	9.39	852	9.86	861	10.36	869	10.92	886	12.08	902	13.14	917	14.27	931	15.28	944	16.30
29792	3200	861	9.69	870	10.23	878	10.68	886	11.23	894	11.77	911	12.99	926	14.09	941	15.24	955	16.34	968	17.40
30723	3300	887	10.59	896	11.12	903	11.59	911	12.16	919	12.69	935	13.94	950	15.07	965	16.24	979	17.44	992	18.51
31654	3400	913	11.54	921	12.04	929	12.57	936	13.13	944	13.66	960	14.93	974	16.14	989	17.31	1003	18.56	1016	19.70
32585	3500	939	12.51	947	12.98	955	13.59	962	14.12	969	14.70	985	15.96	999	17.26	1013	18.46	1027	19.71	1040	20.96

OV = Outlet velocity
SP = Static pressure
CFM = Cubic feet per minute
RPM = Revolution per minute
BHP = Brake horse power

Annexure B

Conversion Factors

Table B.1 Alphabetical listing of common conversions

To convert from	To	Multiply by
Acres	Square feet	43,560
Acres	Square meters	4074
Ampere-hours (absolute)	Coulombs (absolute)	3600
Angstrom units	Inches	3.937×10^{-9}
Atmospheres	Millimeters of mercury at 32 °F	760
Atmospheres	Newtons per square meter	101,325
Atmospheres	Feet of water at 39.1 °F	33.90
Atmospheres	Grams per square centimeter	1033.3
Atmospheres	Inches of mercury at 32 °F	29.921
Atmospheres	Pounds per square foot	2116.2
Bars	Atmospheres	0.9869
Bars	Newtons per square meter	1×10^{5}
Bars	Pounds per square inch	14.504
B.t.u.	Calories (gram	252
B.t.u.	Foot-pounds	777.9
B.t.u.	Horsepower-hours	3.929×10^{-4}
B.t.u.	Joules	1055.1
B.t.u.	Kilowatt-hours	2.930×10^{-4}
B.t.u. per second	Watts	1054.4
Calories, gram	B.t.u.	3.968×10^{-3}
Calories, gram	Foot-pounds	3.087
Calories, gram	Joules	4.1868
Calories, gram	Horsepower-hours	1.5591×10^{-6}
Candle power (spherical)	Lumens	12.556
Centigrade heat units	B.t.u	1.8
Cubic centimeters	Gallons	2.6417×10^{-4}
Cubic feet	Gallons	7.481
Cubic feet	Liters	28.316

(Contd.)

To convert from	To	Multiply by
Degrees	Radians	0.017453
Faradays	Coulombs (abs.)	96,500
Foot-pounds	Kilowatt-hours	3.766×10^{-7}
Foot-pounds per second	Kilowatts	0.0013558
Furlongs	Miles	0.125
Grains	Grams	0.06480
Grains	Pounds	1/7000
Grains per gallon	Parts per million	17.118
Grams per cubic centimeter	Pounds per cubic foot	62.43
Grams per square centimeter	Pounds per square foot	2.0482
Hectares	Acres	2.471
Hectares	Square meters	10,000
Horsepower (British)	B.t.u. per minute	42.42
Horsepower (British)	B.t.u. per hour	2545
Horsepower (British)	Foot-pounds per minute	33,000
Horsepower (British)	Watts	745.7
Joules (absolute)	Calories, gram (mean)	0.2389
Joules (absolute)	Kilowatt-hours	2.7778×10^{-7}
Kilograms force	Newtons	9.807
Kilometers	Miles	0.6214
Knots (international)	Meters per second	0.5144
Knots (nautical miles per hour)	Miles per hour	1.1516
Lumens	Watt	0.001496
Milliliters	Cubic centimeters	1
Radians	Degrees	57.30
Revolutions per minute	Radians per second	0.10472
Tons (refrigeration)	Pounds	12000
Yards	Meters	0.9144

Annexure C

Electrical Tables

Table C.1 Maximum number of PVC insulated 650/1100 V grade aluminium/copper conductor cable in MS conduit conforming to IS: 694–1990

Nominal cross sectional area of conductor in sq mm	20 mm		25 mm		32 mm		38 mm		51 mm		64 mm	
	S	B	S	B	S	B	S	B	S	B	S	B
1	2	3	4	5	6	7	8	9	10	11	12	13
1.5	5	4	10	8	18	12	–	–	–	–	–	–
2.5	5	3	8	6	12	10	–	–	–	–	–	–
4	3	2	6	5	10	8	–	–	–	–	–	–
6	2	–	5	4	8	7	–	–	–	–	–	–
10	2	–	4	3	6	5	8	6	–	–	–	–
16	–	–	2	2	3	3	6	5	10	7	12	8
25	–	–	–	–	3	2	5	3	8	6	9	7
35	–	–	–	–	–	–	3	2	6	5	8	6
50	–	–	–	–	–	–	–	–	5	3	6	5
70	–	–	–	–	–	–	–	–	4	3	5	4

Note
1. The above table shows the maximum capacity of conduits for a simultaneous drawing in of cables.
2. The columns headed 'S' apply to runs of conduits which have distance not exceeding 4.25 m between draw in boxes and which do not deflect from the straight by an angle of more than 15 degrees. The columns headed 'B' apply to runs of conduit, which deflect from the straight by an angle of more than 15 degrees.
3. Conduit sizes are nominal external diameters.

Table C.2 Maximum number of PVC insulated 650/1100 Volt grade aluminium/copper conductor cable in Al channel conforming to IS: 694–1990

Nominal cross sectional area	10/15 mm × 10 mm	20/15 mm × 10 mm	25/15 mm × 16 mm	32 mm × 16 mm	40 mm × 25 mm	40 mm × 40 mm
1.5	3	5	6	8	12	18
2.5	2	4	5	6	9	15
4	2	3	4	5	8	12
6		2	3	4	6	9
10		1	2	3	5	8
16			1	2	4	6
25				1	3	5
35					2	4
50					1	3
70					1	2

Note: Dimensions shown above are outer dimensions of mini trunking.

Table C.3 Aluminium/copper bus-bar sections

Current rating in amps upto	Recommended rectangular cross-section			
	Aluminium		Copper	
	No. of strips/phase	Size in mm	No. of strips/phase	Size in mm
100	1	20 × 5	1	20 x 3
200	1	30 × 5	1	25 x 5
300	1	50 × 5	1	40 x 5
400	1	50 × 6	1	50 x 5
500	1	75 × 6	1	60 x 5
600	1	80 × 6	–	–
800	1	100 × 6	–	–
1000	1	100 × 10	–	–
1200	1	125 × 10	–	–
1600	2	100 × 10	–	–
2000	2	125 × 10	–	–
2500	3	125 × 10	–	–

Note

i. In large bus bars of sizes above 1000 amps, the sections can be accepted in other rectangular cross- sections and numbers also, provided the total cross-sectional area offered is not less than the cross-sectional area shown in the above table against the respective bus bar rating.

ii. With aluminium bus bars, only aluminium wire/solid bar connections shall be made for incoming/outgoing mountings on the switchboards.

iii. With copper bus bars, only copper wire/solid bar connections shall be made for incoming/outgoing mountings on the switchboards.

Table C.4 (i) Marking for AC bus-bars and main connections

	Bus bar and main connections	Colour	Letter/Symbol
i.	Three phase	Red, Yellow, Blue	R.Y.B
	Two phase	Red, Blue	R.B.
	Single phase	Red	R
ii.	Neutral connection	Black	N
iii.	Connection to earth	Green	E
iv.	Phase variable (such as connections to reversible motors)	Grey	Grey

(ii) For DC bus-bars and main connections

	Bus bar and main connections	Colour	Letter/Symbol
i.	Positive	Red	R, or plus
ii.	Negative	Blue	B, or minus
iii.	Neutral connection	Black	N
iv.	Connection to earth	Green	E
v.	Equalizer	Yellow	Y
vi.	Phase variable (such as connections to reversible motors)	Grey	Grey

Note: In the wiring diagram, positive and negative should be indicated by '+' and '–' respectively.

Table C.5. A Earth continuity strip for protective earthing of sub-station equipment

S. no.	Type of installation	Earth electrode	Earth strip from earth electrode to earth bus and loop earthing of equipment
1.	Indoor substation with HT panel, transformer capacity up to 1600 kVa, LT panel, generating set.	Copper plate	25 × 5 mm copper strip
2.	Indoor sub-station with HT panel, transformer capacity above 1600 kVa, LT panel, and generating set.	Copper plate	32 × 5 mm copper strip
3.	HT outdoor substation.	Copper plate	25 × 5 mm copper strip
4.	LT indoor substation with generator.	Copper plate	25 × 5 mm copper strip
5.	LT switch room having main LT switch board.	Copper plate	20 × 3 mm copper strip

B Earth continuity strip for bus trunking and rising main

S. no.	Type of Installation	Material of main conductor	Earth strip
1.	Bus trunking upto 2500 Amp capacity	Copper/Aluminium	2 Nos. 25 × 5 mm copper strip
2.	Bus trunking above 2500 Amp capacity	Copper/Aluminium	2 Nos. 32 × 5 mm copper strip
3.	Bus trunking for connecting generating set and LT panel	Copper/Aluminium	2 Nos. 25 × 5 mm copper strip
4.	Rising main upto 400 Amp capacity	Copper/Aluminium	2 Nos. 20 × 3 mm copper strip
5.	Rising main above 400 Amp and upto 800 Amp capacity	Copper/Aluminium	2 Nos. 20 × 5 mm copper strip

C Neutral earthing of transformers and generators

S. no.	Equipment	Earth electrode	Earth strip from earth station to neutral
1.	Transformer of capacity up to 1600 KVA	Copper plate	25 × 5 mm copper strip
2.	Transformer of capacity above 1600 KVA	Copper plate	0L × 5 mm copper strip
3.	Generating set of all capacity	Copper plate	25 × 5 mm copper strip

Table C.6 Materials and sizes of earth electrodes

Type of electrode	Material	Size
Pipe	GI medium class	40 mm dia 3.45 m long (without any joint)
Plate	(i) GI	60 cm × 60 cm × 6 mm thick
	(ii) Copper	60 cm × 60 cm × 3 mm thick
Strip	(i) GI	100 sq. mm section
	(ii) Copper	40 sq. mm section
Conductor	(i) Copper	4 mm dia (8 SWG)

Note: Galvanization of GI item shall conform to Class IV of IS: 4736–1986.

Table C.7 Chart showing the distance upto which different sizes of UG aluminium conductor cables can be used for different current ratings for 8 volts drop when laid in ground (PVC insulated, PVC sheathed, 3 core or 4 core) when cable grading is 1.1 kV

(Maximum conductor temperature 70 °C)

S. no.	Current Amp	\multicolumn{13}{c}{Distance in meters for the following cable sizes in sq mm}												
		6	10	16	25	35	50	70	95	120	150	185	240	300
1.	5	165	260	415	725	895	1300	1925	2360	3065	3555	4300	5770	6460
2.	10	80	130	205	360	450	650	960	1180	1530	1775	2150	2885	3230
3.	15	55	85	140	240	300	430	640	785	1020	1185	1430	1920	2155
4.	20	40	65	100	180	225	325	480	590	765	890	1075	1440	1615
5.	25	30	50	80	145	180	260	385	470	610	710	860	1150	1290
6.	30	25	40	70	120	150	215	320	390	570	590	715	960	1075
7	40	20	30	50	90	110	160	240	295	380	445	535	720	805
8.	50	–	25	40	70	90	130	190	235	305	355	430	575	645
9.	60	–	–	35	60	75	110	160	195	255	295	355	480	535
10.	70	–	–	30	50	65	90	135	165	215	255	305	410	460
11.	80	–	–	–	45	55	80	120	145	190	220	265	360	405
12.	90	–	–	–	40	50	70	105	130	170	195	235	320	360
13.	100	–	–	–	35	45	65	95	115	150	175	215	290	320
14.	110	–	–	–	–	40	60	85	105	140	160	195	260	290
15.	120	–	–	–	–	35	55	80	95	125	145	180	240	270
16.	130	–	–	–	–	–	45	75	90	115	135	165	220	250
17.	140	–	–	–	–	–	–	70	80	110	125	150	205	230
18.	150	–	–	–	–	–	–	65	75	100	115	140	190	215
19.	160	–	–	–	–	–	–	60	70	95	110	130	180	200
20.	170	–	–	–	–	–	–	55	70	90	105	125	170	190
21.	180	–	–	–	–	–	–	50	65	85	100	120	160	180
22.	190	–	–	–	–	–	–	–	60	80	90	110	150	170
23.	200	–	–	–	–	–	–	–	60	75	90	105	145	160
24.	225	–	–	–	–	–	–	–	–	65	80	95	125	145
25.	250	–	–	–	–	–	–	–	–	–	70	85	115	130
26.	275	–	–	–	–	–	–	–	–	–	–	80	105	115
27.	300	–	–	–	–	–	–	–	–	–	–	70	95	105

Note: For temperature correction please see as detailed below.

1		\multicolumn{4}{l}{*When the voltage drop and length is constant, then to find the size of cable, multiply the current rating with the following factor and then see the size in this table corresponding to the derived value.*}			
Ground temperature		20 °C	25 °C	30 ° C	35 °C
Rating factors		0.95	0.9	0.85	0.8

Table C.8 Current rating (in ground) for XLPE insulated 1.1 kV grade cable

Nominal area of the conductor Sq mm	Aluminium conductor				Copper conductor			
	Single core		Multi core		Single core		Multi core	
	PVC	XLPE	PVC	XLPE	PVC	XLPE	PVC	XLPE
10	51	55	46	50	65	71	60	65
16	66	74	60	68	85	95	77	87
25	86	98	76	90	110	125	99	115
35	100	118	92	108	130	150	120	138
50	120	137	110	126	155	175	145	161
70	140	172	135	158	190	220	175	202
95	175	204	165	187	220	260	210	239
120	195	234	185	215	250	301	240	276
150	220	262	210	240	280	336	270	308
185	240	298	235	273	305	381	300	350
240	270	344	275	316	345	441	345	405
300	295	387	305	355	375	496	385	455
400	325	458	335	420	400	586	425	538
500	345	495	–	–	425	635	–	–
630	390	555	–	–	470	710	–	–
800	440	625	–	–	–	–	–	–
1000	490	685	–	–	–	–	–	–

Rating factors for variation in ambient air temperture

Air temperature (°C)	40	45	50
Rating factor (XLPE)	1.00	0.94	0.88
Rating factor (PVC)	1.00	0.90	0.81

Table C.9 Permissible maximum short-circuit current ratings for XLPE cables

Conductor area sq mm	Short circuit rating for one second duration		Conductor area sq mm	Short circuit rating for one second duration	
	Copper conductors	Aluminium conductors		Copper conductors	Aluminium conductors
	A	B		A	B
16	2,570	1,730	185	28,200	18,700
25	3,970	2,670	240	36,400	24,200
35	5,500	3,690	300	45,300	30,100
50	7,800	5,220	400	60,200	39,900
70	10,850	7,400	500	74,800	49,800
95	14,600	9,740	630	92,700	62,000
120	18,400	12,200	800	–	78,800
150	23,000	15,200	1,000	–	97,800

Table C.10 Class of insulation (for electric motors)

Type	Maximum operating temperature	Materials used
Y	90 °C	Cotton, silk, paper, and similar organic material and combination of such material which are not (impregnated) nor immersed in oil.
A	105 °C	Above materials immersed with varnish or enamel or oil immersed.
E	120 °C	Comprise inorganic materials such as mica, glass fiber asbestos or combination of these materials in built up form with binding cement.
B	130 °C	
F	155 °C	Class B materials when built-up with suitable cement or binder.
H	180 °C	Consists of materials or combination of material also such as mica, glass fiber silicon elastomer with suitable winding, impregnating or coating substances as silicon resins.
C	Above 180 °C	Materials such as mica Porcelain, glass quartz and asbestos with or without an inorganic binder.

Annexure D

List of IS Codes

Table D.1 IS codes and standards

Air-conditioning Equipment

IS 659 : 1964 (reaffirmed 1991)
Safety code for air conditioning (revised) (Amendment 1).
IS 660 : 1963 (reaffirmed 1991)
Safety code of mechanical refrigeration. (revised).
IS 139 : Part I and Part II
Room air conditioners (unitary and split) (2nd revision).
IS 2370 : 1963 (reaffirmed 1991)
Sectional cold rooms. (walk in type).
IS 2371 : 1973
Solid drawn copper alloyed tubes for condensers, evaporators, heaters and cooler tubes
using saline hard water. (withdrawn) (superseded by IS 1545, 1982).
IS 2372 : 1991
Timber for cooling towers (1st revision).
IS 3315 : 1994
Evaporative air coolers. (Desert coolers) (2nd revision).
IS 3615 : 1967 (reaffirmed 1991)
Glossary of terms used in refrigeration and air-conditioning.
IS 4831 : 1968 (reaffirmed 1991)
Recommendation on units and symbols for refrigeration.
IS 5111 : 1993
Testing of refrigeration compressors (1st revision).
IS 6272 : 1987 (reaffirmed 1991)
Industrial cooling fans (Man coolers) (1st revision).
IS 7896 : 1975 (reaffirmed 1991)
Data for outside design conditions for air conditioning for summer months.
IS 8148 : 1976 (reaffirmed 1991)
Packaged air conditioners. (Amendment-1).
IS 8188 : 1976 (reaffirmed 1988)
Code of practice for treatment of water industrial cooling systems.
IS 8362 : 1977 (reaffirmed 1991)
Copper and copper alloy rolled plates for condensers and heat exchangers.
IS 8667 : 1977 (reaffirmed 1993)
Purchasers data sheet for cooling towers for process industry.

IS 9612 : 1980 (reaffirmed 1991)
Aluminium tubes for refrigeration purposes.
IS 10470 : 1983 (reaffirmed 1991)
Air cooled heat exchangers (Amendment-1).
IS 10617 : Part I, Part II and Part III, 1983 (reaffirmed 1991)
Hermetic compressors.
IS 10873 : 1983 (reaffirmed 1993)
Data sheet for air cooled heat exchangers.
IS 11327 : 1985 (reaffirmed 1991)
Requirements for refrigerants condensing units.
IS 11329 : 1985 (reaffirmed 1991)
Finned type heat exchangers for room air conditioners (Amendment-1).
IS 11330 : 1985 (reaffirmed 1991)
Oil separators.
IS 12357 : 1988 (reaffirmed 1993)
Suppliers data sheet for clean air equipments (Laminar flow).
IS 11338 : 1985 (reaffirmed 1991)
Thermostats for use in refrigeration, air conditioners etc.
SP 7 : 1983 Group 4
National Building code.

Electrical Wires and Cables, IT and HT Grade

IS 282 : 1963
Hard drawn copper conductors for overhead power transmission.
IS 398 : 1976 Part I and Part II
Aluminium conductors for overhead transmission purposes.
IS 694 : 1977 Part I and Part II
PVC Insulated cables for voltage upto 1100 V with copper and aluminium conductors respectively.
IS 1554 : 1981 Part I and Part II
PVC insulated (heavy duty) cables for working voltage upto 1.1 kV and upto 11 kV grade respectively.
IS 692 : 1973
Paper insulated, lead sheathed cables for electric supply.
IS 732 : 1000
Code of practice for electrical wiring installation.

Earthing

IS 3043 : 1966
Code of practice for earthing.
IS 3151 : 1965
Earthing transformers.
IS 12776 : 1989
Galvanized stand for earthing.

Fuses

IS 2208 : 1976
HRC fuse links upto 650 V.
IS 2086 : 1963
Carriers and bases used in rewirable type electric fuses upto 650 V.
IS 3106 : 1966
Code of practice for selection, installation and maintenance of fuses (upto 650 V with drawn).

Motors

IS 325 : Specification for 3 phase induction motors.
IS 900 : Code of practice for installation and maintenance of induction motors.

IS 996 : Specification for 1 phase small AC universal motors.

IS 2148 : Specification for flame proof motors.

IS 1231 : Dimensions of 3 phase foot mounted motors.

IS 2223 : Dimensions of 3 phase flange mounted motors.

IS 2253 : Types of constructions and mounting of motors.

IS 2968 : Dimensions of slide rail for electric motors.

IS 4722 : 1968
Rotating electric machines.

IS 1580 : 1983
Service conditions for electrical equipment.

SP : 30 : 1985
National electrical code.

Switch gear and controls

IS 2607 : 1981
Air break isolators for voltage not exceeding 1000 V.

IS 2516 : 1980 Part I and Part II
Circuit breakers AC.

IS 2959 : 1975
Contractors of AC for voltage upto 1100 V.

IS 2675 : 1975
Enclosed distribution fuse boards and cutouts for and upto 1000 V.

IS 4237 : 1982
General requirement for switch gear and control gear for
voltage not exceeding 1000 V AC or 1200 V DC.

IS 8544 : 1977 Part I to Part IV
Motor starter for voltages not exceeding 1000 V.

IS 8623 : 1993 Part I and Part II
Low voltage switch gear and control gear assemblies.

IS 3914 : Code of practice for selection of starters for AC induction motors (withdrawn).

IS 4821 : Specification of cable glands.

IS 10118 : 1982, Part I to Part IV
Code of practice for selection, installation and maintenance of switch gear control gear.

Wiring accessories

IS 3854: 1966
Switches for domestic and similar purposes.

IS 1293 : 1967
3 pin plugs and sockets outlets.

IS 3837 : 1976
Accessories for electrical wiring.

IS 2509 : 1973
Rigid non metallic conduits.

IS 9537 : 1981 Part I to Part IV
Conduits for electrical insulation.

Noise and vibration

IS 2526 : 1963 (reaffirmed 1991)
Code of practice for acoustical design of auditoriums and conference halls. (Amendment 1).

IS 2264 : 1963 (reaffirmed 1994)
Preferred frequencies for acoustical measurements.

IS 3483 : 1965 (reaffirmed 1996)
Code of practice for noise reduction in Industrial buildings.

IS 3932 : 1966
Sound level meter for general purpose use. (Incorporated in IS 9779).

IS 4954 : (reaffirmed 1991)
Recommendations for noise for abatement in town planning.
IS 4758 : 1968 (reaffirmed 1994)
Methods of measurement of noise emitted by machines.
IS 9736 : 1981 (reaffirmed 1991)
Glossary of terms applicable to acoustics in buildings.
IS 9901 : Part I and Part II, Part IX (reaffirmed 1991)
Measurement of sound insulation in building and building elements.
IS 9989 : 1981 (reaffirmed 1991)
Assessment of noise with respect to community response.
IS 9876 : 1981 (reaffirmed 1991)
Guide to the measurement of air borne acoustical noise and evaluation of its effects on man.
IS 10423 : 1982 (reaffirmed 1994)
Personal sound exposure meter.
IS 11446 : 1985 (reaffirmed 1990)
Measurement of air borne noise emitted by compressors units intended for outdoor use.
IS 12998 : Part I and Part II, 1991
Test code and method of measurements of air borne noise emitted by electrical machinery
(superseding IS 6098, 1971).
IS 12710 : 1989
Glossary of terms used in acoustic emission testing.
IS 11050 : Part I, Part II and Part III, 1984 (reaffirmed 1991)
Rating of sound insulation in buildings and of building elements.
IS 14280 : 1995
Mechanical vibration–balancing–shaft and fitment key convention.
IS 14259 : 1995
Vibration and shock –Isolators, procedure of specifying characteristics.
IS 12065 : 1987
Permissible limits of noise level for rotating electrical machines.
IS 1950 : 1962 (reaffirmed 1991)
Code of practice for sound insulation of non industrial building. (Amendment 1).
IS 7194 : 1994
Assessment of noise exposure during work for hearing conversation purpose.
IS 4729 : Measurement and evaluation of vibration for motors. (withdrawn).

Pipe and fittings

IS 638 : 1979 (reaffirmed 1993)
Gaskets.
IS 1239 : Part I and Part II 1990/1992
Mild steel tubes and fittings.
IS 1367: Part I to Part XVIII
Technical supply conditions for threaded steel fasteners.
IS 2501 : 1995
Solid drawn– copper tubes for general engineering purposes. (3rd revision).
IS 3076 : 1985 (reaffirmed 1991)
Low density polyethylene pipes for portable water supplies. (2nd revision, Amendment 1).
IS 5822 : 1994
Code of practice laying of electrically welded steel pipes for water supply (2nd revision).
IS 6630 : 1985 (reaffirmed 1990)
Seamless ferrite alloy steel pipes for high temperature steam service. (1st revision).
IS 6392 : 1971 (reaffirmed 1988)
Steel pipe flanges (Amendment 1).
IS 10773 : 1995
Wrought copper tubes for refrigeration and air conditioning purposes.

Pump and valves

IS 778 : 1984 (reaffirmed 1990)

Copper alloy gate, globe and check valves for water works purposes.

(4th revision, Amendments 2).

IS 779 : 1994

Water meters (domestic type) (6th revision, Amendments 1).

IS 780 : 1984 (reaffirmed 1990)

Sluice valves for water works purposes. (50 to 300 mm size)

(6th revision, Amenndments 3).

IS 781 : 1984 (reaffirmed 1990)

Cast copper alloy, screw down, bib taps and stop valves for water services.

(3rd revision, Amendment 1).

IS 1520 : 1980

Horizontal centrifugal pumps for clear, cold, fresh water (2nd revision).

IS 2685 : 1971 (reaffirmed 1992)

Code of practice for selection, installation and maintenance of sluice valves (1st revision).

IS 2906 : 1984 (reaffirmed 1990)

Specification for sluice valve for water works purposes 350 to 1200 mm size

(3rd revision, Amendments 3).

IS 3233: 1965 (reaffirmed 1992)

Glossary of terms for safety and relief valves and their parts.

IS 4854 : Part I, II 1969, Part III, 1984

Glossary of terms for valves and their parts.

IS 5312 : Part I 1984, Part II, 1986 (reaffirmed 1990, 1991)

Swing check type non return valves.

IS 5659 : 1970 (reaffirmed 1991)

Pump for process water (Amendment 1).

IS 8092 : 1992 (reaffirmed 1990)

Code for inspection of surface quality of steel castings for valves, fittings and other

piping components. (Visual method) (1st revision).

IS 8418 : 1977 (reaffirmed 1990)

Horizontal centrifugal self priming pumps.

IS 8472 : 1977 (reaffirmed 1993)

Regenerative self priming pumps for clear. cold, fresh water. (Amendments 2).

IS 9137 : 1978 (reaffirmed 1993)

Code for acceptance test for centrifugal, mix flow and axial pumps–class C. (Amendments 4).

IS 9542 : 1980 (reaffirmed 1993)

Horizontal centrifugal mono set pumps for clear, cold, fresh water. (Amendment 1).

IS 10596 : Part I, II, III and IV, 1983 (reaffirmed 1993)

Code of practice for selection, installation, operation and maintenance of pumps for industrial applications.

IS 10981: 1983 (reaffirmed 1993)

Class of acceptance test for centrifugal mixed flow and axial flow pumps–Class B.

IS 11745 : 1986 (reaffirmed 1991)

Technical supply conditions for positive displacement pumps-reciprocating.

IS 12969 : 1990

Method of test for quality characteristic of valves.

IS 12992 : 1993, Part I, 1990 Part II

Safety relief valves.

IS 13095 : 1991

Butterfly valve for general purposes.

Refrigerant gas and lubricants

IS 310 : 1954

Method of sampling and test for lubricants. (Part I and Part II)(Superseded by IS 1447 and IS 1448).

IS 718 : 197 (reaffirmed 1991)

Carbon tetra Chloride (2nd revision, Amendments 2).

IS 4578 : 1989

Lubricating oils for refrigeration machinery (1st revision).

IS 6849 : Part I and Part II, 1993

Positive displacement vacuum pumps.

IS 9466: 1980 (reaffirmed 1993)

Viscosity classification of industrial liquid lubricants.

IS 10609 : 1983 (reaffirmed 1991)

Refrigerants–Number–designation.

Safety

IS 954 : 1989

Functional requirements for carbon dioxide tender for fire brigade use. (2nd revision).

IS 1641 : 1988 (reaffirmed 1993)

Code of practice for fire safety of buildings (general) :

General principles of fire grading and classification. (1st revision).

IS 1642 : 1989

Code of practice for fire safety of buildings (general) :

Details of construction (1st revision) (1645 supersedes 1642).

IS 1643 : 1988 (reaffirmed 1993)

Code of practice for fire safety of buildings (general) : Exposure hazard (1st revision).

IS 1644 : 1988 (reaffirmed 1993)

Code of practice for fire safety of buildings (general) : Exit requirements and personal hazard.

IS 1646 : 1982 (reaffirmed 1990)

Code of practice for fire safety of buildings (general) : Electrical installation (1st revision).

IS 3786 : 1983 (reaffirmed 1991)

Methods for computation of frequency and severity rates for industrial injuries and classification of industrial accidents. (1st revision).

IS 3808 : 1979 (reaffirmed 1990)

Method of test non combustibility of building materials (1st revision).

IS 5311 : 1969 (reaffirmed 1990)

Code of safety for carbon tetra chloride.

IS 6382 : 1984 (reaffirmed 1990)

Code of practice for design and installation of fixes carbon dioxide for fire extinguishing system (1st revision).

IS 7969 : 1975 (reaffirmed 1991)

Safety code for handling and storage of building materials (Amendment 1).

Sheet metal work

IS 277 : 1992

Galvanized steel sheet (5th revision, Amendments 2).

IS 513 : 1994

Cold rolled low carbon steel sheets.

IS 655 : 1963 (reaffirmed 1991)

Metal air ducts (revised, Amendments 3).

IS 1079 : 1994

Hot rolled carbon steel sheets.

IS 1977 : 1975 (reaffirmed 1992)

Structural steel (ordinary quality).

IS 2062 : 1992

Steel for general structural purposes.

IS 3768 : 1989 (reaffirmed 1994)

Ventilation ducting–Venyl coated, flexible (1st revision).

IS 7613 : 1975 (reaffirmed 1991)

Method of testing panel type air filters for air conditioning purpose.

Thermal insulation

IS 334 : 1982 (reaffirmed 1991)

Glossary of terms relating to bitumen and tar. (2nd revision).

IS 661 : 1974 (reaffirmed 1990)

Code of practice for thermal insulation of cold storages. (2nd revision, Amendments 2).

IS 1397 : 1990

Kraft paper (2nd revision).

IS 2094 : 1995

Heaters for bitumen and emulsion (2nd revision).

IS 3069 : 1994

Glossary of terms, symbols and units relating to thermal insulation materials (1st revision).

IS 3129 : 1985 (reaffirmed 1990)

Low density particle boards. (1st revision).

IS 3144 : 1992

Mineral wool thermal insulations–Methods of tests. (2nd revision).

IS 3346 : 1980 (reaffirmed 1990)

Method of determination of thermal conductivity of thermal insulation materials.
(2 slab guarded hot plate method) (1st revision).

IS 3677 : 1985 (reaffirmed 1990)

Unbonded rock and slag wool for thermal insulation. (2nd revision (superseding IS 5696).

IS 3690 : 1974 (reaffirmed 1990)

Unbonded glass wool for thermal insulation (1st revision).

IS 3716 : 1978 (reaffirmed 1991)

Application guide for insulation co-ordination (1st revision, Amendments-2).

IS 3792 : 1978 (reaffirmed 1992)

Guide for heat insulation of non industrial buildings (1st revision).

IS 4671 : 1984 (reaffirmed 1990)

Expanded polystyerene for thermal insulation purposes (1st revision).

IS 7240 : 1974

Code of practice for thermal insulation of cold storages.

IS 7413 : 1981

Code of practice for thermal industrial application and finishing of thermal insulating materials at temperature above 40 °C to 700 °C (Superseded by IS 14164, 1994).

IS 8154 : 1993

Preformed calcium silicate insulation (for temperatures upto 650 °C) (1st revision).

IS 8183 : 1993

Bonded mineral wool. (1st revision).

IS 9403 : 1980 (reaffirmed 1990)

Method of test for thermal conductance and transmittance of built up sections by means of guarded hot box.

IS 9428 : 1993

Preformed calcium silicate insulation for temperature upto 950 °C (1st revision).

IS 9489 : 1980 (reaffirmed 1992)

Method of test for thermal conductivity of materials by means of heat flow meter.

IS 9742 : 1993

Sprayed mineral wool insulation. (1st revision, amendment-1).

IS 9842 : 1994

Preformed fibrous pipe insulation (1st revision).

IS 10556 : 1993 (reaffirmed 1990)

Code of practice for storage and handling of insulation materials.

IS 11239 : Part I to Part XIII

Method of test for cellular thermal insulation materials. Dimensions, Dimensional stability, apparent density, water vapour transmission rate, volume percent of open and closed cells, heat distortion temperature, co-efficient of liner thermal expansion at low temperature flame height, time of burning and loss mass, water absorption, flexural strength, compressive strength, horizontal burning characteristics, determination of flammability by oxygen index.

IS 11246 : 1992

Glass fibre reinforced polyester resin (1st revision).

IS 12436 : 1988 (reaffirmed 1994)

Preformed rigid polyurethane foams for thermal insulation.

IS 13204 : 1991

Rigid phenolic foams for thermal insulation.

IS 13205 : 1991

Code of practice for the application of polyurethane insulation by the in-situ pouring method.

Ventilation

IS 2312 : 1976 (reaffirmed 1991)

Propeller type AC ventilation fans. (1st revision, Amendments-7).

IS 3103 : 1975

Code of practice for industrial ventilation (1st revision).

IS 3588 : 1987 (reaffirmed 1991)

Electric axial flow fans. (1st revision).

IS 4894 : 1987 (reaffirmed 1991)

Centrifugal fans (1st revision).

Lifts

IS 4666–1980 specification for electric passenger and goods lifts (first revision).

IS 6383–1971 specification for electric service lifts.

IS 2332–1963 nomenclature of floors and stories.

IS 4289–1967 specification for lift cables.

IS 3043–1966 code of practice for earthing.

IS 1860–1980 code of practice for installation, operation and maintenance of electric passenger and goods lifts (second revision).

IS 6620–1972 code of practice for installation, operation and maintenance of electric service lifts.

IS 2309–1969 code of practice for the protection of building and allied structures against lightning (first revision).

IS 4591–1968 code of practice for installation and maintenance of escalators.

Annexure E

List of International Codes

Table E.1 International codes and practice

Sl. no.	Description	Publisher	Reference
1.	Commercial low pressure, low velocity duct system design	ACCA	Manual Q
2.	HVAC Systems-Applications I Edition, 1986	SMACNA	SMACNA
3.	HVAC Systems-Duct design 1990	SMACNA	SMACNA
4.	HVAC air duct leakage test manual, 1985	SMACNA	SMACNA
5.	HVAC duct construction standards–Metal and flexible I edition, 1985	SMACNA	SMACNA
6.	HVAC duct systems inspection guide, 1989	SMACNA	SMACNA
7.	Rectangular industrial duct construction, 1989	SMACNA	SMACNA
8.	Round industrial duct construction, 1977	SMACNA	SMACNA
9.	Thermoplastic duct (PVC) Construction Manual (Rev. a, 1974)	SMACNA	SMACNA
10.	Gravimetric and Dust spot procedures for testing air cleaning devices used in general ventilation for removing particulate matter	ASHRAE	ANSI/ASHRAE/ 52.1–1992
11.	Method for sodium flame test air filters	BSI	BS 3928
12.	Methods of test for atmospheric dust spot efficiency and synthetic dust weight arrestance	BSI	BS 6540, Part-1
13.	High efficiency, particulate air filter units, 1990	UL	ANSI/UL 586–1990
14.	Method of testing liquid chilling packages	ASHRAE	ASHRAE 30–1978
15.	Absorption water–chilling packages	ARI	ARI 560–82
16.	Centrifugal or rotary screw water chilling packages	ARI	ARI 550–90
17.	Reciprocating water chilling packages	ARI	ARI 590–86
18.	Forced–circulation air cooling and air heating	ARI	ARI 410-91
19.	Methods of testing forced to circulation air cooling and air heating coils	ASHRAE	ASHRAE -33–1978
20.	Thermal environmental conditions for human occupancy	ASHRAE	ANSI/ASHRAE 55–1992
21.	Acceptance test code for water cooling towers, mechanical draft, natural draft, fan assistant type evaluation of results and thermal testing of wet and dry cooling towers. (1990)	CTI	CTI-ATC 105–1990
22.	Code of measurement of sound from water cooling towers.	CTI	CTI-ATC 128–1981
23.	Energy conservation guidelines	SMACNA	SMACNA 1984
24.	Energy recovery equipment and systems, air to air	SMACNA 1991	SMACNA
25.	Retrofit of building energy system and processes	SMACNA	SMACNA 1982

(Contd.)

Sl. no.	Description	Publisher	Reference
26.	Laboratory methods for testing fans for rating	ASHRAE	ANSI/ASHRAE–51 1985ANSI/AMCA-210–1985
27.	Fire dampers, 1990	UL	ANSI-UL-555–1985
28.	Fire protection hand book 17th edition	NFPA	NFPA
29.	Standards of tubular exchanger manufactures association 7th edition, 1988	TEMA	TEMA
30.	Refrigeration piping	ASME	ASME/ANSI B-31.5/1987
31.	Scheme for identification of piping system	ASME	ANSI/ASME A-13.1/1981 (R-1985)
32.	Number designation and safety classification of refrigerants	ASHRAE	ANSI/ASHRAE 34–1992
33.	Refrigerants oil	ASHRAE Guide description lines 3–1990	ASHRAE
34.	Refrigerant recovery recycling equipment, 1989	UL	ANSI/UL/1963–1991
35.	Practices for measurement, testing and balancing of building, heating, ventilation, air conditioning and refrigeration system.	ASHRAE	ANSI/ASHRAE/ 111–1988
36.	HVAC System –Testing, adjusting and balancing	SMACNA	SMACNA
37.	Ventilation for acceptance indoor air quality and balancing, 1983	ASHRAE	ANSI/ASHRAE 62–1989
38.	ASHRAE 90–1–1989 : Energy Efficient Design of New Building except low rise residential buildings.		
39.	ASHRAE 15–1994 : Safety code for Mechanical Refrigeration.		
40.	Thermal DVP Test US Military Standard MIL-STD–282. Federal Standard 209 : Clean Room and clean work station requirements, controlled environment.		
41.	ASHRAE 100–1980 : Energy conservation in existing buildings.		
42.	ASHRAE Guidelines : Guideline 1–1989–Commissioning of HVAC systems.		
43.	Guideline 3–1990–Reducing emission of fully hydro generated Chloro fluoro Carbon (CFC) refrigerants in refrigeration and Air-Conditioning equipment and applications.		

Bibliography

1. *Automatic Sprinkler System Handbook*, National Fire Protection Association, 2013.
2. Boyen JL. *Thermal Energy Recovery*. John Wiley and Sons, Inc., New York, 1980.
3. Cacciola G., G. Restuccia and N Girodano. *Economic Comparison Between Absorptions and Compression Heat Pumps*, 1990.
4. Caravath, James Raley; Lalsingh V R. *Practical Illumination*. McGraw Publishing Co., New York, 1907.
5. Carrier, Water piping systems and pumps, Technical Development Program.
6. *CPWD Specifications for Internal Electrical Installations*. Central Public Works Department, Government of India, New Delhi.
7. *CPWD Specifications for Lifts*. Central Public Works Department, Government of India, New Delhi.
8. *CPWD Specifications for Sub-Stations*. Central Public Works Department, Government of India, New Delhi.
9. Cravath, J V; Lansingh V R. The calculation of Illumination by the flux of light method.
10. Dorgan, Chad B, Steven P. Leight, Charles E. Dorgan. *Application Guide for Absorption Cooling / Refrigeration Using Recovered Heat*. American Society of Heating, Refrigerating and Air-Conditioning Engineers, Inc., Atlanta, 1995.
11. Energy Conservation Building Code, User Guide. Bureau of Energy Efficiency, New Delhi, 2009.
12. *Fire Protection Handbook*, 20th edn, National Fire Protection Association, 2008.
13. Green Building Rating System, (Abridged Reference Guide for New Construction and Major Renovations), Indian Green Building Council, Hyderabad.
14. GRIHA for LD. The Energy and Resource Institute and Association for Development and Research of Sustainable Habitats, India.
15. GRIHA Manual Volume-1 (Introduction to National Rating System-GRIHA). Ministry of New and Renewable Energy and TERI, New Delhi, 2010.
16. *Handbook of Air-Conditioning System Design*. Carrier Corporation. New York: McGraw-Hill, 1965.
17. *Handbook on Fundamentals*, ASHRAE, 2013.
18. *Handbook on HVAC Applications*, ASHRAE, 2011.
19. *Handbook on HVAC System and Equipment*, ASHRAE, 2012.
20. *Handbook on Refrigeration*, ASHRAE, 2010.
21. Harris, KJ. *Consider Absorption Cooling to Tap Cogeneration Waste Heat Power*, 1987.
22. Int-Hout, Dan; Kloostra, Leon. Air Distribution for large spaces.
23. LEED India NC (Reference Guide for New Construction). Indian Green Building Council, Hyderabad.
24. Lorsch, Harold G. *Air-Conditioning System Design Manual*. American Society of Heating, Refrigerating and Air-Conditioning Engineers, Inc., Atlanta, 1993.
25. McLean, Don; Advancements in Roadway Lighting-IESNA BC Chapter, 2009.
26. Mcquiston, Faye C, Jerald D. Parker and Jeffrey D. Spitler. *Heating, Ventilating and Air-Conditioning*. John Wiley and Sons, Inc., New York, 2000.
27. Milwaukee, W I. *Generator Sets, Application and Installation Guide*. Caterpillar: Caterpillar Inc., 1993.
28. *National Building Code of India*, Bureau of Indian Standards, New Delhi, 2005.
29. Pita, Edward G. *Air-Conditioning Principles and Systems*. Prentice-Hall of India, New Delhi, 2003.

30. Sahu, G K. *Pumps.* New Age International Publishers, New Delhi, 2007.
31. Seip, Gunter G. *Siemen's Electrical Installations Handbook,* Parts I & II. Akademia Books International, 1999.
32. Siteco (Osram). Interior Lighting Tools.
33. UNEP. Energy Efficiency Guide for Industry in Asia, 2006.
34. Ventilation for acceptable Indoor Air Quality, Standard 62-2001, ASHRAE.
35. Wulfinghoff, Donald R. Energy Efficiency Maual. Energy Institute Press, Maryland, 1999.

USEFUL WEB SITES

1. www.ashrae.org
2. www.elitesoft.com
3. www.carmelsoft.com
4. www.wrightsoft.com
5. www.cea.nic.in
6. www.eng-tips.com
7. www.mcquay.com
8. www.trane.com
9. www.carrier.com
10. www.mcgrawhill.com
11. www.scribed.com
12. www.engineeringtoolbox.com

Index

A

Absolute viscosity 201
Absorber 114
Accuracy class 174, 176
Accuracy limit factor 175
Actuator 85, 87, 221
Adiabatic 20
Adiabatic humidification 28
Aerofoil 125
Aerofoil Blade Fans 124
AHU 78, 80, 83
Air circuit breaker 158, 159
Air dilution 276
Air handling unit 9
Air insulated bus trunking 173
Air quality sensor 85
Air refrigeration 101
Air sampling smoke detectors 271, 272
Air velocity 75
Ambient temperature 188
Ammonia 106, 113
Apparatus dew point temperature 26
Apparent power 161
Aquifier 218
Aspect ratio 73, 75, 78–80
Auto transformer 191
Automatic power factor control panel 178
Automatic sprinkler 281
Automatic voltage regulator 184–185, 187, 198, 200
Axial fan 123, 124

B

Backward curved fans 123, 125
Balancing valves 95
Ball valve 222

Ballasts 236
Barometric pressure 273
Basement ventilation 134
Bimetallic type 273
Biogas generators 258
Biomass power generators 263
Blowers 9
Body earthing 146
Brakes 301
Branch distribution board 141
Breaking capacity 153
Brush gear 183
Brushless machines 184
Bucholze relay 155, 159
Building management system 258
Burden 174
Bus bar 172
Bus bar chamber 172
Bus trunking 172
Butterfly valves 87, 221
By pass factor 25

C

Cable size 152, 177
Cabling 3
Candela 233, 238, 244, 246
Candela power 248
Candle power 232, 242, 243, 246
Capacitor 161, 168, 171
Capacitor bank 172
Capacitor contactor 169
Capital cost 4
Carnot cycle 101
Cascade condenser 105, 106
Cavitation 215
Centrifugal compressors 106, 111
Centrifugal fan 83, 123, 124

Centrifugal pumps 206
Centripoise 201
Certifying agency 259
CFC gases 257
Characteristic curve 214
Check valves 221
Chemical dehumidifier 28
Chilled water pump 9
Chilled water system 82
Chiller machine 9
Chillers 82
Circuit breaker 141, 155
Circuit wiring 140–141
Clean earth 144
Clean room applications 85
Closed loop 184
Coefficient of performance 101, 116
Coefficient of roughness 203
Combination heat detectors 274
Compact rising main 8
Compost 262
Compound lights 5
Compounding 105
Compression ratio 106
Compressor 106
Computerized flow dynamics 117
Condensation 73, 79–80
Condenser 113
Condenser water pumps 9, 85, 89
Condensing temperature 106
Conductance 22
Conduction 32
Conductivity 22
Connected load 150
Conservation of energy 202
Constant air volume 84
Continuity equation 202

Contour trenching 260
Contract capacity 308, 311
Control and indicating panels 277, 279
Control and indication panel 267
Control valve 85
Controller 302
Convection 32
Cooling tower fans 85
Cooling towers 9, 258, 262
Counter-weights 302, 303
Court building 5
Crossflow fans 125
Current transformer 155, 174, 197

D

Daily range 55
Dalton's law 21
Dampers 128
Darcy-Weisbach equation 95
De-rating factor 172
De-tuned filter circuits 168
Decibel 278, 279
Deep wells 217
Deflector position 290
Dehumidification 26, 29
Delta connection 161
Demand load 152
Density 201
Derate factors 194
Design velocity 77
Detector circuits 279
Dew point 21
Dew point temperature 79
DG set 3
Diesel engine 190
Diesel generator synchronizing panel 178
Diffuser 123
Digital controller 84
Direct on line starting 191
Discharge lamp 236
Distribution boards 140, 143
Distribution pipes 299
Distribution system 141, 150
Diversity factor 4, 150
Door closing time 310
Door opening time 310
Double bus bar panels 178
Double dehumidification 29
Double inlet double width 124
Down comer 13
Down-peak 305, 306
Drawdown 217
Droop 198, 200
Dry air 20
Dry bulb temperature 20, 79
Dual compressor centrifugal 112
Dual-screw 108

Duct design-friction chart 81
Duct size 75
Ducts 132
Dynamic losses 78
Dynamic pressure 120
Dynamic viscosity 201

E

Early suppression fast response
    sprinklers 286
Earth electrodes 144–145
Earth station 148
Earthing 3, 144
Economical consideration 78
Efficiency of refrigeration 101
Electric motor 302
Electric power consumption 150
Electrical distribution system 8, 144
Electrical lag 159
Electrical load 150
Electrical panel 9
Electrical points 3
Electrical wiring 139
Electronic voltage control 184
Elevator 307–308
End feed unit 8
Enthalpy 21, 23, 104
Equal friction method 75, 80
Equivalent area of duct 75
Equivalent length 78, 95
Equivalent pipe length 294
Equivalent rectangular duct 75
Equivalent round duct 126
Erosion control 260
Evaporating temperature 106
Evaporator 104, 114
Evaporator pressure 105
Exciter field 186
Exciter field excitation 185
Exhaust fans 128
Extended coverage sprinklers 286
External light fixtures 3
Extra hazard 285

F

Family curves 222
Fan coil units 80
Fan selection 135
Fan static efficiency 126
Fan static pressure 121
Fan total efficiency 126
Fan total pressure 121
Faucets 262
Fault current 155, 158
Fault levels 152, 157–158
Fault withstand capacity 152, 158
Feeder pillars 3

Fire alarm panel 132
Fire alarm system 266, 278
Fire dampers 132
Fire detector 266
Fire fighting 10, 220
Fixed temperature detector 273
Flame stage 270
Foot candles 234, 246
Forced draft 9
Forward curved fans 123
Four pipe system 92
Frame size 194
Free flow conduits 202
Frequency 165
Friction chart 75
Friction factor 205
Friction head 209, 224, 228
Friction loss 78, 96, 99, 296
Friction loss rate 77
Frictional head 212
Full load current 153
Full type test 178
Fuse 141, 169
Fusible element 273

G

Gauge pressure 119
Geared machine 301
Gearless motors 303
Generating sets 4
Generator 113
Generator alternator 145
Generator efficiency 194
Geothermal 263
Global warming 106
Globe valves 221
Goods lifts 300
Governor 302
Grass paver 261
Green building 258–259, 262–263, 265
Greenhouse gases 262
Grey water 262
Grid demand load 152
Grid diversity factor 4
Grid station 152
Grid sub-station 4
Ground water 216
Group control point wiring 140
Guide rails 302

H

Handling capacity 308–309
Harmonic current 168
Harmonic filter circuit 167
Harmonics 159, 165, 167
Heat detectors 274, 276–277
Heat exchanger 104

Heat island effect 257, 261
Heat load 31, 33, 79
Heat recovery wheel 70
Heat stage 270
Heat transfer 79–80
Heat transfer coefficient 22
Hermetic chillers 112
Hermetic compressor 112
High hazard 287
High intensity sounders 279
High side equipments 9
High tension (HT) panels 3
High velocity system 77
Higher duct velocity 72
Highest reversal floor 310–312
Hoist way 302
Horizontal illuminance 248–251
Horizontal illumination 251
Hospital lifts 300
Humidity ratio 23
HVAC 222
Hydraulic calculations 281, 292–293, 296
Hydraulic elevators 314
Hydraulic lift 300
Hydraulic mean depth 202
Hydraulic method 292
Hydronic piping system 87
Hydronic systems 85–86, 94

I

Illuminance 234–236, 243–244
Illuminance ratio 246
Illumination 235
Impedance 153, 155, 157
Impeller 206–207, 211, 223–224
Impeller blades 125
Impeller trimming 230
Imperviousness 262
In-line fans 125
Incipient stage 269
Indoor environment 265
Inductive load 155, 167
Infrared 232, 272
Initiating device circuit 267
Installation valve 299
Instrument transformers 174
Insulating materials 80
Insulation 79–80
Intelligent building 258
Inter cooling 104–105
Inter floor distance 310
Internal electrical installation 3
Internal wiring 3
Interstage pressure 105
Interval 309
Inverse definite minimum time 159

Inverse square law 234
Ionization detectors 270
IP classification 177
Isentropic compression 102
Isentropic expansion 102
Iso-efficiency curves 211
Isolator 142
Isolux 249
Isolux curves 241, 250
Isolux diagram 249
Isothermal heat absorption 102
Isothermal heat rejection 102

J

Journey time 308

K

Kinematic viscosity 201
Kinetic energy 202, 206, 215, 303
Kinetic head 212

L

Lagging current 196
Lagging reactive power 171
Lambert's cosine law 234
Laminar flow 201–202
Latent heat 106
Latent heat of air 20
Leading current 196
Leading reactive power 171
Life cycle 259
Lift lobby 117
Lift shaft 13
Lift system 308
Lifts 4, 300
Light distribution board 178
Light hazard 284, 299
Light loss factor 244
Light/fan fixtures 3
Lighting circuit 142
Lightning arrester 146
Lightning protection 3
Limit of temperature 175
Load analysis 192
Load factor 172
Locked rotor current 189
Log mean temperature difference 25
Loop earthing 141
Looping system 142
Low intensity sounders 279
Low tension (LT) duct 3
Low velocity system 75
LT cables 3
Lumens 148, 232, 234–235, 246, 248

Luminaires 177
Luminance 235
Luminous efficacy 236
Luminous efficiency 235–236
Luminous flux 232, 234, 240, 244
Luminous intensity 233, 235, 238, 240, 250

M

Machine room 300
Magnetic field 182–185, 189
Main control and indication panel 267
Main distribution boards 3, 141
Main terminal 309
Manometer 73
Manometric head 210
Manual call box 267
Manure 262
Maximum demand 150, 172
Maximum duct velocities 71
Mean temperature difference 32
Mechanical energy 205
Mechanical ventilation 117, 132–133
Micro-irrigation 262
Mimic diagram 267
Miniature circuit breaker 142
Mixed use development 150
Modulating control valves 87
Moist air 20
Moisture sensors 262
Mono-screw 108
Monsoon 80
Motor control center 177
Motor efficiency 189
Mulching 260
Multi-celled tower 89
Multistage pump 208
Multistage vapour compression 104
Multistaging 104–105

N

National building code 117, 128
Natural draft 9
Natural resources 257
Natural ventilation 117
Needle valves 221
Net positive suction head 12, 215
Net-zero 258
Neutral earthing 146
Night sky pollution 262
Non conventional power 258
Non-conventional refrigeration 101
Non-overloading 211
Non-return valves 221
Notification appliance circuit 267
NPSH 210

O

Office building 5
One pipe main 91
Open loop 184
Open wells 217
Optical smoke detectors 270
Ordinary hazard 292
Outdoor environment 265
Outer shell 264
Outlet velocity 126
Output rating 194
Overall heat transfer coefficient 25
Ozone 257
Ozone depleting gases 262
Ozone depletion potential 106
Ozone layer 263

P

Panel boards 177
Panels 177
Parallel circuit 167
Parallel droop 198
Partial pressure 21
Partial type test 178
Passenger lifts 300
Peak solar heat gain 51
Pendant sprinklers 285
Per capita 220
Percentage impedances 158
Performance curves 223–229
Permanent magnet 171, 182
PFC controller 169
Photoelectric detectors 271
Photometer curve 242
Photometry 232
Pipe earth electrode 144
Pipe schedule method 292
Pitot tube 73
Plate earth electrode 144
Plug fans 83
Pneumatic fire detector 273
Point wiring 139, 141
Polar curve 235, 237–238, 240,
    242–243, 246
Polar diagram 238
Potential energy 202
Potential head 212
Potential transformers 155, 176
Power circuit 142
Power control centre 177
Power cum motor control center 178
Power factor 155, 159, 161, 165, 172,
    190, 193, 198
Power factor correction 194
Power-surplus 258
Preliminary cost 3

Pressure 72
Pressure conduits 202
Pressure drop 73, 99
Pressure energy 202, 206
Pressure loss 78, 95
Pressure ratio 105
Pressure relief valves 221
Pressure sensor 84
Pressure wells 217
Pressurizing system 132
Primary chilled water pumps 85
Prime mover 187, 189–190, 194, 205
Probability 309–310
Programmable logic control 178
Propeller fans 124,–125
Protection relays 155
Protective conductor 141
Psychrometric equations 23
Psychrometric processes 25
Public address system 267, 279
Pump capacity 297
Pump house 299
Pumping level 217
Pumping station 218

Q

Quadrature droop 198

R

Radial fans 123
Radiation 31, 232
Radiation detectors 272
Radiation heat 261
Rain water harvesting 258, 262
Range pipes 299
Rate of rise heat detectors 273
Rated burden 174, 176
Rated output 176
Rated short time current 174
Re-cycled water 262
Reactance 155
Reactive current 197
Reactive power 159, 161, 164
Reactor filters 171
Reciprocating compressors 106
Rectangular ducts 73, 77–78, 136
Recycling 264
Refrigerant 101, 106
Refrigerant vapour 104
Refrigerating effect 101, 104
Refrigeration cycle 101
Regenerative heat exchange 30
Relative humidity 23
Relative humidity control 84
Repeater panels 267, 279
Residual drawdown 217
Residual magnetism 184–186

Residual voltage 184
Resistance 155
Resonance 167
Resonant circuit 168
Resonant frequency 167
Response indicator 267
Return air ducts 75, 132
Return duct 132
Reynolds number 202–203, 205
Rising mains 3, 8, 158, 172, 177
Roof extractors 125
Room dew point 73
Rope lift 300
Rotary compressors 108
Rotating armature 184
Rotating field 184
Roughness of surface 75
Round duct 73, 75, 77
Round duct diameter 80
Round equivalent duct 73
Round trip time 308–310
Rousseau diagram 246
Run off 260

S

Safety factor 224
Sandwich insulated bus trunking 174
Screw compressors 108
Scroll compressor 108
Secondary chilled water pumps 85
Sector panel 267
Security lights 5
Sedimentation 260
Self excited 184
Semi-hermetic 113
Sensible cooling 25
Sensible heat of air 20
Separately excited 185
Series circuit 167
Series loop 91
Service lifts 300
Shallow well 217
Short circuit current 152, 183, 186
Short circuit withstand capacity 153, 157
Short-circuit 158, 171
Shunt trip release 177
Sidewall sprinklers 286, 290
Signaling line circuit 267
Single acting compressors 107
Single floor jump time 310
Single stage compression 105
Single stage vapour compression 102
Slip ring induction motors 303
Slip rings 183
Sluice valves 221
Smoke detectors 270, 276–277
Smoldering stage 270

Socket outlet points 140
Soil stabilization 260
Solar gain 51
Solid angle 232–233
Solid waste 257
Sounders 267–278
Space 8
Specific gravity 201
Specific heat 106
Specific humidity 21
Specific speed 208–209
Specific volume 21–22, 106
Specific weight 201–202
Spray density 285
Spring buffers 302
Sprinkler 281, 285, 287, 289, 291,
    293, 297
Sprinkler spacing 288
Squirrel cage motors 303
Staircase pressurization 133
Standby duty 187
Star-delta starter 190
Starting current 189
Static discharge head 209
Static duct pressure 84
Static earthing 144
Static head 214, 228
Static level 217
Static pressure 73, 77, 84, 117, 119
    121–123, 135–136, 292, 296
Static pressure regain 77–78
Static suction head 209
Steady state load 189
Steam jet refrigeration 101
Stefan-Boltzman law 32
Steradian 232
Storage bins 264
STP 258
Strainer 218
Strip earth electrode 144
Sub distribution boards 3, 8
Subcooling 103–104
Submain wiring 140
Substations 3, 150, 152
Suction superheat 103
Superheat 103
Supply air duct 80, 84
Supply pressure 75
Surface area 25
Surface reflections 246
Surface water 216
Switch distribution board 177
Switchgear 143
Switching station 152
Switching unit 8
Symmetrical 238, 241
Synchronous machines 164
Synchronous speed 303

System curve 224
System fault level 155
System head curve 212
System pressure drop 95

T

Talk back unit 267
Tandem motors 303
Temperature difference 22, 61
Temperature sensors 84, 85
Thermal comfort levels 264
Thermal lag 273
Thermal resistance 22
Thermodynamic processes 101
Thermostatic expansion valve 103
Third party evaluators 259
Three pipe system 92
Three-way valve 87
Throttling 214, 230
Thumb rules 3
Time and motion study 307
Top soil 260
Total dynamic discharge head 210
Total dynamic suction head 209
Total fan pressure 123, 136
Total heat of air 20
Total pressure 121, 126
Total pressure loss 75
Total static head 209
Township 4
Trade-off 263
Traffic analysis 307, 310
Transformers 3, 145, 150, 153, 155,
    157–159, 167, 172
Transformer control 186
Transient 187
Transient voltage 190
Transients 165
Tube axial fans 124
Tube well 217
Tuned filter circuit 168
Turbo compressors 111
Turbulent flow 95, 202
Twin control light point 140
Twin screw compressors 110
Two pipe direct return 92
Two pipe reverse 92
Two-way control valve 87
Two-way throttling valve 83

U

Ultraviolet 232, 236, 272
Unbalanced 193
Unified glass rating 240
Unit of refrigeration 101
Up-peak 305–306, 308

Upright sprinklers 288, 290
Usage factor 3
Utilization factor 244, 249

V

Valve authority 86
Valve rangeability 86
Vane axial fans 83
Vapour absorption refrigeration 101, 113
Vapour compression cycle 104
Vapour compression refrigeration 101,
    113
Vapour compression systems 105
Vapour pressure head 215
Variable air volume 84
Variable speed (frequency) drives 82
Vector diagrams 197
Velocity energy 75
Velocity head 209, 224
Velocity pressure 73, 119–120, 122–123,
    126, 136
Ventilation fan 119, 135
Vertical illumination 251
VFD 83
Viscosity 87, 201
Volatile organic compounds 264
Voltage dip 187, 194
Voltage distortion 165
Voltage drop 152, 155, 161
Voltage fluctuation 166
Voltage levels 152
Volumetric efficiency 105–106

W

Waiting interval 308, 310
Waiting time 305
Waste water treatment 262
Water hammer 215
Water harvesting 258
Water supply 10
Water table 216
Water vapour 20
Wavelength 232
Weather maker room 10
Well development 218
Well yield 217
Wet bulb depression 20
Wet bulb temperature 20
Wet riser 13
Whirlpool chamber 207
Whole building performance 263

Z

Zonal flux 249
Zonal flux diagram 248
Zonal panel 267